"How and why is computational statistics taking over the world? In this serious work of synthesis that is also fun to read, Efron and Hastie, two pioneers in the integration of parametric and nonparametric statistical ideas, give their take on the unreasonable effectiveness of statistics and machine learning in the context of a series of clear, historically informed examples."

— Andrew Gelman, *Columbia University*

"This unusual book describes the nature of statistics by displaying multiple examples of the way the field has evolved over the past 60 years, as it has adapted to the rapid increase in available computing power. The authors' perspective is summarized nicely when they say, 'Very roughly speaking, algorithms are what statisticians do, while inference says why they do them.' The book explains this 'why'; that is, it explains the purpose and progress of statistical research, through a close look at many major methods, methods the authors themselves have advanced and studied at great length. Both enjoyable and enlightening, *Computer Age Statistical Inference* is written especially for those who want to hear the big ideas, and see them instantiated through the essential mathematics that defines statistical analysis. It makes a great supplement to the traditional curricula for beginning graduate students."

— Rob Kass, *Carnegie Mellon University*

"This is a terrific book. It gives a clear, accessible, and entertaining account of the interplay between theory and methodological development that has driven statistics in the computer age. The authors succeed brilliantly in locating contemporary algorithmic methodologies for analysis of 'big data' within the framework of established statistical theory."

— Alastair Young, *Imperial College London*

"This is a guided tour of modern statistics that emphasizes the conceptual and computational advances of the last century. Authored by two masters of the field, it offers just the right mix of mathematical analysis and insightful commentary."

— Hal Varian, *Google*

"Efron and Hastie guide us through the maze of breakthrough statistical methodologies following the computing evolution: why they were developed, their properties, and how they are used. Highlighting their origins, the book helps us understand each method's roles in inference and/or prediction. The inference–prediction distinction maintained throughout the book is a welcome and important novelty in the landscape of statistics books."

— Galit Shmueli, *National Tsing Hua University*

"A masterful guide to how the inferential bases of classical statistics can provide a principled disciplinary frame for the data science of the twenty-first century."

— Stephen Stigler, *University of Chicago, author of* Seven Pillars of Statistical Wisdom

"*Computer Age Statistical Inference* offers a refreshing view of modern statistics. Algorithmics are put on equal footing with intuition, properties, and the abstract arguments behind them. The methods covered are indispensable to practicing statistical analysts in today's big data and big computing landscape."

— Robert Gramacy, *The University of Chicago Booth School of Business*

"Every aspiring data scientist should carefully study this book, use it as a reference, and carry it with them everywhere. The presentation through the two-and-a-half-century history of statistical inference provides insight into the development of the discipline, putting data science in its historical place."

— Mark Girolami, *Imperial College London*

"Efron and Hastie are two immensely talented and accomplished scholars who have managed to brilliantly weave the fiber of 250 years of statistical inference into the more recent historical mechanization of computing. This book provides the reader with a mid-level overview of the last 60-some years by detailing the nuances of a statistical community that, historically, has been self-segregated into camps of Bayes, frequentist, and Fisher yet in more recent years has been unified by advances in computing. What is left to be explored is the emergence of, and role that, big data theory will have in bridging the gap between data science and statistical methodology. Whatever the outcome, the authors provide a vision of high-speed computing having tremendous potential to enable the contributions of statistical inference toward methodologies that address both global and societal issues."

— Rebecca Doerge, *Carnegie Mellon University*

"In this book, two masters of modern statistics give an insightful tour of the intertwined worlds of statistics and computation. Through a series of important topics, Efron and Hastie illuminate how modern methods for predicting and understanding data are rooted in both statistical and computational thinking. They show how the rise of computational power has transformed traditional methods and questions, and how it has pointed us to new ways of thinking about statistics."

— David Blei, *Columbia University*

Absolutely brilliant. This beautifully written compendium reviews many big statistical ideas, including the authors' own. A must for anyone engaged creatively in statistics and the data sciences, for repeated use. Efron and Hastie demonstrate the ever-growing power of statistical reasoning, past, present, and future.

— Carl Morris, *Harvard University*

Computer Age Statistical Inference

The twenty-first century has seen a breathtaking expansion of statistical methodology, both in scope and in influence. "Big data," "data science," and "machine learning" have become familiar terms in the news, as statistical methods are brought to bear upon the enormous data sets of modern science and commerce. How did we get here? And where are we going?

This book takes us on an exhilarating journey through the revolution in data analysis following the introduction of electronic computation in the 1950s. Beginning with classical inferential theories – Bayesian, frequentist, Fisherian – individual chapters take up a series of influential topics: survival analysis, logistic regression, empirical Bayes, the jackknife and bootstrap, random forests, neural networks, Markov chain Monte Carlo, inference after model selection, and dozens more. The distinctly modern approach integrates methodology and algorithms with statistical inference. The book ends with speculation on the future direction of statistics and data science.

BRADLEY EFRON is Max H. Stein Professor, Professor of Statistics, and Professor of Biomedical Data Science at Stanford University. He has held visiting faculty appointments at Harvard, UC Berkeley, and Imperial College London. Efron has worked extensively on theories of statistical inference, and is the inventor of the bootstrap sampling technique. He received the National Medal of Science in 2005 and the Guy Medal in Gold of the Royal Statistical Society in 2014.

TREVOR HASTIE is John A. Overdeck Professor, Professor of Statistics, and Professor of Biomedical Data Science at Stanford University. He is coauthor of *Elements of Statistical Learning*, a key text in the field of modern data analysis. He is also known for his work on generalized additive models and principal curves, and for his contributions to the R computing environment. Hastie was awarded the Emmanuel and Carol Parzen prize for Statistical Innovation in 2014.

INSTITUTE OF MATHEMATICAL STATISTICS MONOGRAPHS

IMS Monographs are concise research monographs of high quality on any branch of statistics or probability of sufficient interest to warrant publication as books. Some concern relatively traditional topics in need of up-to-date assessment. Others are on emerging themes. In all cases the objective is to provide a balanced view of the field.

Other Books in the Series

1. *Large-Scale Inference,* by Bradley Efron
2. *Nonparametric Inference on Manifolds,* by Abhishek Bhattacharya and Rabi Battacharya
3. *The Skew-Normal and Related Families*, by Adelchi Azzalini
4. *Case-Control Studies,* by Ruth H. Keogh and D. R. Cox
5. *Computer Age Statistical Inference*, by Bradley Efron and Trevor Hastie

Computer Age Statistical Inference

Algorithms, Evidence, and Data Science

BRADLEY EFRON
Stanford University, California

TREVOR HASTIE
Stanford University, California

CAMBRIDGE
UNIVERSITY PRESS

CAMBRIDGE
UNIVERSITY PRESS

One Liberty Plaza, 20th Floor, New York, NY 10006, USA

Cambridge University Press is part of the University of Cambridge.

It furthers the University's mission by disseminating knowledge in the pursuit of education, learning and research at the highest international levels of excellence.

www.cambridge.org
Information on this title: www.cambridge.org/9781107149892

First published 2016

Printed in the United Kingdom by Clays, St Ives plc

A catalogue record for this publication is available from the British Library

ISBN 978-1-107-14989-2 Hardback

To Donna and Lynda

Contents

Preface xv
Acknowledgments xviii
Notation xix

Part I Classic Statistical Inference 1

1 Algorithms and Inference 3
1.1 A Regression Example 4
1.2 Hypothesis Testing 8
1.3 Notes 11

2 Frequentist Inference 12
2.1 Frequentism in Practice 14
2.2 Frequentist Optimality 18
2.3 Notes and Details 20

3 Bayesian Inference 22
3.1 Two Examples 24
3.2 Uninformative Prior Distributions 28
3.3 Flaws in Frequentist Inference 30
3.4 A Bayesian/Frequentist Comparison List 33
3.5 Notes and Details 36

4 Fisherian Inference and Maximum Likelihood Estimation 38
4.1 Likelihood and Maximum Likelihood 38
4.2 Fisher Information and the MLE 41
4.3 Conditional Inference 45
4.4 Permutation and Randomization 49
4.5 Notes and Details 51

5 Parametric Models and Exponential Families 53

5.1 Univariate Families 54
5.2 The Multivariate Normal Distribution 55
5.3 Fisher's Information Bound for Multiparameter Families 59
5.4 The Multinomial Distribution 61
5.5 Exponential Families 64
5.6 Notes and Details 69

Part II Early Computer-Age Methods 73

6 Empirical Bayes 75
6.1 Robbins' Formula 75
6.2 The Missing-Species Problem 78
6.3 A Medical Example 84
6.4 Indirect Evidence 1 88
6.5 Notes and Details 88

7 James–Stein Estimation and Ridge Regression 91
7.1 The James–Stein Estimator 91
7.2 The Baseball Players 94
7.3 Ridge Regression 97
7.4 Indirect Evidence 2 102
7.5 Notes and Details 104

8 Generalized Linear Models and Regression Trees 108
8.1 Logistic Regression 109
8.2 Generalized Linear Models 116
8.3 Poisson Regression 120
8.4 Regression Trees 124
8.5 Notes and Details 128

9 Survival Analysis and the EM Algorithm 131
9.1 Life Tables and Hazard Rates 131
9.2 Censored Data and the Kaplan–Meier Estimate 134
9.3 The Log-Rank Test 139
9.4 The Proportional Hazards Model 143
9.5 Missing Data and the EM Algorithm 146
9.6 Notes and Details 150

10 The Jackknife and the Bootstrap 155
10.1 The Jackknife Estimate of Standard Error 156
10.2 The Nonparametric Bootstrap 159
10.3 Resampling Plans 162

10.4 The Parametric Bootstrap 169
10.5 Influence Functions and Robust Estimation 174
10.6 Notes and Details 177

11 Bootstrap Confidence Intervals 181
11.1 Neyman's Construction for One-Parameter Problems 181
11.2 The Percentile Method 185
11.3 Bias-Corrected Confidence Intervals 190
11.4 Second-Order Accuracy 192
11.5 Bootstrap-t Intervals 195
11.6 Objective Bayes Intervals and the Confidence Distribution 198
11.7 Notes and Details 204

12 Cross-Validation and C_p Estimates of Prediction Error 208
12.1 Prediction Rules 208
12.2 Cross-Validation 213
12.3 Covariance Penalties 218
12.4 Training, Validation, and Ephemeral Predictors 227
12.5 Notes and Details 230

13 Objective Bayes Inference and MCMC 233
13.1 Objective Prior Distributions 234
13.2 Conjugate Prior Distributions 237
13.3 Model Selection and the Bayesian Information Criterion 243
13.4 Gibbs Sampling and MCMC 251
13.5 Example: Modeling Population Admixture 256
13.6 Notes and Details 261

14 Postwar Statistical Inference and Methodology 264

Part III Twenty-First-Century Topics 269

15 Large-Scale Hypothesis Testing and FDRs 271
15.1 Large-Scale Testing 272
15.2 False-Discovery Rates 275
15.3 Empirical Bayes Large-Scale Testing 278
15.4 Local False-Discovery Rates 282
15.5 Choice of the Null Distribution 286
15.6 Relevance 290
15.7 Notes and Details 294

16 Sparse Modeling and the Lasso 298

16.1 Forward Stepwise Regression 299
16.2 The Lasso 303
16.3 Fitting Lasso Models 308
16.4 Least-Angle Regression 309
16.5 Fitting Generalized Lasso Models 313
16.6 Post-Selection Inference for the Lasso 317
16.7 Connections and Extensions 319
16.8 Notes and Details 321

17 Random Forests and Boosting 324
17.1 Random Forests 325
17.2 Boosting with Squared-Error Loss 333
17.3 Gradient Boosting 338
17.4 Adaboost: the Original Boosting Algorithm 341
17.5 Connections and Extensions 345
17.6 Notes and Details 347

18 Neural Networks and Deep Learning 351
18.1 Neural Networks and the Handwritten Digit Problem 353
18.2 Fitting a Neural Network 356
18.3 Autoencoders 362
18.4 Deep Learning 364
18.5 Learning a Deep Network 368
18.6 Notes and Details 371

19 Support-Vector Machines and Kernel Methods 375
19.1 Optimal Separating Hyperplane 376
19.2 Soft-Margin Classifier 378
19.3 SVM Criterion as Loss Plus Penalty 379
19.4 Computations and the Kernel Trick 381
19.5 Function Fitting Using Kernels 384
19.6 Example: String Kernels for Protein Classification 385
19.7 SVMs: Concluding Remarks 387
19.8 Kernel Smoothing and Local Regression 387
19.9 Notes and Details 390

20 Inference After Model Selection 394
20.1 Simultaneous Confidence Intervals 395
20.2 Accuracy After Model Selection 402
20.3 Selection Bias 408
20.4 Combined Bayes–Frequentist Estimation 412
20.5 Notes and Details 417

21 Empirical Bayes Estimation Strategies 421
21.1 Bayes Deconvolution 421
21.2 g-Modeling and Estimation 424
21.3 Likelihood, Regularization, and Accuracy 427
21.4 Two Examples 432
21.5 Generalized Linear Mixed Models 437
21.6 Deconvolution and f-Modeling 440
21.7 Notes and Details 444

Epilogue 446
References 453
Author Index 463
Subject Index 467

Preface

Statistical inference is an unusually wide-ranging discipline, located as it is at the triple-point of mathematics, empirical science, and philosophy. The discipline can be said to date from 1763, with the publication of Bayes' rule (representing the philosophical side of the subject; the rule's early advocates considered it an argument for the existence of God). The most recent quarter of this 250-year history—from the 1950s to the present—is the "computer age" of our book's title, the time when computation, the traditional bottleneck of statistical applications, became faster and easier by a factor of a million.

The book is an examination of how statistics has evolved over the past sixty years—an aerial view of a vast subject, but seen from the height of a small plane, not a jetliner or satellite. The individual chapters take up a series of influential topics—generalized linear models, survival analysis, the jackknife and bootstrap, false-discovery rates, empirical Bayes, MCMC, neural nets, and a dozen more—describing for each the key methodological developments and their inferential justification.

Needless to say, the role of electronic computation is central to our story. This doesn't mean that every advance was computer-related. A land bridge had opened to a new continent but not all were eager to cross. Topics such as empirical Bayes and James–Stein estimation could have emerged just as well under the constraints of mechanical computation. Others, like the bootstrap and proportional hazards, were pureborn children of the computer age. Almost all topics in twenty-first-century statistics are now computer-dependent, but it will take our small plane a while to reach the new millennium.

Dictionary definitions of statistical inference tend to equate it with the entire discipline. This has become less satisfactory in the "big data" era of immense computer-based processing algorithms. Here we will attempt, not always consistently, to separate the two aspects of the statistical enterprise: algorithmic developments aimed at specific problem areas, for instance random forests for prediction, as distinct from the inferential arguments offered in their support.

Very broadly speaking, algorithms are what statisticians do while inference says why they do them. A particularly energetic brand of the statistical enterprise has flourished in the new century, *data science*, emphasizing algorithmic thinking rather than its inferential justification. The later chapters of our book, where large-scale prediction algorithms such as boosting and deep learning are examined, illustrate the data-science point of view. (See the epilogue for a little more on the sometimes fraught statistics/data science marriage.)

There are no such subjects as Biological Inference or Astronomical Inference or Geological Inference. Why do we need "Statistical Inference"? The answer is simple: the natural sciences have nature to judge the accuracy of their ideas. Statistics operates one step back from Nature, most often interpreting the observations of natural scientists. Without Nature to serve as a disinterested referee, we need a system of mathematical logic for guidance and correction. Statistical inference is that system, distilled from two and a half centuries of data-analytic experience.

The book proceeds historically, in three parts. The great themes of classical inference, Bayesian, frequentist, and Fisherian, reviewed in Part I, were set in place before the age of electronic computation. Modern practice has vastly extended their reach without changing the basic outlines. (An analogy with classical and modern literature might be made.) Part II concerns early computer-age developments, from the 1950s through the 1990s. As a transitional period, this is the time when it is easiest to see the effects, or noneffects, of fast computation on the progress of statistical methodology, both in its theory and practice. Part III, "Twenty-First-Century topics," brings the story up to the present. Ours is a time of enormously ambitious algorithms ("machine learning" being the somewhat disquieting catchphrase). Their justification is the ongoing task of modern statistical inference.

Neither a catalog nor an encyclopedia, the book's topics were chosen as apt illustrations of the interplay between computational methodology and inferential theory. Some missing topics that might have served just as well include time series, general estimating equations, causal inference, graphical models, and experimental design. In any case, there is no implication that the topics presented here are the only ones worthy of discussion.

Also underrepresented are asymptotics and decision theory, the "math stat" side of the field. Our intention was to maintain a technical level of discussion appropriate to Masters'-level statisticians or first-year PhD students. Inevitably, some of the presentation drifts into more difficult waters, more from the nature of the statistical ideas than the mathematics. Readers who find our aerial view circling too long over some topic shouldn't hesitate to move ahead in the book. For the most part, the chapters can be read independently of each other (though there is a connecting overall theme). This comment applies especially to nonstatisticians who have picked up the book because of interest in some particular topic, say survival analysis or boosting.

Useful disciplines that serve a wide variety of demanding clients run the risk of losing their center. Statistics has managed, for the most part, to maintain its philosophical cohesion despite a rising curve of outside demand. The center of the field has in fact moved in the past sixty years, from its traditional home in mathematics and logic toward a more computational focus. Our book traces that movement on a topic-by-topic basis. An answer to the intriguing question "What happens next?" won't be attempted here, except for a few words in the epilogue, where the rise of data science is discussed.

Acknowledgments

We are indebted to Cindy Kirby for her skillful work in the preparation of this book, and Galit Shmueli for her helpful comments on an earlier draft. At Cambridge University Press, a huge thank you to Steven Holt for his excellent copy editing, Clare Dennison for guiding us through the production phase, and to Diana Gillooly, our editor, for her unfailing support.

Bradley Efron
Trevor Hastie
Department of Statistics
Stanford University
May 2016

Notation

Throughout the book the numbered † sign indicates a technical note or reference element which is elaborated on at the end of the chapter. There, next to the number, the page number of the referenced location is given in parenthesis. For example, `lowess` in the notes on page 11 was referenced via a $\dagger_1$ on page 6. Matrices such as $\boldsymbol{\Sigma}$ are represented in bold font, as are certain vectors such as $\boldsymbol{y}$, a data vector with n elements. Most other vectors, such as coefficient vectors, are typically not bold. We use a dark green `typewriter` font to indicate data set names such as `prostate`, variable names such as `prog` from data sets, and `R` commands such as `glmnet` or `locfdr`. No bibliographic references are given in the body of the text; important references are given in the endnotes of each chapter.

Part I

Classic Statistical Inference

1

Algorithms and Inference

Statistics is the science of learning from experience, particularly experience that arrives a little bit at a time: the successes and failures of a new experimental drug, the uncertain measurements of an asteroid's path toward Earth. It may seem surprising that any one theory can cover such an amorphous target as "learning from experience." In fact, there are *two* main statistical theories, Bayesianism and frequentism, whose connections and disagreements animate many of the succeeding chapters.

First, however, we want to discuss a less philosophical, more operational division of labor that applies to both theories: between the *algorithmic* and *inferential* aspects of statistical analysis. The distinction begins with the most basic, and most popular, statistical method, averaging. Suppose we have observed numbers $x_1, x_2, \ldots, x_n$ applying to some phenomenon of interest, perhaps the automobile accident rates in the $n = 50$ states. The *mean*

$$\bar{x} = \sum_{i=1}^{n} x_i / n \tag{1.1}$$

summarizes the results in a single number.

How accurate is that number? The textbook answer is given in terms of the *standard error*,

$$\widehat{\text{se}} = \left[\sum_{i=1}^{n} (x_i - \bar{x})^2 \Big/ (n(n-1)) \right]^{1/2}. \tag{1.2}$$

Here *averaging* (1.1) is the algorithm, while the standard error provides an inference of the algorithm's accuracy. It is a surprising, and crucial, aspect of statistical theory that the same data that supplies an estimate can also assess its accuracy.[1]

[1] "Inference" concerns more than accuracy: speaking broadly, algorithms say what the statistician does while inference says why he or she does it.

Of course, $\widehat{\text{se}}$ (1.2) is itself an algorithm, which could be (and is) subject to further inferential analysis concerning *its* accuracy. The point is that the algorithm comes first and the inference follows at a second level of statistical consideration. In practice this means that algorithmic invention is a more free-wheeling and adventurous enterprise, with inference playing catch-up as it strives to assess the accuracy, good or bad, of some hot new algorithmic methodology.

If the inference/algorithm race is a tortoise-and-hare affair, then modern electronic computation has bred a bionic hare. There are two effects at work here: computer-based technology allows scientists to collect enormous data sets, orders of magnitude larger than those that classic statistical theory was designed to deal with; huge data demands new methodology, and the demand is being met by a burst of innovative computer-based statistical algorithms. When one reads of "big data" in the news, it is usually these algorithms playing the starring roles.

Our book's title, *Computer Age Statistical Inference*, emphasizes the tortoise's side of the story. The past few decades have been a golden age of statistical methodology. It hasn't been, quite, a golden age for statistical inference, but it has not been a dark age either. The efflorescence of ambitious new algorithms has forced an evolution (though not a revolution) in inference, the theories by which statisticians choose among competing methods. The book traces the interplay between methodology and inference as it has developed since the 1950s, the beginning of our discipline's computer age. As a preview, we end this chapter with two examples illustrating the transition from classic to computer-age practice.

1.1 A Regression Example

Figure 1.1 concerns a study of kidney function. Data points (x_i, y_i) have been observed for $n = 157$ healthy volunteers, with x_i the ith volunteer's **age** in years, and y_i a composite measure "**tot**" of overall function. Kidney function generally declines with **age**, as evident in the downward scatter of the points. The rate of decline is an important question in kidney transplantation: in the past, potential donors past **age** 60 were prohibited, though, given a shortage of donors, this is no longer enforced.

The solid line in Figure 1.1 is a *linear regression*

$$y = \hat{\beta}_0 + \hat{\beta}_1 x \tag{1.3}$$

fit to the data by *least squares*, that is by minimizing the sum of squared

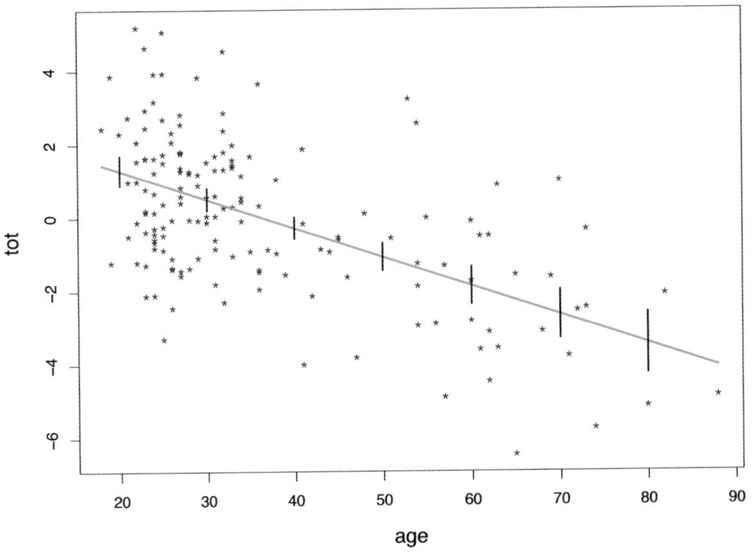

Figure 1.1 Kidney fitness `tot` vs `age` for 157 volunteers. The line is a linear regression fit, showing ±2 standard errors at selected values of `age`.

deviations

$$\sum_{i=1}^{n}(y_i - \beta_0 - \beta_1 x_i)^2 \tag{1.4}$$

over all choices of (β_0, β_1). The least squares algorithm, which dates back to Gauss and Legendre in the early 1800s, gives $\hat{\beta}_0 = 2.86$ and $\hat{\beta}_1 = -0.079$ as the least squares estimates. We can read off of the fitted line an estimated value of kidney fitness for any chosen `age`. The top line of Table 1.1 shows estimate 1.29 at `age` 20, down to −3.43 at `age` 80.

How accurate are these estimates? This is where inference comes in: an extended version of formula (1.2), also going back to the 1800s, provides the standard errors, shown in line 2 of the table. The vertical bars in Figure 1.1 are ± two standard errors, giving them about 95% chance of containing the true expected value of `tot` at each `age`.

That 95% coverage depends on the validity of the linear regression model (1.3). We might instead try a quadratic regression $y = \hat{\beta}_0 + \hat{\beta}_1 x + \hat{\beta}_2 x^2$, or a cubic, etc., all of this being well within the reach of pre-computer statistical theory.

Table 1.1 *Regression analysis of the kidney data; (1) linear regression estimates; (2) their standard errors; (3)* `lowess` *estimates; (4) their bootstrap standard errors.*

`age`	20	30	40	50	60	70	80
1. linear regression	1.29	.50	−.28	−1.07	−1.86	−2.64	−3.43
2. std error	.21	.15	.15	.19	.26	.34	.42
3. lowess	1.66	.65	−.59	−1.27	−1.91	−2.68	−3.50
4. bootstrap std error	.71	.23	.31	.32	.37	.47	.70

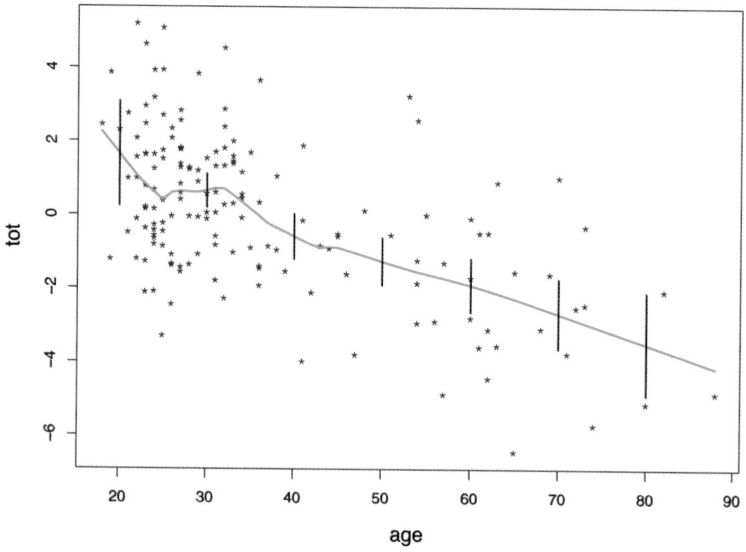

Figure 1.2 Local polynomial `lowess(x,y,1/3)` fit to the kidney-fitness data, with ±2 bootstrap standard deviations.

A modern computer-based algorithm `lowess` produced the somewhat
†1 bumpy regression curve in Figure 1.2. The `lowess`† [2] algorithm moves its attention along the x-axis, fitting local polynomial curves of differing degrees to nearby (x, y) points. (The 1/3 in the call[3] `lowess(x,y,1/3)`

[2] Here and throughout the book, the numbered † sign indicates a technical note or reference element which is elaborated on at the end of the chapter.

[3] Here and in all our examples we are employing the language `R`, itself one of the key developments in computer-based statistical methodology.

determines the definition of local.) Repeated passes over the x-axis refine the fit, reducing the effects of occasional anomalous points. The fitted curve in Figure 1.2 is nearly linear at the right, but more complicated at the left where points are more densely packed. It is flat between ages 25 and 35, a potentially important difference from the uniform decline portrayed in Figure 1.1.

There is no formula such as (1.2) to infer the accuracy of the `lowess` curve. Instead, a computer-intensive inferential engine, the *bootstrap*, was used to calculate the error bars in Figure 1.2. A bootstrap data set is produced by resampling 157 pairs (x_i, y_i) from the original 157 *with replacement*, so perhaps (x_1, y_1) might show up twice in the bootstrap sample, (x_2, y_2) might be missing, (x_3, y_3) present once, etc. Applying `lowess` to the bootstrap sample generates a bootstrap replication of the original calculation.

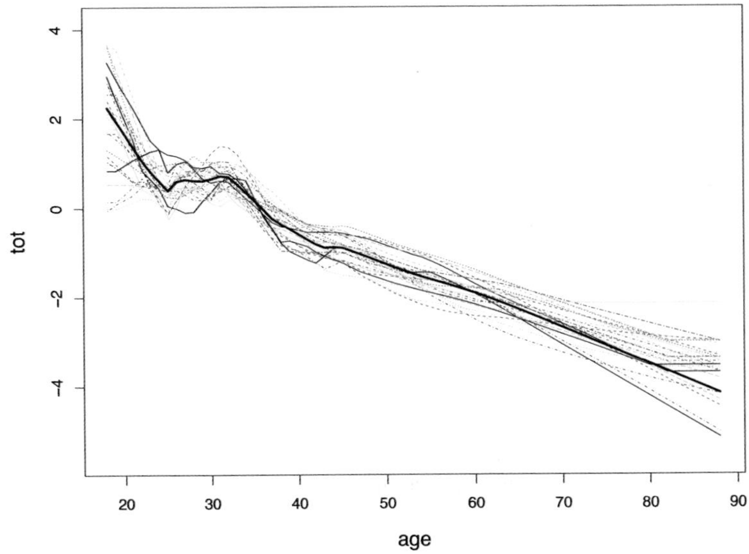

Figure 1.3 25 bootstrap replications of `lowess(x,y,1/3)`.

Figure 1.3 shows the first 25 (of 250) bootstrap `lowess` replications bouncing around the original curve from Figure 1.2. The variability of the replications at any one `age`, the *bootstrap standard deviation*, determined the original curve's accuracy. How and why the bootstrap works is discussed in Chapter 10. It has the great virtue of assessing estimation accuracy for *any* algorithm, no matter how complicated. The price is a hundred- or thousand-fold increase in computation, unthinkable in 1930, but routine now.

The bottom two lines of Table 1.1 show the `lowess` estimates and their standard errors. We have paid a price for the increased flexibility of `lowess`, its standard errors roughly doubling those for linear regression.

1.2 Hypothesis Testing

Our second example concerns the march of methodology and inference for *hypothesis testing* rather than estimation: 72 leukemia patients, 45 with **ALL** (acute lymphoblastic leukemia) and 27 with **AML** (acute myeloid leukemia, a worse prognosis) have each had genetic activity measured for a panel of 7,128 genes. The histograms in Figure 1.4 compare the genetic activities in the two groups for gene 136.

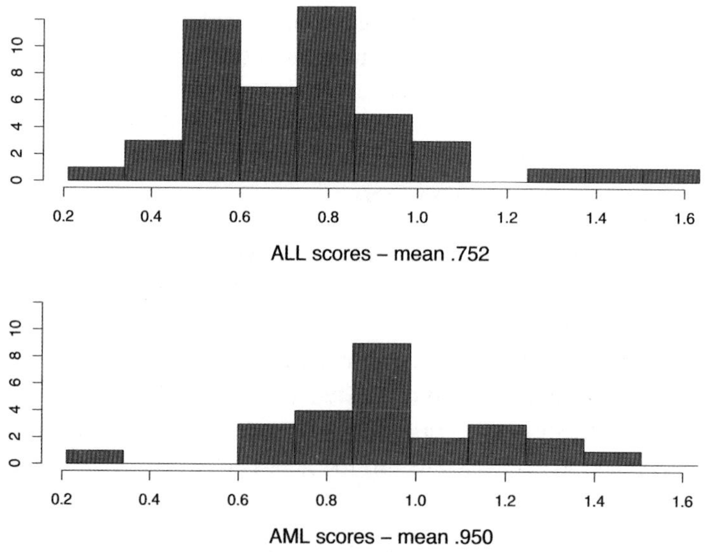

Figure 1.4 Scores for gene 136, leukemia data. Top **ALL** ($n = 47$), bottom **AML** ($n = 25$). A two-sample t-statistic $= 3.01$ with p-value $= .0036$.

The **AML** group appears to show greater activity, the mean values being

$$\overline{\text{ALL}} = 0.752 \quad \text{and} \quad \overline{\text{AML}} = 0.950. \tag{1.5}$$

Is the perceived difference genuine, or perhaps, as people like to say, "a statistical fluke"? The classic answer to this question is via a *two-sample t-statistic*,

$$t = \frac{\overline{\texttt{AML}} - \overline{\texttt{ALL}}}{\widehat{\text{sd}}}, \tag{1.6}$$

where $\widehat{\text{sd}}$ is an estimate of the numerator's standard deviation.[4]

Dividing by $\widehat{\text{sd}}$ allows us (under Gaussian assumptions discussed in Chapter 5) to compare the observed value of t with a standard "null" distribution, in this case a Student's t distribution with 70 degrees of freedom. We obtain $t = 3.01$ from (1.6), which would classically be considered very strong evidence that the apparent difference (1.5) is genuine; in standard terminology, "with two-sided significance level 0.0036."

A small significance level (or "p-value") is a statement of statistical surprise: something very unusual has happened if in fact there is no difference in gene 136 expression levels between **ALL** and **AML** patients. We are less surprised by $t = 3.01$ if gene 136 is just one candidate out of thousands that might have produced "interesting" results.

That is the case here. Figure 1.5 shows the histogram of the two-sample t-statistics for the panel of 7128 genes. Now $t = 3.01$ looks less unusual; 400 other genes have t exceeding 3.01, about 5.6% of them.

This doesn't mean that gene 136 is "significant at the 0.056 level." There are two powerful complicating factors:

1. Large numbers of candidates, 7128 here, will produce some large t-values even if there is really no difference in genetic expression between **ALL** and **AML** patients.
2. The histogram implies that in this study there is something wrong with the theoretical null distribution ("Student's t with 70 degrees of freedom"), the smooth curve in Figure 1.5. It is much too narrow at the center, where presumably most of the genes are reporting unsignificant results.

We will see in Chapter 15 that a low *false-discovery rate*, i.e., a low chance of crying wolf over an innocuous gene, requires t exceeding 6.16 in the **ALL/AML** study. Only 47 of the 7128 genes make the cut. False-discovery-rate theory is an impressive advance in statistical inference, incorporating Bayesian, frequentist, and empirical Bayesian (Chapter 6) el-

[4] Formally, a standard error is the standard deviation of a summary statistic, and $\widehat{\text{sd}}$ might better be called $\widehat{\text{se}}$, but we will follow the distinction less than punctiliously here.

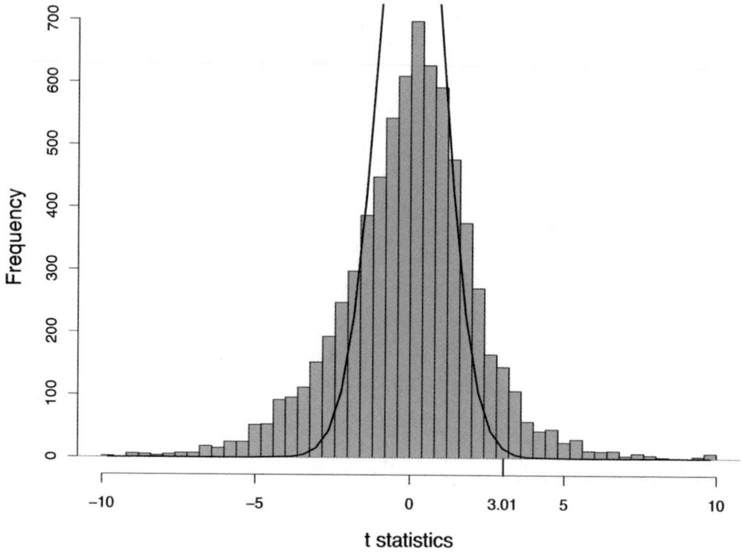

Figure 1.5 Two-sample t-statistics for 7128 genes, leukemia data. The smooth curve is the theoretical null density for the t-statistic.

ements. It was a *necessary* advance in a scientific world where computer-based technology routinely presents thousands of comparisons to be evaluated at once.

There is one more thing to say about the algorithm/inference statistical cycle. Important new algorithms often arise outside the world of professional statisticians: neural nets, support vector machines, and boosting are three famous examples. None of this is surprising. New sources of data, satellite imagery for example, or medical microarrays, inspire novel methodology from the observing scientists. The early literature tends toward the enthusiastic, with claims of enormous applicability and power.

In the second phase, statisticians try to locate the new methodology within the framework of statistical theory. In other words, they carry out the statistical inference part of the cycle, placing the new methodology within the known Bayesian and frequentist limits of performance. (Boosting offers a nice example, Chapter 17.) This is a healthy chain of events, good both for the hybrid vigor of the statistics profession and for the further progress of algorithmic technology.

1.3 Notes

Legendre published the least squares algorithm in 1805, causing Gauss to state that he had been using the method in astronomical orbit-fitting since 1795. Given Gauss' astonishing production of major mathematical advances, this says something about the importance attached to the least squares idea. Chapter 8 includes its usual algebraic formulation, as well as Gauss' formula for the standard errors, line 2 of Table 1.1.

Our division between algorithms and inference brings to mind Tukey's explanatory/confirmatory system. However the current algorithmic world is often bolder in its claims than the word "exploratory" implies, while to our minds "inference" conveys something richer than mere confirmation.

†1 [p. 6] **`lowess`** was devised by William Cleveland (Cleveland, 1981) and is available in the R statistical computing language. It is applied to the kidney data in Efron (2004). The kidney data originated in the nephrology laboratory of Dr. Brian Myers, Stanford University, and is available from this book's web site.

2

Frequentist Inference

Before the computer age there was the calculator age, and before "big data" there were small data sets, often a few hundred numbers or fewer, laboriously collected by individual scientists working under restrictive experimental constraints. Precious data calls for maximally efficient statistical analysis. A remarkably effective theory, feasible for execution on mechanical desk calculators, was developed beginning in 1900 by Pearson, Fisher, Neyman, Hotelling, and others, and grew to dominate twentieth-century statistical practice. The theory, now referred to as *classical*, relied almost entirely on frequentist inferential ideas. This chapter sketches a quick and simplified picture of frequentist inference, particularly as employed in classical applications.

We begin with another example from Dr. Myers' nephrology laboratory: 211 kidney patients have had their *glomerular filtration rates* measured, with the results shown in Figure 2.1; **gfr** is an important indicator of kidney function, with low values suggesting trouble. (It is a key component of **tot** in Figure 1.1.) The mean and standard error (1.1)–(1.2) are $\bar{x} = 54.25$ and $\widehat{\text{se}} = 0.95$, typically reported as

$$54.25 \pm 0.95; \tag{2.1}$$

± 0.95 denotes a frequentist inference for the accuracy of the estimate $\bar{x} = 54.25$, and suggests that we shouldn't take the ".25" very seriously, even the "4" being open to doubt. Where the inference comes from and what exactly it means remains to be said.

Statistical inference usually begins with the assumption that some probability model has produced the observed data $\boldsymbol{x}$, in our case the vector of $n = 211$ **gfr** measurements $\boldsymbol{x} = (x_1, x_2, \ldots, x_n)$. Let $\boldsymbol{X} = (X_1, X_2, \ldots, X_n)$ indicate n independent draws from a probability distribution F, written

$$F \to \boldsymbol{X}, \tag{2.2}$$

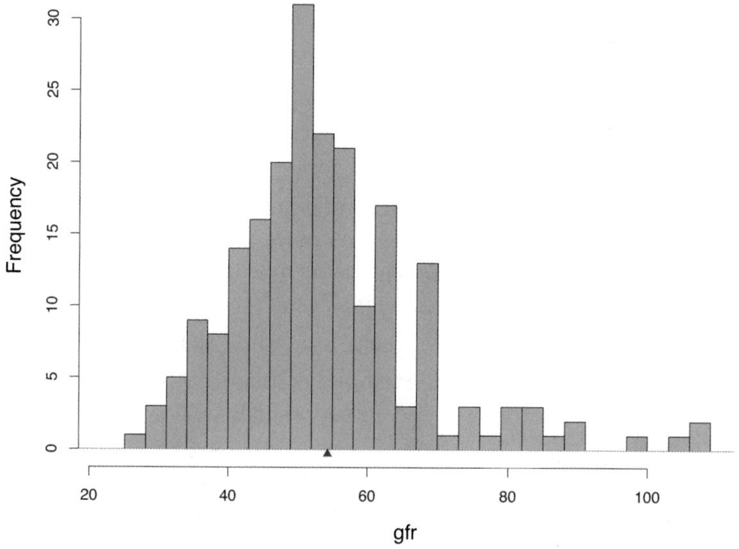

Figure 2.1 Glomerular filtration rates for 211 kidney patients; mean 54.25, standard error .95.

F being the underlying distribution of possible **gfr** scores here. A realization $\boldsymbol{X} = \boldsymbol{x}$ of (2.2) has been observed, and the statistican wishes to *infer* some property of the unknown distribution F.

Suppose the desired property is the *expectation* of a single random draw X from F, denoted

$$\theta = E_F\{X\} \tag{2.3}$$

(which also equals the expectation of the average $\bar{X} = \sum X_i/n$ of random vector (2.2)[1]). The obvious estimate of θ is $\hat{\theta} = \bar{x}$, the sample average. If n were enormous, say 10^{10}, we would expect $\hat{\theta}$ to nearly equal θ, but otherwise there is room for error. How much error is the inferential question.

The estimate $\hat{\theta}$ is calculated from $\boldsymbol{x}$ according to some known algorithm, say

$$\hat{\theta} = t(\boldsymbol{x}), \tag{2.4}$$

$t(\boldsymbol{x})$ in our example being the averaging function $\bar{x} = \sum x_i/n$; $\hat{\theta}$ is a

[1] The fact that $E_F\{\bar{X}\}$ equals $E_F\{X\}$ is a crucial, though easily proved, probabilistic result.

realization of

$$\hat{\Theta} = t(\boldsymbol{X}), \tag{2.5}$$

the output of $t(\cdot)$ applied to a theoretical sample $\boldsymbol{X}$ from F (2.2). We have chosen $t(\boldsymbol{X})$, we hope, to make $\hat{\Theta}$ a good estimator of θ, the desired property of F.

We can now give a first definition of frequentist inference: *the accuracy of an observed estimate $\hat{\theta} = t(\boldsymbol{x})$ is the probabilistic accuracy of $\hat{\Theta} = t(\boldsymbol{X})$ as an estimator of θ*. This may seem more a tautology than a definition, but it contains a powerful idea: $\hat{\theta}$ is just a single number but $\hat{\Theta}$ takes on a range of values whose spread can define measures of accuracy.

Bias and variance are familiar examples of frequentist inference. Define μ to be the expectation of $\hat{\Theta} = t(\boldsymbol{X})$ under model (2.2),

$$\mu = E_F\{\hat{\Theta}\}. \tag{2.6}$$

Then the bias and variance attributed to estimate $\hat{\theta}$ of parameter θ are

$$\text{bias} = \mu - \theta \quad \text{and} \quad \text{var} = E_F\left\{(\hat{\Theta} - \mu)^2\right\}. \tag{2.7}$$

Again, what keeps this from tautology is the attribution to the single number $\hat{\theta}$ of the probabilistic properties of $\hat{\Theta}$ following from model (2.2). If all of this seems too obvious to worry about, the Bayesian criticisms of Chapter 3 may come as a shock.

Frequentism is often defined with respect to "an infinite sequence of future trials." We imagine hypothetical data sets $\boldsymbol{X}^{(1)}, \boldsymbol{X}^{(2)}, \boldsymbol{X}^{(3)}, \ldots$ generated by the same mechanism as $\boldsymbol{x}$ providing corresponding values $\hat{\Theta}^{(1)}$, $\hat{\Theta}^{(2)}, \hat{\Theta}^{(3)}, \ldots$ as in (2.5). The frequentist principle is then to attribute for $\hat{\theta}$ the accuracy properties of the ensemble of $\hat{\Theta}$ values.[2] If the $\hat{\Theta}$s have empirical variance of, say, 0.04, then $\hat{\theta}$ is claimed to have standard error $0.2 = \sqrt{0.04}$, etc. This amounts to a more picturesque restatement of the previous definition.

2.1 Frequentism in Practice

Our working definition of frequentism is that *the probabilistic properties of a procedure of interest are derived and then applied verbatim to the procedure's output for the observed data*. This has an obvious defect: it requires calculating the properties of estimators $\hat{\Theta} = t(\boldsymbol{X})$ obtained from

[2] In essence, frequentists ask themselves "What would I see if I reran the same situation again (and again and again...)?"

the true distribution F, even though F is unknown. Practical frequentism uses a collection of more or less ingenious devices to circumvent the defect.

1. The plug-in principle. A simple formula relates the standard error of $\bar{X} = \sum X_i/n$ to $\text{var}_F(X)$, the variance of a single X drawn from F,

$$\text{se}\left(\bar{X}\right) = \left[\text{var}_F(X)/n\right]^{1/2}. \tag{2.8}$$

But having observed $\boldsymbol{x} = (x_1, x_2, \ldots, x_n)$ we can estimate $\text{var}_F(X)$ without bias by

$$\widehat{\text{var}}_F = \sum (x_i - \bar{x})^2 / (n-1). \tag{2.9}$$

Plugging formula (2.9) into (2.8) gives $\widehat{\text{se}}$ (1.2), the usual estimate for the standard error of an average $\bar{x}$. In other words, the frequentist accuracy estimate for $\bar{x}$ is itself estimated from the observed data.[3]

2. Taylor-series approximations. Statistics $\hat{\theta} = t(\boldsymbol{x})$ more complicated than $\bar{x}$ can often be related back to the plug-in formula by local linear approximations, sometimes known as the "delta method."† For example, †1
$\hat{\theta} = \bar{x}^2$ has $d\hat{\theta}/d\bar{x} = 2\bar{x}$. Thinking of $2\bar{x}$ as a constant gives

$$\text{se}\left(\bar{x}^2\right) \doteq 2\left|\bar{x}\right| \widehat{\text{se}}, \tag{2.10}$$

with $\widehat{\text{se}}$ as in (1.2). Large sample calculations, as sample size n goes to infinity, validate the delta method which, fortunately, often performs well in small samples.

3. Parametric families and maximum likelihood theory. Theoretical expressions for the standard error of a maximum likelihood estimate (MLE) are discussed in Chapters 4 and 5, in the context of parametric families of distributions. These combine Fisherian theory, Taylor-series approximations, and the plug-in principle in an easy-to-apply package.

4. Simulation and the bootstrap. Modern computation has opened up the possibility of numerically implementing the "infinite sequence of future trials" definition, except for the infinite part. An estimate $\hat{F}$ of F, perhaps the MLE, is found, and values $\hat{\Theta}^{(k)} = t(\boldsymbol{X}^{(k)})$ simulated from $\hat{F}$ for $k = 1, 2, \ldots, B$, say $B = 1000$. The empirical standard deviation of the $\hat{\Theta}$s is then the frequentist estimate of standard error for $\hat{\theta} = t(\boldsymbol{x})$, and similarly with other measures of accuracy.

This is a good description of the bootstrap, Chapter 10. (Notice that

[3] The most familiar example is the observed proportion p of heads in n flips of a coin having true probability π: the actual standard error is $[\pi(1-\pi)/n]^{1/2}$ but we can only report the plug-in estimate $[p(1-p)/n]^{1/2}$.

Table 2.1 *Three estimates of location for the* `gfr` *data, and their estimated standard errors; last two standard errors using the bootstrap,* $B = 1000$.

	Estimate	Standard error
mean	54.25	.95
25% Winsorized mean	52.61	.78
median	52.24	.87

here the plugging-in, of $\hat{F}$ for F, comes *first* rather than at the end of the process.) The classical methods 1–3 above are restricted to estimates $\hat{\theta} = t(\boldsymbol{x})$ that are smoothly defined functions of various sample means. Simulation calculations remove this restriction. Table 2.1 shows three "location" estimates for the `gfr` data, the mean, the 25% Winsorized mean,[4] and the median, along with their standard errors, the last two computed by the bootstrap. A happy feature of computer-age statistical inference is the tremendous expansion of useful and usable statistics $t(\boldsymbol{x})$ in the statistician's working toolbox, the `lowess` algorithm in Figures 1.2 and 1.3 providing a nice example.

5. Pivotal statistics. A pivotal statistic $\hat{\theta} = t(\boldsymbol{x})$ is one whose distribution does *not* depend upon the underlying probability distribution F. In such a case the theoretical distribution of $\hat{\Theta} = t(\boldsymbol{X})$ applies exactly to $\hat{\theta}$, removing the need for devices 1–4 above. The classic example concerns Student's two-sample t-test.

In a two-sample problem the statistician observes two sets of numbers,

$$\boldsymbol{x}_1 = (x_{11}, x_{12}, \dots, x_{1n_1}) \quad \boldsymbol{x}_2 = (x_{21}, x_{22}, \dots, x_{2n_2}), \tag{2.11}$$

and wishes to test the *null hypothesis* that they come from the same distribution (as opposed to, say, the second set tending toward larger values than the first). It is assumed that the distributuion F_1 for $\boldsymbol{x}_1$ is *normal*, or *Gaussian*,

$$X_{1i} \overset{\text{ind}}{\sim} \mathcal{N}(\mu_1, \sigma^2), \qquad i = 1, 2, \dots, n_1, \tag{2.12}$$

the notation indicating n_1 independent draws from a normal distribution[5]

[4] All observations below the 25th percentile of the 211 observations are moved up to that point, similarly those above the 75th percentile are moved down, and finally the mean is taken.

[5] Each draw having probability density $(2\pi\sigma^2)^{-1/2} \exp\{-0.5 \cdot (x - \mu_1)^2/\sigma^2\}$.

with expectation μ_1 and variance σ^2. Likewise

$$X_{2i} \overset{\text{ind}}{\sim} \mathcal{N}(\mu_2, \sigma^2) \qquad i = 1, 2, \ldots, n_2. \tag{2.13}$$

We wish to test the null hypothesis

$$H_0 : \mu_1 = \mu_2. \tag{2.14}$$

The obvious test statistic $\hat{\theta} = \bar{x}_2 - \bar{x}_1$, the difference of the means, has distribution

$$\hat{\theta} \sim \mathcal{N}\left(0, \sigma^2 \left(\tfrac{1}{n_1} + \tfrac{1}{n_2}\right)\right) \tag{2.15}$$

under H_0. We could plug in the unbiased estimate of σ^2,

$$\hat{\sigma}^2 = \left[\sum_1^{n_1} (x_{1i} - \bar{x}_1)^2 + \sum_1^{n_2} (x_{2i} - \bar{x}_2)^2\right] \Big/ (n_1 + n_2 - 2), \tag{2.16}$$

but Student provided a more elegant solution: instead of $\hat{\theta}$, we test H_0 using the two-sample t-statistic

$$t = \frac{\bar{x}_2 - \bar{x}_1}{\widehat{\text{sd}}}, \qquad \text{where } \widehat{\text{sd}} = \hat{\sigma}\left(\tfrac{1}{n_1} + \tfrac{1}{n_2}\right)^{1/2}. \tag{2.17}$$

Under H_0, t is pivotal, having the same distribution (Student's t distribution with $n_1 + n_2 - 2$ degrees of freedom), no matter what the value of the "nuisance parameter" σ.

For $n_1 + n_2 - 2 = 70$, as in the leukemia example (1.5)–(1.6), Student's distribution gives

$$\Pr_{H_0}\{-1.99 \le t \le 1.99\} = 0.95. \tag{2.18}$$

The hypothesis test that rejects H_0 if $|t|$ exceeds 1.99 has probability exactly 0.05 of mistaken rejection. Similarly,

$$\bar{x}_2 - \bar{x}_1 \pm 1.99 \tag{2.19}$$

is an exact 0.95 *confidence interval* for the difference $\mu_2 - \mu_1$, covering the true value in 95% of repetitions of probability model (2.12)–(2.13).[6]

[6] Occasionally, one sees frequentism defined in careerist terms, e.g., "A statistician who always rejects null hypotheses at the 95% level will over time make only 5% errors of the first kind." This is not a comforting criterion for the statistician's clients, who are interested in their own situations, not everyone else's. Here we are only assuming hypothetical repetitions of the specific problem at hand.

What might be called the *strong definition of frequentism* insists on exact frequentist correctness under experimental repetitions. Pivotality, unfortunately, is unavailable in most statistical situations. Our looser definition of frequentism, supplemented by devices such as those above,[7] presents a more realistic picture of actual frequentist practice.

2.2 Frequentist Optimality

The popularity of frequentist methods reflects their relatively modest mathematical modeling assumptions: only a probability model F (more exactly a family of probabilities, Chapter 3) and an algorithm of choice $t(\boldsymbol{x})$. This flexibility is also a defect in that the principle of frequentist correctness doesn't help with the choice of algorithm. Should we use the sample mean to estimate the location of the `gfr` distribution? Maybe the 25% Winsorized mean would be better, as Table 2.1 suggests.

The years 1920–1935 saw the development of two key results on *frequentist optimality*, that is, finding the *best* choice of $t(\boldsymbol{x})$ given model F. The first of these was Fisher's theory of maximum likelihood estimation and the Fisher information bound: in parametric probability models of the type discussed in Chapter 4, the MLE is the optimum estimate in terms of minimum (asymptotic) standard error.

In the same spirit, the Neyman–Pearson lemma provides an optimum hypothesis-testing algorithm. This is perhaps the most elegant of frequentist constructions. In its simplest formulation, the NP lemma assumes we are trying to decide between two possible probability density functions for the observed data $\boldsymbol{x}$, a null hypothesis density $f_0(\boldsymbol{x})$ and an alternative density $f_1(\boldsymbol{x})$. A testing rule $t(\boldsymbol{x})$ says which choice, 0 or 1, we will make having observed data $\boldsymbol{x}$. Any such rule has two associated frequentist error probabilities: choosing f_1 when actually f_0 generated $\boldsymbol{x}$, and vice versa,

$$\begin{aligned} \alpha &= \Pr_{f_0}\{t(\boldsymbol{x}) = 1\}, \\ \beta &= \Pr_{f_1}\{t(\boldsymbol{x}) = 0\}. \end{aligned} \tag{2.20}$$

Let $L(\boldsymbol{x})$ be the *likelihood ratio*,

$$L(\boldsymbol{x}) = f_1(\boldsymbol{x})/f_0(\boldsymbol{x}) \tag{2.21}$$

[7] The list of devices is not complete. Asymptotic calculations play a major role, as do more elaborate combinations of pivotality and the plug-in principle; see the discussion of approximate bootstrap confidence intervals in Chapter 11.

and define the testing rule $t_c(\boldsymbol{x})$ by

$$t_c(\boldsymbol{x}) = \begin{cases} 1 & \text{if } \log L(\boldsymbol{x}) \geq c \\ 0 & \text{if } \log L(\boldsymbol{x}) < c. \end{cases} \tag{2.22}$$

There is one such rule for each choice of the cutoff c. The Neyman–Pearson lemma says that only rules of form (2.22) can be optimum; for any other rule $t(\boldsymbol{x})$ there will be a rule $t_c(\boldsymbol{x})$ having smaller errors of both kinds,[8]

$$\alpha_c < \alpha \quad \text{and} \quad \beta_c < \beta. \tag{2.23}$$

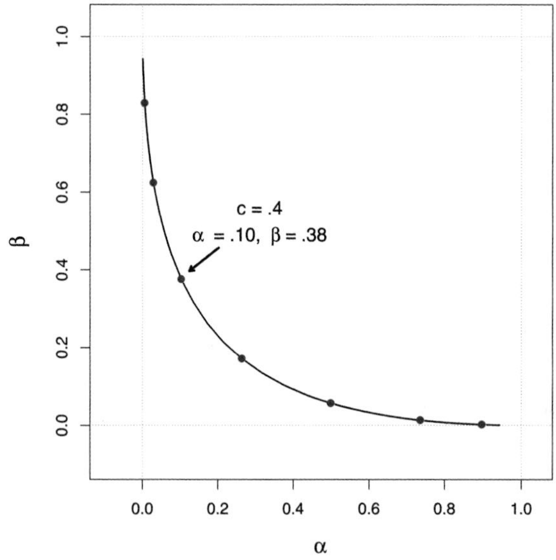

Figure 2.2 Neyman–Pearson alpha–beta curve for $f_0 \sim \mathcal{N}(0, 1)$, $f_1 \sim \mathcal{N}(.5, 1)$, and sample size $n = 10$. Red dots correspond to cutoffs $c = .8, .6, .4, \ldots, -.4$.

Figure 2.2 graphs (α_c, β_c) as a function of the cutoff c, for the case where $\boldsymbol{x} = (x_1, x_2, \ldots, x_{10})$ is obtained by independent sampling from a normal distribution, $\mathcal{N}(0, 1)$ for f_0 versus $\mathcal{N}(0.5, 1)$ for f_1. The NP lemma says that any rule not of form (2.22) must have its (α, β) point lying above the curve.

[8] Here we are ignoring some minor definitional difficulties that can occur if f_0 and f_1 are discrete.

Frequentist optimality theory, both for estimation and for testing, anchored statistical practice in the twentieth century. The larger data sets and more complicated inferential questions of the current era have strained the capabilities of that theory. Computer-age statistical inference, as we will see, often displays an unsettling ad hoc character. Perhaps some contemporary Fishers and Neymans will provide us with a more capacious optimality theory equal to the challenges of current practice, but for now that is only a hope.

Frequentism cannot claim to be a seamless philosophy of statistical inference. Paradoxes and contradictions abound within its borders, as will be shown in the next chapter. That being said, frequentist methods have a natural appeal to working scientists, an impressive history of successful application, and, as our list of five "devices" suggests, the capacity to encourage clever methodology. The story that follows is not one of abandonment of frequentist thinking, but rather a broadening of connections with other methods.

2.3 Notes and Details

The name "frequentism" seems to have been suggested by Neyman as a statistical analogue of Richard von Mises' frequentist theory of probability, the connection being made explicit in his 1977 paper, "Frequentist probability and frequentist statistics." "Behaviorism" might have been a more descriptive name[9] since the theory revolves around the long-run behavior of statistics $t(\boldsymbol{x})$, but in any case "frequentism" has stuck, replacing the older (sometimes disparaging) term "objectivism." Neyman's attempt at a complete frequentist theory of statistical inference, "inductive behavior," is not much quoted today, but can claim to be an important influence on Wald's development of decision theory.

R. A. Fisher's work on maximum likelihood estimation is featured in Chapter 4. Fisher, arguably the founder of frequentist optimality theory, was not a pure frequentist himself, as discussed in Chapter 4 and Efron (1998), "R. A. Fisher in the 21st Century." (Now that we are well into the twenty-first century, the author's talents as a prognosticator can be frequentistically evaluated.)

†1 [p. 15] *Delta method.* The delta method uses a first-order Taylor series to approximate the variance of a function $s(\hat{\theta})$ of a statistic $\hat{\theta}$. Suppose $\hat{\theta}$ has mean/variance (θ, σ^2), and consider the approximation $s(\hat{\theta}) \approx s(\theta) +$

[9] That name is already spoken for in the psychology literature.

$s'(\theta)(\hat{\theta} - \theta)$. Hence $\text{var}\{s(\hat{\theta})\} \approx |s'(\theta)|^2\sigma^2$. We typically plug-in $\hat{\theta}$ for θ, and use an estimate for σ^2.

3

Bayesian Inference

The human mind is an inference machine: "It's getting windy, the sky is darkening, I'd better bring my umbrella with me." Unfortunately, it's not a very dependable machine, especially when weighing complicated choices against past experience. *Bayes' theorem* is a surprisingly simple mathematical guide to accurate inference. The theorem (or "rule"), now 250 years old, marked the beginning of statistical inference as a serious scientific subject. It has waxed and waned in influence over the centuries, now waxing again in the service of computer-age applications.

Bayesian inference, if not directly opposed to frequentism, is at least orthogonal. It reveals some worrisome flaws in the frequentist point of view, while at the same time exposing itself to the criticism of dangerous overuse. The struggle to combine the virtues of the two philosophies has become more acute in an era of massively complicated data sets. Much of what follows in succeeding chapters concerns this struggle. Here we will review some basic Bayesian ideas and the ways they impinge on frequentism.

The fundamental unit of statistical inference both for frequentists and for Bayesians is a *family* of probability densities

$$\mathcal{F} = \left\{ f_\mu(x);\; x \in \mathcal{X},\; \mu \in \Omega \right\}; \tag{3.1}$$

x, the observed data, is a point[1] in the *sample space* $\mathcal{X}$, while the unobserved parameter μ is a point in the *parameter space* Ω. The statistician observes x from $f_\mu(x)$, and infers the value of μ.

Perhaps the most familiar case is the normal family

$$f_\mu(x) = \frac{1}{\sqrt{2\pi}} e^{-\frac{1}{2}(x-\mu)^2} \tag{3.2}$$

[1] Both x and μ may be scalars, vectors, or more complicated objects. Other names for the generic "x" and "μ" occur in specific situations, for instance $\boldsymbol{x}$ for x in Chapter 2. We will also call $\mathcal{F}$ a "family of probability distributions."

(more exactly, the one-dimensional normal translation family[2] with variance 1), with both $\mathcal{X}$ and Ω equaling $\mathcal{R}^1$, the entire real line $(-\infty, \infty)$. Another central example is the Poisson family

$$f_\mu(x) = e^{-\mu}\mu^x/x!, \tag{3.3}$$

where $\mathcal{X}$ is the nonnegative integers $\{0, 1, 2, \dots\}$ and Ω is the nonnegative real line $(0, \infty)$. (Here the "density" (3.3) specifies the atoms of probability on the discrete points of $\mathcal{X}$.)

Bayesian inference requires one crucial assumption in addition to the probability family $\mathcal{F}$, the knowledge of a *prior density*

$$g(\mu), \qquad \mu \in \Omega; \tag{3.4}$$

$g(\mu)$ represents prior information concerning the parameter μ, available to the statistician before the observation of x. For instance, in an application of the normal model (3.2), it could be known that μ is positive, while past experience shows it never exceeding 10, in which case we might take $g(\mu)$ to be the uniform density $g(\mu) = 1/10$ on the interval $[0, 10]$. Exactly what constitutes "prior knowledge" is a crucial question we will consider in ongoing discussions of Bayes' theorem.

Bayes' theorem is a rule for combining the prior knowledge in $g(\mu)$ with the current evidence in x. Let $g(\mu|x)$ denote the *posterior density* of μ, that is, our update of the prior density $g(\mu)$ after taking account of observation x. Bayes' rule provides a simple expression for $g(\mu|x)$ in terms of $g(\mu)$ and $\mathcal{F}$.

$$\textbf{Bayes' Rule:} \quad g(\mu|x) = g(\mu) f_\mu(x)/f(x), \qquad \mu \in \Omega, \tag{3.5}$$

where $f(x)$ is the *marginal density* of x,

$$f(x) = \int_\Omega f_\mu(x) g(\mu)\, d\mu. \tag{3.6}$$

(The integral in (3.6) would be a sum if Ω were discrete.) The Rule is a straightforward exercise in conditional probability,[3] and yet has far-reaching and sometimes surprising consequences.

In Bayes' formula (3.5), x is fixed at its observed value while μ varies over Ω, just the opposite of frequentist calculations. We can emphasize this

[2] Standard notation is $x \sim \mathcal{N}(\mu, \sigma^2)$ for a normal distribution with expectation μ and variance σ^2, so (3.2) has $x \sim \mathcal{N}(\mu, 1)$.

[3] $g(\mu|x)$ is the ratio of $g(\mu) f_\mu(x)$, the joint probability of the pair (μ, x), and $f(x)$, the marginal probability of x.

by rewriting (3.5) as

$$g(\mu|x) = c_x L_x(\mu) g(\mu), \tag{3.7}$$

where $L_x(\mu)$ is the *likelihood function*, that is, $f_\mu(x)$ with x fixed and μ varying. Having computed $L_x(\mu)g(\mu)$, the constant c_x can be determined numerically from the requirement that $g(\mu|x)$ integrate to 1, obviating the calculation of $f(x)$ (3.6).

Note Multiplying the likelihood function by any fixed constant c_0 has no effect on (3.7) since c_0 can be absorbed into c_x. So for the Poisson family (3.3) we can take $L_x(\mu) = e^{-\mu}\mu^x$, ignoring the $x!$ factor, which acts as a constant in Bayes' rule. The luxury of ignoring factors depending only on x often simplifies Bayesian calculations.

For any two points μ_1 and μ_2 in Ω, the ratio of posterior densities is, by division in (3.5),

$$\frac{g(\mu_1|x)}{g(\mu_2|x)} = \frac{g(\mu_1)}{g(\mu_2)} \frac{f_{\mu_1}(x)}{f_{\mu_2}(x)} \tag{3.8}$$

(no longer involving the marginal density $f(x)$), that is, "the posterior odds ratio is the prior odds ratio times the likelihood ratio," a memorable restatement of Bayes' rule.

3.1 Two Examples

A simple but genuine example of Bayes' rule in action is provided by the story of the *Physicist's Twins*: thanks to sonograms, a physicist found out she was going to have twin boys. "What is the probability my twins will be *Identical*, rather than *Fraternal*?" she asked. The doctor answered that one-third of twin births were Identicals, and two-thirds Fraternals.

In this situation μ, the unknown parameter (or "state of nature") is either *Identical* or *Fraternal* with prior probability 1/3 or 2/3; X, the possible sonogram results for twin births, is either *Same Sex* or *Different Sexes*, and $x =$ *Same Sex* was observed. (We can ignore sex since that does not affect the calculation.) A crucial fact is that identical twins are always same-sex while fraternals have probability 0.5 of same or different, so *Same Sex* in the sonogram is twice as likely if the twins are Identical. Applying Bayes' rule in ratio form (3.8) answers the physicist's question:

$$\begin{aligned}\frac{g(\text{Identical} \mid \text{Same})}{g(\text{Fraternal} \mid \text{Same})} &= \frac{g(\text{Identical})}{g(\text{Fraternal})} \cdot \frac{f_{\text{Identical}}(\text{Same})}{f_{\text{Fraternal}}(\text{Same})} \\ &= \frac{1/3}{2/3} \cdot \frac{1}{1/2} = 1.\end{aligned} \tag{3.9}$$

That is, the posterior odds are even, and the physicist's twins have equal probabilities 0.5 of being Identical or Fraternal.[4] Here the doctor's prior odds ratio, 2 to 1 in favor of Fraternal, is balanced out by the sonogram's likelihood ratio of 2 to 1 in favor of Identical.

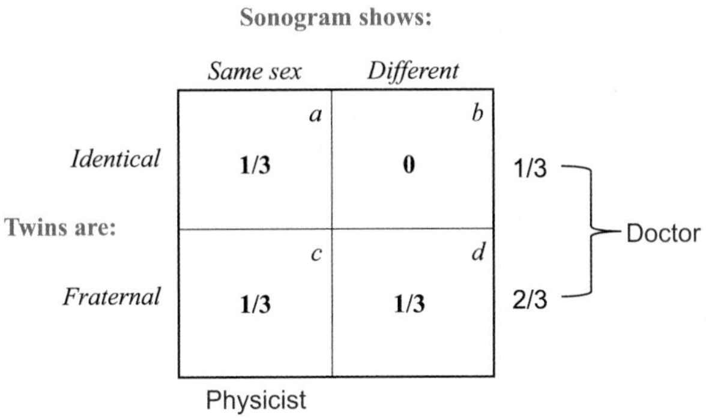

Figure 3.1 Analyzing the twins problem.

There are only four possible combinations of parameter μ and outcome x in the twins problem, labeled *a, b, c,* and *d* in Figure 3.1. Cell *b* has probability 0 since Identicals cannot be of Different Sexes. Cells *c* and *d* have equal probabilities because of the random sexes of Fraternals. Finally, $a + b$ must have total probability 1/3, and $c + d$ total probability 2/3, according to the doctor's prior distribution. Putting all this together, we can fill in the probabilities for all four cells, as shown. The physicist knows she is in the first column of the table, where the conditional probabilities of Identical or Fraternal are equal, just as provided by Bayes' rule in (3.9).

Presumably the doctor's prior distribution came from some enormous state or national database, say three million previous twin births, one million Identical pairs and two million Fraternals. We deduce that cells *a, c,* and *d* must have had one million entries each in the database, while cell *b* was empty. Bayes' rule can be thought of as a *big book* with one page

[4] They turned out to be Fraternal.

for each possible outcome x. (The book has only two pages in Figure 3.1.) The physicist turns to the page "Same Sex" and sees two million previous twin births, half Identical and half Fraternal, correctly concluding that the odds are equal in her situation.

Given any prior distribution $g(\mu)$ and any family of densities $f_\mu(x)$, Bayes' rule will always provide a version of the big book. That doesn't mean that the book's contents will always be equally convincing. The prior for the twins problems was based on a large amount of relevant previous experience. Such experience is most often unavailable. Modern Bayesian practice uses various strategies to construct an appropriate "prior" $g(\mu)$ in the absence of prior experience, leaving many statisticians unconvinced by the resulting Bayesian inferences. Our second example illustrates the difficulty.

Table 3.1 *Scores from two tests taken by 22 students,* `mechanics` *and* `vectors`.

	1	2	3	4	5	6	7	8	9	10	11
`mechanics`	7	44	49	59	34	46	0	32	49	52	44
`vectors`	51	69	41	70	42	40	40	45	57	64	61

	12	13	14	15	16	17	18	19	20	21	22
`mechanics`	36	42	5	22	18	41	48	31	42	46	63
`vectors`	59	60	30	58	51	63	38	42	69	49	63

Table 3.1 shows the scores on two tests, `mechanics` and `vectors`, achieved by $n = 22$ students. The sample correlation coefficient between the two scores is $\hat{\theta} = 0.498$,

$$\hat{\theta} = \sum_{i=1}^{22}(m_i - \bar{m})(v_i - \bar{v}) \Bigg/ \left[\sum_{i=1}^{22}(m_i - \bar{m})^2 \sum_{i=1}^{22}(v_i - \bar{v})^2\right]^{1/2}, \quad (3.10)$$

with m and v short for `mechanics` and `vectors`, $\bar{m}$ and $\bar{v}$ their averages. We wish to assign a Bayesian measure of posterior accuracy to the true correlation coefficient θ, "true" meaning the correlation for the hypothetical population of all students, of which we observed only 22.

If we assume that the joint (m, v) distribution is bivariate normal (as discussed in Chapter 5), then the density of $\hat{\theta}$ as a function of θ has a
†1 known form,†

$$f_\theta\left(\hat{\theta}\right) = \frac{(n-2)(1-\theta^2)^{(n-1)/2}\left(1-\hat{\theta}^2\right)^{(n-4)/2}}{\pi} \int_0^\infty \frac{dw}{\left(\cosh w - \theta\hat{\theta}\right)^{n-1}}. \tag{3.11}$$

In terms of our general Bayes notation, parameter μ is θ, observation x is $\hat{\theta}$, and family $\mathcal{F}$ is given by (3.11), with both Ω and $\mathcal{X}$ equaling the interval $[-1, 1]$. Formula (3.11) looks formidable to the human eye but not to the computer eye, which makes quick work of it.

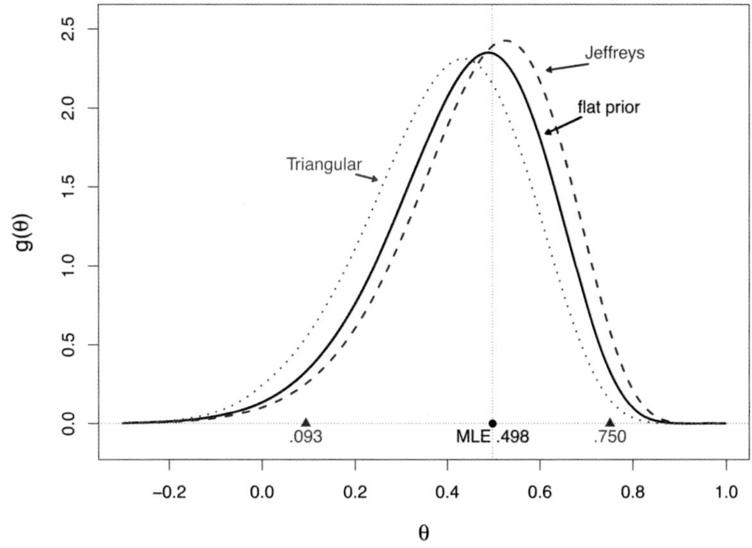

Figure 3.2 Student scores data; posterior density of correlation θ for three possible priors.

In this case, as in the majority of scientific situations, we don't have a trove of relevant past experience ready to provide a prior $g(\theta)$. One expedient, going back to Laplace, is the "principle of insufficient reason," that is, we take θ to be uniformly distributed over Ω,

$$g(\theta) = \tfrac{1}{2} \qquad \text{for } -1 \leq \theta \leq 1, \tag{3.12}$$

a "flat prior." The solid black curve in Figure 3.2 shows the resulting posterior density (3.5), which is just the likelihood $f_\theta(0.498)$ plotted as a function of θ (and scaled to have integral 1).

Jeffreys' prior,

$$g^{\text{Jeff}}(\theta) = 1/(1 - \theta^2), \tag{3.13}$$

yields posterior density $g(\theta|\hat{\theta})$ shown by the dashed red curve. It suggests somewhat bigger values for the unknown parameter θ. Formula (3.13) arises from a theory of "uninformative priors" discussed in the next section, an improvement on the principle of insufficient reason; (3.13) is an *improper density* in that $\int_{-1}^{1} g(\theta)\, d\theta = \infty$, but it still provides proper posterior densities when deployed in Bayes' rule (3.5).

The dotted blue curve in Figure 3.2 is posterior density $g(\theta|\hat{\theta})$ obtained from the triangular-shaped prior

$$g(\theta) = 1 - |\theta|. \tag{3.14}$$

This is a primitive example of a *shrinkage* prior, one designed to favor smaller values of θ. Its effect is seen in the leftward shift of the posterior density. Shrinkage priors will play a major role in our discussion of large-scale estimation and testing problems, where we are hoping to find a few large effects hidden among thousands of negligible ones.

3.2 Uninformative Prior Distributions

Given a convincing prior distribution, Bayes' rule is easier to use and produces more satisfactory inferences than frequentist methods. The dominance of frequentist practice reflects the scarcity of useful prior information in day-to-day scientific applications. But the Bayesian impulse is strong, and almost from its inception 250 years ago there have been proposals for the construction of "priors" that permit the use of Bayes' rule in the absence of relevant experience.

One approach, perhaps the most influential in current practice, is the employment of *uninformative priors*. "Uninformative" has a positive connotation here, implying that the use of such a prior in Bayes' rule does not tacitly bias the resulting inference. Laplace's principle of insufficient reason, i.e., assigning uniform prior distributions to unknown parameters, is an obvious attempt at this goal. Its use went unchallenged for more than a century, perhaps because of Laplace's influence more than its own virtues.

Venn (of the Venn diagram) in the 1860s, and Fisher in the 1920s, attacking the routine use of Bayes' theorem, pointed out that Laplace's principle could not be applied consistently. In the student correlation example, for instance, a uniform prior distribution for θ would not be uniform if we changed parameters to $\gamma = e^{\theta}$; posterior probabilities such as

$$\Pr\left\{\theta > 0|\hat{\theta}\right\} = \Pr\left\{\gamma > 1|\hat{\theta}\right\} \tag{3.15}$$

would depend on whether θ or γ was taken to be uniform a priori. Neither choice then could be considered uninformative.

A more sophisticated version of Laplace's principle was put forward by Jeffreys beginning in the 1930s. It depends, interestingly enough, on the frequentist notion of *Fisher information* (Chapter 4). For a *one-parameter family* $f_\mu(x)$, where the parameter space Ω is an interval of the real line $\mathcal{R}^1$, the Fisher information is defined to be

$$\mathcal{I}_\mu = E_\mu\left\{\left(\frac{\partial}{\partial\mu}\log f_\mu(x)\right)^2\right\}. \tag{3.16}$$

(For the Poisson family (3.3), $\partial/\partial\mu(\log f_\mu(x)) = x/\mu - 1$ and $\mathcal{I}_\mu = 1/\mu$.) The Jeffreys' prior $g^{\text{Jeff}}(\mu)$ is by definition

$$g^{\text{Jeff}}(\mu) = \mathcal{I}_\mu^{1/2}. \tag{3.17}$$

Because $1/\mathcal{I}_\mu$ equals, approximately, the variance σ_μ^2 of the MLE $\hat{\mu}$, an equivalent definition is

$$g^{\text{Jeff}}(\mu) = 1/\sigma_\mu. \tag{3.18}$$

Formula (3.17) does in fact transform correctly under parameter changes, avoiding the Venn–Fisher criticism.† It is known that $\hat{\theta}$ in family (3.11) has approximate standard deviation †2

$$\sigma_\theta = c(1 - \theta^2), \tag{3.19}$$

yielding Jeffreys' prior (3.13) from (3.18), the constant factor c having no effect on Bayes' rule (3.5)–(3.6).

The red triangles in Figure 3.2 indicate the "95% credible interval" [0.093, 0.750] for θ, based on Jeffreys' prior. That is, the posterior probability $0.093 \le \theta \le 0.750$ equals 0.95,

$$\int_{0.093}^{0.750} g^{\text{Jeff}}\left(\theta|\hat{\theta}\right)\, d\theta = 0.95, \tag{3.20}$$

with probability 0.025 for $\theta < 0.093$ or $\theta > 0.750$. It is not an accident that this nearly equals the standard Neyman 95% confidence interval based on $f_\theta(\hat{\theta})$ (3.11). Jeffreys' prior tends to induce this nice connection between the Bayesian and frequentist worlds, at least in one-parameter families.

Multiparameter probability families, Chapter 4, make everything more difficult. Suppose, for instance, the statistician observes 10 independent versions of the normal model (3.2), with possibly different values of μ,

$$x_i \overset{\text{ind}}{\sim} \mathcal{N}(\mu_i, 1) \qquad \text{for } i = 1, 2, \dots, 10, \tag{3.21}$$

in standard notation. Jeffreys' prior is flat for any one of the 10 problems, which is reasonable for dealing with them separately, but the joint Jeffreys' prior

$$g(\mu_1, \mu_2, \dots, \mu_{10}) = \text{constant}, \tag{3.22}$$

also flat, can produce disastrous overall results, as discussed in Chapter 13.

Computer-age applications are often more like (3.21) than (3.11), except with hundreds or thousands of cases rather than 10 to consider simultaneously. Uninformative priors of many sorts, including Jeffreys', are highly popular in current applications, as we will discuss. This leads to an interplay between Bayesian and frequentist methodology, the latter intended to control possible biases in the former, exemplifying our general theme of computer-age statistical inference.

3.3 Flaws in Frequentist Inference

Bayesian statistics provides an internally consistent ("coherent") program of inference. The same cannot be said of frequentism. The apocryphal story of the *meter reader* makes the point: an engineer measures the voltages on a batch of 12 tubes, using a voltmeter that is normally calibrated,

$$x \sim \mathcal{N}(\mu, 1), \tag{3.23}$$

x being any one measurement and μ the true batch voltage. The measurements range from 82 to 99, with an average of $\bar{x} = 92$, which he reports
†3 back as an unbiased estimate of μ.†

The next day he discovers a glitch in his voltmeter such that any voltage exceeding 100 would have been reported as $x = 100$. His frequentist statistician tells him that $\bar{x} = 92$ is no longer unbiased for the true expectation μ since (3.23) no longer completely describes the probability family. (The statistician says that 92 is a little too small.) The fact that the glitch didn't affect any of the actual measurements doesn't let him off the hook; $\bar{x}$ would not be unbiased for μ in future realizations of $\bar{X}$ from the actual probability model.

A Bayesian statistician comes to the meter reader's rescue. For any prior density $g(\mu)$, the posterior density $g(\mu|\boldsymbol{x}) = g(\mu)f_\mu(\boldsymbol{x})/f(\boldsymbol{x})$, where $\boldsymbol{x}$ is the vector of 12 measurements, depends only on the data $\boldsymbol{x}$ actually observed, and *not on other potential data sets* $\boldsymbol{X}$ *that might have been seen.* The flat Jeffreys' prior $g(\mu) =$ constant yields posterior expectation $\bar{x} = 92$ for μ, irrespective of whether or not the glitch would have affected readings above 100.

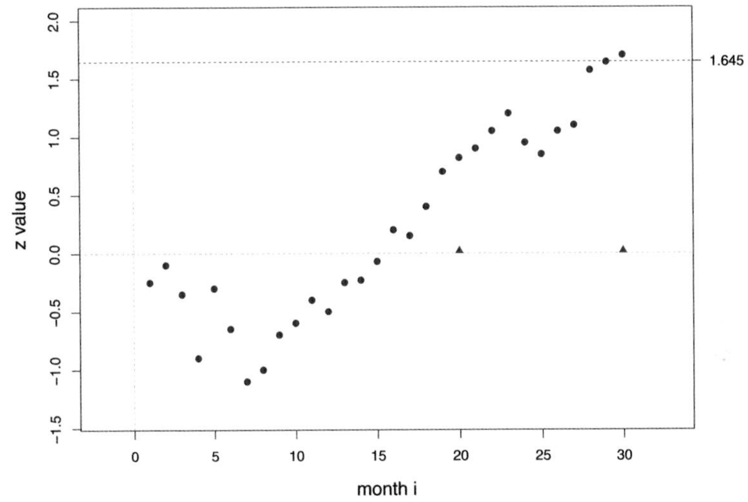

Figure 3.3 Z-values against null hypothesis $\mu = 0$ for months 1 through 30.

A less contrived version of the same phenomenon is illustrated in Figure 3.3. An ongoing experiment is being run. Each month i an independent normal variate is observed,

$$x_i \sim \mathcal{N}(\mu, 1), \tag{3.24}$$

with the intention of testing the null hypothesis $H_0 : \mu = 0$ versus the alternative $\mu > 0$. The plotted points are test statistics

$$Z_i = \sum_{j=1}^{i} x_j \Big/ \sqrt{i}\,, \tag{3.25}$$

a "z-value" based on all the data up to month i,

$$Z_i \sim \mathcal{N}\left(\sqrt{i}\,\mu, 1\right). \tag{3.26}$$

At month 30, the scheduled end of the experiment, $Z_{30} = 1.66$, just exceeding 1.645, the upper 95% point for a $\mathcal{N}(0, 1)$ distribution. Victory! The investigators get to claim "significant" rejection of H_0 at level 0.05.

Unfortunately, it turns out that the investigators broke protocol and peeked at the data at month 20, in the hope of being able to stop an expensive experiment early. This proved a vain hope, $Z_{20} = 0.79$ not being anywhere near significance, so they continued on to month 30 as originally planned. This means they effectively used the stopping rule "stop and declare significance if either Z_{20} or Z_{30} exceeds 1.645." Some computation shows that this rule had probability 0.083, not 0.05, of rejecting H_0 if it were true. Victory has turned into defeat according to the honored frequentist 0.05 criterion.

Once again, the Bayesian statistician is more lenient. The likelihood function for the full data set $\boldsymbol{x} = (x_1, x_2, \dots, x_{30})$,

$$L_{\boldsymbol{x}}(\mu) = \prod_{i=1}^{30} e^{-\frac{1}{2}(x_i - \mu)^2}, \tag{3.27}$$

is the same irrespective of whether or not the experiment *might have* stopped early. The stopping rule doesn't affect the posterior distribution $g(\mu|\boldsymbol{x})$, which depends on $\boldsymbol{x}$ only through the likelihood (3.7).

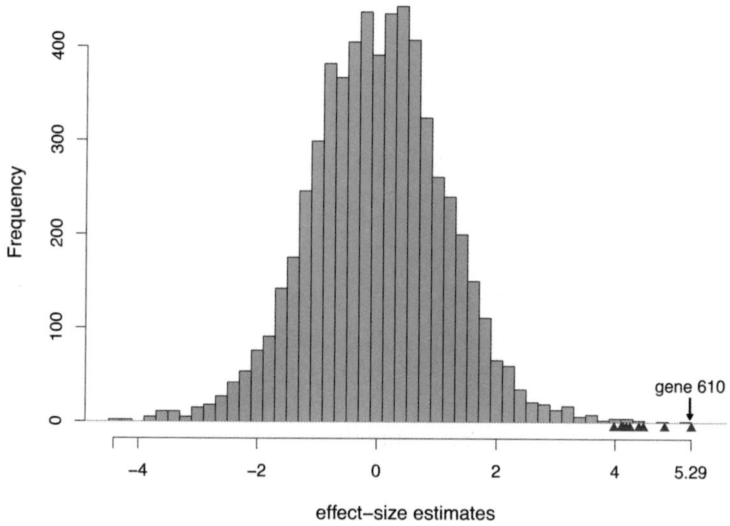

Figure 3.4 Unbiased effect-size estimates for 6033 genes, prostate cancer study. The estimate for gene 610 is $x_{610} = 5.29$. What is its effect size?

The lenient nature of Bayesian inference can look less benign in multiparameter settings. Figure 3.4 concerns a prostate cancer study comparing 52 patients with 50 healthy controls. Each man had his genetic activity measured for a panel of $N = 6033$ genes. A statistic x was computed for each gene,[5] comparing the patients with controls, say† †4

$$x_i \sim \mathcal{N}(\mu_i, 1) \qquad i = 1, 2, \ldots, N, \tag{3.28}$$

where μ_i represents the *true effect size* for gene i. Most of the genes, probably not being involved in prostate cancer, would be expected to have effect sizes near 0, but the investigators hoped to spot a few large μ_i values, either positive or negative.

The histogram of the 6033 x_i values does in fact reveal some large values, $x_{610} = 5.29$ being the winner. Question: what estimate should we give for μ_{610}? Even though x_{610} was individually unbiased for μ_{610}, a frequentist would (correctly) worry that focusing attention on the *largest* of 6033 values would produce an upward bias, and that our estimate should downwardly correct 5.29. "Selection bias," "regression to the mean," and "the winner's curse" are three names for this phenomenon.

Bayesian inference, surprisingly, is immune to selection bias.† Irrespective of whether gene 610 was prespecified for particular attention or only came to attention as the "winner," the Bayes' estimate for μ_{610} given all the data stays the same. This isn't obvious, but follows from the fact that any data-based selection process does not affect the likelihood function in (3.7). †5

What *does* affect Bayesian inference is the prior $g(\boldsymbol{\mu})$ for the full vector $\boldsymbol{\mu}$ of 6033 effect sizes. The flat prior, $g(\boldsymbol{\mu})$ constant, results in the dangerous overestimate $\hat{\mu}_{610} = x_{610} = 5.29$. A more appropriate uninformative prior appears as part of the empirical Bayes calculations of Chapter 15 (and gives $\hat{\mu}_{610} = 4.11$). The operative point here is that there is a price to be paid for the desirable properties of Bayesian inference. Attention shifts from choosing a good frequentist procedure to choosing an appropriate prior distribution. This can be a formidable task in high-dimensional problems, the very kinds featured in computer-age inference.

3.4 A Bayesian/Frequentist Comparison List

Bayesians and frequentists start out on the same playing field, a family of probability distributions $f_\mu(x)$ (3.1), but play the game in orthogonal

[5] The statistic was the two-sample t-statistic (2.17) transformed to normality (3.28); see the endnotes.

directions, as indicated schematically in Figure 3.5: Bayesian inference proceeds vertically, with x fixed, according to the posterior distribution $g(\mu|x)$, while frequentists reason horizontally, with μ fixed and x varying. Advantages and disadvantages accrue to both strategies, some of which are compared next.

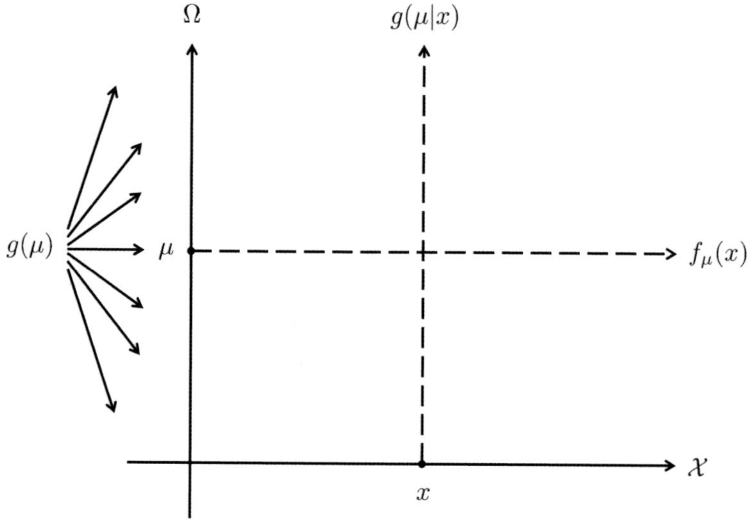

Figure 3.5 Bayesian inference proceeds vertically, given x; frequentist inference proceeds horizontally, given μ.

- Bayesian inference requires a prior distribution $g(\mu)$. When past experience provides $g(\mu)$, as in the twins example, there is every good reason to employ Bayes' theorem. If not, techniques such as those of Jeffreys still permit the use of Bayes' rule, but the results lack the full logical force of the theorem; the Bayesian's right to ignore selection bias, for instance, must then be treated with caution.
- Frequentism replaces the choice of a prior with the choice of a method, or algorithm, $t(x)$, designed to answer the specific question at hand. This adds an arbitrary element to the inferential process, and can lead to meter-reader kinds of contradictions. Optimal choice of $t(x)$ reduces arbitrary behavior, but computer-age applications typically move outside the safe waters of classical optimality theory, lending an ad-hoc character to frequentist analyses.
- Modern data-analysis problems are often approached via a favored meth-

odology, such as logistic regression or regression trees in the examples of Chapter 8. This plays into the methodological orientation of frequentism, which is more flexible than Bayes' rule in dealing with specific algorithms (though one always hopes for a reasonable Bayesian justification for the method at hand).

- Having chosen $g(\mu)$, only a single probability distribution $g(\mu|x)$ is in play for Bayesians. Frequentists, by contrast, must struggle to balance the behavior of $t(x)$ over a family of possible distributions, since μ in Figure 3.5 is unknown. The growing popularity of Bayesian applications (usually begun with uninformative priors) reflects their simplicity of application and interpretation.
- The simplicity argument cuts both ways. The Bayesian essentially bets it all on the choice of his or her prior being correct, or at least not harmful. Frequentism takes a more defensive posture, hoping to do well, or at least not poorly, whatever μ might be.
- A Bayesian analysis answers *all* possible questions at once, for example, estimating $E\{\text{gfr}\}$ or $\Pr\{\text{gfr} < 40\}$ or anything else relating to Figure 2.1. Frequentism focuses on the problem at hand, requiring different estimators for different questions. This is more work, but allows for more intense inspection of particular problems. In situation (2.9) for example, estimators of the form

$$\sum (x_i - \bar{x})^2/(n - c) \tag{3.29}$$

  might be investigated for different choices of the constant c, hoping to reduce expected mean-squared error.
- The simplicity of the Bayesian approach is especially appealing in dynamic contexts, where data arrives sequentially and updating one's beliefs is a natural practice. Bayes' rule was used to devastating effect before the 2012 US presidential election, updating sequential polling results to correctly predict the outcome in all 50 states. Bayes' theorem is an excellent tool in general for combining statistical evidence from disparate sources, the closest frequentist analog being maximum likelihood estimation.
- In the absence of genuine prior information, a whiff of subjectivity[6] hangs over Bayesian results, even those based on uninformative priors. Classical frequentism claimed for itself the high ground of scientific objectivity, especially in contentious areas such as drug testing and approval, where skeptics as well as friends hang on the statistical details.

Figure 3.5 is soothingly misleading in its schematics: μ and x will

[6] Here we are not discussing the important subjectivist school of Bayesian inference, of Savage, de Finetti, and others, covered in Chapter 13.

typically be high-dimensional in the chapters that follow, sometimes *very* high-dimensional, straining to the breaking point both the frequentist and the Bayesian paradigms. Computer-age statistical inference at its most successful *combines* elements of the two philosophies, as for instance in the empirical Bayes methods of Chapter 6, and the lasso in Chapter 16. There are two potent arrows in the statistician's philosophical quiver, and faced, say, with 1000 parameters and 1,000,000 data points, there's no need to go hunting armed with just one of them.

3.5 Notes and Details

Thomas Bayes, if transferred to modern times, might well be employed as a successful professor of mathematics. Actually, he was a mid-eighteenth-century nonconformist English minister with substantial mathematical interests. Richard Price, a leading figure of letters, science, and politics, had Bayes' theorem published in the 1763 *Transactions of the Royal Society* (two years after Bayes' death), his interest being partly theological, with the rule somehow proving the existence of God. Bellhouse's (2004) biography includes some of Bayes' other mathematical accomplishments.

Harold Jeffreys was another part-time statistician, working from his day job as the world's premier geophysicist of the inter-war period (and fierce opponent of the theory of continental drift). What we called *uninformative* priors are also called *noninformative* or *objective*. Jeffreys' brand of Bayesianism had a dubious reputation among Bayesians in the period 1950–1990, with preference going to subjective analysis of the type advocated by Savage and de Finetti. The introduction of *Markov chain Monte Carlo* methodology was the kind of technological innovation that changes philosophies. MCMC (Chapter 13), being very well suited to Jeffreys-style analysis of Big Data problems, moved Bayesian statistics out of the textbooks and into the world of computer-age applications. Berger (2006) makes a spirited case for the objective Bayes approach.

†1 [p. 26] *Correlation coefficient density.* Formula (3.11) for the correlation coefficient density was R. A. Fisher's debut contribution to the statistics literature. Chapter 32 of Johnson and Kotz (1970b) gives several equivalent forms. The constant c in (3.19) is often taken to be $(n-3)^{-1/2}$, with n the sample size.

†2 [p. 29] *Jeffreys' prior and transformations.* Suppose we change parameters from μ to $\tilde{\mu}$ in a smoothly differentiable way. The new family $\tilde{f}_{\tilde{\mu}}(x)$

satisfies

$$\frac{\partial}{\partial \tilde{\mu}} \log \tilde{f}_{\tilde{\mu}}(x) = \frac{\partial \mu}{\partial \tilde{\mu}} \frac{\partial}{\partial \mu} \log f_\mu(x). \tag{3.30}$$

Then $\tilde{\mathcal{I}}_{\tilde{\mu}} = \left(\frac{\partial \mu}{\partial \tilde{\mu}}\right)^2 \mathcal{I}_\mu$ (3.16) and $\tilde{g}^{\text{Jeff}}(\tilde{\mu}) = \left|\frac{\partial \mu}{\partial \tilde{\mu}}\right| g^{\text{Jeff}}(\mu)$. But this just says that $g^{\text{Jeff}}(\mu)$ transforms correctly to $\tilde{g}^{\text{Jeff}}(\tilde{\mu})$.

†3 [p. 30] The *meter-reader* fable is taken from Edwards' (1992) book *Likelihood*, where he credits John Pratt. It nicely makes the point that frequentist inferences, which are calibrated in terms of possible observed data sets X, may be inappropriate for the actual observation x. This is the difference between working in the horizontal and vertical directions of Figure 3.5.

†4 [p. 33] *Two-sample t-statistic.* Applied to gene i's data in the prostate study, the two-sample t-statistic t_i (2.17) has theoretical null hypothesis distribution t_{100}, a Student's t distribution with 100 degrees of freedom; x_i in (3.28) is $\Phi^{-1} F_{100}(t_i)$, where Φ and F_{100} are the cumulative distribution functions of standard normal and t_{100} variables. Section 7.4 of Efron (2010) motivates approximation (3.28).

†5 [p. 33] *Selection bias.* Senn (2008) discusses the immunity of Bayesian inferences to selection bias and other "paradoxes," crediting Phil Dawid for the original idea. The article catches the possible uneasiness of following Bayes' theorem too literally in applications.

The 22 students in Table 3.1 were randomly selected from a larger data set of 88 in Mardia *et al.* (1979) (which gave $\hat{\theta} = 0.553$). Welch and Peers (1963) initiated the study of priors whose credible intervals, such as $[0.093, 0.750]$ in Figure 3.2, match frequentist confidence intervals. In one-parameter problems, Jeffreys' priors provide good matches, but not ususally in multiparameter situations. In fact, no single multiparameter prior can give good matches for all one-parameter subproblems, a source of tension between Bayesian and frequentist methods revisited in Chapter 11.

4

Fisherian Inference and Maximum Likelihood Estimation

Sir Ronald Fisher was arguably the most influential anti-Bayesian of all time, but that did not make him a conventional frequentist. His key data-analytic methods—analysis of variance, significance testing, and maximum likelihood estimation—were almost always applied frequentistically. Their Fisherian rationale, however, often drew on ideas neither Bayesian nor frequentist in nature, or sometimes the two in combination. Fisher's work held a central place in twentieth-century applied statistics, and some of it, particularly maximum likelihood estimation, has moved forcefully into computer-age practice. This chapter's brief review of Fisherian methodology sketches parts of its unique philosophical structure, while concentrating on those topics of greatest current importance.

4.1 Likelihood and Maximum Likelihood

Fisher's seminal work on estimation focused on the likelihood function, or more exactly its logarithm. For a family of probability densities $f_\mu(x)$ (3.1), the *log likelihood function* is

$$l_x(\mu) = \log\{f_\mu(x)\}, \tag{4.1}$$

the notation $l_x(\mu)$ emphasizing that the parameter vector μ is varying while the observed data vector x is fixed. The *maximum likelihood estimate* (MLE) is the value of μ in parameter space Ω that maximizes $l_x(\mu)$,

$$\text{MLE}: \quad \hat{\mu} = \arg\max_{\mu \in \Omega}\{l_x(\mu)\}. \tag{4.2}$$

It can happen that $\hat{\mu}$ doesn't exist or that there are multiple maximizers, but here we will assume the usual case where $\hat{\mu}$ exists uniquely. More careful references are provided in the endnotes.

Definition (4.2) is extended to provide maximum likelihood estimates for a function $\theta = T(\mu)$ of μ according to the simple plug-in rule

$$\hat{\theta} = T(\hat{\mu}), \tag{4.3}$$

most often with θ being a scalar parameter of particular interest, such as the regression coefficient of an important covariate in a linear model.

Maximum likelihood estimation came to dominate classical applied estimation practice. Less dominant now, for reasons we will be investigating in subsequent chapters, the MLE algorithm still has iconic status, being often the method of first choice in any novel situation. There are several good reasons for its ubiquity.

1 The MLE algorithm is *automatic*: in theory, and almost in practice, a single numerical algorithm produces $\hat{\mu}$ without further statistical input. This contrasts with unbiased estimation, for instance, where each new situation requires clever theoretical calculations.
2 The MLE enjoys excellent frequentist properties. In large-sample situations, maximum likelihood estimates tend to be nearly unbiased, with the least possible variance. Even in small samples, MLEs are usually quite efficient, within say a few percent of the best possible performance.
3 The MLE also has reasonable Bayesian justification. Looking at Bayes' rule (3.7),

$$g(\mu|x) = c_x g(\mu) e^{l_x(\mu)}, \tag{4.4}$$

we see that $\hat{\mu}$ is the maximizer of the posterior density $g(\mu|x)$ if the prior $g(\mu)$ is flat, that is, constant. Because the MLE depends on the family $\mathcal{F}$ only through the likelihood function, anomalies of the meter-reader type are averted.

Figure 4.1 displays two maximum likelihood estimates for the `gfr` data of Figure 2.1. Here the data[1] is the vector $\boldsymbol{x} = (x_1, x_2, \ldots, x_n)$, $n = 211$. We assume that $\boldsymbol{x}$ was obtained as a random sample of size n from a density $f_\mu(x)$,

$$x_i \overset{\text{iid}}{\sim} f_\mu(x) \qquad \text{for } i = 1, 2, \ldots, n, \tag{4.5}$$

"iid" abbreviating "independent and identically distributed." Two families are considered for the component density $f_\mu(x)$, the *normal*, with $\mu = (\theta, \sigma)$,

$$f_\mu(x) = \frac{1}{\sqrt{2\pi\sigma^2}} e^{-\frac{1}{2}\left(\frac{x-\theta}{\sigma}\right)^2}, \tag{4.6}$$

[1] Now $\boldsymbol{x}$ is what we have been calling "x" before, while we will henceforth use x as a symbol for the individual components of $\boldsymbol{x}$.

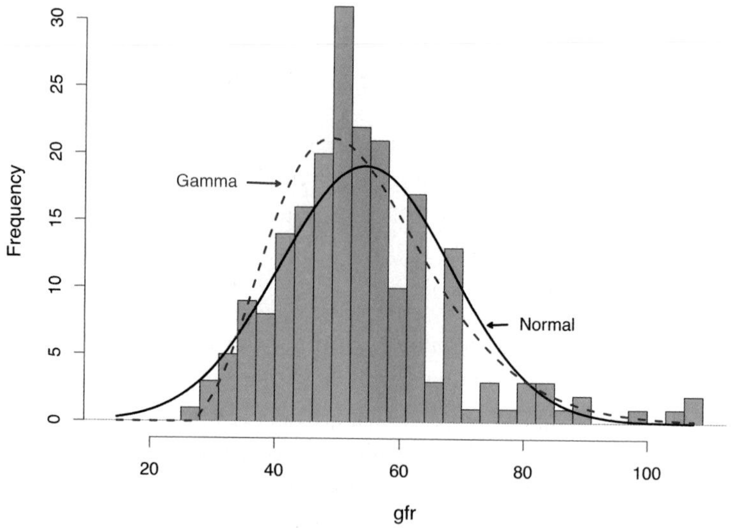

Figure 4.1 Glomerular filtration data of Figure 2.1 and two maximum-likelihood density estimates, normal (solid black), and gamma (dashed blue).

and the gamma,[2] with $\mu = (\lambda, \sigma, \nu)$,

$$f_\mu(x) = \frac{(x-\lambda)^{\nu-1}}{\sigma^2\Gamma(\nu)} e^{-\frac{x-\lambda}{\sigma}} \qquad \text{(for } x \geq \lambda, 0 \text{ otherwise).} \tag{4.7}$$

Since

$$f_\mu(\boldsymbol{x}) = \prod_{i=1}^{n} f_\mu(x_i) \tag{4.8}$$

under iid sampling, we have

$$l_{\boldsymbol{x}}(\mu) = \sum_{i=1}^{n} \log f_\mu(x_i) = \sum_{i=1}^{n} l_{x_i}(\mu). \tag{4.9}$$

Maximum likelihood estimates were found by maximizing $l_{\boldsymbol{x}}(\mu)$. For the normal model (4.6),

$$\left(\hat{\theta}, \hat{\sigma}\right) = (54.3, 13.7) = \left(\bar{x}, \left[\sum (x_i - \bar{x})^2 / n\right]^{1/2}\right). \tag{4.10}$$

[2] The gamma distribution is usually defined with $\lambda = 0$ as the lower limit of x. Here we are allowing the lower limit λ to vary as a free parameter.

There is no closed-form solution for gamma model (4.7), where numerical maximization gave

$$\left(\hat{\lambda}, \hat{\sigma}, \hat{\nu}\right) = (21.4, 5.47, 6.0). \tag{4.11}$$

The plotted curves in Figure 4.1 are the two MLE densities $f_{\hat{\mu}}(x)$. The gamma model gives a better fit than the normal, but neither is really satisfactory. (A more ambitious maximum likelihood fit appears in Figure 5.7.)

Most MLEs require numerical minimization, as for the gamma model. When introduced in the 1920s, maximum likelihood was criticized as computationally difficult, invidious comparisons being made with the older method of moments, which relied only on sample moments of various kinds.

There is a downside to maximum likelihood estimation that remained nearly invisible in classical applications: it is dangerous to rely upon in problems involving large numbers of parameters. If the parameter vector μ has 1000 components, each component individually may be well estimated by maximum likelihood, while the MLE $\hat{\theta} = T(\hat{\mu})$ for a quantity of particular interest can be grossly misleading.

For the prostate data of Figure 3.4, model (3.6) gives MLE $\hat{\mu}_i = x_i$ for each of the 6033 genes. This seems reasonable, but if we are interested in the maximum coordinate value

$$\theta = T(\mu) = \max_i \{\mu_i\}, \tag{4.12}$$

the MLE is $\hat{\theta} = 5.29$, almost certainly a flagrant overestimate. "Regularized" versions of maximum likelihood estimation more suitable for high-dimensional applications play an important role in succeeding chapters.

4.2 Fisher Information and the MLE

Fisher was not the first to suggest the maximum likelihood algorithm for parameter estimation. His paradigm-shifting work concerned the favorable inferential properties of the MLE, and in particular its achievement of the Fisher information bound. Only a brief heuristic review will be provided here, with more careful derivations referenced in the endnotes.

We begin[3] with a one-parameter family of densities

$$\mathcal{F} = \{f_\theta(x),\ \theta \in \Omega,\ x \in \mathcal{X}\}, \tag{4.13}$$

[3] The multiparameter case is considered in the next chapter.

where Ω is an interval of the real line, possibly infinite, while the sample space $\mathcal{X}$ may be multidimensional. (As in the Poisson example (3.3), $f_\theta(x)$ can represent a discrete density, but for convenience we assume here the continuous case, with the probability of set A equaling $\int_A f_\theta(x)\,dx$, etc.) The log likelihood function is $l_x(\theta) = \log f_\theta(x)$ and the MLE $\hat{\theta} = \arg\max\{l_x(\theta)\}$, with θ replacing μ in (4.1)–(4.2) in the one-dimensional case.

Dots will indicate differentiation with respect to θ, e.g., for the *score function*

$$\dot{l}_x(\theta) = \frac{\partial}{\partial\theta}\log f_\theta(x) = \dot{f}_\theta(x)/f_\theta(x). \tag{4.14}$$

The score function has expectation 0,

$$\begin{aligned}\int_{\mathcal{X}} \dot{l}_x(\theta) f_\theta(x)\,dx = \int_{\mathcal{X}} \dot{f}_\theta(x)\,dx &= \frac{\partial}{\partial\theta}\int_{\mathcal{X}} f_\theta(x)\,dx \\ &= \frac{\partial}{\partial\theta} 1 = 0,\end{aligned} \tag{4.15}$$

where we are assuming the regularity conditions necessary for differentiating under the integral sign at the third step.

The *Fisher information* $\mathcal{I}_\theta$ is defined to be the variance of the score function,

$$\mathcal{I}_\theta = \int_{\mathcal{X}} \dot{l}_x(\theta)^2 f_\theta(x)\,dx, \tag{4.16}$$

the notation

$$\dot{l}_x(\theta) \sim (0, \mathcal{I}_\theta) \tag{4.17}$$

indicating that $\dot{l}_x(\theta)$ has mean 0 and variance $\mathcal{I}_\theta$. The term "information" is well chosen. The main result for maximum likelihood estimation, sketched next, is that the MLE $\hat{\theta}$ has an approximately normal distribution with mean θ and variance $1/\mathcal{I}_\theta$,

$$\hat{\theta} \mathrel{\dot{\sim}} \mathcal{N}(\theta, 1/\mathcal{I}_\theta), \tag{4.18}$$

and that no "nearly unbiased" estimator of θ can do better. In other words, bigger Fisher information implies smaller variance for the MLE.

The second derivative of the log likelihood function

$$\ddot{l}_x(\theta) = \frac{\partial^2}{\partial\theta^2}\log f_\theta(x) = \frac{\ddot{f}_\theta(x)}{f_\theta(x)} - \left(\frac{\dot{f}_\theta(x)}{f_\theta(x)}\right)^2 \tag{4.19}$$

has expectation

$$E_\theta\left\{\ddot{l}_x(\theta)\right\} = -\mathcal{I}_\theta \tag{4.20}$$

(the $\ddot{f}_\theta(x)/f_\theta(x)$ term having expectation 0 as in (4.15)). We can write

$$-\ddot{l}_x(\theta) \sim (\mathcal{I}_\theta, \mathcal{J}_\theta), \tag{4.21}$$

where $\mathcal{J}_\theta$ is the variance of $\ddot{l}_x(\theta)$.

Now suppose that $\boldsymbol{x} = (x_1, x_2, \ldots, x_n)$ is an iid sample from $f_\theta(x)$, as in (4.5), so that the total score function $\dot{l}_{\boldsymbol{x}}(\theta)$, as in (4.9), is

$$\dot{l}_{\boldsymbol{x}}(\theta) = \sum_{i=1}^{n} \dot{l}_{x_i}(\theta), \tag{4.22}$$

and similarly

$$-\ddot{l}_{\boldsymbol{x}}(\theta) = \sum_{i=1}^{n} -\ddot{l}_{x_i}(\theta). \tag{4.23}$$

The MLE $\hat{\theta}$ based on the full sample $\boldsymbol{x}$ satisfies the maximizing condition $\dot{l}_{\boldsymbol{x}}(\hat{\theta}) = 0$. A first-order Taylor series gives the approximation

$$0 = \dot{l}_{\boldsymbol{x}}\left(\hat{\theta}\right) \doteq \dot{l}_{\boldsymbol{x}}(\theta) + \ddot{l}_{\boldsymbol{x}}(\theta)\left(\hat{\theta} - \theta\right), \tag{4.24}$$

or

$$\hat{\theta} \doteq \theta + \frac{\dot{l}_{\boldsymbol{x}}(\theta)/n}{-\ddot{l}_{\boldsymbol{x}}(\theta)/n}. \tag{4.25}$$

Under reasonable regularity conditions, (4.17) and the central limit theorem imply that

$$\dot{l}_{\boldsymbol{x}}(\theta)/n \;\dot{\sim}\; \mathcal{N}(0, \mathcal{I}_\theta/n), \tag{4.26}$$

while the law of large numbers has $-\ddot{l}_{\boldsymbol{x}}(\theta)/n$ approaching the constant $\mathcal{I}_\theta$ (4.21).

Putting all of this together, (4.25) produces Fisher's fundamental theorem for the MLE, that in large samples

$$\hat{\theta} \;\dot{\sim}\; \mathcal{N}\left(\theta, 1/(n\mathcal{I}_\theta)\right). \tag{4.27}$$

This is the same as result (4.18) since the total Fisher information in an iid sample (4.5) is $n\mathcal{I}_\theta$, as can be seen by taking expectations in (4.23).

In the case of normal sampling,

$$x_i \stackrel{\text{iid}}{\sim} \mathcal{N}(\theta, \sigma^2) \qquad \text{for } i = 1, 2, \ldots, n, \tag{4.28}$$

with σ^2 known, we compute the log likelihood

$$l_{\boldsymbol{x}}(\theta) = -\frac{1}{2}\sum_{i=1}^{n}\frac{(x_i-\theta)^2}{\sigma^2} - \frac{n}{2}\log(2\pi\sigma^2). \tag{4.29}$$

This gives

$$\dot{l}_{\boldsymbol{x}}(\theta) = \frac{1}{\sigma^2}\sum_{i=1}^{n}(x_i-\theta) \quad \text{and} \quad -\ddot{l}_{\boldsymbol{x}}(\theta) = \frac{n}{\sigma^2}, \tag{4.30}$$

yielding the familiar result $\hat{\theta} = \bar{x}$ and, since $\mathcal{I}_\theta = 1/\sigma^2$,

$$\hat{\theta} \sim \mathcal{N}(\theta, \sigma^2/n) \tag{4.31}$$

from (4.27).

This brings us to an aspect of Fisherian inference neither Bayesian nor frequentist. Fisher believed there was a "logic of inductive inference" that would produce the *correct* answer to any statistical question, in the same way ordinary logic solves deductive problems. His principal tactic was to logically reduce a complicated inferential question to a simple form where the solution should be obvious to all.

Fisher's favorite target for the obvious was (4.31), where a single scalar observation $\hat{\theta}$ is normally distributed around the unknown parameter of interest θ, with known variance σ^2/n. Then everyone should agree in the absence of prior information that $\hat{\theta}$ is the best estimate of θ, that θ has about 95% chance of lying in the interval $\hat{\theta} \pm 1.96\hat{\sigma}/\sqrt{n}$, etc.

Fisher was astoundingly resourceful at reducing statistical problems to the form (4.31). Sufficiency, efficiency, conditionality, and ancillarity were all brought to bear, with the maximum likelihood approximation (4.27) being the most influential example. Fisher's logical system is not in favor these days, but its conclusions remain as staples of conventional statistical practice.

Suppose that $\tilde{\theta} = t(\boldsymbol{x})$ is any *unbiased* estimate of θ based on an iid sample $\boldsymbol{x} = (x_1, x_2, \ldots, x_n)$ from $f_\theta(x)$. That is,

$$\theta = E_\theta\{t(\boldsymbol{x})\}. \tag{4.32}$$

Then the *Cramér–Rao lower bound*, described in the endnotes, says that
†1 the variance of $\tilde{\theta}$ exceeds the Fisher information bound (4.27),†

$$\mathrm{var}_\theta\left\{\tilde{\theta}\right\} \geq 1/(n\mathcal{I}_\theta). \tag{4.33}$$

A loose interpretation is that the MLE has variance at least as small as the best unbiased estimate of θ. The MLE is generally not unbiased, but its bias is small (of order $1/n$, compared with standard deviation of order $1/\sqrt{n}$), making the comparison with unbiased estimates and the Cramér–Rao bound appropriate.

4.3 Conditional Inference

A simple example gets across the idea of conditional inference: an i.i.d. sample

$$x_i \overset{\text{iid}}{\sim} \mathcal{N}(0, 1), \qquad i = 1, 2, \dots, n, \tag{4.34}$$

has produced estimate $\hat{\theta} = \bar{x}$. The investigators originally disagreed on an affordable sample size n and flipped a fair coin to decide,

$$n = \begin{cases} 25 & \text{probability } 1/2 \\ 100 & \text{probability } 1/2; \end{cases} \tag{4.35}$$

$n = 25$ won. Question: What is the standard deviation of $\bar{x}$?

If you answered $1/\sqrt{25} = 0.2$ then you, like Fisher, are an advocate of *conditional inference*. The *unconditional* frequentist answer says that $\bar{x}$ could have been $\mathcal{N}(0, 1/100)$ or $\mathcal{N}(0, 1/25)$ with equal probability, yielding standard deviation $[(0.01 + 0.04)/2]^{1/2} = 0.158$. Some less obvious (and less trivial) examples follow in this section, and in Chapter 9, where conditional inference plays a central role.

The data for a typical regression problem consists of pairs (x_i, y_i), $i = 1, 2, \dots, n$, where x_i is a p-dimensional vector of covariates for the ith subject and y_i is a scalar response. In Figure 1.1, x_i is `age` and y_i the kidney fitness measure `tot`. Let $\boldsymbol{x}$ be the $n \times p$ matrix having x_i as its ith row, and $\boldsymbol{y}$ the vector of responses. A regression algorithm uses $\boldsymbol{x}$ and $\boldsymbol{y}$ to construct a function $r_{\boldsymbol{x},\boldsymbol{y}}(x)$ predicting y for any value of x, as in (1.3), where $\hat{\beta}_0$ and $\hat{\beta}_1$ were obtained using least squares.

How accurate is $r(\boldsymbol{x}, \boldsymbol{y})$? This question is usually answered under the assumption that $\boldsymbol{x}$ is fixed, not random: in other words, by *conditioning on the observed value of* $\boldsymbol{x}$. The standard errors in the second line of Table 1.1 are conditional in this sense; they are frequentist standard deviations of $\hat{\beta}_0 + \hat{\beta}_1 x$, assuming that the 157 values for `age` are fixed as observed. (A *correlation* analysis between `age` and `tot` would *not* make this assumption.)

Fisher argued for conditional inference on two grounds.

1 *More relevant inferences.* The conditional standard deviation in situation (4.35) seems obviously more relevant to the accuracy of the observed $\hat{\theta}$ for estimating θ. It is less obvious in the regression example, though arguably still the case.
2 *Simpler inferences.* Conditional inferences are often simpler to execute and interpret. This is the case with regression, where the statistician doesn't have to worry about correlation relationships among the covariates, and also with our next example, a Fisherian classic.

Table 4.1 shows the results of a randomized trial on 45 ulcer patients, comparing `new` and `old` surgical treatments. Was the `new` surgery significantly better? Fisher argued for carrying out the hypothesis test conditional on the marginals of the table $(16, 29, 21, 24)$. With the marginals fixed, the number y in the upper left cell determines the other three cells by subtraction. We need only test whether the number $y = 9$ is too big under the null hypothesis of no treatment difference, instead of trying to test the numbers in all four cells.[4]

Table 4.1 *Forty-five ulcer patients randomly assigned to either* `new` *or* `old` *surgery, with results evaluated as either* `success` *or* `failure`. *Was the* `new` *surgery significantly better?*

	`success`	`failure`	
`new`	9	12	21
`old`	7	17	24
	16	29	45

An ancillary statistic (again, Fisher's terminology) is one that contains no direct information by itself, but does determine the conditioning framework for frequentist calculations. Our three examples of ancillaries were the sample size n, the covariate matrix $\boldsymbol{x}$, and the table's marginals. "Contains no information" is a contentious claim. More realistically, the two advantages of conditioning, relevance and simplicity, are thought to outweigh the loss of information that comes from treating the ancillary statistic as nonrandom. Chapter 9 makes this case specifically for standard survival analysis methods.

[4] Section 9.3 gives the details of such tests; in the surgery example, the difference was not significant.

Our final example concerns the accuracy of a maximum likelihood estimate $\hat{\theta}$. Rather than

$$\hat{\theta} \dot{\sim} \mathcal{N}\left(\theta, 1/\left(n\mathcal{I}_{\hat{\theta}}\right)\right), \tag{4.36}$$

the plug-in version of (4.27), Fisher suggested using

$$\hat{\theta} \dot{\sim} \mathcal{N}\left(\theta, 1/I(\boldsymbol{x})\right), \tag{4.37}$$

where $I(\boldsymbol{x})$ is the *observed Fisher information*

$$I(\boldsymbol{x}) = -\ddot{l}_{\boldsymbol{x}}\left(\hat{\theta}\right) = -\frac{\partial^2}{\partial\theta^2} l_{\boldsymbol{x}}(\theta)\bigg|_{\hat{\theta}}. \tag{4.38}$$

The expectation of $I(\boldsymbol{x})$ is $n\mathcal{I}_\theta$, so in large samples the distribution (4.37) converges to (4.36). Before convergence, however, Fisher suggested that (4.37) gives a better idea of $\hat{\theta}$'s accuracy.

As a check, a simulation was run involving i.i.d. samples $\boldsymbol{x}$ of size $n = 20$ drawn from a Cauchy density

$$f_\theta(x) = \frac{1}{\pi} \frac{1}{1 + (x - \theta)^2}. \tag{4.39}$$

10,000 samples $\boldsymbol{x}$ of size $n = 20$ were drawn (with $\theta = 0$) and the observed information bound $1/I(\boldsymbol{x})$ computed for each. The 10,000 $\hat{\theta}$ values were grouped according to deciles of $1/I(\boldsymbol{x})$, and the observed empirical variance of $\hat{\theta}$ within each group was then calculated.

This amounts to calculating a somewhat crude estimate of the conditional variance of the MLE $\hat{\theta}$, given the observed information bound $1/I(\boldsymbol{x})$. Figure 4.2 shows the results. We see that the conditional variance is close to $1/I(\boldsymbol{x})$, as Fisher predicted. The conditioning effect is quite substantial; the unconditional variance $1/n\mathcal{I}_\theta$ is 0.10 here, while the conditional variance ranges from 0.05 to 0.20.

The observed Fisher information $I(\boldsymbol{x})$ acts as an approximate ancillary, enjoying both of the virtues claimed by Fisher: it is more relevant than the unconditional information $n\mathcal{I}_{\hat{\theta}}$, and it is usually easier to calculate. Once $\hat{\theta}$ has been found, $I(\boldsymbol{x})$ is obtained by numerical second differentiation. Unlike $\mathcal{I}_\theta$, no probability calculations are required.

There is a strong Bayesian current flowing here. A narrow peak for the log likelihood function, i.e., a large value of $I(\boldsymbol{x})$, also implies a narrow posterior distribution for θ given $\boldsymbol{x}$. Conditional inference, of which Figure 4.2 is an evocative example, helps counter the central Bayesian criticism of frequentist inference: that the frequentist properties relate to data sets possibly much different than the one actually observed. The maximum

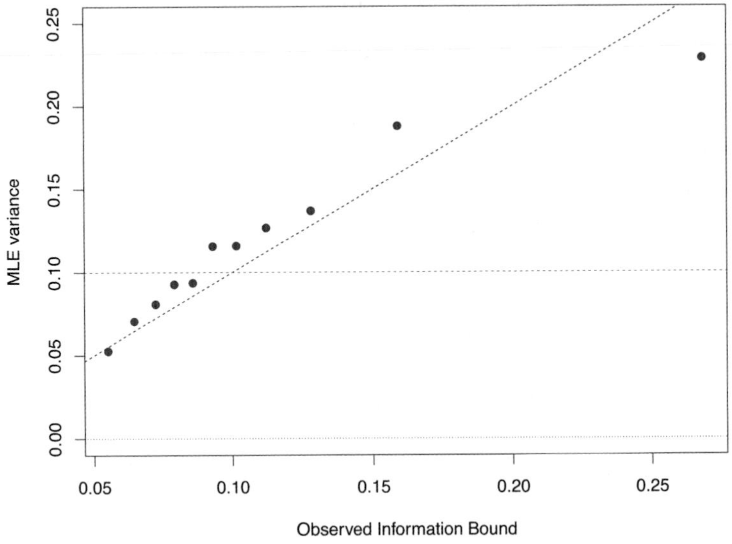

Figure 4.2 Conditional variance of MLE for Cauchy samples of size 20, plotted versus the observed information bound $1/I(\boldsymbol{x})$. Observed information bounds are grouped by quantile intervals for variance calculations (in percentages): (0–5), (5–15), ..., (85–95), (95–100). The broken red horizontal line is the unconditional variance $1/n\mathcal{I}_\theta$.

likelihood algorithm can be interpreted both vertically and horizontally in Figure 3.5, acting as a connection between the Bayesian and frequentist worlds.

The equivalent of result (4.37) for multiparameter families, Section 5.3,

$$\hat{\mu} \dot{\sim} \mathcal{N}_p\left(\mu, I(\boldsymbol{x})^{-1}\right), \tag{4.40}$$

plays an important role in succeeding chapters, with $-I(\boldsymbol{x})$ the $p \times p$ matrix of second derivatives

$$I(\boldsymbol{x}) = -\ddot{l}_{\boldsymbol{x}}(\mu) = -\left[\frac{\partial^2}{\partial\mu_i\,\partial\mu_j}\log f_\mu(\boldsymbol{x})\right]_{\hat{\mu}}. \tag{4.41}$$

4.4 Permutation and Randomization

Fisherian methodology faced criticism for its overdependence on normal sampling assumptions. Consider the comparison between the 47 **ALL** and 25 **AML** patients in the gene 136 leukemia example of Figure 1.1. The two-sample t-statistic (1.6) had value 3.13, with two-sided significance level 0.0025 according to a Student-t null distribution with 70 degrees of freedom. All of this depended on the Gaussian, or normal, assumptions (2.12)–(2.13).

As an alternative significance-level calculation, Fisher suggested using permutations of the 72 data points. The 72 values are *randomly* divided into disjoint sets of size 47 and 25, and the two-sample t-statistic (2.17) is recomputed. This is done some large number B times, yielding permutation t-values $t_1^*, t_2^*, \ldots, t_B^*$. The two-sided permutation significance level for the original value t is then the proportion of the t_i^* values exceeding t in absolute value,

$$\#\{|t_i^*| \geq t\}/B. \tag{4.42}$$

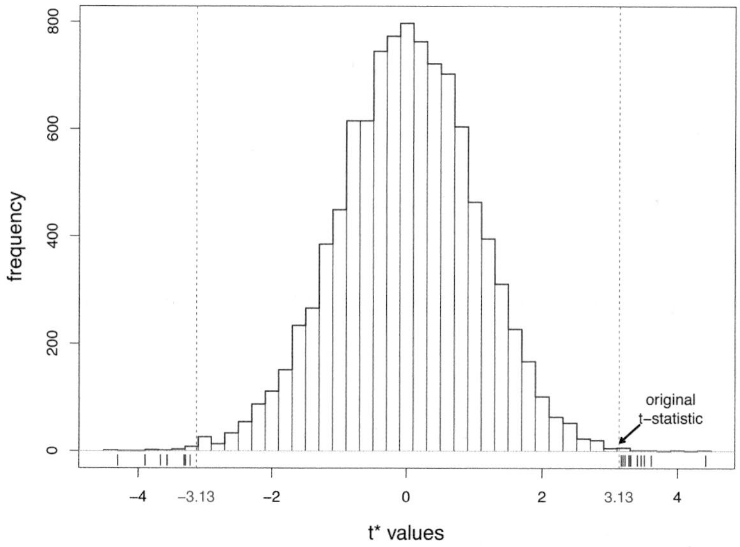

Figure 4.3 10,000 permutation t^*-values for testing **ALL** vs **AML**, for gene 136 in the `leukemia` data of Figure 1.3. Of these, 22 t^*-values (red ticks) exceeded in absolute value the observed t-statistic 3.13, giving permutation significance level 0.0022.

Figure 4.3 shows the histogram of $B = 10{,}000$ t_i^* values for the gene 136 data in Figure 1.3: 22 of these exceeded $t = 3.13$ in absolute value, yielding significance level 0.0022 against the null hypothesis of no **ALL/AML** difference, remarkably close to the normal-theory significance level 0.0025. (We were a little lucky here.)

Why should we believe the permutation significance level (4.42)? Fisher provided two arguments.

- Suppose we assume as a null hypothesis that the $n = 72$ observed measurements $\boldsymbol{x}$ are an iid sample obtained from the *same* distribution $f_\mu(x)$,

$$x_i \overset{\text{iid}}{\sim} f_\mu(x) \qquad \text{for } i = 1, 2, \ldots, n. \tag{4.43}$$

  (There is no normal assumption here, say that $f_\mu(x)$ is $\mathcal{N}(\theta, \sigma^2)$.)

  Let $\boldsymbol{o}$ indicate the *order statistic* of $\boldsymbol{x}$, i.e., the 72 numbers ordered from smallest to largest, with their **AML** or **ALL** labels removed. Then it can be shown that all $72!/(47!25!)$ ways of obtaining $\boldsymbol{x}$ by dividing $\boldsymbol{o}$ into disjoint subsets of sizes 47 and 25 are equally likely under null hypothesis (4.43). A small value of the permutation significance level (4.42) indicates that the actual division of **AML/ALL** measurements was *not* random, but rather resulted from negation of the null hypothesis (4.43). This might be considered an example of Fisher's logic of inductive inference, where the conclusion "should be obvious to all." It is certainly an example of conditional inference, now with conditioning used to avoid specific assumptions about the sampling density $f_\mu(x)$.
- In experimental situations, Fisher forcefully argued for *randomization*, that is for randomly assigning the experimental units to the possible treatment groups. Most famously, in a clinical trial comparing drug A with drug B, each patient should be randomly assigned to A or B.

  Randomization greatly strengthens the conclusions of a permutation test. In the **AML/ALL** gene 136 situation, where randomization wasn't feasible, we wind up almost certain that the **AML** group has systematically larger numbers, but cannot be certain that it is the different disease states causing the difference. Perhaps the **AML** patients are older, or heavier, or have more of some other characteristic affecting gene 136. Experimental randomization *almost* guarantees that age, weight, etc., will be well-balanced between the treatment groups. Fisher's RCT (randomized clinical trial) was and is the gold standard for statistical inference in medical trials.

Permutation testing is frequentistic: a statistician following the procedure has 5% chance of rejecting a valid null hypothesis at level 0.05, etc.

Randomization inference is somewhat different, amounting to a kind of forced frequentism, with the statistician imposing his or her preferred probability mechanism upon the data. Permutation methods are enjoying a healthy computer-age revival, in contexts far beyond Fisher's original justification for the t-test, as we will see in Chapter 15.

4.5 Notes and Details

On a linear scale that puts Bayesian on the left and frequentist on the right, Fisherian inference winds up somewhere in the middle. Fisher rejected Bayesianism early on, but later criticized as "wooden" the hard-line frequentism of the Neyman–Wald decision-theoretic school. Efron (1998) locates Fisher along the Bayes–frequentist scale for several different criteria; see in particular Figure 1 of that paper.

Bayesians, of course, believe there is only one true logic of inductive inference. Fisher disagreed. His most ambitious attempt to "enjoy the Bayesian omelette without breaking the Bayesian eggs"[5] was *fiducial inference*. The simplest example concerns the normal translation model $x \sim \mathcal{N}(\theta, 1)$, where $\theta - x$ has a standard $\mathcal{N}(0, 1)$ distribution, the fiducial distribution of θ given x then being $\mathcal{N}(x, 1)$. Among Fisher's many contributions, fiducial inference was the only outright popular bust. Nevertheless the idea has popped up again in the current literature under the name "confidence distribution;" see Efron (1993) and Xie and Singh (2013). A brief discussion appears in Chapter 11.

†1 [p. 44] For an unbiased estimator $\tilde{\theta} = t(\boldsymbol{x})$ (4.32), we have

$$\int_{\boldsymbol{\mathcal{X}}} t(\boldsymbol{x}) \dot{l}_{\boldsymbol{x}}(\theta) f_\theta(\boldsymbol{x})\, d\boldsymbol{x} = \int_{\boldsymbol{\mathcal{X}}} t(\boldsymbol{x}) \dot{f}_\theta(\boldsymbol{x})\, d\boldsymbol{x} = \frac{\partial}{\partial\theta} \int_{\boldsymbol{\mathcal{X}}} t(\boldsymbol{x}) f_\theta(\boldsymbol{x})\, d\boldsymbol{x} = \frac{\partial}{\partial\theta}\theta = 1. \tag{4.44}$$

Here $\boldsymbol{\mathcal{X}}$ is $\mathcal{X}^n$, the sample space of $\boldsymbol{x} = (x_1, x_2, \ldots, x_n)$, and we are assuming the conditions necessary for differentiating under the integral sign; (4.44) gives $\int (t(\boldsymbol{x}) - \theta) \dot{l}_{\boldsymbol{x}}(\theta) f_\theta(\boldsymbol{x})\, d\boldsymbol{x} = 1$ (since $\dot{l}_{\boldsymbol{x}}(\theta)$ has expectation

[5] Attributed to the important Bayesian theorist L. J. Savage.

0), and then, applying the *Cauchy–Schwarz inequality*,

$$\left[\int_{\mathcal{X}} (t(x) - \theta)\, \dot{l}_x(\theta)\, f_\theta(x)\, dx\right]^2 \\ \leq \left[\int_{\mathcal{X}} (t(x) - \theta)^2\, f_\theta(x)\right]\left[\int_{\mathcal{X}} \dot{l}_x(\theta)^2\, f_\theta(x)\, dx\right], \tag{4.45}$$

or

$$1 \leq \text{var}_\theta\left\{\tilde{\theta}\right\} \mathcal{I}_\theta. \tag{4.46}$$

This verifies the Cramér–Rao lower bound (4.33): the optimal variance for an unbiased estimator is one over the Fisher information.

Optimality results are a sign of scientific maturity. Fisher information and its estimation bound mark the transition of statistics from a collection of ad-hoc techniques to a coherent discipline. (We have lost some ground recently, where, as discussed in Chapter 1, ad-hoc algorithmic coinages have outrun their inferential justification.) Fisher's information bound was a major mathematical innovation, closely related to and predating, Heisenberg's uncertainty principle and Shannon's information bound; see Dembo *et al.* (1991).

Unbiased estimation has strong appeal in statistical applications, where "biased," its opposite, carries a hint of self-interested data manipulation. In large-scale settings, such as the prostate study of Figure 3.4, one can, however, strongly argue for biased estimates. We saw this for gene 610, where the usual unbiased estimate $\hat{\mu}_{610} = 5.29$ is almost certainly too large. Biased estimation will play a major role in our subsequent chapters.

Maximum likelihood estimation is effectively unbiased in most situations. Under repeated sampling, the expected mean squared error

$$\text{MSE} = E\left\{\left(\hat{\theta} - \theta\right)^2\right\} = \text{variance} + \text{bias}^2 \tag{4.47}$$

has order-of-magnitude variance $= O(1/n)$ and bias$^2 = O(1/n^2)$, the latter usually becoming negligible as sample size n increases. (Important exceptions, where bias *is* substantial, can occur if $\hat{\theta} = T(\hat{\mu})$ when $\hat{\mu}$ is high-dimensional, as in the James–Stein situation of Chapter 7.) Section 10 of Efron (1975) provides a detailed analysis.

Section 9.2 of Cox and Hinkley (1974) gives a careful and wide-ranging account of the MLE and Fisher information. Lehmann (1983) covers the same ground, somewhat more technically, in his Chapter 6.

5

Parametric Models and Exponential Families

We have been reviewing classic approaches to statistical inference—frequentist, Bayesian, and Fisherian—with an eye toward examining their strengths and limitations in modern applications. Putting philosophical differences aside, there is a common methodological theme in classical statistics: a strong preference for low-dimensional parametric models; that is, for modeling data-analysis problems using parametric families of probability densities (3.1),

$$\mathcal{F} = \left\{ f_\mu(x);\, x \in \mathcal{X}, \mu \in \Omega \right\}, \tag{5.1}$$

where the dimension of parameter μ is small, perhaps no greater than 5 or 10 or 20. The inverted nomenclature "nonparametric" suggests the predominance of classical parametric methods.

Two words explain the classic preference for parametric models: mathematical tractability. In a world of sliderules and slow mechanical arithmetic, mathematical formulation, by necessity, becomes the computational tool of choice. Our new computation-rich environment has unplugged the mathematical bottleneck, giving us a more realistic, flexible, and far-reaching body of statistical techniques. But the classic parametric families still play an important role in computer-age statistics, often assembled as small parts of larger methodologies (as with the generalized linear models of Chapter 8). This chapter[1] presents a brief review of the most widely used parametric models, ending with an overview of exponential families, the great connecting thread of classical theory and a player of continuing importance in computer-age applications.

[1] This chapter covers a large amount of technical material for use later, and may be reviewed lightly at first reading.

5.1 Univariate Families

Univariate parametric families, in which the sample space $\mathcal{X}$ of observation x is a subset of the real line $\mathcal{R}^1$, are the building blocks of most statistical analyses. Table 5.1 names and describes the five most familiar univariate families: normal, Poisson, binomial, gamma, and beta. (The chi-squared distribution with n degrees of freedom χ^2_n is also included since it is distributed as $2 \cdot \text{Gam}(n/2, 1)$.) The normal distribution $\mathcal{N}(\mu, \sigma^2)$ is a scaled version of the $\mathcal{N}(\mu, 1)$ distribution[2] used in (3.27),

$$\mathcal{N}(\mu, \sigma^2) \sim \sigma \mathcal{N}(\mu, 1). \tag{5.2}$$

Table 5.1 *Five familiar univariate densities, and their sample spaces $\mathcal{X}$, parameter spaces Ω, and expectations and variances; chi-squared distribution with n degrees of freedom is* $2\,\text{Gam}(n/2, 1)$.

Name, Notation	Density	$\mathcal{X}$	Ω	Expectation, Variance
Normal $\mathcal{N}(\mu, \sigma^2)$	$\frac{1}{\sigma\sqrt{2\pi}} e^{-\frac{1}{2}\left(\frac{x-\mu}{\sigma}\right)^2}$	$\mathcal{R}^1$	$\mu \in \mathcal{R}^1$ $\sigma^2 > 0$	μ σ^2
Poisson $\text{Poi}(\mu)$	$\frac{e^{-\mu}\mu^x}{x!}$	$\{0, 1, \ldots\}$	$\mu > 0$	μ μ
Binomial $\text{Bi}(n, \pi)$	$\frac{n!}{x!(n-x)!}\pi^x(1-\pi)^{n-x}$	$\{0, 1, \ldots, n\}$	$0 < \pi < 1$	$n\pi$ $n\pi(1-\pi)$
Gamma $\text{Gam}(\nu, \sigma)$	$\frac{x^{\nu-1}e^{-x/\sigma}}{\sigma^\nu \Gamma(\nu)}$	$x \geq 0$	$\nu > 0$ $\sigma > 0$	$\sigma\nu$ $\sigma^2\nu$
Beta $\text{Be}(\nu_1, \nu_2)$	$\frac{\Gamma(\nu_1+\nu_2)}{\Gamma(\nu_1)\Gamma(\nu_2)} x^{\nu_1-1}(1-x)^{\nu_2-1}$	$0 \leq x \leq 1$	$\nu_1 > 0$ $\nu_2 > 0$	$\nu_1/(\nu_1+\nu_2)$ $\frac{\nu_1\nu_2}{(\nu_1+\nu_2)^2(\nu_1+\nu_2+1)}$

Relationships abound among the table's families. For instance, independent gamma variables $\text{Gam}(\nu_1, \sigma)$ and $\text{Gam}(\nu_2, \sigma)$ yield a beta variate according to

$$\text{Be}(\nu_1, \nu_2) \sim \frac{\text{Gam}(\nu_1, \sigma)}{\text{Gam}(\nu_1, \sigma) + \text{Gam}(\nu_2, \sigma)}. \tag{5.3}$$

The binomial and Poisson are particularly close cousins. A $\text{Bi}(n, \pi)$ distribution (the number of heads in n independent flips of a coin with probabil-

[2] The notation in (5.2) indicates that if $X \sim \mathcal{N}(\mu, \sigma^2)$ and $Y \sim \mathcal{N}(\mu, 1)$ then X and σY have the same distribution.

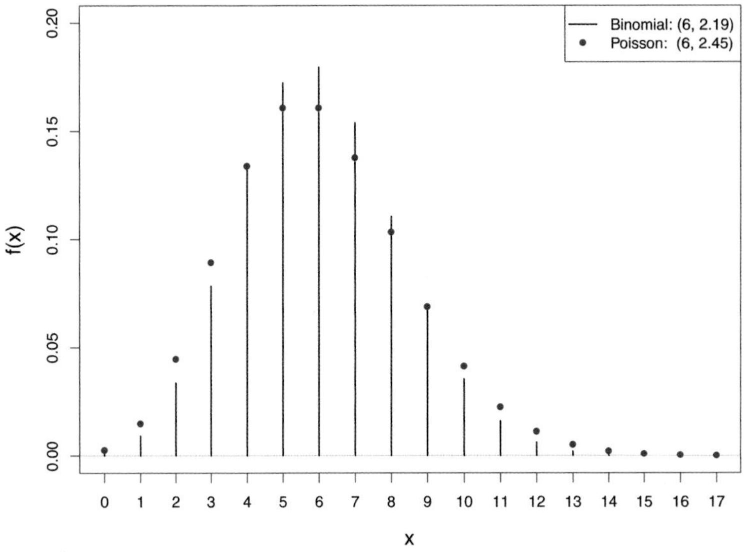

Figure 5.1 Comparison of the binomial distribution Bi(30, 0.2) (black lines) with the Poisson Poi(6) (red dots). In the legend we show the mean and standard deviation for each distribution.

ity of heads π) approaches a Poi($n\pi$) distribution,

$$\text{Bi}(n,\pi) \,\dot{\sim}\, \text{Poi}(n\pi) \tag{5.4}$$

as n grows large and p small, the notation $\dot{\sim}$ indicating approximate equality of the two distributions. Figure 5.1 shows the approximation already working quite effectively for $n = 30$ and $\pi = 0.2$.

The five families in Table 5.1 have five different sample spaces, making them appropriate in different situations. Beta distributions, for example, are natural candidates for modeling continuous data on the unit interval [0, 1]. Choices of the two parameters (ν_1, ν_2) provide a variety of possible shapes, as illustrated in Figure 5.2. Later we will discuss general exponential families, unavailable in classical theory, that greatly expand the catalog of possible shapes.

5.2 The Multivariate Normal Distribution

Classical statistics produced a less rich catalog of multivariate distributions, ones where the sample space $\mathcal{X}$ exists in $\mathcal{R}^p$, p-dimensional Eu-

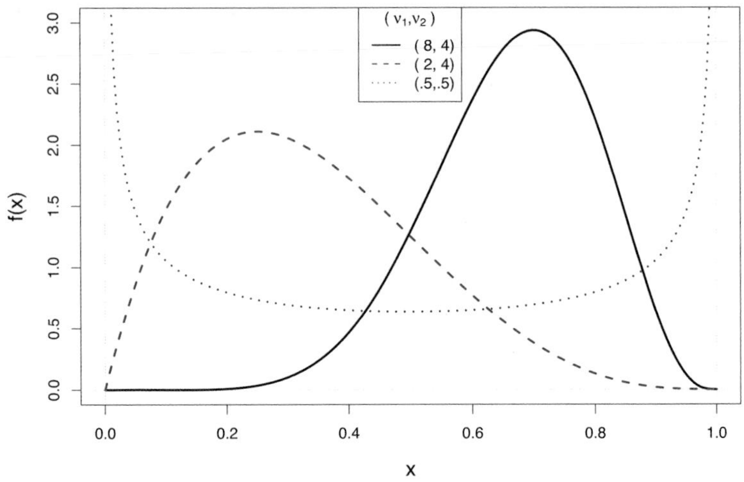

Figure 5.2 Three beta densities, with (ν_1, ν_2) indicated.

clidean space, $p > 1$. By far the greatest amount of attention focused on the multivariate normal distribution.

A random vector $x = (x_1, x_2, \ldots, x_p)'$, normally distributed or not, has *mean vector*

$$\mu = E\{x\} = \big(E\{x_1\}, E\{x_2\}, \ldots, E\{x_p\}\big)' \tag{5.5}$$

and $p \times p$ *covariance matrix*[3]

$$\boldsymbol{\Sigma} = E\left\{(x-\mu)(x-\mu)'\right\} = \left(E\left\{(x_i-\mu_i)(x_j-\mu_j)\right\}\right). \tag{5.6}$$

(The outer product uv' of vectors u and v is the matrix having elements $u_i v_j$.) We will use the convenient notation

$$x \sim (\mu, \boldsymbol{\Sigma}) \tag{5.7}$$

for (5.5) and (5.6), reducing to the familiar form $x \sim (\mu, \sigma^2)$ in the univariate case.

Denoting the entries of $\boldsymbol{\Sigma}$ by σ_{ij}, for i and j equaling $1, 2, \ldots, p$, the diagonal elements are variances,

$$\sigma_{ii} = \text{var}(x_i). \tag{5.8}$$

[3] The notation $\boldsymbol{\Sigma} = (\sigma_{ij})$ defines the ijth element of a matrix.

The off-diagonal elements relate to the correlations between the coordinates of x,

$$\operatorname{cor}(x_i, x_j) = \frac{\sigma_{ij}}{\sqrt{\sigma_{ii}\sigma_{jj}}}. \tag{5.9}$$

The multivariate normal distribution extends the univariate definition $\mathcal{N}(\mu, \sigma^2)$ in Table 5.1. To begin with, let $z = (z_1, z_2, \ldots, z_p)'$ be a vector of p independent $\mathcal{N}(0, 1)$ variates, with probability density function

$$f(z) = (2\pi)^{-\frac{p}{2}} e^{-\frac{1}{2}\sum_1^p z_i^2} = (2\pi)^{-\frac{p}{2}} e^{-\frac{1}{2}z'z} \tag{5.10}$$

according to line 1 of Table 5.1.

The multivariate normal family is obtained by linear transformations of z: let μ be a p-dimensional vector and $\boldsymbol{T}$ a $p \times p$ nonsingular matrix, and define the random vector

$$x = \mu + \boldsymbol{T} z. \tag{5.11}$$

Following the usual rules of probability transformations yields the density of x,

$$f_{\mu,\boldsymbol{\Sigma}}(x) = \frac{(2\pi)^{-p/2}}{|\boldsymbol{\Sigma}|^{1/2}} e^{-\frac{1}{2}(x-\mu)'\boldsymbol{\Sigma}^{-1}(x-\mu)}, \tag{5.12}$$

where $\boldsymbol{\Sigma}$ is the $p \times p$ symmetric positive definite matrix

$$\boldsymbol{\Sigma} = \boldsymbol{T}\boldsymbol{T}' \tag{5.13}$$

and $|\boldsymbol{\Sigma}|$ its determinant;[†] $f_{\mu,\boldsymbol{\Sigma}}(x)$, the p-dimensional multivariate normal †1
distribution with mean μ and covariance $\boldsymbol{\Sigma}$, is denoted

$$x \sim \mathcal{N}_p(\mu, \boldsymbol{\Sigma}). \tag{5.14}$$

Figure 5.3 illustrates the bivariate normal distribution with $\mu = (0, 0)'$ and $\boldsymbol{\Sigma}$ having $\sigma_{11} = \sigma_{22} = 1$ and $\sigma_{12} = 0.5$ (so $\operatorname{cor}(x_1, x_2) = 0.5$). The bell-shaped mountain on the left is a plot of density (5.12). The right panel shows a scatterplot of 2000 points drawn from this distribution. Concentric ellipses illustrate curves of constant density,

$$(x - \mu)'\boldsymbol{\Sigma}^{-1}(x - \mu) = \text{constant}. \tag{5.15}$$

Classical multivariate analysis was the study of the multivariate normal distribution, both of its probabilistic and statistical properties. The notes reference some important (and lengthy) multivariate texts. Here we will just recall a couple of results useful in the chapters to follow.

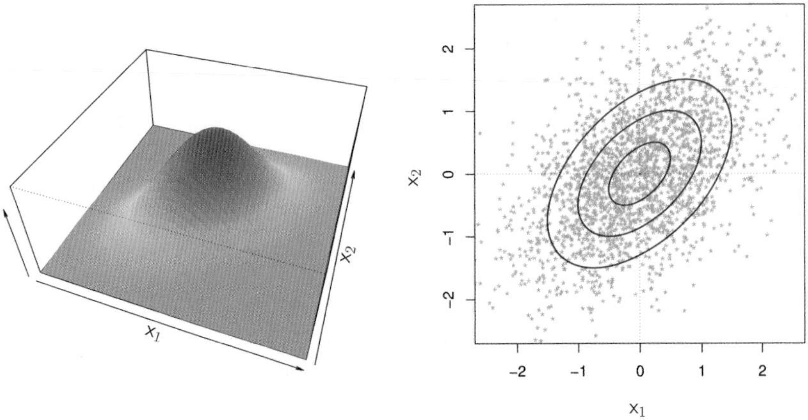

Figure 5.3 *Left:* bivariate normal density, with $\text{var}(x_1) = \text{var}(x_2) = 1$ and $\text{cor}(x_1, x_2) = 0.5$. *Right:* sample of 2000 (x_1, x_2) pairs from this bivariate normal density.

Suppose that $x = (x_1, x_2, \ldots, x_p)'$ is partitioned into

$$x_{(1)} = (x_1, x_2, \ldots, x_{p_1})' \quad \text{and} \quad x_{(2)} = (x_{p_1+1}, x_{p_1+2}, \ldots, x_{p_1+p_2})', \tag{5.16}$$

$p_1 + p_2 = p$, with μ and $\mathbf{\Sigma}$ similarly partitioned,

$$\begin{pmatrix} x_{(1)} \\ x_{(2)} \end{pmatrix} \sim \mathcal{N}_p \left(\begin{pmatrix} \mu_{(1)} \\ \mu_{(2)} \end{pmatrix}, \begin{pmatrix} \mathbf{\Sigma}_{11} & \mathbf{\Sigma}_{12} \\ \mathbf{\Sigma}_{21} & \mathbf{\Sigma}_{22} \end{pmatrix} \right) \tag{5.17}$$

(so $\mathbf{\Sigma}_{11}$ is $p_1 \times p_1$, $\mathbf{\Sigma}_{12}$ is $p_1 \times p_2$, etc.). Then the conditional distribution
†2 of $x_{(2)}$ given $x_{(1)}$ is itself normal,†

$$x_{(2)} | x_{(1)} \sim \mathcal{N}_{p_2} \left(\mu_{(2)} + \mathbf{\Sigma}_{21} \mathbf{\Sigma}_{11}^{-1} (x_{(1)} - \mu_{(1)}), \mathbf{\Sigma}_{22} - \mathbf{\Sigma}_{21} \mathbf{\Sigma}_{11}^{-1} \mathbf{\Sigma}_{12} \right). \tag{5.18}$$

If $p_1 = p_2 = 1$, then (5.18) reduces to

$$x_2 | x_1 \sim \mathcal{N} \left(\mu_2 + \frac{\sigma_{12}}{\sigma_{11}} (x_1 - \mu_1), \ \sigma_{22} - \frac{\sigma_{12}^2}{\sigma_{11}} \right); \tag{5.19}$$

here σ_{12}/σ_{11} is familiar as the linear regression coefficient of x_2 as a function of x_1, while $\sigma_{12}^2/\sigma_{11}\sigma_{22}$ equals $\text{cor}(x_1, x_2)^2$, the squared proportion R^2 of the variance of x_2 explained by x_1. Hence we can write the (unexplained) variance term in (5.19) as $\sigma_{22}(1 - R^2)$.

Bayesian statistics also makes good use of the normal family. It helps to begin with the univariate case $x \sim \mathcal{N}(\mu, \sigma^2)$, where now we assume that the expectation vector itself has a normal prior distribution $\mathcal{N}(M, A)$:

$$\mu \sim \mathcal{N}(M, A) \quad \text{and} \quad x|\mu \sim \mathcal{N}(\mu, \sigma^2). \tag{5.20}$$

Bayes' theorem and some algebra show that the posterior distribution of μ having observed x is normal,† †3

$$\mu|x \sim \mathcal{N}\left(M + \frac{A}{A+\sigma^2}(x - M), \frac{A\sigma^2}{A+\sigma^2}\right). \tag{5.21}$$

The posterior expectation $\hat{\mu}_{\text{Bayes}} = M + (A/(A+\sigma^2))(x-M)$ is a *shrinkage estimator* of μ: if, say, A equals σ^2, then $\hat{\mu}_{\text{Bayes}} = M + (x - M)/2$ is shrunk half the way back from the unbiased estimate $\hat{\mu} = x$ toward the prior mean M, while the posterior variance $\sigma^2/2$ of $\hat{\mu}_{\text{Bayes}}$ is only one-half that of $\hat{\mu}$.

The multivariate version of the Bayesian setup (5.20) is

$$\mu \sim \mathcal{N}_p(M, \boldsymbol{A}) \quad \text{and} \quad x|\mu \sim \mathcal{N}_p(\mu, \boldsymbol{\Sigma}), \tag{5.22}$$

now with M and μ p-vectors, and $\boldsymbol{A}$ and $\boldsymbol{\Sigma}$ positive definite $p \times p$ matrices. As indicated in the notes, the posterior distribution of μ given x is then

$$\mu|x \sim \mathcal{N}_p\left(M + (\boldsymbol{A}^{-1} + \boldsymbol{\Sigma}^{-1})^{-1}\boldsymbol{\Sigma}^{-1}(x - M), (\boldsymbol{A}^{-1} + \boldsymbol{\Sigma}^{-1})^{-1}\right), \tag{5.23}$$

which reduces to (5.21) when $p = 1$.

5.3 Fisher's Information Bound for Multiparameter Families

The multivariate normal distribution plays its biggest role in applications as a large-sample approximation for maximum likelihood estimates. We suppose that the parametric family of densities $\{f_\mu(x)\}$, normal or not, is smoothly defined in terms of its p-dimensional parameter vector μ. (In terms of (5.1), Ω is a subset of $\mathcal{R}^p$.)

The MLE definitions and results are direct analogues of the single-parameter calculations beginning at (4.14) in Chapter 4. The *score function* $\dot{l}_x(\mu)$ is now defined as the gradient of $\log\{f_\mu(x)\}$,

$$\dot{l}_x(\mu) = \nabla_\mu \left\{\log f_\mu(x)\right\} = \left(\ldots, \frac{\partial \log f_\mu(x)}{\partial \mu_i}, \ldots\right)', \tag{5.24}$$

the p-vector of partial derivatives of $\log f_\mu(x)$ with respect to the coordinates of μ. It has mean zero,

$$E_\mu\left\{\dot{l}_x(\mu)\right\} = 0 = (0, 0, 0, \ldots, 0)'. \tag{5.25}$$

By definition, the Fisher information matrix $\boldsymbol{\mathcal{I}}_\mu$ for μ is the $p \times p$ covariance matrix of $\dot{l}_x(\mu)$; using outer product notation,

$$\boldsymbol{\mathcal{I}}_\mu = E_\mu\left\{\dot{l}_x(\mu)\dot{l}_x(\mu)'\right\} = \left(E_\mu\left\{\frac{\partial \log f_\mu(x)}{\partial \mu_i}\frac{\partial \log f_\mu(x)}{\partial \mu_j}\right\}\right). \tag{5.26}$$

The key result is that the MLE $\hat{\mu} = \arg\max_\mu\{f_\mu(x)\}$ has an approximately normal distribution with covariance matrix $\boldsymbol{\mathcal{I}}_\mu^{-1}$,

$$\hat{\mu} \dot{\sim} \mathcal{N}_p(\mu, \boldsymbol{\mathcal{I}}_\mu^{-1}). \tag{5.27}$$

Approximation (5.27) is justified by large-sample arguments, say with x an iid sample in $\mathcal{R}^p$, $(x_1, x_2, \ldots, x_n)$, n going to infinity.

Suppose the statistician is particularly interested in μ_1, the first coordinate of μ. Let $\mu_{(2)} = (\mu_2, \mu_3, \ldots, \mu_p)$ denote the other $p-1$ coordinates of μ, which are now "nuisance parameters" as far as the estimation of μ_1 goes. According to (5.27), the MLE $\hat{\mu}_1$, which is the first coordinate of $\hat{\mu}$, has

$$\hat{\mu}_1 \dot{\sim} \mathcal{N}\left(\mu_1, (\boldsymbol{\mathcal{I}}_\mu^{-1})_{11}\right), \tag{5.28}$$

where the notation indicates the upper leftmost entry of $\boldsymbol{\mathcal{I}}_\mu^{-1}$.

We can partition the information matrix $\boldsymbol{\mathcal{I}}_\mu$ into the two parts corresponding to μ_1 and $\mu_{(2)}$,

$$\boldsymbol{\mathcal{I}}_\mu = \begin{pmatrix} \mathcal{I}_{\mu 11} & \mathcal{I}_{\mu 1(2)} \\ \mathcal{I}_{\mu(2)1} & \boldsymbol{\mathcal{I}}_{\mu(22)} \end{pmatrix} \tag{5.29}$$

(with $\mathcal{I}_{\mu 1(2)} = \mathcal{I}'_{\mu(2)1}$ of dimension $1 \times (p-1)$ and $\boldsymbol{\mathcal{I}}_{\mu(22)}$ $(p-1)\times(p-1)$).
†4 The endnotes show that†

$$(\boldsymbol{\mathcal{I}}_\mu^{-1})_{11} = \left(\mathcal{I}_{\mu 11} - \mathcal{I}_{\mu 1(2)}\boldsymbol{\mathcal{I}}_{\mu(22)}^{-1}\mathcal{I}_{\mu(2)1}\right)^{-1}. \tag{5.30}$$

The subtracted term on the right side of (5.30) is nonnegative, implying that

$$(\boldsymbol{\mathcal{I}}_\mu^{-1})_{11} \geq \mathcal{I}_{\mu 11}^{-1}. \tag{5.31}$$

If $\mu_{(2)}$ were known to the statistician, rather than requiring estimation, then $f_{\mu_1\mu_{(2)}}(x)$ would be a one-parameter family, with Fisher information $\mathcal{I}_{\mu 11}$ for estimating μ_1, giving

$$\hat{\mu}_1 \dot{\sim} \mathcal{N}(\mu_1, \mathcal{I}_{\mu 11}^{-1}). \tag{5.32}$$

Comparing (5.28) with (5.32), (5.31) shows that the variance of the MLE $\hat{\mu}_1$ must always increase[4] in the presence of nuisance parameters.[†] †5

Maximum likelihood, and in fact any form of unbiased or nearly unbiased estimation, pays a nuisance tax for the presence of "other" parameters. Modern applications often involve thousands of *others*; think of regression fits with too many predictors. In some circumstances, biased estimation methods can reverse the situation, using the others to actually improve estimation of a target parameter; see Chapter 6 on empirical Bayes techniques, and Chapter 16 on ℓ_1 regularized regression models.

5.4 The Multinomial Distribution

Second in the small catalog of well-known classic multivariate distributions is the multinomial. The multinomial applies to situations in which the observations take on only a finite number of discrete values, say L of them. The 2×2 ulcer surgery of Table 4.1 is repeated in Table 5.2, now with the cells labeled $1, 2, 3,$ and 4. Here there are $L = 4$ possible outcomes for each patient: (`new`, `success`), (`new`, `failure`), (`old`, `success`), (`old`, `failure`).

Table 5.2 *The ulcer study of Table 4.1, now with the cells numbered 1 through 4 as shown.*

	`success`	`failure`
`new`	1 **9**	2 **12**
`old`	3 **7**	4 **17**

A number n of cases has been observed, $n = 45$ in Table 5.2. Let $\boldsymbol{x} = (x_1, x_2, \dots, x_L)$ be the vector of counts for the L possible outcomes,

$$x_l = \#\{\text{cases having outcome } l\}, \tag{5.33}$$

$\boldsymbol{x} = (9, 12, 7, 17)'$ for the ulcer data. It is convenient to code the outcomes in terms of the coordinate vectors $\boldsymbol{e}_l$ of length L,

$$\boldsymbol{e}_l = (0, 0, \dots, 0, 1, 0, \dots, 0)', \tag{5.34}$$

with a 1 in the lth place.

[4] Unless $\mathcal{I}_{\mu 1(2)}$ is a vector of zeros, a condition that amounts to approximate independence of $\hat{\mu}_1$ and $\hat{\mu}_{(2)}$.

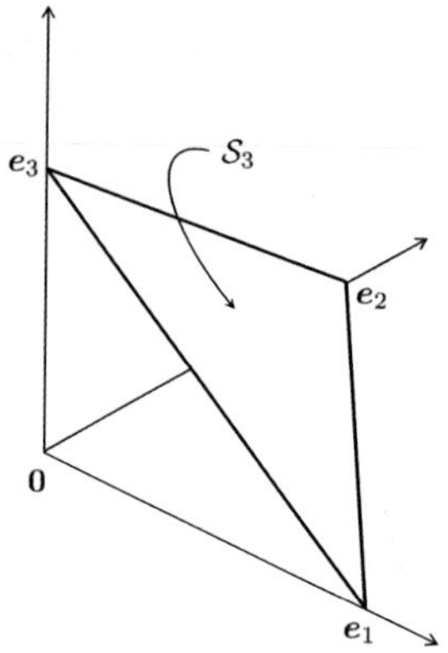

Figure 5.4 The simplex $\mathcal{S}_3$ is an equilateral triangle set at an angle to the coordinate axes in $\mathcal{R}^3$.

The multinomial probability model assumes that the n cases are independent of each other, with each case having probability π_l for outcome e_l,

$$\pi_l = \Pr\{e_l\}, \qquad l = 1, 2, \dots, L. \tag{5.35}$$

Let

$$\boldsymbol{\pi} = (\pi_1, \pi_2, \dots, \pi_L)' \tag{5.36}$$

indicate the vector of probabilities. The count vector $\boldsymbol{x}$ then follows the *multinomial distribution*,

$$f_{\boldsymbol{\pi}}(\boldsymbol{x}) = \left(\frac{n!}{x_1!x_2!\dots x_L!} \right) \prod_{l=1}^{L} \pi_l^{x_l}, \tag{5.37}$$

denoted

$$\boldsymbol{x} \sim \text{Mult}_L(n, \boldsymbol{\pi}) \tag{5.38}$$

(for n observations, L outcomes, probability vector $\boldsymbol{\pi}$).

The parameter space Ω for $\boldsymbol{\pi}$ is the *simplex* $\mathcal{S}_L$,

$$\mathcal{S}_L = \left\{ \boldsymbol{\pi} : \pi_l \geq 0 \text{ and } \sum_{l=1}^{L} \pi_l = 1 \right\}. \tag{5.39}$$

Figure 5.4 shows $\mathcal{S}_3$, an equilateral triangle sitting at an angle to the coordinate axes $\boldsymbol{e}_1, \boldsymbol{e}_2,$ and $\boldsymbol{e}_3$. The midpoint of the triangle $\boldsymbol{\pi} = (1/3, 1/3, 1/3)$

corresponds to a multinomial distribution putting equal probability on the three possible outcomes.

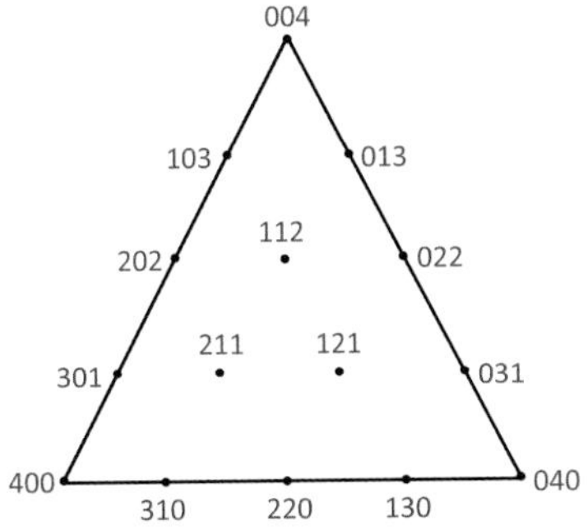

Figure 5.5 Sample space $\mathcal{X}$ for $\boldsymbol{x} \sim \text{Mult}_3(4, \boldsymbol{\pi})$; numbers indicate (x_1, x_2, x_3).

The sample space $\mathcal{X}$ for $\boldsymbol{x}$ is the subset of $n\mathcal{S}_L$ (the set of nonnegative vectors summing to n) having integer components. Figure 5.5 illustrates the case $n = 4$ and $L = 3$, now with the triangle of Figure 5.4 multiplied by 4 and set flat on the page. The point 121 indicates $\boldsymbol{x} = (1, 2, 1)$, with probability $12 \cdot \pi_1 \pi_2^2 \pi_3$ according to (5.37), etc.

In the *dichotomous* case, $L = 2$, the multinomial distribution reduces to the binomial, with (π_1, π_2) equaling $(\pi, 1 - \pi)$ in line 3 of Table 5.1, and (x_1, x_2) equaling $(x, n - x)$. The mean vector and covariance matrix of $\text{Mult}_L(n, \boldsymbol{\pi})$, for any value of L, are† †6

$$\boldsymbol{x} \sim \left(n\boldsymbol{\pi}, n\left[\text{diag}(\boldsymbol{\pi}) - \boldsymbol{\pi}\boldsymbol{\pi}'\right]\right) \tag{5.40}$$

(diag($\boldsymbol{\pi}$) is the diagonal matrix with diagonal elements π_l), so $\text{var}(x_l) = n\pi_l(1 - \pi_l)$ and covariance $(x_l, x_j) = -n\pi_l\pi_j$; (5.40) generalizes the binomial mean and variance $(n\pi, n\pi(1 - \pi))$.

There is a useful relationship between the multinomial distribution and the Poisson. Suppose $S_1, S_2, \ldots, S_L$ are independent Poissons having possibly different parameters,

$$S_l \overset{\text{ind}}{\sim} \text{Poi}(\mu_l), \qquad l = 1, 2, \ldots, L, \tag{5.41}$$

or, more concisely,

$$\boldsymbol{S} \sim \text{Poi}(\boldsymbol{\mu}) \tag{5.42}$$

with $\boldsymbol{S} = (S_1, S_2, \ldots, S_L)'$ and $\boldsymbol{\mu} = (\mu_1, \mu_2, \ldots, \mu_L)'$, the independence

being assumed in notation (5.42). Then the conditional distribution of $\boldsymbol{S}$
†7 given the sum $S_+ = \sum S_l$ is multinomial,†

$$\boldsymbol{S}|S_+ \sim \text{Mult}_L(S_+, \boldsymbol{\mu}/\mu_+), \tag{5.43}$$

$\mu_+ = \sum \mu_l$.

Going in the other direction, suppose $N \sim \text{Poi}(n)$. Then the unconditional or marginal distribution of $\text{Mult}_L(N, \boldsymbol{\pi})$ is Poisson,

$$\text{Mult}_L(N, \boldsymbol{\pi}) \sim \text{Poi}(n\boldsymbol{\pi}) \qquad \text{if } N \sim \text{Poi}(n). \tag{5.44}$$

Calculations involving $\boldsymbol{x} \sim \text{Mult}_L(n, \boldsymbol{\pi})$ are sometimes complicated by the multinomial's correlations. The approximation $\boldsymbol{x} \mathrel{\dot{\sim}} \text{Poi}(n\boldsymbol{\pi})$ removes the correlations and is usually quite accurate if n is large.

There is one more important thing to say about the multinomial family: it contains *all* distributions on a sample space $\mathcal{X}$ composed of L discrete categories. In this sense it is a model for *nonparametric* inference on $\mathcal{X}$. The nonparametric bootstrap calculations of Chapter 10 use the multinomial in this way. Nonparametrics, and the multinomial, have played a larger role in the modern environment of large, difficult to model, data sets.

5.5 Exponential Families

Classic parametric families dominated statistical theory and practice for a century and more, with an enormous catalog of their individual properties—means, variances, tail areas, etc.—being compiled. A surprise, though a slowly emerging one beginning in the 1930s, was that all of them were examples of a powerful general construction: *exponential families*. What follows here is a brief introduction to the basic theory, with further development to come in subsequent chapters.

To begin with, consider the Poisson family, line 2 of Table 5.1. The ratio of Poisson densities at two parameter values μ and μ_0 is

$$\frac{f_\mu(x)}{f_{\mu_0}(x)} = e^{-(\mu-\mu_0)} \left(\frac{\mu}{\mu_0}\right)^x, \tag{5.45}$$

which can be re-expressed as

$$f_\mu(x) = e^{\alpha x - \psi(\alpha)} f_{\mu_0}(x), \tag{5.46}$$

where we have defined

$$\alpha = \log\{\mu/\mu_0\} \quad \text{and} \quad \psi(\alpha) = \mu_0(e^\alpha - 1). \tag{5.47}$$

Looking at (5.46), we can describe the Poisson family in three steps.

1 Start with any one Poisson distribution $f_{\mu_0}(x)$.
2 For any value of $\mu > 0$ let $\alpha = \log\{\mu/\mu_0\}$ and calculate

$$\tilde{f}_\mu(x) = e^{\alpha x} f_{\mu_0}(x) \qquad \text{for } x = 0, 1, 2, \ldots. \tag{5.48}$$

3 Finally, divide $\tilde{f}_\mu(x)$ by $\exp(\psi(\alpha))$ to get the Poisson density $f_\mu(x)$.

In other words, we "tilt" $f_{\mu_0}(x)$ with the exponential factor $e^{\alpha x}$ to get $\tilde{f}_\mu(x)$, and then renormalize $\tilde{f}_\mu(x)$ to sum to 1. Notice that (5.46) gives $\exp(-\psi(\alpha))$ as the renormalizing constant since

$$e^{\psi(\alpha)} = \sum_0^\infty e^{\alpha x} f_{\mu_0}(x). \tag{5.49}$$

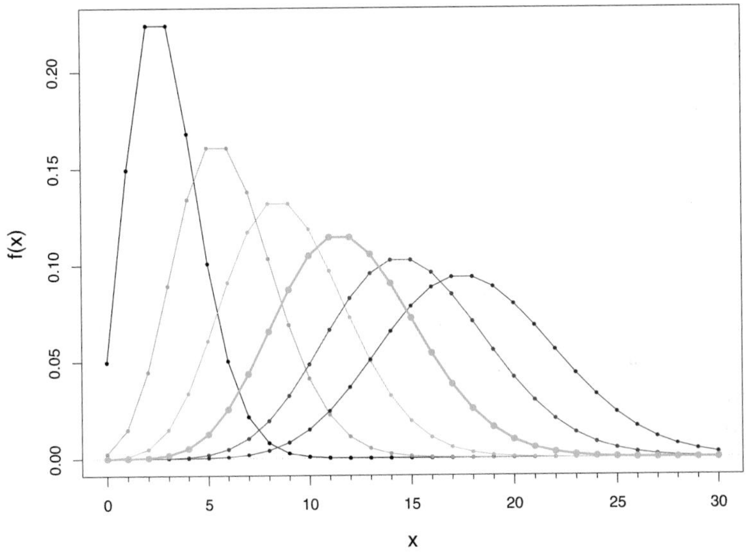

Figure 5.6 Poisson densities for $\mu = 3, 6, 9, 12, 15, 18$; heavy green curve with dots for $\mu = 12$.

Figure 5.6 graphs the Poisson density $f_\mu(x)$ for $\mu = 3, 6, 9, 12, 15, 18$. Each Poisson density is a renormalized exponential tilt of any other Poisson density. So for instance $f_6(x)$ is obtained from $f_{12}(x)$ via the tilt $e^{\alpha x}$ with $\alpha = \log\{6/12\} = -0.693$.[5]

[5] Alternate expressions for $f_\mu(x)$ as an exponential family are available, for example $\exp(\alpha x - \psi(\alpha)) f_0(x)$, where $\alpha = \log \mu$, $\psi(\alpha) = \exp(\alpha)$, and $f_0(x) = 1/x!$. (It isn't necessary for $f_0(x)$ to be a member of the family.)

The Poisson is a *one-parameter exponential family*, in that α and x in expression (5.46) are one-dimensional. A *p-parameter exponential family* has the form

$$f_\alpha(x) = e^{\alpha' y - \psi(\alpha)} f_0(x) \qquad \text{for } \alpha \in A, \tag{5.50}$$

where α and y are p-vectors and A is contained in $\mathcal{R}^p$. Here α is the "canonical" or "natural" parameter vector and $y = t(x)$ is the "sufficient statistic" vector. The normalizing function $\psi(\alpha)$, which makes $f_\alpha(x)$ integrate (or sum) to one, satisfies

$$e^{\psi(\alpha)} = \int_{\mathcal{X}} e^{\alpha' y} f_0(x)\, dx, \tag{5.51}$$

and it can be shown that the parameter space A for which the integral is
†8 finite is a convex set† in $\mathcal{R}^p$. As an example, the gamma family on line 4 of Table 5.1 is a two-parameter exponential family, with α and $y = t(x)$ given by

$$(\alpha_1, \alpha_2) = \left(-\frac{1}{\sigma}, \nu\right), \qquad (y_1, y_2) = (x, \log x), \tag{5.52}$$

and

$$\begin{aligned} \psi(\alpha) &= \nu \log \sigma + \log \Gamma(\nu) \\ &= -\alpha_2 \log\{-\alpha_1\} + \log\{\Gamma(\alpha_2)\}. \end{aligned} \tag{5.53}$$

The parameter space A is $\{\alpha_1 < 0 \text{ and } \alpha_2 > 0\}$.

Why are we interested in exponential tilting rather than some other transformational form? The answer has to do with repeated sampling. Suppose $\boldsymbol{x} = (x_1, x_2, \ldots, x_n)$ is an iid sample from a p-parameter exponential family (5.50). Then, letting $y_i = t(x_i)$ denote the sufficient vector corresponding to x_i,

$$\begin{aligned} f_\alpha(\boldsymbol{x}) &= \prod_{i=1}^{n} e^{\alpha' y_i - \psi(\alpha)} f_0(x_i) \\ &= e^{n(\alpha' \bar{y} - \psi(\alpha))} f_0(\boldsymbol{x}), \end{aligned} \tag{5.54}$$

where $\bar{y} = \sum_1^n y_i / n$. This is still a p-parameter exponential family, now with natural parameter $n\alpha$, sufficient statistic $\bar{y}$, and normalizer $n\psi(\alpha)$. No matter how large n may be, the statistician can still compress all the inferential information into a p-dimensional statistic $\bar{y}$. Only exponential families enjoy this property.

Even though they were discovered and developed in quite different contexts, and at quite different times, all of the distributions discussed in this chapter exist in exponential families. This isn't quite the coincidence it seems. Mathematical tractability was the prized property of classic parametric distributions, and tractability was greatly facilitated by exponential structure, even if that structure went unrecognized.

In one-parameter exponential families, the normalizer $\psi(\alpha)$ is also known as the *cumulant generating function*. Derivatives of $\psi(\alpha)$ yield the cumulants of y,[6] the first two giving the mean and variance† †9

$$\dot{\psi}(\alpha) = E_\alpha\{y\} \quad \text{and} \quad \ddot{\psi}(\alpha) = \text{var}_\alpha\{y\}. \tag{5.55}$$

Similarly, in p-parametric families

$$\dot{\psi}(\alpha) = (\dots \partial\psi/\partial\alpha_j \dots)' = E_\alpha\{y\} \tag{5.56}$$

and

$$\ddot{\psi}(\alpha) = \left(\frac{\partial^2\psi(\alpha)}{\partial\alpha_j\,\partial\alpha_k}\right) = \text{cov}_\alpha\{y\}. \tag{5.57}$$

The p-dimensional *expectation parameter*, denoted

$$\beta = E_\alpha\{y\}, \tag{5.58}$$

is a one-to-one function of the natural parameter α. Let V_α indicate the $p \times p$ covariance matrix,

$$V_\alpha = \text{cov}_\alpha(y). \tag{5.59}$$

Then the $p \times p$ derivate matrix of β with respect to α is

$$\frac{d\beta}{d\alpha} = (\partial\beta_j/\partial\alpha_k) = V_\alpha, \tag{5.60}$$

this following from (5.56)–(5.57), the inverse mapping being $d\alpha/d\beta = V_\alpha^{-1}$. As a one-parameter example, the Poisson in Table 5.1 has $\alpha = \log\mu$, $\beta = \mu$, $y = x$, and $d\beta/d\alpha = 1/(d\alpha/d\beta) = \mu = V_\alpha$.

The maximum likelihood estimate for the expectation parameter β is simply y (or $\bar{y}$ under repeated sampling (5.54)), which makes it immediate to calculate in most situations.† Less immediate is the MLE for the natural parameter α: the one-to-one mapping $\beta = \dot{\psi}(\alpha)$ (5.56) has inverse $\alpha = \dot{\psi}^{-1}(\beta)$, so †10

$$\hat{\alpha} = \dot{\psi}^{-1}(y), \tag{5.61}$$

[6] The simplified dot notation leads to more compact expressions: $\dot{\psi}(\alpha) = d\psi(\alpha)/d\alpha$ and $\ddot{\psi}(\alpha) = d^2\psi(\alpha)/d\alpha^2$.

e.g., $\hat{\alpha} = \log y$ for the Poisson. The trouble is that $\dot{\psi}^{-1}(\cdot)$ is usually unavailable in closed form. Numerical approximation algorithms are necessary to calculate $\hat{\alpha}$ in most cases.

All of the classic exponential families have closed-form expressions for $\psi(\alpha)$ (and $f_\alpha(x)$), yielding pleasant formulas for the mean β and covariance V_α, (5.56)–(5.57). Modern computational technology allows us to work with general exponential families, designed for specific tasks, without concern for mathematical tractability.

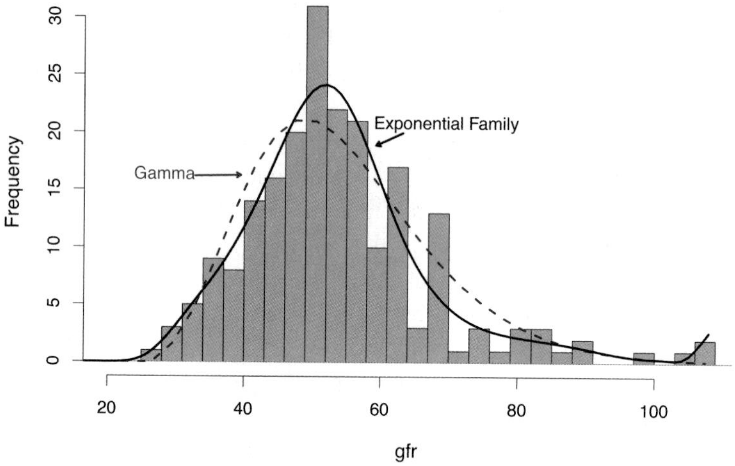

Figure 5.7 A seven-parameter exponential family fit to the **gfr** data of Figure 2.1 (solid) compared with gamma fit of Figure 4.1 (dashed).

As an example we again consider fitting the **gfr** data of Figure 2.1. For our exponential family of possible densities we take $f_0(x) \equiv 1$, and sufficient statistic vector

$$y(x) = (x, x^2, \ldots, x^7), \tag{5.62}$$

so $\alpha' y$ in (5.50) can represent all 7th-order polynomials in x, the **gfr** measurement.[7] (Stopping at power 2 gives the $\mathcal{N}(\mu, \sigma^2)$ family, which we already know fits poorly from Figure 4.1.) The heavy curve in Figure 5.7 shows the MLE fit $f_{\hat{\alpha}}(x)$ now following the **gfr** histogram quite closely. Chapter 15 discusses "Lindsey's method," a simplified algorithm for calculating the MLE $\hat{\alpha}$.

[7] Any intercept in the polynomial is absorbed into the $\psi(\alpha)$ term in (5.57).

A more exotic example concerns the generation of random graphs on a fixed set of N nodes. Each possible graph has a certain total number E of edges, and T of triangles. A popular choice for generating such graphs is the two-parameter exponential family having $y = (E, T)$, so that larger values of α_1 and α_2 yield more connections.

5.6 Notes and Details

The notion of *sufficient statistics*, ones that contain all available inferential information, was perhaps Fisher's happiest contribution to the classic corpus. He noticed that in the exponential family form (5.50), the fact that the parameter α interacts with the data x only through the factor $\exp(\alpha' y)$ makes $y(x)$ sufficient for estimating α. In 1935–36, a trio of authors, working independently in different countries, Pitman, Darmois, and Koopmans, showed that exponential families are the only ones that enjoy fixed-dimensional sufficient statistics under repeated independent sampling. Until the late 1950s such distributions were called Pitman–Darmois–Koopmans families, the long name suggesting infrequent usage.

Generalized linear models, Chapter 8, show the continuing impact of sufficiency on statistical practice. Peter Bickel has pointed out that *data compression*, a lively topic in areas such as image transmission, is a modern, less stringent, version of sufficiency.

Our only nonexponential family so far was (4.39), the Cauchy translational model. Efron and Hinkley (1978) analyze the Cauchy family in terms of *curved exponential families*, a generalization of model (5.50).

Properties of classical distributions (lots of properties and lots of distributions) are covered in Johnson and Kotz's invaluable series of reference books, 1969–1972. Two classic multivariate analysis texts are Anderson (2003) and Mardia *et al.* (1979).

†1 [p. 57] *Formula* (5.12). From $z = \boldsymbol{T}^{-1}(x - \mu)$ we have $dz/dx = \boldsymbol{T}^{-1}$ and

$$f_{\mu,\boldsymbol{\Sigma}}(x) = f(z)|\boldsymbol{T}^{-1}| = (2\pi)^{-\frac{p}{2}}|\boldsymbol{T}^{-1}|e^{-\frac{1}{2}(x-\mu)'\boldsymbol{T}^{-1\prime}\boldsymbol{T}^{-1}(x-\mu)}, \tag{5.63}$$

so (5.12) follows from $\boldsymbol{T}\boldsymbol{T}' = \boldsymbol{\Sigma}$ and $|\boldsymbol{T}| = |\boldsymbol{\Sigma}|^{1/2}$.

†2 [p. 58] *Formula* (5.18). Let $\boldsymbol{\Lambda} = \boldsymbol{\Sigma}^{-1}$ be partitioned as in (5.17). Then

$$\begin{pmatrix} \boldsymbol{\Lambda}_{11} & \boldsymbol{\Lambda}_{12} \\ \boldsymbol{\Lambda}_{21} & \boldsymbol{\Lambda}_{22} \end{pmatrix} = \begin{pmatrix} \left(\boldsymbol{\Sigma}_{11} - \boldsymbol{\Sigma}_{12}\boldsymbol{\Sigma}_{22}^{-1}\boldsymbol{\Sigma}_{21}\right)^{-1} & -\boldsymbol{\Sigma}_{11}^{-1}\boldsymbol{\Sigma}_{12}\boldsymbol{\Lambda}_{22} \\ -\boldsymbol{\Sigma}_{22}^{-1}\boldsymbol{\Sigma}_{21}\boldsymbol{\Lambda}_{11} & \left(\boldsymbol{\Sigma}_{22} = \boldsymbol{\Sigma}_{21}\boldsymbol{\Sigma}_{11}^{-1}\boldsymbol{\Sigma}_{12}\right)^{-1} \end{pmatrix}, \tag{5.64}$$

direct multiplication showing that $\boldsymbol{\Lambda}\boldsymbol{\Sigma} = \boldsymbol{I}$, the identity matrix. If $\boldsymbol{\Sigma}$ is symmetric then $\mathbf{\Lambda}_{21} = \mathbf{\Lambda}'_{12}$. By redefining x to be $x - \mu$ we can set $\mu_{(1)}$ and $\mu_{(2)}$ equal to zero in (5.18). The quadratic form in the exponent of (5.12) is

$$(x'_{(1)}, x'_{(2)})\mathbf{\Lambda}\left(x_{(1)}, x_{(2)}\right) = x'_{(2)}\mathbf{\Lambda}_{22}x_{(2)} + 2x'_{(1)}\mathbf{\Lambda}_{12}x_{(2)} + x'_{(1)}\mathbf{\Lambda}_{11}x_{(1)}. \tag{5.65}$$

But, using (5.64), this matches the quadratic form from (5.18),

$$\left(x_{(2)} - \mathbf{\Sigma}_{21}\mathbf{\Sigma}_{11}^{-1}x_{(1)}\right)'\mathbf{\Lambda}_{22}\left(x_{(2)} - \mathbf{\Sigma}_{21}\mathbf{\Sigma}_{11}^{-1}x_{(1)}\right) \tag{5.66}$$

except for an added term that does *not* involve $x_{(2)}$. For a multivariate normal distribution, this is sufficient to show that the conditional distribution of $x_{(2)}$ given $x_{(1)}$ is indeed (5.18) (see †3).

†3 [p. 59] *Formulas* (5.21) *and* (5.23). Suppose that the continuous univariate random variable z has density of the form

$$f(z) = c_0 e^{-\frac{1}{2}Q(z)}, \qquad \text{where } Q(z) = az^2 + 2bz + c_1, \tag{5.67}$$

a, b, c_0 and c_1 constants, $a > 0$. Then, by "completing the square,"

$$f(z) = c_2 e^{-\frac{1}{2}a\left(z - \frac{b}{a}\right)^2}, \tag{5.68}$$

and we see that $z \sim \mathcal{N}(b/a, 1/a)$. The key point is that form (5.67) specifies z as normal, with mean and variance uniquely determined by a and b. The multivariate version of this fact was used in the derivation of formula (5.18).

By redefining μ and x as $\mu - M$ and $x - M$, we can take $M = 0$ in (5.21). Setting $B = A/(A + \sigma^2)$, density (5.21) for $\mu|x$ is of form (5.67), with

$$Q(\mu) = \frac{\mu^2}{B\sigma^2} - \frac{2x\mu}{\sigma^2} + \frac{Bx^2}{\sigma^2}. \tag{5.69}$$

But Bayes' rule says that the density of $\mu|x$ is proportional to $g(\mu)f_\mu(x)$, also of form (5.67), now with

$$Q(\mu) = \left(\frac{1}{A} + \frac{1}{\sigma^2}\right)\mu^2 - \frac{2x\mu}{\sigma^2} + \frac{x^2}{\sigma^2}. \tag{5.70}$$

A little algebra shows that the quadratic and linear coefficients of μ match in (5.69)–(5.70), verifying (5.21).

The multivariate result (5.23) follows by a similar argument, making use of the matrix identity

$$\left(A^{-1} + \mathbf{\Sigma}^{-1}\right)^{-1} = A - A(A + \mathbf{\Sigma})^{-1}A = \mathbf{\Sigma} - \mathbf{\Sigma}(A + \mathbf{\Sigma})^{-1}\mathbf{\Sigma}. \tag{5.71}$$

†4 [p. 60] *Formula* (5.30). This is the matrix identity (5.64), now with $\mathbf{\Sigma}$ equaling $\mathcal{I}_\mu$.

†5 [p. 61] *Multivariate Gaussian and nuisance parameters.* The cautionary message here—that increasing the number of unknown nuisance parameters decreases the accuracy of the estimate of interest—can be stated more positively: if some nuisance parameters are actually known, then the MLE of the parameter of interest becomes more accurate. Suppose, for example, we wish to estimate μ_1 from a sample of size n in a bivariate normal model $x \sim \mathcal{N}_2(\mu, \mathbf{\Sigma})$ (5.14). The MLE $\bar{x}_1$ has variance σ_{11}/n in notation (5.19). But if μ_2 is known then the MLE of μ_1 becomes $\bar{x}_1 - (\sigma_{12}/\sigma_{22})(\bar{x}_2 - \mu_2)$ with variance $(\sigma_{11}/n) \cdot (1 - \rho^2)$, ρ being the correlation $\sigma_{12}/\sqrt{\sigma_{11}\sigma_{22}}$.

†6 [p. 63] *Formula* (5.40). $\boldsymbol{x} = \sum_{i=1}^n \boldsymbol{x}_i$, where the $\boldsymbol{x}_i$ are iid observations having $\Pr\{\boldsymbol{x}_i = \boldsymbol{e}_i\} = \pi_l$, as in (5.35). The mean and covariance of each $\boldsymbol{x}_i$ are

$$E\{\boldsymbol{x}_i\} = \sum_1^L \pi_l \boldsymbol{e}_l = \boldsymbol{\pi} \tag{5.72}$$

and

$$\begin{aligned} \operatorname{cov}\{\boldsymbol{x}_i\} &= E\{\boldsymbol{x}_i \boldsymbol{x}_i'\} - E\{\boldsymbol{x}_i\}E\{\boldsymbol{x}_i'\} = \sum \pi_l \boldsymbol{e}_l \boldsymbol{e}_l' - \boldsymbol{\pi}\boldsymbol{\pi}' \\ &= \operatorname{diag}(\boldsymbol{\pi}) - \boldsymbol{\pi}\boldsymbol{\pi}'. \end{aligned} \tag{5.73}$$

Formula (5.40) follows from $E\{\boldsymbol{x}\} = \sum E\{\boldsymbol{x}_i\}$ and $\operatorname{cov}(\boldsymbol{x}) = \sum \operatorname{cov}(\boldsymbol{x}_i)$.

†7 [p. 64] *Formula* (5.43). The densities of $\boldsymbol{S}$ (5.42) and $S_+ = \sum S_l$ are

$$f_{\boldsymbol{\mu}}(\boldsymbol{S}) = \prod_{l=1}^L e^{-\mu_l} \mu_l^{x_l}(x_l) \quad \text{and} \quad f_{\mu_+}(S_+) = e^{-\mu_+} \mu_+^{S_+}/S_+!. \tag{5.74}$$

The conditional density of $\boldsymbol{S}$ given S_+ is the ratio

$$f_{\boldsymbol{\mu}}(\boldsymbol{S}|S_+) = \left(\frac{S!}{\prod_1^L x_l!}\right) \prod_{l=1}^L \left(\frac{\mu_l}{\mu_+}\right)^{x_l}, \tag{5.75}$$

which is (5.43).

†8 [p. 66] *Formula* (5.51) *and the convexity of* A. Suppose α_1 and α_2 are any two points in A, i.e., values of α having the integral in (5.51) finite. For any value of c in the interval [0, 1], and any value of y, we have

$$ce^{\alpha_1' y} + (1-c)e^{\alpha_2' y} \geq e^{[c\alpha_1 + (1-c)\alpha_2]' y} \tag{5.76}$$

because of the convexity in c of the function on the right (verified by showing that its second derivative is positive). Integrating both sides of (5.76) over $\mathcal{X}$ with respect to $f_0(x)$ shows that the integral on the right must be finite: that is, $c\alpha_1 + (1-c)\alpha_2$ is in A, verifying A's convexity.

†9 [p. 67] *Formula* (5.55). In the univariate case, differentiating both sides of (5.51) with respect to α gives

$$\dot{\psi}(\alpha)e^{\psi(\alpha)} = \int_{\mathcal{X}} y e^{\alpha y} f_0(x)\, dx; \tag{5.77}$$

dividing by $e^{\psi(\alpha)}$ shows that $\dot{\psi}(\alpha) = E_\alpha\{y\}$. Differentiating (5.77) again gives

$$\left(\ddot{\psi}(\alpha) + \dot{\psi}(\alpha)^2\right) e^{\psi(\alpha)} = \int_{\mathcal{X}} y^2 e^{\alpha y} f_0(x)\, dx, \tag{5.78}$$

or

$$\ddot{\psi}(\alpha) = E_\alpha\{y^2\} - E_\alpha\{y\}^2 = \text{var}_\alpha\{y\}. \tag{5.79}$$

Successive derivatives of $\psi(\alpha)$ yield the higher cumulants of y, its skewness, kurtosis, etc.

†10 [p. 67] *MLE for* β. The gradient with respect to α of $\log f_\alpha(y)$ (5.50) is

$$\nabla_\alpha \left(\alpha' y - \psi(\alpha)\right) = y - \dot{\psi}(\alpha) = y - E_\alpha\{y^*\}, \tag{5.80}$$

(5.56), where y^* represents a hypothetical realization $y(x^*)$ drawn from $f_\alpha(\cdot)$. We achieve the MLE $\hat{\alpha}$ at $\nabla_{\hat{\alpha}} = 0$, or

$$E_{\hat{\alpha}}\{y^*\} = y. \tag{5.81}$$

In other words the MLE $\hat{\alpha}$ is the value of α that makes the expectation $E_\alpha\{y^*\}$ match the observed y. Thus (5.58) implies that the MLE of parameter β is y.

Part II

Early Computer-Age Methods

6

Empirical Bayes

The constraints of slow mechanical computation molded classical statistics into a mathematically ingenious theory of sharply delimited scope. Emerging after the Second World War, electronic computation loosened the computational stranglehold, allowing a more expansive and useful statistical methodology.

Some revolutions start slowly. The journals of the 1950s continued to emphasize classical themes: pure mathematical development typically centered around the normal distribution. Change came gradually, but by the 1990s a new statistical technology, computer enabled, was firmly in place. Key developments from this period are described in the next several chapters. The ideas, for the most part, would not startle a pre-war statistician, but their computational demands, factors of 100 or 1000 times those of classical methods, would. More factors of a thousand lay ahead, as will be told in Part III, the story of statistics in the twenty-first century.

Empirical Bayes methodology, this chapter's topic, has been a particularly slow developer despite an early start in the 1940s. The roadblock here was not so much the computational demands of the theory as a lack of appropriate data sets. Modern scientific equipment now provides ample grist for the empirical Bayes mill, as will be illustrated later in the chapter, and more dramatically in Chapters 15–21.

6.1 Robbins' Formula

Table 6.1 shows one year of claims data for a European automobile insurance company; 7840 of the 9461 policy holders made no claims during the year, 1317 made a single claim, 239 made two claims each, etc., with Table 6.1 continuing to the one person who made seven claims. Of course the insurance company is concerned about the claims each policy holder will make in the *next* year.

Bayes' formula seems promising here. We suppose that x_k, the number

Table 6.1 *Counts y_x of number of claims x made in a single year by 9461 automobile insurance policy holders. Robbins' formula* (6.7) *estimates the number of claims expected in a succeeding year, for instance* 0.168 *for a customer in the $x = 0$ category. Parametric maximum likelihood analysis based on a gamma prior gives less noisy estimates.*

Claims x	0	1	2	3	4	5	6	7
Counts y_x	7840	1317	239	42	14	4	4	1
Formula (6.7)	.168	.363	.527	1.33	1.43	6.00	1.25	
Gamma MLE	.164	.398	.633	.87	1.10	1.34	1.57	

of claims to be made in a single year by policy holder k, follows a Poisson distribution with parameter θ_k,

$$\Pr\{x_k = x\} = p_{\theta_k}(x) = e^{-\theta_k}\theta_k^x/x!, \tag{6.1}$$

for $x = 0, 1, 2, 3, \dots$; θ_k is the expected value of x_k. A good customer, from the company's point of view, has a small value of θ_k, though in any one year his or her actual number of accidents x_k will vary randomly according to probability density (6.1).

Suppose we knew the prior density $g(\theta)$ for the customers' θ values. Then Bayes' rule (3.5) would yield

$$E\{\theta|x\} = \frac{\int_0^\infty \theta p_\theta(x)g(\theta)\,d\theta}{\int_0^\infty p_\theta(x)g(\theta)\,d\theta} \tag{6.2}$$

for the expected value of θ of a customer observed to make x claims in a single year. This would answer the insurance company's question of what number of claims X to expect the next year from the same customer, since $E\{\theta|x\}$ is also $E\{X|x\}$ (θ being the expectation of X).

Formula (6.2) is just the ticket if the prior $g(\theta)$ is known to the company, but what if it is not? A clever rewriting of (6.2) provides a way forward. Using (6.1), (6.2) becomes

$$\begin{aligned} E\{\theta|x\} &= \frac{\int_0^\infty \left[e^{-\theta}\theta^{x+1}/x!\right] g(\theta)\,d\theta}{\int_0^\infty \left[e^{-\theta}\theta^x/x!\right] g(\theta)\,d\theta} \\ &= \frac{(x+1)\int_0^\infty \left[e^{-\theta}\theta^{x+1}/(x+1)!\right] g(\theta)\,d\theta}{\int_0^\infty \left[e^{-\theta}\theta^x/x!\right] g(\theta)\,d\theta}. \end{aligned} \tag{6.3}$$

The *marginal density* of x, integrating $p_\theta(x)$ over the prior $g(\theta)$, is

$$f(x) = \int_0^\infty p_\theta(x)g(\theta)\,d\theta = \int_0^\infty \left[e^{-\theta}\theta^x/x!\right] g(\theta)\,d\theta. \tag{6.4}$$

Comparing (6.3) with (6.4) gives *Robbins' formula*,

$$E\{\theta|x\} = (x+1)f(x+1)/f(x). \tag{6.5}$$

The surprising and gratifying fact is that, even with no knowledge of the prior density $g(\theta)$, the insurance company can estimate $E\{\theta|x\}$ (6.2) from formula (6.5). The obvious estimate of the marginal density $f(x)$ is the proportion of total counts in category x,

$$\hat{f}(x) = y_x/N, \quad \text{with } N = \textstyle\sum_x y_x, \text{ the total count,} \tag{6.6}$$

$\hat{f}(0) = 7840/9461$, $\hat{f}(1) = 1317/9461$, etc. This yields an empirical version of Robbins' formula,

$$\hat{E}\{\theta|x\} = (x+1)\hat{f}(x+1)/\hat{f}(x) = (x+1)y_{x+1}/y_x, \tag{6.7}$$

the final expression not requiring N. Table 6.1 gives $\hat{E}\{\theta|0\} = 0.168$: customers who made zero claims in one year had expectation 0.168 of a claim the next year; those with one claim had expectation 0.363, and so on.

Robbins' formula came as a surprise[1] to the statistical world of the 1950s: the expectation $E\{\theta_k|x_k\}$ for a single customer, unavailable without the prior $g(\theta)$, somehow becomes available in the context of a large study. The terminology *empirical Bayes* is apt here: Bayesian formula (6.5) for a single subject is estimated empirically (i.e., frequentistically) from a collection of similar cases. The crucial point, and the surprise, is that *large data sets of parallel situations carry within them their own Bayesian information.* Large parallel data sets are a hallmark of twenty-first-century scientific investigation, promoting the popularity of empirical Bayes methods.

Formula (6.7) goes awry at the right end of Table 6.1, where it is destabilized by small count numbers. A parametric approach gives more dependable results: now we assume that the prior density $g(\theta)$ for the customers' θ_k values has a gamma form (Table 5.1)

$$g(\theta) = \frac{\theta^{\nu-1}e^{-\theta/\sigma}}{\sigma^\nu\Gamma(\nu)}, \qquad \text{for } \theta \geq 0, \tag{6.8}$$

but with parameters ν and σ unknown. Estimates $(\hat{\nu}, \hat{\sigma})$ are obtained by

[1] Perhaps it shouldn't have; estimation methods similar to (6.7) were familiar in the actuarial literature.

maximum likelihood fitting to the counts y_x, yielding a parametrically es-
†1 timated marginal density†

$$\hat{f}(x) = f_{\hat{\nu},\hat{\sigma}}(x), \tag{6.9}$$

or equivalently $\hat{y}_x = N f_{\hat{\nu},\hat{\sigma}}(x)$.

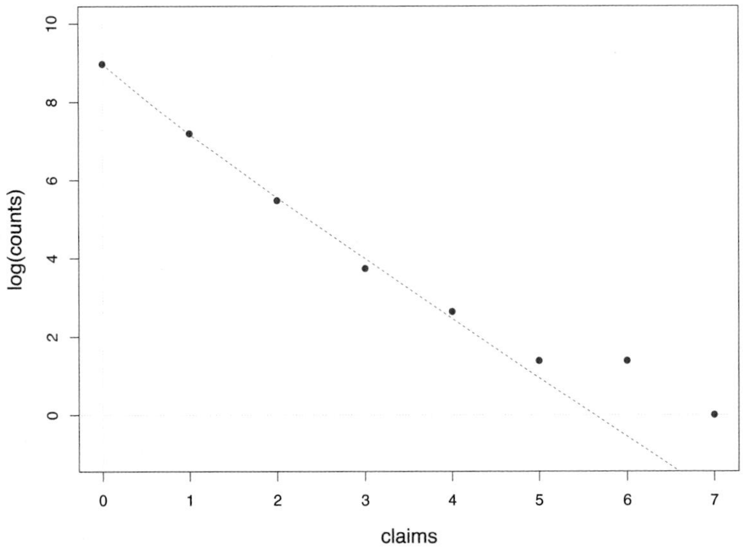

Figure 6.1 Auto accident data; log(counts) vs claims for 9461 auto insurance policies. The dashed line is a gamma MLE fit.

The bottom row of Table 6.1 gives parametric estimates $E_{\hat{\nu},\hat{\sigma}}\{\theta|x\} = (x+1)\hat{y}_{x+1}/\hat{y}_x$, which are seen to be less eccentric for large x. Figure 6.1 compares (on the log scale) the raw counts y_x with their parametric cousins $\hat{y}_x$.

6.2 The Missing-Species Problem

The very first empirical Bayes success story related to the butterfly data of Table 6.2. Even in the midst of World War II Alexander Corbet, a leading naturalist, had been trapping butterflies for two years in Malaysia (then Malaya): 118 species were so rare that he had trapped only one speciman each, 74 species had been trapped twice each, Table 6.2 going on to show that three species were trapped 44 times each, and so on. Some of the more common species had appeared hundreds of times each, but of course Corbet was interested in the rarer specimans.

Table 6.2 *Butterfly data; number y of species seen x times each in two years of trapping; 118 species trapped just once, 74 trapped twice each, etc.*

x	1	2	3	4	5	6	7	8	9	10	11	12
y	118	74	44	24	29	22	20	19	20	15	12	14

x	13	14	15	16	17	18	19	20	21	22	23	24
y	6	12	6	9	9	6	10	10	11	5	3	3

Corbet then asked a seemingly impossible question: if he trapped for one additional year, how many new species would he expect to capture? The question relates to the *absent* entry in Table 6.2, $x = 0$, the species that haven't been seen yet. Do we really have any evidence at all for answering Corbet? Fortunately he asked the right man: R. A. Fisher, who produced a surprisingly satisfying solution for the "missing-species problem."

Suppose there are S species in all, seen or unseen, and that x_k, the number of times species k is trapped in one time unit,[2] follows a Poisson distribution with parameter θ_k as in (6.1),

$$x_k \sim \text{Poi}(\theta_k), \qquad \text{for } k = 1, 2, \ldots, S. \tag{6.10}$$

The entries in Table 6.2 are

$$y_x = \#\{x_k = x\}, \qquad \text{for } x = 1, 2, \ldots, 24, \tag{6.11}$$

the number of species trapped exactly x times each.

Now consider a further trapping period of t time units, $t = 1/2$ in Corbet's question, and let $x_k(t)$ be the number of times species k is trapped in the new period. Fisher's key assumption is that

$$x_k(t) \sim \text{Poi}(\theta_k t) \tag{6.12}$$

independently of x_k. That is, any one species is trapped independently over time[3] at a rate proportional to its parameter θ_k.

The probability that species k is *not* seen in the initial trapping period

[2] One time unit equals two years in Corbet's situation.
[3] This is the definition of a *Poisson process*.

but *is* seen in the new period, that is $x_k = 0$ and $x_k(t) > 0$, is

$$e^{-\theta_k}\left(1 - e^{-\theta_k t}\right), \tag{6.13}$$

so that $E(t)$, the expected number of new species seen in the new trapping period, is

$$E(t) = \sum_{k=1}^{S} e^{-\theta_k}\left(1 - e^{-\theta_k t}\right). \tag{6.14}$$

It is convenient to write (6.14) as an integral,

$$E(t) = S \int_0^\infty e^{-\theta}\left(1 - e^{-\theta t}\right) g(\theta)\, d\theta, \tag{6.15}$$

where $g(\theta)$ is the "empirical density" putting probability $1/S$ on each of the θ_k values. (Later we will think of $g(\theta)$ as a continuous prior density on the possible θ_k values.)

Expanding $1 - e^{-\theta t}$ gives

$$E(t) = S \int_0^\infty e^{-\theta}\left[\theta t - (\theta t)^2/2! + (\theta t)^3/3! + \cdots\right] g(\theta)\, d\theta. \tag{6.16}$$

Notice that the expected value e_x of y_x is the sum of the probabilities of being seen exactly x times in the initial period,

$$\begin{aligned} e_x = E\{y_x\} &= \sum_{k=1}^{S} e^{-\theta_k}\theta_k^x/x! \\ &= S \int_0^\infty \left[e^{-\theta}\theta^x/x!\right] g(\theta)\, d\theta. \end{aligned} \tag{6.17}$$

Comparing (6.16) with (6.17) provides a surprising result,

$$E(t) = e_1 t - e_2 t^2 + e_3 t^3 - \cdots. \tag{6.18}$$

We don't know the e_x values but, as in Robbins' formula, we can estimate them by the y_x values, yielding an answer to Corbet's question,

$$\hat{E}(t) = y_1 t - y_2 t^2 + y_3 t^3 - \cdots. \tag{6.19}$$

Corbet specified $t = 1/2$, so[4]

$$\begin{aligned} \hat{E}(1/2) &= 118(1/2) - 74(1/2)^2 + 44(1/2)^3 - \cdots \\ &= 45.2. \end{aligned} \tag{6.20}$$

[4] This may have been discouraging; there were no new trapping results reported.

Table 6.3 *Expectation* (6.19) *and its standard error* (6.21) *for the number of new species captured in t additional fractional units of trapping time.*

t	0.0	0.1	0.2	0.3	0.4	0.5	0.6	0.7	0.8	0.9	1.0
$E(t)$	0	11.10	20.96	29.79	37.79	45.2	52.1	58.9	65.6	71.6	75.0
$\widehat{\text{sd}}(t)$	0	2.24	4.48	6.71	8.95	11.2	13.4	15.7	17.9	20.1	22.4

Formulas (6.18) and (6.19) do not require the butterflies to arrive independently. If we are willing to add the assumption that the x_k's are mutually independent, we can calculate† †2

$$\widehat{\text{sd}}(t) = \left(\sum_{x=1}^{24} y_x t^{2x} \right)^{1/2} \tag{6.21}$$

as an approximate standard error for $\hat{E}(t)$. Table 6.3 shows $\hat{E}(t)$ and $\widehat{\text{sd}}(t)$ for $t = 0, 0.1, 0.2, \ldots, 1$; in particular,

$$\hat{E}(0.5) = 45.2 \pm 11.2. \tag{6.22}$$

Formula (6.19) becomes unstable for $t > 1$. This is our price for substituting the nonparametric estimates y_x for e_x in (6.18). Fisher actually answered Corbet using a parametric empirical Bayes model in which the prior $g(\theta)$ for the Poisson parameters θ_k (6.12) was assumed to be of the gamma form (6.8). It can be shown† that then $E(t)$ (6.15) is given by †3

$$E(t) = e_1 \left\{1 - (1 + \gamma t)^{-\nu}\right\} / (\gamma \nu), \tag{6.23}$$

where $\gamma = \sigma/(1 + \sigma)$. Taking $\hat{e}_1 = y_1$, maximum likelihood estimation gave

$$\hat{\nu} = 0.104 \quad \text{and} \quad \hat{\sigma} = 89.79. \tag{6.24}$$

Figure 6.2 shows that the parametric estimate of $E(t)$ (6.23) using $\hat{e}_1$, $\hat{\nu}$, and $\hat{\sigma}$ is just slightly greater than the nonparametric estimate (6.19) over the range $0 \le t \le 1$. Fisher's parametric estimate, however, gives reasonable results for $t > 1$, $\hat{E}(2) = 123$ for instance, for a future trapping period of 2 units (4 years). "Reasonable" does not necessarily mean dependable. The gamma prior is a mathematical convenience, not a fact of nature; projections into the far future fall into the category of educated guessing.

The missing-species problem encompasses more than butterflies. There are 884,647 words in total in the recognized Shakespearean canon, of which 14,376 are so rare they appear just once each, 4343 appear twice each, etc.,

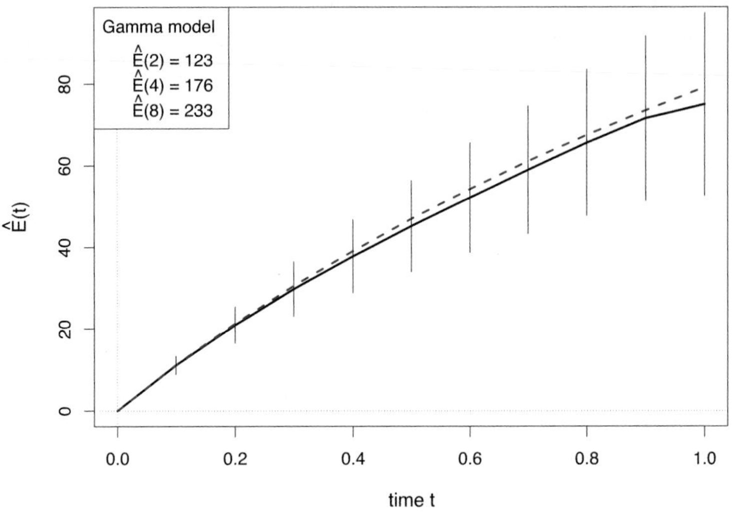

Figure 6.2 Butterfly data; expected number of new species in t units of additional trapping time. Nonparametric fit (solid) $\pm$ 1 standard deviation; gamma model (dashed).

Table 6.4 *Shakespeare's word counts; 14,376 distinct words appeared once each in the canon, 4343 distinct words twice each, etc. The canon has 884,647 words in total, counting repeats.*

	1	2	3	4	5	6	7	8	9	10
0+	14376	4343	2292	1463	1043	837	638	519	430	364
10+	305	259	242	223	187	181	179	130	127	128
20+	104	105	99	112	93	74	83	76	72	63
30+	73	47	56	59	53	45	34	49	45	52
40+	49	41	30	35	37	21	41	30	28	19
50+	25	19	28	27	31	19	19	22	23	14
60+	30	19	21	18	15	10	15	14	11	16
70+	13	12	10	16	18	11	8	15	12	7
80+	13	12	11	8	10	11	7	12	9	8
90+	4	7	6	7	10	10	15	7	7	5

as in Table 6.4, which goes on to the five words appearing 100 times each. All told, 31,534 distinct words appear (including those that appear more than 100 times each), this being the observed size of Shakespeare's vocabulary. But what of the words Shakespeare knew but didn't use? These are the "missing species" in Table 6.4.

Suppose another quantity of previously unknown Shakespeare manuscripts was discovered, comprising $884647 \cdot t$ words (so $t = 1$ would represent a new canon just as large as the old one). How many previously unseen distinct words would we expect to discover?

Employing formulas (6.19) and (6.21) gives

$$11430 \pm 178 \tag{6.25}$$

for the expected number of distinct new words if $t = 1$. This is a very conservative lower bound on how many words Shakespeare knew but didn't use. We can imagine t rising toward infinity, revealing ever more unseen vocabulary. Formula (6.19) fails for $t > 1$, and Fisher's gamma assumption is just that, but more elaborate empirical Bayes calculations give a firm lower bound of $35,000+$ on Shakespeare's unseen vocabulary, exceeding the visible portion!

Missing mass is an easier version of the missing-species problem, in which we only ask for the proportion of the total sum of θ_k values corresponding to the species that went unseen in the original trapping period,

$$M = \sum_{\text{unseen}} \theta_k \Big/ \sum_{\text{all}} \theta_k. \tag{6.26}$$

The numerator has expectation

$$\sum_{\text{all}} \theta_k e^{-\theta_k} = S \int_0^\infty \theta e^{-\theta} g(\theta) = e_1 \tag{6.27}$$

as in (6.15), while the expectation of the denominator is

$$\sum_{\text{all}} \theta_k = \sum_{\text{all}} E\{x_s\} = E\left\{\sum_{\text{all}} x_s\right\} = E\{N\}, \tag{6.28}$$

where N is the number of species observed. The obvious missing-mass estimate is then

$$\hat{M} = y_1/N. \tag{6.29}$$

For the Shakespeare data,

$$\hat{M} = 14376/884647 = 0.016. \tag{6.30}$$

We have seen most of Shakespeare's vocabulary, as weighted by his usage, though not by his vocabulary count.

All of this seems to live in the rarefied world of mathematical abstraction, but in fact some previously unknown Shakespearean work *might* have been discovered in 1985. A short poem, “Shall I die?,” was found in the archives of the Bodleian Library and, controversially, attributed to Shakespeare by some but not all experts.

The poem of 429 words provided a new “trapping period” of length only

$$t = 429/884647 = 4.85 \cdot 10^{-4}, \tag{6.31}$$

and a prediction from (6.19) of

$$E\{t\} = 6.97 \tag{6.32}$$

new “species,” i.e., distinct words not appearing in the canon. In fact there were nine such words in the poem. Similar empirical Bayes predictions for the number of words appearing once each in the canon, twice each, etc., showed reasonable agreement with the poem’s counts, but not enough to stifle doubters. “Shall I die?” is currently grouped with other canonical apocrypha by a majority of experts.

6.3 A Medical Example

The reader may have noticed that our examples so far have not been particularly computer intensive; all of the calculations could have been (and originally were) done by hand.[5] This section discusses a medical study where the empirical Bayes analysis is more elaborate.

Cancer surgery sometimes involves the removal of surrounding lymph nodes as well as the primary target at the site. Figure 6.3 concerns $N = 844$ surgeries, each reporting

$$n = \#\text{ nodes removed} \quad \text{and} \quad x = \#\text{ nodes found positive}, \tag{6.33}$$

“positive” meaning malignant. The ratios

$$p_k = x_k/n_k, \qquad k = 1, 2, \ldots, N, \tag{6.34}$$

are described in the histogram. A large proportion of them, $340/844$ or 40%, were zero, the remainder spreading unevenly between zero and one. The denominators n_k ranged from 1 to 69, with a mean of 19 and standard deviation of 11.

We suppose that each patient has some true probability of a node being

[5] Not so collecting the data. Corbet’s work was pre-computer but Shakespeare’s word counts were done electronically. Twenty-first-century scientific technology excels at the production of the large parallel-structured data sets conducive to empirical Bayes analysis.

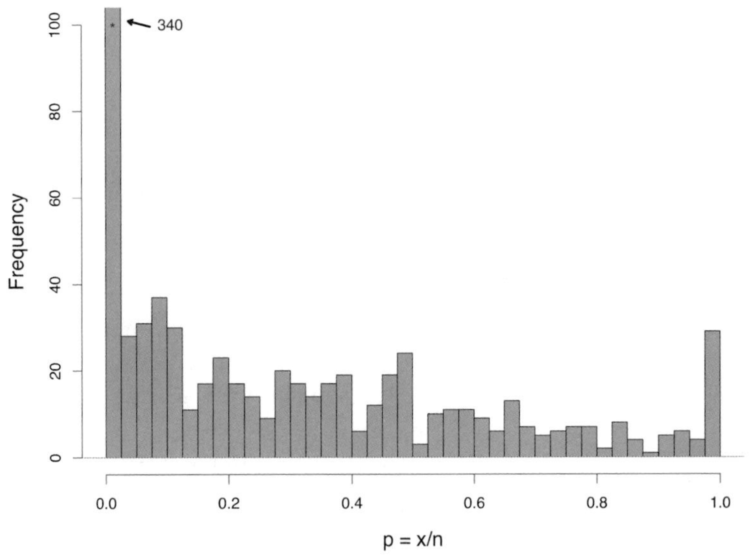

Figure 6.3 Nodes study; ratio $p = x/n$ for 844 patients; $n =$ number of nodes removed, $x =$ number positive.

positive, say probability θ_k for patient k, and that his or her nodal results occur independently of each other, making x_k binomial,

$$x_k \sim \text{Bi}(n_k, \theta_k). \tag{6.35}$$

This gives $p_k = x_k/n_k$ with mean and variance

$$p_k \sim (\theta_k, \theta_k(1-\theta_k)/n_k), \tag{6.36}$$

so that θ_k is estimated more accurately when n_k is large.

A Bayesian analysis would begin with the assumption of a prior density $g(\theta)$ for the θ_k values,

$$\theta_k \sim g(\theta), \qquad \text{for } k = 1, 2, \ldots, N = 844. \tag{6.37}$$

We don't know $g(\theta)$, but the parallel nature of the nodes data set—844 similar cases—suggests an empirical Bayes approach. As a first try for the nodes study, we assume that $\log\{g(\theta)\}$ is a fourth-degree polynomial in θ,

$$\log\{g_\alpha(\theta)\} = a_0 + \sum_{j=1}^{4} \alpha_j \theta^j; \tag{6.38}$$

$g_\alpha(\theta)$ is determined by the parameter vector $\alpha = (\alpha_1, \alpha_2, \alpha_3, \alpha_4)$ since, given α, a_0 can be calculated from the requirement that

$$\int_0^1 g_\alpha(\theta)\, d\theta = 1 = \int_0^1 \exp\left\{a_0 + \sum_1^4 \alpha_j \theta^j\right\} d\theta. \tag{6.39}$$

For a given choice of α, let $f_\alpha(x_k)$ be the marginal probability of the observed value x_k for patient k,

$$f_\alpha(x_k) = \int_0^1 \binom{n_k}{x_k} \theta^{x_k}(1-\theta)^{n_k - x_k} g_\alpha(\theta)\, d\theta. \tag{6.40}$$

The maximum likelihood estimate of α is the maximizer

$$\hat{\alpha} = \arg\max_\alpha \left\{\sum_{k=1}^N \log f_\alpha(x_k)\right\}. \tag{6.41}$$

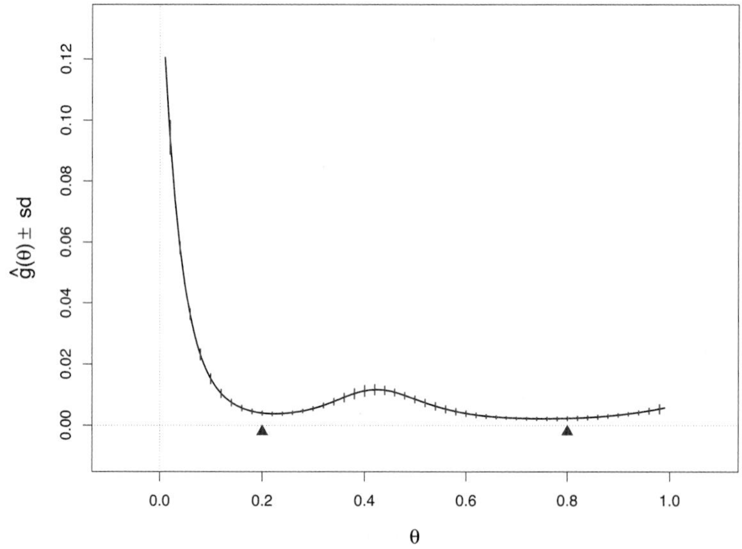

Figure 6.4 Estimated prior density $g(\theta)$ for the nodes study; 59% of patients have $\theta \le 0.2$, 7% have $\theta \ge 0.8$.

Figure 6.4 graphs $g_{\hat{\alpha}}(\theta)$, the empirical Bayes estimate for the prior distribution of the θ_k values. The huge spike at zero in Figure 6.3 is now reduced: $\Pr\{\theta_k \le 0.01\} = 0.12$ compared with the 38% of the p_k values less than 0.01. Small θ values are still the rule though, for instance

$$\int_0^{0.20} g_{\hat{\alpha}}(\theta)\, d\theta = 0.59 \text{ compared with } \int_{0.80}^{1.00} g_{\hat{\alpha}}(\theta)\, d\theta = 0.07. \quad (6.42)$$

The vertical bars in Figure 6.4 indicate $\pm$ one standard error for the estimation of $g(\theta)$. The curve seems to have been estimated very accurately, at least if we assume the adequacy of model (6.37). Chapter 21 describes the computations involved in Figure 6.4.

The posterior distribution of θ_k given x_k and n_k is estimated according to Bayes' rule (3.5) to be

$$\hat{g}(\theta|x_k, n_k) = g_{\hat{\alpha}}(\theta)\binom{n_k}{x_k}\theta^{x_k}(1-\theta)^{n_k-x_k} \Big/ f_{\hat{\alpha}}(x_k), \quad (6.43)$$

with $f_{\hat{\alpha}}(x_k)$ from (6.40).

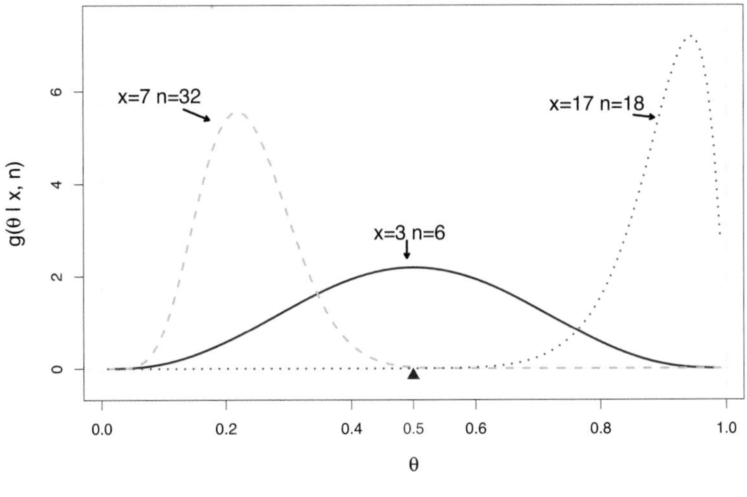

Figure 6.5 Empirical Bayes posterior densities of θ for three patients, given x = number of positive nodes, n = number of nodes.

Figure 6.5 graphs $\hat{g}(\theta|x_k, n_k)$ for three choices of (x_k, n_k): $(7, 32)$, $(3, 6)$, and $(17, 18)$. If we take $\theta \geq 0.50$ as indicating poor prognosis (and suggesting more aggressive follow-up therapy), then the first patient is almost surely on safe ground, the third patient almost surely needs more follow-up therapy and the situation of the second is uncertain.

6.4 Indirect Evidence 1

A good definition of a statistical argument is one in which many small pieces of evidence, often contradictory, are combined to produce an overall conclusion. In the clinical trial of a new drug, for instance, we don't expect the drug to cure every patient, or the placebo to always fail, but eventually perhaps we will obtain convincing evidence of the new drug's efficacy.

The clinical trial is collecting *direct* statistical evidence, in which each subject's success or failure bears directly upon the question of interest. Direct evidence, interpreted by frequentist methods, was the dominant mode of statistical application in the twentieth century, being strongly connected to the idea of scientific objectivity.

Bayesian inference provides a theoretical basis for incorporating *indirect* evidence, for example the doctor's prior experience with twin sexes in Section 3.1. The assertion of a prior density $g(\theta)$ amounts to a claim for the relevance of past data to the case at hand.

Empirical Bayes removes the Bayes scaffolding. In place of a reassuring prior $g(\theta)$, the statistician must put his or her faith in the relevance of the "other" cases in a large data set to the case of direct interest. For the second patient in Figure 6.5, the direct estimate of his θ value is $\hat{\theta} = 3/6 = 0.50$. The empirical Bayes estimate is a little less,

$$\hat{\theta}^{\text{EB}} = \int_0^1 \theta \hat{g}(\theta | x_k = 3, n_k = 6) = 0.446. \tag{6.44}$$

A small difference, but we will see bigger ones in succeeding chapters.

The changes in twenty-first-century statistics have largely been demand driven, responding to the massive data sets enabled by modern scientific equipment. Philosophically, as opposed to methodologically, the biggest change has been the increased acceptance of indirect evidence, especially as seen in empirical Bayes and objective ("uninformative") Bayes applications. *False-discovery rates*, Chapter 15, provide a particularly striking shift from direct to indirect evidence in hypothesis testing. Indirect evidence in estimation is the subject of our next chapter.

6.5 Notes and Details

Robbins (1956) introduced the term "empirical Bayes" as well as rule (6.7) as part of a general theory of empirical Bayes estimation. 1956 was also the publication year for Good and Toulmin's solution (6.19) to the missing-species problem. Good went out of his way to credit his famous Bletchley colleague Alan Turing for some of the ideas. The auto accident data is taken from Table 3.1 of Carlin and Louis (1996), who provide a more complete discussion. Empirical Bayes estimates such as 11430 in (6.25) do not depend on independence among the "species," but accuracies such as ± 178 do; and similarly for the error bars in Figures 6.2 and 6.4.

Corbet's enormous efforts illustrate the difficulties of amassing large data sets in pre-computer times. *Dependable* data is still hard to come by, but these days it is often the statistician's job to pry it out of enormous databases. Efron and Thisted (1976) apply formula (6.19) to the Shakespeare word counts, and then use linear programming methods to bound Shakespeare's unseen vocabulary from below at 35,000 words. (Shakespeare was actually less "wordy" than his contemporaries, Marlow and Donne.) "Shall I die," the possibly Shakespearean poem recovered in 1985, is analyzed by a variety of empirical Bayes techniques in Thisted and Efron (1987). Comparisons are made with other Elizabethan authors, none of whom seem likely candidates for authorship.

The Shakespeare word counts are from Spevack's (1968) concordance. (The first concordance was compiled by hand in the mid 1800s, listing every word Shakespeare wrote and where it appeared, a full life's labor.)

The nodes example, Figure 6.3, is taken from Gholami *et al.* (2015).

†1 [p. 78] *Formula* (6.9). For any positive numbers c and d we have

$$\int_0^\infty \theta^{c-1} e^{-\theta/d} \, d\theta = d^c \Gamma(c), \tag{6.45}$$

so combining gamma prior (6.8) with Poisson density (6.1) gives marginal density

$$\begin{aligned} f_{\nu,\sigma}(x) &= \frac{\int_0^\infty \theta^{\nu+x-1} e^{-\theta/\gamma} \, d\theta}{\sigma^\nu \Gamma(\nu) x!} \\ &= \frac{\gamma^{\nu+x} \Gamma(\nu + x)}{\sigma^\nu \Gamma(\nu) x!}, \end{aligned} \tag{6.46}$$

where $\gamma = \sigma/(1+\sigma)$. Assuming independence among the counts y_x (which is exactly true if the customers act independently of each other and N, the total number of them, is itself Poisson), the log likelihood function for the accident data is

$$\sum_{x=0}^{x_{\max}} y_x \log \{ f_{\nu,\sigma}(x) \} . \tag{6.47}$$

Here $x_{\max}$ is some notional upper bound on the maximum possible number of accidents for a single customer; since $y_x = 0$ for $x > 7$ the choice of $x_{\max}$ is irrelevant. The values $(\hat{\nu}, \hat{\sigma})$ in (6.8) maximize (6.47).

†2 [p. 81] *Formula* (6.21). If $N = \sum y_x$, the total number trapped, is assumed to be Poisson, and if the N observed values x_k are mutually independent, then a useful property of the Poisson distribution implies that the counts y_x are themselves approximately independent Poisson variates

$$y_x \overset{\text{ind}}{\sim} \text{Poi}(e_x), \qquad \text{for } x = 0, 1, 2, \ldots, \tag{6.48}$$

in notation (6.17). Formula (6.19) and $\text{var}\{y_x\} = e_x$ then give

$$\text{var}\left\{\hat{E}(t)\right\} = \sum_{x \geq 1} e_x t^{2x}. \tag{6.49}$$

Substituting y_x for e_x produces (6.21). Section 11.5 of Efron (2010) shows that (6.49) is an upper bound on $\text{var}\{\hat{E}(t)\}$ if N is considered fixed rather than Poisson.

†3 [p. 81] *Formula* (6.23). Combining the case $x = 1$ in (6.17) with (6.15) yields

$$E(t) = \frac{e_1 \left[\int_0^\infty e^{-\theta} g(\theta)\, d\theta - \int_0^\infty e^{-\theta(1+t)} g(\theta)\, d\theta\right]}{\int_0^\infty \theta e^{-\theta} g(\theta)\, d\theta}. \tag{6.50}$$

Substituting the gamma prior (6.8) for $g(\theta)$, and using (6.45) three times, gives formula (6.23).

7

James–Stein Estimation and Ridge Regression

If Fisher had lived in the era of “apps,” maximum likelihood estimation might have made him a billionaire. Arguably the twentieth century’s most influential piece of applied mathematics, maximum likelihood continues to be a prime method of choice in the statistician’s toolkit. Roughly speaking, maximum likelihood provides nearly unbiased estimates of nearly minimum variance, and does so in an automatic way.

That being said, maximum likelihood estimation has shown itself to be an inadequate and dangerous tool in many twenty-first-century applications. Again speaking roughly, unbiasedness can be an unaffordable luxury when there are hundreds or thousands of parameters to estimate at the same time.

The James–Stein estimator made this point dramatically in 1961, and made it in the context of just a few unknown parameters, not hundreds or thousands. It begins the story of *shrinkage estimation*, in which deliberate biases are introduced to improve overall performance, at a possible danger to individual estimates. Chapters 7 and 21 will carry on the story in its modern implementations.

7.1 The James–Stein Estimator

Suppose we wish to estimate a single parameter μ from observation x in the Bayesian situation

$$\mu \sim \mathcal{N}(M, A) \quad \text{and} \quad x|\mu \sim \mathcal{N}(\mu, 1), \tag{7.1}$$

in which case μ has posterior distribution

$$\mu|x \sim \mathcal{N}\left(M + B(x - M), B\right) \qquad [B = A/(A+1)] \tag{7.2}$$

as given in (5.21) (where we take $\sigma^2 = 1$ for convenience). The Bayes estimator of μ,

$$\hat{\mu}^{\text{Bayes}} = M + B(x - M), \tag{7.3}$$

has expected squared error

$$E\left\{\left(\hat{\mu}^{\text{Bayes}} - \mu\right)^2\right\} = B, \tag{7.4}$$

compared with 1 for the MLE $\hat{\mu}^{\text{MLE}} = x$,

$$E\left\{\left(\hat{\mu}^{\text{MLE}} - \mu\right)^2\right\} = 1. \tag{7.5}$$

If, say, $A = 1$ in (7.1) then $B = 1/2$ and $\hat{\mu}^{\text{Bayes}}$ has only half the risk of the MLE.

The same calculation applies to a situation where we have N independent versions of (7.1), say

$$\boldsymbol{\mu} = (\mu_1, \mu_2, \ldots, \mu_N)' \quad \text{and} \quad \boldsymbol{x} = (x_1, x_2, \ldots, x_N)', \tag{7.6}$$

with

$$\mu_i \sim \mathcal{N}(M, A) \quad \text{and} \quad x_i | \mu_i \sim \mathcal{N}(\mu_i, 1), \tag{7.7}$$

independently for $i = 1, 2, \ldots, N$. (Notice that the μ_i differ from each other, and that this situation is not the same as (5.22)–(5.23).) Let $\hat{\boldsymbol{\mu}}^{\text{Bayes}}$ indicate the vector of individual Bayes estimates $\hat{\mu}_i^{\text{Bayes}} = M + B(x_i - M)$,

$$\hat{\boldsymbol{\mu}}^{\text{Bayes}} = \boldsymbol{M} + B(\boldsymbol{x} - \boldsymbol{M}), \qquad \left[\boldsymbol{M} = (M, M, \ldots, M)'\right], \tag{7.8}$$

and similarly

$$\hat{\boldsymbol{\mu}}^{\text{MLE}} = \boldsymbol{x}.$$

Using (7.4) the total squared error risk of $\hat{\boldsymbol{\mu}}^{\text{Bayes}}$ is

$$E\left\{\left\|\hat{\boldsymbol{\mu}}^{\text{Bayes}} - \boldsymbol{\mu}\right\|^2\right\} = E\left\{\sum_{i=1}^{N}\left(\hat{\mu}_i^{\text{Bayes}} - \mu_i\right)^2\right\} = N \cdot B \tag{7.9}$$

compared with

$$E\left\{\left\|\hat{\boldsymbol{\mu}}^{\text{MLE}} - \boldsymbol{\mu}\right\|^2\right\} = N. \tag{7.10}$$

Again, $\hat{\boldsymbol{\mu}}^{\text{Bayes}}$ has only B times the risk of $\hat{\boldsymbol{\mu}}^{\text{MLE}}$.

This is fine if we know M and A (or equivalently M and B) in (7.1). If not, we might try to estimate them from $\boldsymbol{x} = (x_1, x_2, \ldots, x_N)$. Marginally, (7.7) gives

$$x_i \overset{\text{ind}}{\sim} \mathcal{N}(M, A + 1). \tag{7.11}$$

Then $\hat{M} = \bar{x}$ is an unbiased estimate of M. Moreover,

$$\hat{B} = 1 - (N - 3)/S \qquad \left[S = \sum_{i=1}^{N}(x - \bar{x})^2\right] \tag{7.12}$$

unbiasedly estimates B, as long as $N > 3$.[†] The James–Stein estimator is †1
the plug-in version of (7.3),

$$\hat{\mu}_i^{\text{JS}} = \hat{M} + \hat{B}\left(x_i - \hat{M}\right) \qquad \text{for } i = 1, 2, \ldots, N, \tag{7.13}$$

or equivalently $\hat{\boldsymbol{\mu}}^{\text{JS}} = \hat{\boldsymbol{M}} + \hat{B}(\boldsymbol{x} - \hat{\boldsymbol{M}})$, with $\hat{\boldsymbol{M}} = (\hat{M}, \hat{M}, \ldots, \hat{M})'$.

At this point the terminology "empirical Bayes" seems especially apt: Bayesian model (7.7) leads to the Bayes estimator (7.8), which itself is estimated empirically (i.e., frequentistically) from all the data $\boldsymbol{x}$, and then applied to the individual cases. Of course $\hat{\boldsymbol{\mu}}^{\text{JS}}$ cannot perform as well as the actual Bayes' rule $\hat{\boldsymbol{\mu}}^{\text{Bayes}}$, but the increased risk is surprisingly modest.
The expected squared risk of $\hat{\boldsymbol{\mu}}^{\text{JS}}$ under model (7.7) is[†] †2

$$E\left\{\left\|\hat{\boldsymbol{\mu}}^{\text{JS}} - \boldsymbol{\mu}\right\|^2\right\} = NB + 3(1 - B). \tag{7.14}$$

If, say, $N = 20$ and $A = 1$, then (7.14) equals 11.5, compared with true Bayes risk 10 from (7.9), much less than risk 20 for $\hat{\boldsymbol{\mu}}^{\text{MLE}}$.

A defender of maximum likelihood might respond that none of this is surprising: Bayesian model (7.7) specifies the parameters μ_i to be clustered more or less closely around a central point M, while $\hat{\boldsymbol{\mu}}^{\text{MLE}}$ makes no such assumption, and cannot be expected to perform as well. Wrong! Removing the Bayesian assumptions does not rescue $\hat{\boldsymbol{\mu}}^{\text{MLE}}$, as James and Stein proved in 1961:

James–Stein Theorem *Suppose that*

$$x_i | \mu_i \sim \mathcal{N}(\mu_i, 1) \tag{7.15}$$

independently for $i = 1, 2, \ldots, N$, *with* $N \geq 4$. *Then*

$$E\left\{\left\|\hat{\boldsymbol{\mu}}^{\text{JS}} - \boldsymbol{\mu}\right\|^2\right\} < N = E\left\{\left\|\hat{\boldsymbol{\mu}}^{\text{MLE}} - \boldsymbol{\mu}\right\|^2\right\} \tag{7.16}$$

for all choices of $\boldsymbol{\mu} \in \mathcal{R}^N$. (The expectations in (7.16) are with $\boldsymbol{\mu}$ fixed and $\boldsymbol{x}$ varying according to (7.15).)

In the language of decision theory, equation (7.16) says that $\hat{\boldsymbol{\mu}}^{\text{MLE}}$ is
inadmissable:[†] its total squared error risk exceeds that of $\hat{\boldsymbol{\mu}}^{\text{JS}}$ no matter †3
what $\boldsymbol{\mu}$ may be. This is a strong frequentist form of defeat for $\hat{\boldsymbol{\mu}}^{\text{MLE}}$, not depending on Bayesian assumptions.

The James–Stein theorem came as a rude shock to the statistical world of 1961. First of all, the defeat came on MLE's home field: normal observations with squared error loss. Fisher's "logic of inductive inference," Chapter 4, claimed that $\hat{\mu}^{\text{MLE}} = x$ was the obviously correct estimator in the univariate case, an assumption tacitly carried forward to multiparameter linear regression problems, where versions of $\hat{\boldsymbol{\mu}}^{\text{MLE}}$ were predominant. There are still some good reasons for sticking with $\hat{\boldsymbol{\mu}}^{\text{MLE}}$ in low-dimensional problems, as discussed in Section 7.4. But shrinkage estimation, as exemplified by the James–Stein rule, has become a necessity in the high-dimensional situations of modern practice.

7.2 The Baseball Players

The James–Stein theorem doesn't say by how much $\hat{\boldsymbol{\mu}}^{\text{JS}}$ beats $\hat{\boldsymbol{\mu}}^{\text{MLE}}$. If the improvement were infinitesimal nobody except theorists would be interested. In favorable situations the gains can in fact be substantial, as suggested by (7.14). One such situation appears in Table 7.1. The batting averages[1] of 18 Major League players have been observed over the 1970 season. The column labeled `MLE` reports the player's observed average over his first 90 at bats; `TRUTH` is the average over the remainder of the 1970 season (370 further at bats on average). We would like to predict `TRUTH` from the early-season observations.

The column labeled `JS` in Table 7.1 is from a version of the James–Stein estimator applied to the 18 MLE numbers. We suppose that each player's `MLE` value p_i (his batting average in the first 90 tries) is a binomial proportion,

$$p_i \sim \text{Bi}(90, P_i)/90. \tag{7.17}$$

Here P_i is his *true average*, how he would perform over an infinite number of tries; $\texttt{TRUTH}_i$ is itself a binomial proportion, taken over an average of 370 more tries per player.

At this point there are two ways to proceed. The simplest uses a normal approximation to (7.17),

$$p_i \dot{\sim} \mathcal{N}(P_i, \sigma_0^2), \tag{7.18}$$

where σ_0^2 is the binomial variance

$$\sigma_0^2 = \bar{p}(1-\bar{p})/90, \tag{7.19}$$

with $\bar{p} = 0.254$ the average of the p_i values. Letting $x_i = p_i/\sigma_0$, applying (7.13), and transforming back to $\hat{p}_i^{\text{JS}} = \sigma_0 \hat{\mu}_i^{\text{JS}}$, gives James–Stein estimates

$$\hat{p}_i^{\text{JS}} = \bar{p} + \left[1 - \frac{(N-3)\sigma_0^2}{\sum (p_i - \bar{p})^2}\right](p_i - \bar{p}). \tag{7.20}$$

[1] Batting average = # hits /# at bats, that is, the success rate. For example, Player 1 hits successfully 31 times in his first 90 tries, for batting average $31/90 = 0.345$. This data is based on 1970 Major League performances, but is partly artificial; see the endnotes.

Table 7.1 *Eighteen baseball players;* `MLE` *is batting average in first 90 at bats;* `TRUTH` *is average in remainder of 1970 season; James–Stein estimator* `JS` *is based on arcsin transformation of MLEs. Sum of squared errors for predicting* `TRUTH`*:* `MLE` *.0425,* `JS` *.0218.*

Player	MLE	JS	TRUTH	x
1	.345	.283	.298	11.96
2	.333	.279	.346	11.74
3	.322	.276	.222	11.51
4	.311	.272	.276	11.29
5	.289	.265	.263	10.83
6	.289	.264	.273	10.83
7	.278	.261	.303	10.60
8	.255	.253	.270	10.13
9	.244	.249	.230	9.88
10	.233	.245	.264	9.64
11	.233	.245	.264	9.64
12	.222	.242	.210	9.40
13	.222	.241	.256	9.39
14	.222	.241	.269	9.39
15	.211	.238	.316	9.14
16	.211	.238	.226	9.14
17	.200	.234	.285	8.88
18	.145	.212	.200	7.50

A second approach begins with the *arcsin transformation*

$$x_i = 2(n+0.5)^{1/2} \sin^{-1}\left[\left(\frac{np_i + 0.375}{n+0.75}\right)^{1/2}\right], \tag{7.21}$$

$n = 90$ (column labeled `x` in Table 7.1), a classical device that produces approximate normal deviates of variance 1,

$$x_i \mathrel{\dot\sim} \mathcal{N}(\mu_i, 1), \tag{7.22}$$

where μ_i is transformation (7.21) applied to $\texttt{TRUTH}_i$. Using (7.13) gives $\hat\mu_i^{\text{JS}}$, which is finally inverted back to the binomial scale,

$$\hat p_i^{\text{JS}} = \frac{1}{n}\left[\frac{n+0.75}{n+0.5}\left(\frac{\sin \hat\mu_i^{\text{JS}}}{2}\right)^2 - 0.375\right]. \tag{7.23}$$

Formulas (7.20) and (7.23) yielded nearly the same estimates for the baseball players; the `JS` column in Table 7.1 is from (7.23). James and Stein's theorem requires normality, but the James–Stein estimator often works perfectly well in less ideal situations. That is the case in Table 7.1:

$$\sum_{i=1}^{18} (\texttt{MLE}_i - \texttt{TRUTH}_i)^2 = 0.0425 \quad \text{while} \quad \sum_{i=1}^{18} (\texttt{JS}_i - \texttt{TRUTH}_i)^2 = 0.0218. \tag{7.24}$$

In other words, the James–Stein estimator reduced total predictive squared error by about 50%.

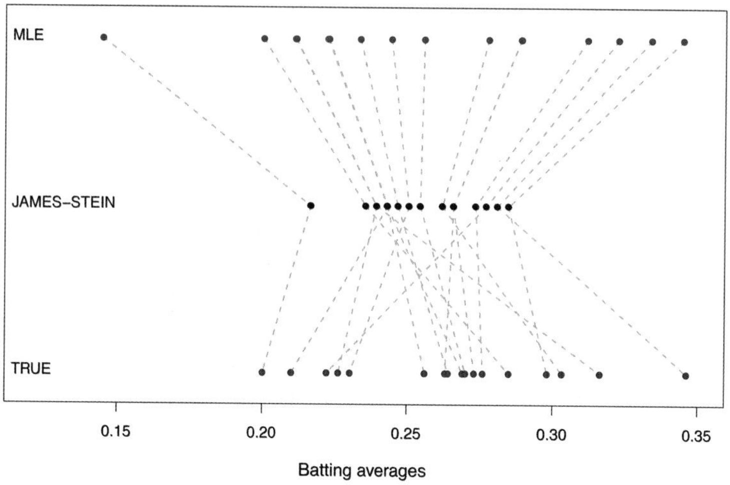

Figure 7.1 Eighteen baseball players; top line MLE, middle James–Stein, bottom true values. Only 13 points are visible, since there are ties.

The James–Stein rule describes a *shrinkage estimator*, each MLE value x_i being shrunk by factor $\hat{B}$ toward the grand mean $\hat{M} = \bar{x}$ (7.13). ($\hat{B} = 0.34$ in (7.20).) Figure 7.1 illustrates the shrinking process for the baseball players.

To see why shrinking might make sense, let us return to the original Bayes model (7.8) and take $M = 0$ for simplicity, so that the x_i are marginally $\mathcal{N}(0, A + 1)$ (7.11). Even though each x_i is unbiased for its parameter μ_i, as a group they are "overdispersed,"

$$E\left\{\sum_{i=1}^{N} x_i^2\right\} = N(A + 1) \quad \text{compared with} \quad E\left\{\sum_{i=1}^{N} \mu_i^2\right\} = NA. \tag{7.25}$$

The sum of squares of the MLEs exceeds that of the true values by expected amount N; shrinkage improves group estimation by removing the excess.

In fact the James–Stein rule *overshrinks* the data, as seen in the bottom two lines of Figure 7.1, a property it inherits from the underlying Bayes model: the Bayes estimates $\hat{\mu}_i^{\text{Bayes}} = Bx_i$ have

$$E\left\{\sum_{i=1}^{N}\left(\hat{\mu}_i^{\text{Bayes}}\right)^2\right\} = NB^2(A+1) = NA\frac{A}{A+1}, \tag{7.26}$$

overshrinking $E(\sum \mu_i^2) = NA$ by factor $A/(A+1)$. We could use the less extreme shrinking rule $\tilde{\mu}_i = \sqrt{B}x_i$, which gives the correct expected sum of squares NA, but a larger expected sum of squared estimation errors $E\{\sum(\tilde{\mu}_i - \mu_i)^2|\boldsymbol{x}\}$.

The most extreme shrinkage rule would be "all the way," that is, to

$$\hat{\mu}_i^{\text{NULL}} = \bar{x} \qquad \text{for } i = 1, 2, \dots, N, \tag{7.27}$$

NULL indicating that in a classical sense we have accepted the null hypothesis of no differences among the μ_i values. (This gave $\sum(P_i - \bar{p})^2 = 0.0266$ for the baseball data (7.24).) The James–Stein estimator is a data-based rule for compromising between the null hypothesis of no differences and the MLE's tacit assumption of no relationship at all among the μ_i values. In this sense it blurs the classical distinction between hypothesis testing and estimation.

7.3 Ridge Regression

Linear regression, perhaps the most widely used estimation technique, is based on a version of $\hat{\boldsymbol{\mu}}^{\text{MLE}}$. In the usual notation, we observe an n-dimensional vector $\boldsymbol{y} = (y_1, y_2, \dots, y_n)'$ from the linear model

$$\boldsymbol{y} = \boldsymbol{X}\beta + \boldsymbol{\epsilon}. \tag{7.28}$$

Here $\boldsymbol{X}$ is a known $n \times p$ *structure matrix*, β is an unknown p-dimensional parameter vector, while the *noise vector* $\boldsymbol{\epsilon} = (\epsilon_1, \epsilon_2, \dots, \epsilon_n)'$ has its components uncorrelated and with constant variance σ^2,

$$\boldsymbol{\epsilon} \sim (\boldsymbol{0}, \sigma^2\boldsymbol{I}), \tag{7.29}$$

where $\boldsymbol{I}$ is the $n \times n$ identity matrix. Often $\boldsymbol{\epsilon}$ is assumed to be multivariate normal,

$$\boldsymbol{\epsilon} \sim \mathcal{N}_n(\boldsymbol{0}, \sigma^2\boldsymbol{I}), \tag{7.30}$$

but that is not required for most of what follows.

The *least squares estimate* $\hat{\beta}$, going back to Gauss and Legendre in the early 1800s, is the minimizer of the total sum of squared errors,

$$\hat{\beta} = \arg\min_{\beta} \left\{ \|\boldsymbol{y} - \boldsymbol{X}\beta\|^2 \right\}. \tag{7.31}$$

It is given by

$$\hat{\beta} = \boldsymbol{S}^{-1}\boldsymbol{X}'\boldsymbol{y}, \tag{7.32}$$

where $\boldsymbol{S}$ is the $p \times p$ inner product matrix

$$\boldsymbol{S} = \boldsymbol{X}'\boldsymbol{X}; \tag{7.33}$$

$\hat{\beta}$ is unbiased for β and has covariance matrix $\sigma^2\boldsymbol{S}^{-1}$,

$$\hat{\beta} \sim \left(\beta, \sigma^2\boldsymbol{S}^{-1}\right). \tag{7.34}$$

In the normal case (7.30) $\hat{\beta}$ is the MLE of β. Before 1950 a great deal of effort went into designing matrices $\boldsymbol{X}$ such that $\boldsymbol{S}^{-1}$ could be feasibly calculated, which is now no longer a concern.

A great advantage of the linear model is that it reduces the number of unknown parameters to p (or $p + 1$ including σ^2), no matter how large n may be. In the kidney data example of Section 1.1, $n = 157$ while $p = 2$. In modern applications, however, p has grown larger and larger, sometimes into the thousands or more, as we will see in Part III, causing statisticians again to confront the limitations of high-dimensional unbiased estimation.

Ridge regression is a shrinkage method designed to improve the estima-
†4 tion of β in linear models. By transformations† we can *standardize* (7.28) so that the columns of $\boldsymbol{X}$ each have mean 0 and sum of squares 1, that is,

$$S_{ii} = 1 \qquad \text{for } i = 1, 2, \ldots, p. \tag{7.35}$$

(This puts the regression coefficients $\beta_1, \beta_2, \ldots, \beta_p$ on comparable scales.) For convenience, we also assume $\bar{y} = 0$. A ridge regression estimate $\hat{\beta}(\lambda)$ is defined, for $\lambda \geq 0$, to be

$$\hat{\beta}(\lambda) = (\boldsymbol{S} + \lambda\boldsymbol{I})^{-1}\boldsymbol{X}'\boldsymbol{y} = (\boldsymbol{S} + \lambda\boldsymbol{I})^{-1}\boldsymbol{S}\hat{\beta} \tag{7.36}$$

(using (7.32)); $\hat{\beta}(\lambda)$ is a shrunken version of $\hat{\beta}$, the bigger λ the more extreme the shrinkage: $\hat{\beta}(0) = \hat{\beta}$ while $\hat{\beta}(\infty)$ equals the vector of zeros.

Ridge regression effects can be quite dramatic. As an example, consider the diabetes data, partially shown in Table 7.2, in which 10 prediction variables measured at baseline—`age`, `sex`, `bmi` (body mass index), `map` (mean arterial blood pressure), and six blood serum measurements—have

Table 7.2 *First 7 of $n = 442$ patients in the diabetes study; we wish to predict disease progression at one year "prog" from the 10 baseline measurements* `age`, `sex`, ..., `glu`.

age	sex	bmi	map	tc	ldl	hdl	tch	ltg	glu	prog
59	1	32.1	101	157	93.2	38	4	2.11	87	151
48	0	21.6	87	183	103.2	70	3	1.69	69	75
72	1	30.5	93	156	93.6	41	4	2.03	85	141
24	0	25.3	84	198	131.4	40	5	2.12	89	206
50	0	23.0	101	192	125.4	52	4	1.86	80	135
23	0	22.6	89	139	64.8	61	2	1.82	68	97
36	1	22.0	90	160	99.6	50	3	1.72	82	138
⋮	⋮	⋮	⋮	⋮	⋮	⋮	⋮	⋮	⋮	⋮

been obtained for $n = 442$ patients. We wish to use the 10 variables to predict `prog`, a quantitative assessment of disease progression one year after baseline. In this case $\boldsymbol{X}$ is the 442×10 matrix of standardized predictor variables, and $\boldsymbol{y}$ is `prog` with its mean subtracted off.

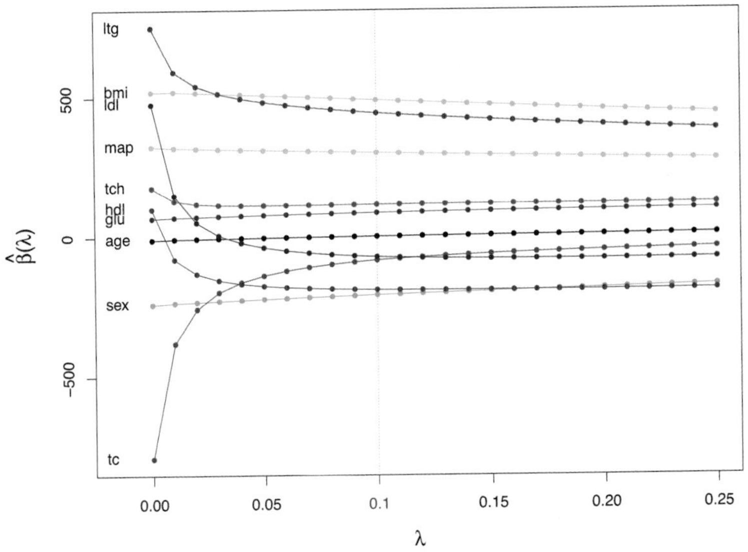

Figure 7.2 Ridge coefficient trace for the standardized diabetes data.

Table 7.3 *Ordinary least squares estimate $\hat{\beta}(0)$ compared with ridge regression estimate $\hat{\beta}(0.1)$ with $\lambda = 0.1$. The columns sd(0) and sd(0.1) are their estimated standard errors. (Here σ was taken to be 54.1, the usual OLS estimate based on model (7.28).)*

	$\hat{\beta}(0)$	$\hat{\beta}(0.1)$	sd(0)	sd(0.1)
`age`	−10.0	1.3	59.7	52.7
`sex`	−239.8	−207.2	61.2	53.2
`bmi`	519.8	489.7	66.5	56.3
`map`	324.4	301.8	65.3	55.7
`tc`	−792.2	−83.5	416.2	43.6
`ldl`	476.7	−70.8	338.6	52.4
`hdl`	101.0	−188.7	212.3	58.4
`tch`	177.1	115.7	161.3	70.8
`ltg`	751.3	443.8	171.7	58.4
`glu`	67.6	86.7	65.9	56.6

Figure 7.2 vertically plots the 10 coordinates of $\hat{\beta}(\lambda)$ as the ridge parameter λ increases from 0 to 0.25. Four of the coefficients change rapidly at first. Table 7.3 compares $\hat{\beta}(0)$, that is the usual estimate $\hat{\beta}$, with $\hat{\beta}(0.1)$. Positive coefficients predict increased disease progression. Notice that **`ldl`**, the "bad cholesterol" measurement, goes from being a strongly positive predictor in $\hat{\beta}$ to a mildly negative one in $\hat{\beta}(0.1)$.

There is a Bayesian rationale for ridge regression. Assume that the noise vector $\boldsymbol{\epsilon}$ is normal as in (7.30), so that

$$\hat{\beta} \sim \mathcal{N}_p\left(\beta, \sigma^2 \boldsymbol{S}^{-1}\right) \tag{7.37}$$

rather than just (7.34). Then the Bayesian prior

$$\beta \sim \mathcal{N}_p\left(\mathbf{0}, \frac{\sigma^2}{\lambda}\boldsymbol{I}\right) \tag{7.38}$$

makes

$$E\left\{\beta|\hat{\beta}\right\} = (\boldsymbol{S} + \lambda \boldsymbol{I})^{-1}\boldsymbol{S}\hat{\beta}, \tag{7.39}$$

the same as the ridge regression estimate $\hat{\beta}(\lambda)$ (using (5.23) with $M = 0$, $\boldsymbol{A} = (\sigma^2/\lambda)\boldsymbol{I}$, and $\boldsymbol{\Sigma} = (\boldsymbol{S}/\sigma^2)^{-1}$). Ridge regression amounts to an increased prior belief that β lies near 0.

†5 The last two columns of Table 7.3 compare the standard deviations† of $\hat{\beta}$ and $\hat{\beta}(0.1)$. Ridging has greatly reduced the variability of the estimated

regression coefficients. This does *not* guarantee that the corresponding estimate of $\boldsymbol{\mu} = \boldsymbol{X}\beta$,

$$\hat{\boldsymbol{\mu}}(\lambda) = \boldsymbol{X}\hat{\beta}(\lambda), \tag{7.40}$$

will be more accurate than the ordinary least squares estimate $\hat{\boldsymbol{\mu}} = \boldsymbol{X}\hat{\beta}$. We have (deliberately) introduced bias, and the squared bias term counteracts some of the advantage of reduced variability. The C_p calculations of Chapter 12 suggest that the two effects nearly offset each other for the diabetes data. However, if interest centers on the coefficients of β, then ridging can be crucial, as Table 7.3 emphasizes.

By current standards, $p = 10$ is a small number of predictors. Data sets with p in the thousands, and more, will show up in Part III. In such situations the scientist is often looking for a few interesting predictor variables hidden in a sea of uninteresting ones: the prior belief is that most of the β_i values lie near zero. Biasing the maximum likelihood estimates $\hat{\beta}_i$ toward zero then becomes a necessity.

There is still another way to motivate the ridge regression estimator $\hat{\beta}(\lambda)$:

$$\hat{\beta}(\lambda) = \arg\min_{\beta}\{\|\boldsymbol{y} - \boldsymbol{X}\beta\|^2 + \lambda\|\beta\|^2\}. \tag{7.41}$$

Differentiating the term in brackets with respect to β shows that $\hat{\beta}(\lambda) = (\boldsymbol{S} + \lambda\boldsymbol{I})^{-1}\boldsymbol{X}'\boldsymbol{y}$ as in (7.36). If $\lambda = 0$ then (7.41) describes the ordinary least squares algorithm; $\lambda > 0$ *penalizes* choices of β having $\|\beta\|$ large, biasing $\hat{\beta}(\lambda)$ toward the origin.

Various terminologies are used to describe algorithms such as (7.41): *penalized least squares*; *penalized likelihood*; *maximized a-posteriori probability* (MAP);†and, generically, *regularization* describes almost any method †6
that tamps down statistical variability in high-dimensional estimation or prediction problems.

A wide variety of penalty terms are in current use, the most influential one involving the "ℓ_1 norm" $\|\beta\|_1 = \sum_1^p |\beta_j|$,

$$\tilde{\beta}(\lambda) = \arg\min_{\beta}\{\|\boldsymbol{y} - \boldsymbol{X}\beta\|^2 + \lambda\|\beta\|_1\}, \tag{7.42}$$

the so-called *lasso* estimator, Chapter 16. Despite the Bayesian provenance, most regularization research is carried out frequentistically, with various penalty terms investigated for their probabilistic behavior regarding estimation, prediction, and variable selection.

If we apply the James–Stein rule to the normal model (7.37), we get a
different shrinkage rule† for $\hat{\beta}$, say $\tilde{\beta}^{\text{JS}}$, †7

$$\tilde{\beta}^{\text{JS}} = \left[1 - \frac{(p-2)\sigma^2}{\hat{\beta}^{-1}\boldsymbol{S}\hat{\beta}}\right]\hat{\beta}. \tag{7.43}$$

Letting $\tilde{\boldsymbol{\mu}}^{\text{JS}} = \boldsymbol{X}\tilde{\beta}^{\text{JS}}$ be the corresponding estimator of $\boldsymbol{\mu} = E\{\boldsymbol{y}\}$ in (7.28), the James–Stein Theorem guarantees that

$$E\left\{\left\|\tilde{\boldsymbol{\mu}}^{\text{JS}} - \boldsymbol{\mu}\right\|^2\right\} < p\sigma^2 \tag{7.44}$$

no matter what β is, as long as $p \geq 3$. There is no such guarantee for ridge regression, and no foolproof way to choose the ridge parameter λ.[2] On the other hand, $\tilde{\beta}^{\text{JS}}$ does not stabilize the coordinate standard deviations, as in the sd(0.1) column of Table 7.3. The main point here is that at present there is no optimality theory for shrinkage estimation. Fisher provided an elegant theory for optimal unbiased estimation. It remains to be seen whether biased estimation can be neatly codified.

7.4 Indirect Evidence 2

There is a downside to shrinkage estimation, which we can examine by returning to the baseball data of Table 7.1. One thousand simulations were run, each one generating simulated batting averages

$$p_i^* \sim \text{Bi}(90, \texttt{TRUTH}_i)/90 \qquad i = 1, 2, \ldots, 18. \tag{7.45}$$

These gave corresponding James–Stein (JS) estimates (7.20), with $\sigma_0^2 = \bar{p}^*(1 - \bar{p}^*)/90$.

Table 7.4 shows the root mean square error for the MLE and JS estimates over 1000 simulations for each of the 18 players,

$$\left[\sum_{j=1}^{1000}(p_{ij}^* - \texttt{TRUTH}_i)^2\right]^{1/2} \quad \text{and} \quad \left[\sum_{j=1}^{1000}(\hat{p}_{ij}^{*\,\text{JS}} - \texttt{TRUTH}_i)^2\right]^{1/2}. \tag{7.46}$$

As foretold by the James–Stein Theorem, the JS estimates are easy victors in terms of total squared error (summing over all 18 players). However, $\hat{p}_i^{*\,\text{JS}}$ loses to $\hat{p}_i^{*\,\text{MLE}} = p_i^*$ for 4 of the 18 players, losing badly in the case of player 2.

Histograms comparing the 1000 simulations of p_i^* with those of $\hat{p}_i^{*\,\text{JS}}$ for player 2 appear in Figure 7.3. Strikingly, all 1000 of the $\hat{p}_{2j}^{*\,\text{JS}}$ values lie

[2] Of course we are assumimg σ^2 is known in (7.43); if it is estimated, some of the improvement erodes away.

Table 7.4 *Simulation study comparing root mean square errors for MLE and JS estimators (7.20) as estimates of* **TRUTH**. *Total mean square errors .0384 (***MLE***) and .0235 (***JS***). Asterisks indicate four players for whom* **rmsJS** *exceeded* **rmsMLE**; *these have two largest and two smallest* **TRUTH** *values (player 2 is Clemente). Column* **rmsJS1** *is for the limited translation version of* **JS** *that bounds shrinkage to within one standard deviation of the* **MLE**.

Player	TRUTH	rmsMLE	rmsJS	rmsJS1
1	.298	.046	.033	.032
2	.346*	.049	.077	.056
3	.222	.044	.042	.038
4	.276	.048	.015	.023
5	.263	.047	.011	.020
6	.273	.046	.014	.021
7	.303	.047	.037	.035
8	.270	.049	.012	.022
9	.230	.044	.034	.033
10	.264	.047	.011	.021
11	.264	.047	.012	.020
12	.210*	.043	.053	.044
13	.256	.045	.014	.020
14	.269	.048	.012	.021
15	.316*	.048	.049	.043
16	.226	.045	.038	.036
17	.285	.046	.022	.026
18	.200*	.043	.062	.048

below $\mathbf{TRUTH}_2 = 0.346$. Player 2 could have had a legitimate complaint if the James–Stein estimate were used to set his next year's salary.

The four losing cases for $\hat{p}_i^{*\,\text{JS}}$ are the players with the two largest and two smallest values of the **TRUTH**. Shrinkage estimators work against cases that are genuinely outstanding (in a positive or negative sense). Player 2 was Roberto Clemente. A better informed Bayesian, that is, a baseball fan, would know that Clemente had led the league in batting over the previous several years, and shouldn't be thrown into a shrinkage pool with 17 ordinary hitters.

Of course the James–Stein estimates *were* more accurate for 14 of the 18 players. Shrinkage estimation tends to produce better results *in general*, at the possible expense of extreme cases. Nobody cares much about Cold War batting averages, but if the context were the efficacies of 18 new anticancer drugs the stakes would be higher.

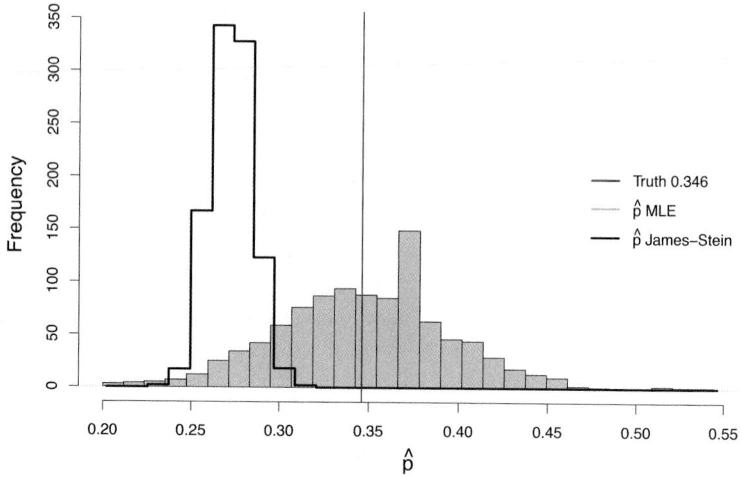

Figure 7.3 Comparing MLE estimates (solid) with JS estimates (line) for Clemente; 1000 simulations, 90 at bats each.

Compromise methods are available. The `rmsJS1` column of Table 7.4 refers to a *limited translation* version of $\hat{p}_i^{\text{JS}}$ in which shrinkage is not allowed to diverge more than one σ_0 unit from $\hat{p}_i$; in formulaic terms,

$$\hat{p}_i^{\text{JS}\,1} = \min\left\{\max\left(\hat{p}_i^{\text{JS}}, \hat{p}_i - \sigma_0\right), \hat{p}_i + \sigma_0\right\}. \tag{7.47}$$

This mitigates the Clemente problem while still gaining most of the shrinkage advantages.

The use of indirect evidence amounts to *learning from the experience of others*, each batter learning from the 17 others in the baseball examples. "Which others?" is a key question in applying computer-age methods. Chapter 15 returns to the question in the context of false-discovery rates.

7.5 Notes and Details

The Bayesian motivation emphasized in Chapters 6 and 7 is anachronistic: originally the work emerged mainly from frequentist considerations and was justified frequentistically, as in Robbins (1956). Stein (1956) proved the inadmissibility of $\hat{\boldsymbol{\mu}}^{\text{MLE}}$, the neat version of $\hat{\boldsymbol{\mu}}^{\text{JS}}$ appearing in James and Stein (1961) (Willard James was Stein's graduate student); $\hat{\boldsymbol{\mu}}^{\text{JS}}$ is itself inadmissable, being everywhere improvable by changing $\hat{B}$ in (7.13) to $\max(\hat{B}, 0)$. This in turn is inadmissable, but further gains tend to the minuscule.

In a series of papers in the early 1970s, Efron and Morris emphasized the empirical Bayes motivation of the James–Stein rule, Efron and Morris (1972) giving the limited translation version (7.47). The baseball data in its original form appears in Table 1.1 of Efron (2010). Here the original 45 at bats recorded for each player have been artificially augmented by adding 45 binomial draws, $\text{Bi}(45, \texttt{TRUTH}_i)$ for player i. This gives a somewhat less optimistic view of the James–Stein rule's performance.

"Stein's paradox in statistics," Efron and Morris' title for their 1977 *Scientific American* article, catches the statistics world's sense of discomfort with the James–Stein theorem. Why should our estimate for Player A go up or down depending on the other players' performances? This is the question of direct versus indirect evidence, raised again in the context of hypothesis testing in Chapter 15. Unbiased estimation has great scientific appeal, so the argument is by no means settled.

Ridge regression was introduced into the statistics literature by Hoerl and Kennard (1970). It appeared previously in the numerical analysis literature as Tikhonov regularization.

†1 [p. 93] *Formula* (7.12). If Z has a chi-squared distribution with ν degrees of freedom, $Z \sim \chi^2_\nu$ (that is, $Z \sim \text{Gam}(\nu/2, 1)$ in Table 5.1), it has density

$$f(z) = \frac{z^{\nu/2-1}e^{-z/2}}{2^{\nu/2}\Gamma(\nu/2)} \qquad \text{for } z \geq 0, \tag{7.48}$$

yielding

$$E\left\{\frac{1}{z}\right\} = \int_0^\infty \frac{z^{\nu/2-2}e^{-z/2}}{2^{\nu/2}\Gamma(\nu/2)}\,dz = \frac{2^{\nu/2-1}}{2^{\nu/2}}\frac{\Gamma(\nu/2-1)}{\Gamma(\nu/2)} = \frac{1}{\nu-2}. \tag{7.49}$$

But standard results, starting from (7.11), show that $S \sim (A+1)\chi^2_{N-1}$. With $\nu = N-1$ in (7.49),

$$E\left\{\frac{N-3}{S}\right\} = \frac{1}{A+1}, \tag{7.50}$$

verifying (7.12).

†2 [p. 93] *Formula* (7.14). First consider the simpler situation where M in (7.11) is known to equal zero, in which case the James–Stein estimator is

$$\hat{\mu}_i^{\text{JS}} = \hat{B}x_i \qquad \text{with } \hat{B} = 1 - (N-2)/S, \tag{7.51}$$

where $S = \sum_1^N x_i^2$. For convenient notation let

$$\hat{C} = 1 - \hat{B} = (N-2)/S \quad \text{and} \quad C = 1 - B = 1/(A+1). \tag{7.52}$$

The conditional distribution $\mu_i | \boldsymbol{x} \sim \mathcal{N}(Bx_i, B)$ gives

$$E\left\{\left(\hat{\mu}_i^{\text{JS}} - \mu_i\right)^2 \middle| \boldsymbol{x}\right\} = B + (\hat{C} - C)^2 x_i^2, \tag{7.53}$$

and, adding over the N coordinates,

$$E\left\{\left\|\hat{\boldsymbol{\mu}}^{\text{JS}} - \boldsymbol{\mu}\right\|^2 \middle| \boldsymbol{x}\right\} = NB + (\hat{C} - C)^2 S. \tag{7.54}$$

The marginal distribution $S \sim (A+1)\chi_N^2$ and (7.49) yields, after a little calculation,

$$E\left\{(\hat{C} - C)^2 S\right\} = 2(1 - B), \tag{7.55}$$

and so

$$E\left\{\left\|\hat{\boldsymbol{\mu}}^{\text{JS}} - \boldsymbol{\mu}\right\|^2\right\} = NB + 2(1 - B). \tag{7.56}$$

By orthogonal transformations, in situation (7.7), where M is not assumed to be zero, $\hat{\boldsymbol{\mu}}^{\text{JS}}$ can be represented as the sum of two parts: a JS estimate in $N - 1$ dimensions but with $M = 0$ as in (7.51), and a MLE estimate of the remaining one coordinate. Using (7.56) this gives

$$\begin{aligned} E\left\{\left\|\hat{\boldsymbol{\mu}}^{\text{JS}} - \boldsymbol{\mu}\right\|^2\right\} &= (N - 1)B + 2(1 - B) + 1 \\ &= NB + 3(1 - B), \end{aligned} \tag{7.57}$$

which is (7.14).

†3 [p. 93] *The James–Stein Theorem.* Stein (1981) derived a simpler proof of the JS Theorem that appears in Section 1.2 of Efron (2010).

†4 [p. 98] *Transformations to form* (7.35). The linear regression model (7.28) is *equivariant* under scale changes of the variables $\boldsymbol{x}_j$. What this means is that the space of fits using linear combinations of the $\boldsymbol{x}_j$ is the same as the space of linear combinations using scaled versions $\tilde{\boldsymbol{x}}_j = \boldsymbol{x}_j / s_j$, with $s_j > 0$. Furthermore, the least squares fits are the same, and the coefficient estimates map in the obvious way: $\hat{\tilde{\beta}}_j = s_j \hat{\beta}_j$.

Not so for ridge regression. Changing the scales of the columns of $\boldsymbol{X}$ will generally lead to different fits. Using the penalty version (7.41) of ridge regression, we see that the penalty term $\|\beta\|^2 = \sum_j \beta_j^2$ treats all the coefficients as equals. This penalty is most natural if all the variables are measured on the same scale. Hence we typically use for s_j the standard deviation of variable $\boldsymbol{x}_j$, which leads to (7.35). Furthermore, with ridge regression we typically do not penalize the intercept. This can be achieved by *centering* and scaling each of the variables, $\tilde{\boldsymbol{x}}_j = (\boldsymbol{x}_j - \mathbf{1}\bar{x}_j)/s_j$, where

$$\bar{x}_j = \sum_{i=1}^{n} x_{ij}/n \quad \text{and} \quad s_j = \left[\sum (x_{ij} - \bar{x}_j)^2\right]^{1/2}, \tag{7.58}$$

with $\mathbf{1}$ the n-vector of 1s. We now work with $\tilde{\boldsymbol{X}} = (\tilde{\boldsymbol{x}}_1, \tilde{\boldsymbol{x}}_2, \ldots, \tilde{\boldsymbol{x}}_p)$ rather than $\boldsymbol{X}$, and the intercept is estimated separately as $\bar{y}$.

†5 [p. 100] *Standard deviations in Table 7.3.* From the first equality in (7.36) we calculate the covariance matrix of $\hat{\beta}(\lambda)$ to be

$$\text{Cov}_\lambda = \sigma^2 (\boldsymbol{S} + \lambda \boldsymbol{I})^{-1} \boldsymbol{S} (\boldsymbol{S} + \lambda \boldsymbol{I}). \tag{7.59}$$

The entries sd(0.1) in Table 7.3 are square roots of the diagonal elements of Cov_λ, substituting the ordinary least squares estimate $\hat{\sigma} = 54.1$ for σ^2.

†6 [p. 101] *Penalized likelihood and MAP.* With σ^2 fixed and known in the normal linear model $\boldsymbol{y} \sim \mathcal{N}_n(\boldsymbol{X}\beta, \sigma^2 \boldsymbol{I})$, minimizing $\|\boldsymbol{y} - \boldsymbol{X}\beta\|^2$ is the same as maximizing the log density function

$$\log f_\beta(y) = -\frac{1}{2}\|\boldsymbol{y} - \boldsymbol{X}\beta\|^2 + \text{constant}. \tag{7.60}$$

In this sense, the term $\lambda\|\beta\|^2$ in (7.41) *penalizes* the likelihood $\log f_\beta(y)$ connected with β in proportion to the magnitude $\|\beta\|^2$. Under the prior distribution (7.38), the log posterior density of β given $\boldsymbol{y}$ (the log of (3.5)) is

$$-\frac{1}{2\sigma^2}\left\{\|\boldsymbol{y} - \boldsymbol{X}\beta\|^2 + \lambda\|\beta\|^2\right\}, \tag{7.61}$$

plus a term that doesn't depend on β. That makes the maximizer of (7.41) also the maximizer of the posterior density of β given $\boldsymbol{y}$, or the MAP.

†7 [p. 101] *Formula* (7.43). Let $\gamma = (\boldsymbol{S}^{1/2}/\sigma)\beta$ and $\hat{\gamma} = (\boldsymbol{S}^{1/2}/\sigma)\hat{\beta}$ in (7.37), where $\boldsymbol{S}^{1/2}$ is a matrix square root of $\boldsymbol{S}$, $(\boldsymbol{S}^{1/2})^2 = \boldsymbol{S}$. Then

$$\hat{\gamma} \sim \mathcal{N}_p(\gamma, \boldsymbol{I}), \tag{7.62}$$

and the $M = 0$ form of the James–Stein rule (7.51) is

$$\hat{\gamma}^{\text{JS}} = \left[1 - \frac{p-2}{\|\hat{\gamma}\|^2}\right]\hat{\gamma}. \tag{7.63}$$

Transforming back to the β scale gives (7.43).

8

Generalized Linear Models and Regression Trees

Indirect evidence is not the sole property of Bayesians. Regression models are the frequentist method of choice for incorporating the experience of "others." As an example, Figure 8.1 returns to the kidney fitness data of Section 1.1. A potential new donor, aged 55, has appeared, and we wish to assess his kidney fitness without subjecting him to an arduous series of medical tests. Only one of the 157 previously tested volunteers was age 55, his **tot** score being −0.01 (the upper large dot in Figure 8.1). Most applied statisticians, though, would prefer to read off the height of the least squares regression line at age = 55 (the green dot on the regression line), $\widehat{\texttt{tot}} = -1.46$. The former is the only direct evidence we have, while the

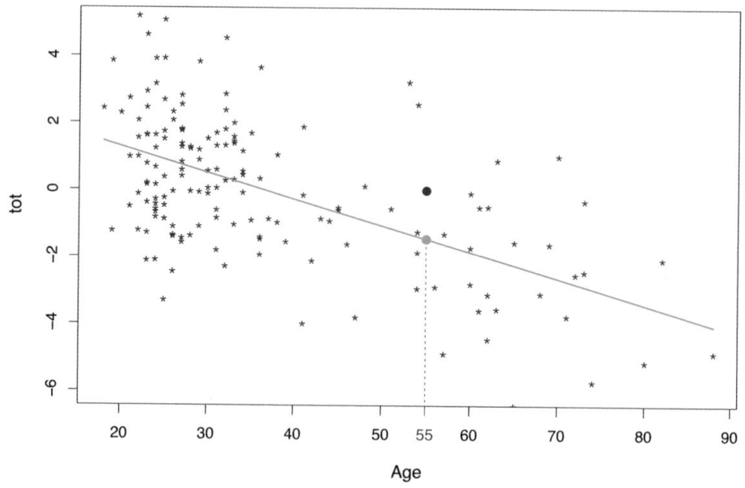

Figure 8.1 Kidney data; a new volunteer donor is aged 55. Which prediction is preferred for his kidney function?

regression line lets us incorporate indirect evidence for age 55 from all 157 previous cases.

Increasingly aggressive use of regression techniques is a hallmark of modern statistical practice, "aggressive" applying to the number and type of predictor variables, the coinage of new methodology, and the sheer size of the target data sets. Generalized linear models, this chapter's main topic, have been the most pervasively influential of the new methods. The chapter ends with a brief review of regression trees, a completely different regression methodology that will play an important role in the prediction algorithms of Chapter 17.

8.1 Logistic Regression

An experimental new anti-cancer drug called `Xilathon` is under development. Before human testing can begin, animal studies are needed to determine safe dosages. To this end, a *bioassay* or dose–response experiment was carried out: 11 groups of $n = 10$ mice each were injected with increasing amounts of `Xilathon`, dosages coded[1] $1, 2, \ldots, 11$.

Let

$$y_i = \#\text{ mice dying in } i\text{th group.} \tag{8.1}$$

The points in Figure 8.2 show the proportion of deaths

$$p_i = y_i/10, \tag{8.2}$$

lethality generally increasing with dose. The counts y_i are modeled as independent binomials,

$$y_i \overset{\text{ind}}{\sim} \text{Bi}(n_i, \pi_i) \qquad \text{for } i = 1, 2, \ldots, N, \tag{8.3}$$

$N = 11$ and all n_i equaling 10 here; π_i is the true death rate in group i, estimated unbiasedly by p_i, the direct evidence for π_i. The regression curve in Figure 8.2 uses *all* the doses to give a better picture of the true dose–response relation.

Logistic regression is a specialized technique for regression analysis of count or proportion data. The *logit* parameter λ is defined as

$$\lambda = \log\left\{\frac{\pi}{1-\pi}\right\}, \tag{8.4}$$

[1] Dose would usually be labeled on a log scale, each one, say, 50% larger than its predecessor.

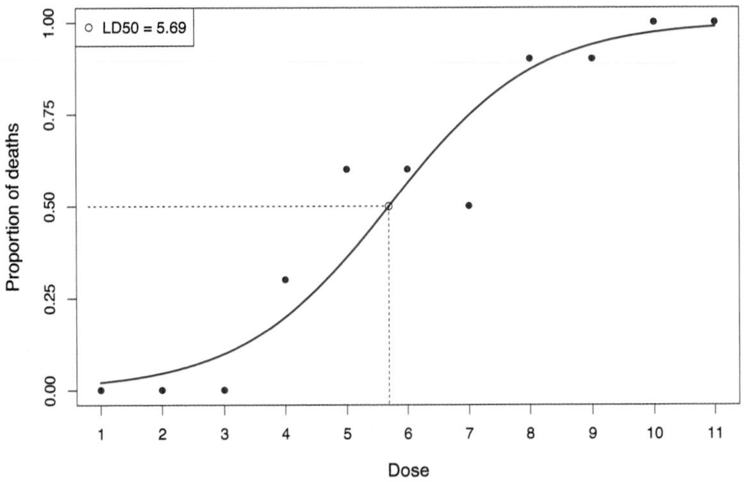

Figure 8.2 Dose–response study; groups of 10 mice exposed to increasing doses of experimental drug. The points are the observed proportions that died in each group. The fitted curve is the maximum-likelihoood estimate of the linear logistic regression model. The open circle on the curve is the LD50, the estimated dose for 50% mortality.

with λ increasing from $-\infty$ to ∞ as π increases from 0 to 1. A linear logistic regression dose–response analysis begins with binomial model (8.3), and assumes that the logit is a linear function of dose,

$$\lambda_i = \log \left\{ \frac{\pi_i}{1 - \pi_i} \right\} = \alpha_0 + \alpha_1 x_i. \tag{8.5}$$

Maximum likelihood gives estimates $(\hat{\alpha}_0, \hat{\alpha}_1)$, and fitted curve

$$\hat{\lambda}(x) = \hat{\alpha}_0 + \hat{\alpha}_1 x. \tag{8.6}$$

Since the inverse transformation of (8.4) is

$$\pi = \left(1 + e^{-\lambda}\right)^{-1} \tag{8.7}$$

we obtain from (8.6) the linear logistic regression curve

$$\hat{\pi}(x) = \left(1 + e^{-(\hat{\alpha}_0 + \hat{\alpha}_1 x)}\right)^{-1} \tag{8.8}$$

pictured in Figure 8.2.

Table 8.1 compares the standard deviation of the estimated regression

Table 8.1 *Standard deviation estimates for $\hat{\pi}(x)$ in Figure 8.1. The first row is for the linear logistic regression fit* (8.8)*; the second row is based on the individual binomial estimates p_i.*

x	1	2	3	4	5	6	7	8	9	10	11
sd $\hat{\pi}(x)$	.015	.027	.043	.061	.071	.072	.065	.050	.032	.019	.010
sd p_i	.045	.066	.094	.126	.152	.157	.138	.106	.076	.052	.035

curve (8.8) at $x = 1, 2, \ldots, 11$ (as discussed in the next section) with the usual binomial standard deviation estimate $[p_i(1-p_i)/10]^{1/2}$ obtained by considering the 11 doses separately.[2] Regression has reduced error by better than 50%, the price being possible bias if model (8.5) goes seriously wrong.

One advantage of the logit transformation is that λ isn't restricted to the range $[0, 1]$, so model (8.5) never verges on forbidden territory. A better reason has to do with the exploitation of exponential family properties. We can rewrite the density function for $\text{Bi}(n, y)$ as

$$\binom{n}{y}\pi^y(1-\pi)^{n-y} = e^{\lambda y - n\psi(\lambda)}\binom{n}{y} \tag{8.9}$$

with λ the logit parameter (8.4) and

$$\psi(\lambda) = \log\{1 + e^{\lambda}\}; \tag{8.10}$$

(8.9) is a one-parameter exponential family[3] as described in Section 5.5, with λ the natural parameter, called α there.

Let $\boldsymbol{y} = (y_1, y_2, \ldots, y_N)$ denote the full data set, $N = 11$ in Figure 8.2. Using (8.5), (8.9), and the independence of the y_i gives the probability density of $\boldsymbol{y}$ as a function of (α_0, α_1),

$$\begin{aligned} f_{\alpha_0,\alpha_1}(\boldsymbol{y}) &= \prod_{i=1}^{N} e^{\lambda_i y_i - n_i\psi(\lambda_i)}\binom{n_i}{y_i} \\ &= e^{\alpha_0 S_0 + \alpha_1 S_1} \cdot e^{-\sum_1^N n_i\psi(\alpha_0+\alpha_1 x_i)} \cdot \prod_1^N \binom{n_i}{y_i}, \end{aligned} \tag{8.11}$$

[2] For the separate-dose standard error, p_i was taken equal to the fitted value from the curve in Figure 8.2.

[3] It is not necessary for $f_{\mu_0}(x)$ in (5.46) on page 64 to be a probability density function, only that it not depend on the parameter μ.

where

$$S_0 = \sum_{i=1}^{N} y_i \quad \text{and} \quad S_1 = \sum_{i=1}^{N} x_i y_i. \tag{8.12}$$

Formula (8.11) expresses $f_{\alpha_0,\alpha_1}(\boldsymbol{y})$ as the product of three factors,

$$f_{\alpha_0,\alpha_1}(\boldsymbol{y}) = g_{\alpha_0,\alpha_1}(S_0, S_1)h(\alpha_0, \alpha_1)j(\boldsymbol{y}), \tag{8.13}$$

only the first of which involves both the parameters and the data. This im-
†1 plies that (S_0, S_1) is a *sufficient statistic*:† no matter how large N might be (later we will have N in the thousands), just the two numbers (S_0, S_1) contain all of the experiment's information. Only the logistic parameterization (8.4) makes this happen.[4]

A more intuitive picture of logistic regression depends on $D(p_i, \hat{\pi}_i)$, the *deviance* between an observed proportion p_i (8.2) and an estimate $\hat{\pi}_i$,

$$D(p_i, \hat{\pi}_i) = 2n_i \left[p_i \log\left(\frac{p_i}{\hat{\pi}_i}\right) + (1 - p_i)\log\left(\frac{1 - p_i}{1 - \hat{\pi}_i}\right)\right]. \tag{8.14}$$

The deviance[5] is zero if $\hat{\pi}_i = p_i$, otherwise it increases as $\hat{\pi}_i$ departs further from p_i.

The logistic regression MLE value $(\hat{\alpha}_0, \hat{\alpha}_1)$ also turns out to be the choice of (α_0, α_1) minimizing the total deviance between the N points p_i and their corresponding estimates $\hat{\pi}_i = \pi_{\hat{\alpha}_0,\hat{\alpha}_1}(x_i)$ (8.8):

$$(\hat{\alpha}_0, \hat{\alpha}_1) = \underset{(\alpha_0,\alpha_1)}{\arg\min} \sum_{i=1}^{N} D\left(p_i, \pi_{\alpha_0,\alpha_1}(x_i)\right). \tag{8.15}$$

The solid line in Figure 8.2 is the linear logistic curve coming closest to the 11 points, when distance is measured by total deviance. In this way the 200-year-old notion of least squares is generalized to binomial regression, as discussed in the next section. A more sophisticated notion of distance between data and models is one of the accomplishments of modern statistics.

Table 8.2 reports on the data for a more structured logistic regression analysis. Human muscle cell colonies were infused with mouse nuclei in five different ratios, cultured over time periods ranging from one to five

[4] Where the name "logistic regression" comes from is explained in the endnotes, along with a description of its nonexponential family predecessor *probit analysis*.

[5] Deviance is analogous to squared error in ordinary regression theory, as discussed in what follows. It is twice the "Kullback–Leibler distance," the preferred name in the information-theory literature.

Table 8.2 *Cell infusion data; human cell colonies infused with mouse nuclei in five ratios over 1 to 5 days and observed to see whether they did or did not thrive. Green numbers are estimates $\hat{\pi}_{ij}$ from the logistic regression model. For example, 5 of 31 colonies in the lowest ratio/days category thrived, with observed proportion $5/31 = 0.16$, and logistic regression estimate $\hat{\pi}_{11} = 0.11$.*

		Time 1	2	3	4	5
Ratio	1	5/31 .11	3/28 .25	20/45 .42	24/47 .54	29/35 .75
	2	15/77 .24	36/78 .45	43/71 .64	56/71 .74	66/74 .88
	3	48/126 .38	68/116 .62	145/171 .77	98/119 .85	114/129 .93
	4	29/92 .32	35/52 .56	57/85 .73	38/50 .81	72/77 .92
	5	11/53 .18	20/52 .37	20/48 .55	40/55 .67	52/61 .84

days, and observed to see whether they thrived. For example, of the 126 colonies having the third ratio and shortest time period, 48 thrived.

Let π_{ij} denote the true probability of thriving for ratio i during time period j, and λ_{ij} its logit $\log\{\pi_{ij}/(1-\pi_{ij})\}$. A two-way additive logistic regression was fit to the data,[6]

$$\lambda_{ij} = \mu + \alpha_i + \beta_j, \qquad i = 1, 2, \ldots, 5, \; j = 1, 2, \ldots, 5. \tag{8.16}$$

The green numbers in Table 8.2 show the maximum likelihood estimates

$$\hat{\pi}_{ij} = 1 \Big/ \left[1 + e^{-\left(\hat{\mu} + \hat{\alpha}_i + \hat{\beta}_j\right)} \right]. \tag{8.17}$$

Model (8.16) has nine free parameters (taking into account the constraints $\sum \alpha_i = \sum \beta_j = 0$ necessary to avoid definitional difficulties) compared with just two in the dose–response experiment. The count can easily go much higher these days.

Table 8.3 reports on a 57-variable logistic regression applied to the `spam` data. A researcher (named George) labeled $N = 4601$ of his email mes-

[6] Using the statistical computing language **R**; see the endnotes.

Table 8.3 *Logistic regression analysis of the* **spam** *data, model* (8.17)*; estimated regression coefficients, standard errors, and* $z = \text{estimate}/\text{se}$*, for 57 keyword predictors. The notation* **char$** *means the relative number of times* **$** *appears, etc. The last three entries measure characteristics such as length of capital-letter strings. The word* **george** *is special, since the recipient of the email is named George, and the goal here is to build a customized spam filter.*

	Estimate	se	z-value		Estimate	se	z-value
intercept	−12.27	1.99	−6.16	lab	−1.48	.89	−1.66
make	−.12	.07	−1.68	labs	−.15	.14	−1.05
address	−.19	.09	−2.10	telnet	−.07	.19	−.35
all	.06	.06	1.03	857	.84	1.08	.78
3d	3.14	2.10	1.49	data	−.41	.17	−2.37
our	.38	.07	5.52	415	.22	.53	.42
over	.24	.07	3.53	85	−1.09	.42	−2.61
remove	.89	.13	6.85	technology	.37	.12	2.99
internet	.23	.07	3.39	1999	.02	.07	.26
order	.20	.08	2.58	parts	−.13	.09	−1.41
mail	.08	.05	1.75	pm	−.38	.17	−2.26
receive	−.05	.06	−.86	direct	−.11	.13	−.84
will	−.12	.06	−1.87	cs	−16.27	9.61	−1.69
people	−.02	.07	−.35	meeting	−2.06	.64	−3.21
report	.05	.05	1.06	original	−.28	.18	−1.55
addresses	.32	.19	1.70	project	−.98	.33	−2.97
free	.86	.12	7.13	re	−.80	.16	−5.09
business	.43	.10	4.26	edu	−1.33	.24	−5.43
email	.06	.06	1.03	table	−.18	.13	−1.40
you	.14	.06	2.32	conference	−1.15	.46	−2.49
credit	.53	.27	1.95	char;	−.31	.11	−2.92
your	.29	.06	4.62	char(	−.05	.07	−.75
font	.21	.17	1.24	char_	−.07	.09	−.78
000	.79	.16	4.76	char!	.28	.07	3.89
money	.19	.07	2.63	char$	1.31	.17	7.55
hp	−3.21	.52	−6.14	char#	1.03	.48	2.16
hpl	−.92	.39	−2.37	cap.ave	.38	.60	.64
george	−39.62	7.12	−5.57	cap.long	1.78	.49	3.62
650	.24	.11	2.24	cap.tot	.51	.14	3.75

sages as either **spam** or **ham** (nonspam[7]), say

$$y_i = \begin{cases} 1 & \text{if email } i \text{ is } \textbf{spam} \\ 0 & \text{if email } i \text{ is } \textbf{ham} \end{cases} \tag{8.18}$$

[7] "Ham" refers to "nonspam" or good email; this is a playful connection to the processed

(40% of the messages were **spam**). The $p = 57$ predictor variables represent the most frequently used words and tokens in George's corpus of email (excluding trivial words such as articles), and are in fact the relative frequencies of these chosen words in each email (standardized by the length of the email). The goal of the study was to predict whether future emails are **spam** or **ham** using these keywords; that is, to build a customized *spam filter*.

Let x_{ij} denote the relative frequency of keyword j in email i, and π_i represent the probability that email i is **spam**. Letting λ_i be the logit transform $\log\{\pi_i/(1-\pi_i)\}$, we fit the additive logistic model

$$\lambda_i = \alpha_0 + \sum_{j=1}^{57} \alpha_i x_{ij}. \tag{8.19}$$

Table 8.3 shows $\hat{\alpha}_i$ for each word—for example, -0.12 for **make**—as well as the estimated standard error and the *z-value*: estimate/se.

It looks like certain words, such as **free** and **your**, are good **spam** predictors. However, the table as a whole has an unstable appearance, with occasional very large estimates $\hat{\alpha}_i$ accompanied by very large standard deviations.[8] The dangers of high-dimensional maximum likelihood estimation are apparent here. Some sort of shrinkage estimation is called for, as discussed in Chapter 16.

Regression analysis, either in its classical form or in modern formulations, requires covariate information x to put the various cases into some sort of geometrical relationship. Given such information, regression is the statistician's most powerful tool for bringing "other" results to bear on a case of primary interest: for instance, the age-55 volunteer in Figure 8.1.

Empirical Bayes methods do not require covariate information but may be improvable if it exists. If, for example, the player's age were an important covariate in the baseball example of Table 7.1, we might first regress the MLE values on age, and then shrink them toward the regression line rather than toward the grand mean $\bar{p}$ as in (7.19). In this way, two different sorts of indirect evidence would be brought to bear on the estimation of each player's ability.

spam that was fake ham during WWII, and has been adopted by the machine-learning community.

8 The 4601×57 $\boldsymbol{X}$ matrix (x_{ij}) was standardized, so disparate scalings are not the cause of these discrepancies. Some of the features have mostly "zero" observations, which may account for their unstable estimation.

8.2 Generalized Linear Models[9]

Logistic regression is a special case of *generalized linear models* (GLMs), a key 1970s methodology having both algorithmic and inferential influence. GLMs extend ordinary linear regression, that is least squares curve-fitting, to situations where the response variables are binomial, Poisson, gamma, beta, or in fact any exponential family form.

We begin with a one-parameter exponential family,

$$\left\{f_\lambda(y) = e^{\lambda y - \gamma(\lambda)} f_0(y),\ \lambda \in \Lambda\right\}, \tag{8.20}$$

as in (5.46) (now with α and x replaced by λ and y, and $\psi(\alpha)$ replaced by $\gamma(\lambda)$, for clearer notation in what follows). Here λ is the *natural parameter* and y the *sufficient statistic*, both being one-dimensional in usual applications; λ takes its values in an interval of the real line. Each coordinate y_i of an observed data set $\boldsymbol{y} = (y_1, y_2, \dots, y_i, \dots, y_N)'$ is assumed to come from a member of family (8.20),

$$y_i \sim f_{\lambda_i}(\cdot) \text{ independently for } i = 1, 2, \dots, N. \tag{8.21}$$

Table 8.4 lists λ and y for the first four families in Table 5.1, as well as their deviance and normalizing functions.

By itself, model (8.21) requires N parameters $\lambda_1, \lambda_2, \dots, \lambda_N$, usually too many for effective individual estimation. A key GLM tactic is to specify the λs in terms of a linear regression equation. Let $\boldsymbol{X}$ be an $N \times p$ "structure matrix," with ith row say x_i', and α an unknown vector of p parameters; the N-vector $\boldsymbol{\lambda} = (\lambda_1, \lambda_2, \dots, \lambda_N)'$ is then specified by

$$\boldsymbol{\lambda} = \boldsymbol{X}\alpha. \tag{8.22}$$

In the dose–response experiment of Figure 8.2 and model (8.5), $\boldsymbol{X}$ is $N \times 2$ with ith row $(1, x_i)$ and parameter vector $\alpha = (\alpha_0, \alpha_1)$.

The probability density function $f_\alpha(\boldsymbol{y})$ of the data vector $\boldsymbol{y}$ is

$$f_\alpha(\boldsymbol{y}) = \prod_{i=1}^{N} f_{\lambda_i}(y_i) = e^{\sum_1^N (\lambda_i y_i - \gamma(\lambda_i))} \prod_{i=1}^{N} f_0(y_i), \tag{8.23}$$

which can be written as

$$f_\alpha(\boldsymbol{y}) = e^{\alpha' z - \psi(\alpha)} f_0(\boldsymbol{y}), \tag{8.24}$$

[9] Some of the more technical points raised in this section are referred to in later chapters, and can be scanned or omitted at first reading.

Table 8.4 *Exponential family form for first four cases in Table 5.1; natural parameter λ, sufficient statistic y, deviance* (8.31) *between family members f_1 and f_2, $D(f_1, f_2)$, and normalizing function $\gamma(\lambda)$.*

	λ	y	$D(f_1, f_2)$	$\gamma(\lambda)$
1. *Normal* $\mathcal{N}(\mu, \sigma^2)$, σ^2 known	μ/σ^2	x	$\left(\frac{\mu_1-\mu_2}{\sigma}\right)^2$	$\sigma^2\lambda^2/2$
2. *Poisson* Poi(μ)	$\log\mu$	x	$2\mu_1\left[\left(\frac{\mu_2}{\mu_1}-1\right)-\log\frac{\mu_2}{\mu_1}\right]$	e^λ
3. *binomial* Bi(n,π)	$\log\frac{\pi}{1-\pi}$	x	$2n\left[\pi_1\log\frac{\pi_1}{\pi_2}+(1-\pi_1)\log\frac{1-\pi_1}{1-\pi_2}\right]$	$n\log(1+e^\lambda)$
4. *Gamma* Gam(ν,σ), ν known	$-1/\sigma$	x	$2\nu\left[\left(\frac{\sigma_1}{\sigma_2}-1\right)-\log\frac{\sigma_1}{\sigma_2}\right]$	$-\nu\log(-\lambda)$

where

$$z = \boldsymbol{X}'\boldsymbol{y} \quad \text{and} \quad \psi(\alpha) = \sum_{i=1}^{N} \gamma(\boldsymbol{x}_i\alpha), \tag{8.25}$$

a p-parameter exponential family (5.50), with natural parameter vector α and sufficient statistic vector z. The main point is that all the information from a p-parameter GLM is summarized in the p-dimensional vector z, no matter how large N may be, making it easier both to understand and to analyze.

We have now reduced the N-parameter model (8.20)–(8.21) to the p-parameter exponential family (8.24), with p usually much smaller than N, in this way avoiding the difficulties of high-dimensional estimation. The moments of the one-parameter constituents (8.20) determine the estimation properties in model (8.22)–(8.24). Let $(\mu_\lambda, \sigma_\lambda^2)$ denote the expectation and variance of univariate density $f_\lambda(y)$ (8.20),

$$y \sim (\mu_\lambda, \sigma_\lambda^2), \tag{8.26}$$

for instance $(\mu_\lambda, \sigma_\lambda^2) = (e^\lambda, e^\lambda)$ for the Poisson. The N-vector $\boldsymbol{y}$ obtained from GLM (8.22) then has mean vector and covariance matrix

$$\boldsymbol{y} \sim (\boldsymbol{\mu}(\alpha), \boldsymbol{\Sigma}(\alpha)), \tag{8.27}$$

where $\boldsymbol{\mu}(\alpha)$ is the vector with ith component μ_{λ_i} with $\lambda_i = x_i'\alpha$, and $\boldsymbol{\Sigma}(\alpha)$ is the $N \times N$ diagonal matrix having diagonal elements $\sigma^2_{\lambda_i}$.

The maximum likelihood estimate $\hat{\alpha}$ of the parameter vector α can be
†2 shown to satisfy the simple equation†

$$\boldsymbol{X}'[\boldsymbol{y} - \boldsymbol{\mu}(\hat{\alpha})] = 0. \tag{8.28}$$

For the normal case where $y_i \sim \mathcal{N}(\mu_i, \sigma^2)$ in (8.21), that is, for ordinary linear regression, $\boldsymbol{\mu}(\hat{\alpha}) = \boldsymbol{X}\alpha$ and (8.28) becomes $\boldsymbol{X}'(\boldsymbol{y} - \boldsymbol{X}\alpha) = 0$, with the familiar solution

$$\hat{\alpha} = (\boldsymbol{X}'\boldsymbol{X})^{-1}\boldsymbol{X}'\boldsymbol{y}; \tag{8.29}$$

otherwise, $\boldsymbol{\mu}(\alpha)$ is a nonlinear function of α, and (8.28) must be solved by numerical iteration. This is made easier by the fact that, for GLMs, $\log f_\alpha(\boldsymbol{y})$, the likelihood function we wish to maximize, is a *concave func-*
†3 *tion of* α. The MLE $\hat{\alpha}$ has approximate expectation and covariance†

$$\hat{\alpha} \dot{\sim} \left(\alpha, \left(\boldsymbol{X}'\boldsymbol{\Sigma}(\alpha)\boldsymbol{X}\right)^{-1}\right), \tag{8.30}$$

†4 similar to the exact OLS result $\hat{\alpha} \sim (\alpha, \sigma^2(\boldsymbol{X}'\boldsymbol{X})^{-1})$.†

Generalizing the binomial definition (8.14), the *deviance* between densities $f_1(y)$ and $f_2(y)$ is defined to be

$$D(f_1, f_2) = 2\int_{\mathcal{Y}} f_1(y)\log\left\{\frac{f_1(y)}{f_2(y)}\right\} dy, \tag{8.31}$$

the integral (or sum for discrete distributions) being over their common sample space $\mathcal{Y}$. $D(f_1, f_2)$ is always nonnegative, equaling zero only if f_1 and f_2 are the same; in general $D(f_1, f_2)$ does not equal $D(f_2, f_1)$. Deviance does not depend on how the two densities are named, for example (8.14) having the same expression as the *Binomial* entry in Table 8.4.

In what follows it will sometimes be useful to label the family (8.20) by its *expectation parameter* $\mu = E_\lambda\{y\}$ rather than by the natural parameter λ:

$$f_\mu(y) = e^{\lambda y - \gamma(\lambda)} f_0(y), \tag{8.32}$$

meaning the same thing as (8.20), only the names attached to the individual family members being changed. In this notation it is easy to show a fundamental result sometimes known as

†5 **Hoeffding's Lemma**† *The maximum likelihood estimate of* μ *given* y *is* y *itself, and the log likelihood* $\log f_\mu(y)$ *decreases from its maximum* $\log f_y(y)$ *by an amount that depends on the deviance* $D(y, \mu)$,

$$f_\mu(y) = f_y(y)e^{-D(y,\mu)/2}. \tag{8.33}$$

Returning to the GLM framework (8.21)–(8.22), parameter vector α gives $\boldsymbol{\lambda}(\alpha) = \boldsymbol{X}\alpha$, which in turn gives the vector of expectation parameters

$$\boldsymbol{\mu}(\alpha) = (\ldots \mu_i(\alpha) \ldots)', \tag{8.34}$$

for instance $\mu_i(\alpha) = \exp\{\lambda_i(\alpha)\}$ for the Poisson family. Multiplying Hoeffding's lemma (8.33) over the N cases $\boldsymbol{y} = (y_1, y_2, \ldots, y_N)'$ yields

$$f_\alpha(\boldsymbol{y}) = \prod_{i=1}^{N} f_{\mu_i(\alpha)}(y_i) = \left[\prod_{i=1}^{N} f_{y_i}(y_i)\right] e^{-\sum_1^N D(y_i,\mu_i(\alpha))}. \tag{8.35}$$

This has an important consequence: *the MLE $\hat{\alpha}$ is the choice of α that minimizes the total deviance $\sum_1^N D(y_i, \mu_i(\alpha))$.* As in Figure 8.2, GLM maximum likelihood fitting is "least total deviance" in the same way that ordinary linear regression is least sum of squares.

••————••————••————••

The inner circle of Figure 8.3 represents normal theory, the preferred venue of classical applied statistics. Exact inferences—t-tests, F distributions, most of multivariate analysis—were feasible within the circle. Outside the circle was a general theory based mainly on asymptotic (large-sample) approximations involving Taylor expansions and the central limit theorem.

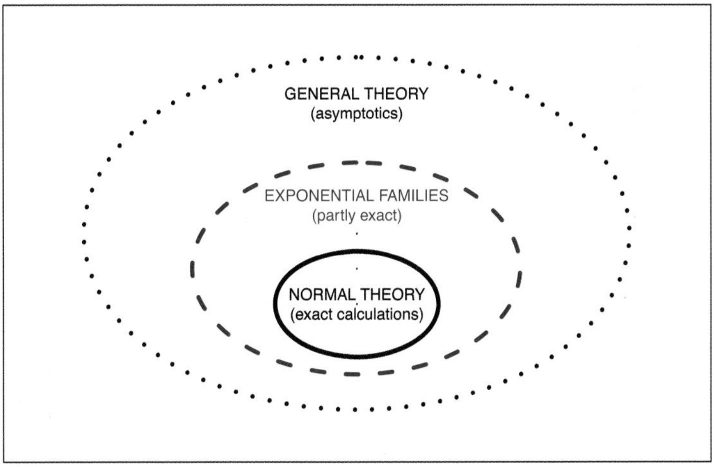

Figure 8.3 Three levels of statistical modeling.

A few useful exact results lay outside the normal theory circle, relating to a few special families: the binomial, Poisson, gamma, beta, and others less well known. Exponential family theory, the second circle in Figure 8.3, unified the special cases into a coherent whole. It has a "partly exact" flavor, with some ideal counterparts to normal theory—convex likelihood surfaces, least deviance regression—but with some approximations necessary, as in (8.30). Even the approximations, though, are often more convincing than those of general theory, exponential families' fixed-dimension sufficient statistics making the asymptotics more transparent.

Logistic regression has banished its predecessors (such as probit analysis) almost entirely from the field, and not only because of estimating efficiencies and computational advantages (which are actually rather modest), but also because it is seen as a clearer analogue to ordinary least squares, our 200-year-old dependable standby. GLM research development has been mostly frequentist, but with a substantial admixture of likelihood-based reasoning, and a hint of Fisher's "logic of inductive inference."

Helping the statistician choose between competing methodologies is the job of statistical inference. In the case of generalized linear models the choice has been made, at least partly, in terms of aesthetics as well as philosophy.

8.3 Poisson Regression

The third most-used member of the GLM family, after normal theory least squares and logistic regression, is Poisson regression. N independent Poisson variates are observed,

$$y_i \overset{\text{ind}}{\sim} \text{Poi}(\mu_i), \qquad i = 1, 2, \dots, N, \tag{8.36}$$

where $\lambda_i = \log \mu_i$ is assumed to follow a linear model,

$$\boldsymbol{\lambda}(\alpha) = \boldsymbol{X}\alpha, \tag{8.37}$$

where $\boldsymbol{X}$ is a known $N \times p$ structure matrix and α an unknown p-vector of regression coefficients. That is, $\lambda_i = x_i'\alpha$ for $i = 1, 2, \dots, N$, where x_i' is the ith row of $\boldsymbol{X}$.

In the chapters that follow we will see Poisson regression come to the rescue in what at first appear to be awkward data-analytic situations. Here we will settle for an example involving density estimation from a spatially truncated sample.

†6 Table 8.5 shows galaxy counts † from a small portion of the sky: 486 galaxies have had their redshifts r and apparent magnitudes m measured.

Table 8.5 *Counts for a truncated sample of 486 galaxies, binned by redshift and magnitude.*

magnitude (dimmer) ↑ \ `redshift` (farther) ⟶	1	2	3	4	5	6	7	8	9	10	11	12	13	14	15
18	1	6	6	3	1	4	6	8	8	20	10	7	16	9	4
17	3	2	3	4	0	5	7	6	6	7	5	7	6	8	5
16	3	2	3	3	3	2	9	9	6	3	5	4	5	2	1
15	1	1	4	3	4	3	2	3	8	9	4	3	4	1	1
14	1	3	2	3	3	4	5	7	6	7	3	4	0	0	1
13	3	2	4	5	3	6	4	3	2	2	5	1	0	0	0
12	2	0	2	4	5	4	2	3	3	0	1	2	0	0	1
11	4	1	1	4	7	3	3	1	2	0	1	1	0	0	0
10	1	0	0	2	2	2	1	2	0	0	0	1	2	0	0
9	1	1	0	2	2	2	0	0	0	0	1	0	0	0	0
8	1	0	0	0	1	1	0	0	0	0	1	1	0	0	0
7	0	1	0	1	1	0	0	0	0	0	0	0	0	0	0
6	0	0	3	1	1	0	0	0	0	0	0	0	0	0	0
5	0	3	1	1	0	0	0	0	0	0	0	0	0	0	0
4	0	0	1	1	1	0	0	0	0	0	0	0	0	0	0
3	0	1	0	0	0	0	0	0	0	0	0	0	0	0	0
2	0	1	0	0	0	0	0	0	0	0	0	0	0	0	0
1	0	1	0	0	0	0	0	0	0	0	0	0	0	0	0

Distance from earth is an increasing function of r, while apparent brightness is a decreasing function[10] of m. In this survey, counts were limited to galaxies having

$$1.22 \le r \le 3.32 \quad \text{and} \quad 17.2 \le m \le 21.5, \tag{8.38}$$

the upper limit reflecting the difficulty of measuring very dim galaxies.

The range of $\log r$ has been divided into 15 equal intervals and likewise 18 equal intervals for m. Table 8.5 gives the counts of the 486 galaxies in the $18 \times 15 = 270$ bins. (The lower right corner of the table is empty because distant galaxies always appear dim.) The multinomial/Poisson connection (5.44) helps motivate model (8.36), picturing the table as a multinomial observation on 270 categories, in which the sample size N was itself Poisson.

We can imagine Table 8.5 as a small portion of a much more extensive table, hypothetically available if the data were *not* truncated. Experience suggests that we might then fit an appropriate bivariate normal density to the data, as in Figure 5.3. It seems like it might be awkward to fit part of a bivariate normal density to truncated data, but Poisson regression offers an easy solution.

[10] An object of the second magnitude is less bright than one of the first, and so on, a classification system owing to the Greeks.

Let $\boldsymbol{r}$ be the 270-vector listing the values of r in each bin of the table (in column order), and likewise $\boldsymbol{m}$ for the 270 m values—for instance $\boldsymbol{m} = (18, 17, \ldots, 1)$ repeated 15 times—and define the 270×5 matrix $\boldsymbol{X}$ as

$$\boldsymbol{X} = [\boldsymbol{r}, \boldsymbol{m}, \boldsymbol{r}^2, \boldsymbol{rm}, \boldsymbol{m}^2], \tag{8.39}$$

where $\boldsymbol{r}^2$ is the vector whose components are the square of $\boldsymbol{r}$'s, etc. The log density of a bivariate normal distribution in (r, m) is of the form $\alpha_1 r + \alpha_2 m + \alpha_3 r^2 + \alpha_4 rm + \alpha_5 m^2$, agreeing with $\log \mu_i = x_i' \alpha$ as specified by (8.39). We can use a Poisson GLM, with y_i the ith bin's count, to estimate the portion of our hypothesized bivariate normal distribution in the truncation region (8.38).

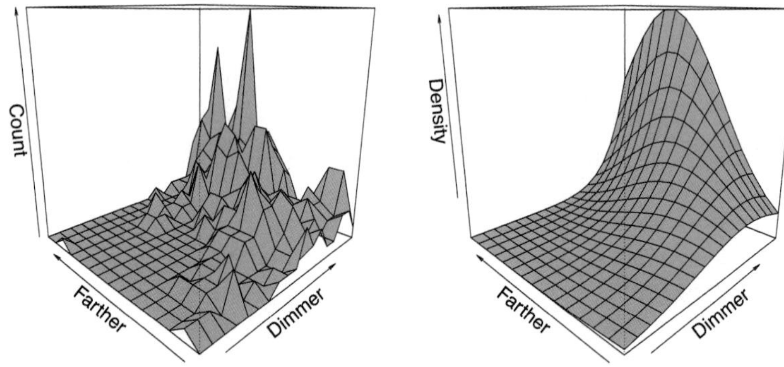

Figure 8.4 *Left* galaxy data; binned counts. *Right* Poisson GLM density estimate.

The left panel of Figure 8.4 is a perspective picture of the raw counts in Table 8.5. On the right is the fitted density from the Poisson regression. Irrespective of density estimation, Poisson regression has done a useful job of smoothing the raw bin counts.

Contours of equal value of the fitted log density

$$\hat{\alpha}_0 + \hat{\alpha}_1 r + \hat{\alpha}_2 m + \hat{\alpha}_3 r^2 + \hat{\alpha}_4 rm + \hat{\alpha}_5 m^2 \tag{8.40}$$

are shown in Figure 8.5. One can imagine the contours as truncated portions of ellipsoids, of the type shown in Figure 5.3. The right panel of Figure 8.4 makes it clear that we are nowhere near the center of the hypothetical bivariate normal density, which must lie well beyond our dimness limit.

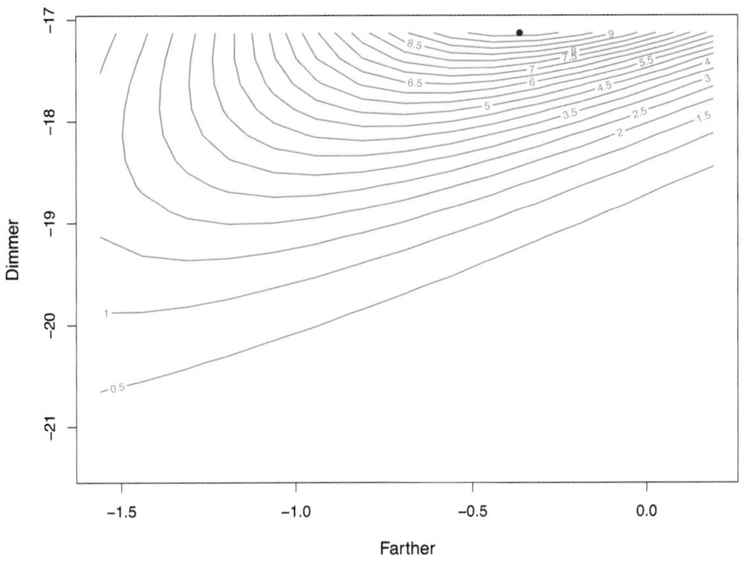

Figure 8.5 Contour curves for Poisson GLM density estimate for the galaxy data. The red dot shows the point of maximum density.

The Poisson *deviance residual* Z between an observed count y and a fitted value $\hat{\mu}$ is

$$Z = \operatorname{sign}(y - \hat{\mu}) D(y, \hat{\mu})^{1/2}, \tag{8.41}$$

with D the Poisson deviance from Table 8.4. Z_{jk}, the deviance residual between the count y_{ij} in the ijth bin of Table 8.5 and the fitted value $\hat{\mu}_{jk}$ from the Poisson GLM, was calculated for all 270 bins. Standard frequentist GLM theory says that $S = \sum_{jk} Z_{jk}^2$ should be about 270 if the bivariate normal model (8.39) is correct.[11] Actually the fit was poor: $S = 610$.

In practice we might try adding columns to $\boldsymbol{X}$ in (8.39), e.g., $\boldsymbol{rm}^2$ or $\boldsymbol{r}^2\boldsymbol{m}^2$, improving the fit where it was worst, near the boundaries of the table. Chapter 12 demonstrates some other examples of Poisson density estimation. In general, Poisson GLMs reduce density estimation to regression model fitting, a familiar and flexible inferential technology.

[11] This is a modern version of the classic chi-squared goodness-of-fit test.

8.4 Regression Trees

The data set $\boldsymbol{d}$ for a regression problem typically consists of N pairs (x_i, y_i),

$$\boldsymbol{d} = \{(x_i, y_i),\ i = 1, 2, \dots, N\}, \tag{8.42}$$

where x_i is a vector of *predictors*, or "covariates," taking its value in some space $\mathcal{X}$, and y_i is the *response*, assumed to be univariate in what follows. The regression algorithm, perhaps a Poisson GLM, inputs $\boldsymbol{d}$ and outputs a *rule* $r_{\boldsymbol{d}}(x)$: for any value of x in $\mathcal{X}$, $r_{\boldsymbol{d}}(x)$ produces an estimate $\hat{y}$ for a possible future value of y,

$$\hat{y} = r_{\boldsymbol{d}}(x). \tag{8.43}$$

In the logistic regression example (8.8), $r_{\boldsymbol{d}}(x)$ is $\hat{\pi}(x)$.

There are three principal uses for the rule $r_{\boldsymbol{d}}(x)$.

1 For *prediction*: Given a new observation of x, but not of its corresponding y, we use $\hat{y} = r_{\boldsymbol{d}}(x)$ to predict y. In the `spam` example, the 57 keywords of an incoming message could be used to predict whether or not it is spam.[12] (See Chapter 12.)

2 For *estimation*: The rule $r_{\boldsymbol{d}}(x)$ describes a "regression surface" $\hat{S}$ over $\mathcal{X}$,

$$\hat{S} = \{r_{\boldsymbol{d}}(x),\ x \in \mathcal{X}\}. \tag{8.44}$$

The right panel of Figure 8.4 shows $\hat{S}$ for the galaxy example. $\hat{S}$ can be thought of as estimating S, the *true* regression surface, often defined in the form of conditional expectation,

$$S = \{E\{y|x\},\ x \in \mathcal{X}\}. \tag{8.45}$$

(In a dichotomous situation where y is coded as 0 or 1, $S = \{\Pr\{y = 1|x\},\ x \in \mathcal{X}\}$.)

For estimation, but not necessarily for prediction, we want $\hat{S}$ to accurately portray S. The right panel of Figure 8.4 shows the estimated galaxy density still increasing monotonically in `dimmer` at the top end of the truncation region, but not so in `farther`, perhaps an important clue for directing future search counts.[13] The flat region in the kidney function regression curve of Figure 1.2 makes almost no difference to prediction, but is of scientific interest if accurate.

[12] Prediction of dichotomous outcomes is often called "classification."

[13] Physicists call a regression-based search for new objects "bump hunting."

3 For *explanation*: The 10 predictors for the diabetes data of Section 7.3, **age**, **sex**, **bmi**,..., were selected by the researcher in the hope of explaining the etiology of diabetes progression. The relative contribution of the different predictors to $r_d(x)$ is then of interest. *How* the regression surface is composed is of prime concern in this use, but not in use 1 or 2 above.

The three different uses of $r_d(x)$ raise different inferential questions. Use 1 calls for estimates of prediction error. In a dichotomous situation such as the **spam** study, we would want to know both error probabilities

$$\Pr\{\hat{y} = \texttt{spam}|y = \texttt{ham}\} \quad \text{and} \quad \Pr\{\hat{y} = \texttt{ham}|y = \texttt{spam}\}. \tag{8.46}$$

For estimation, the accuracy of $r_d(x)$ as a function of x, perhaps in standard deviation terms,

$$\text{sd}(x) = \text{sd}(\hat{y}|x), \tag{8.47}$$

would tell how closely $\hat{S}$ approximates S. Use 3, explanation, requires more elaborate inferential tools, saying for example which of the regression coefficients α_i in (8.19) can safely be set to zero.

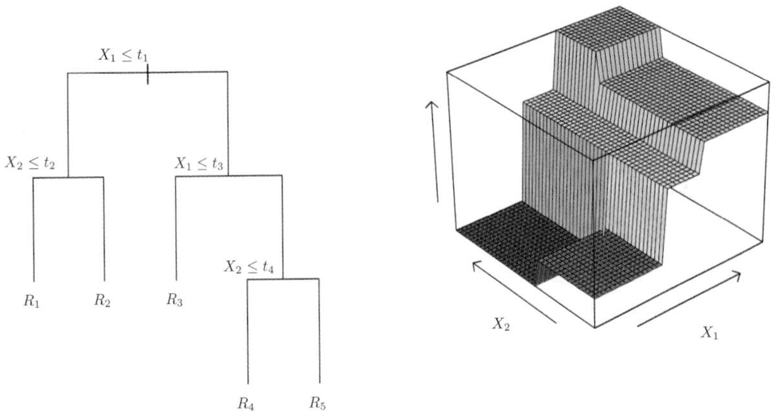

Figure 8.6 *Left* a hypothetical regression tree based on two predictors X_1 and X_2. *Right* corresponding regression surface.

Regression trees use a simple but intuitively appealing technique to form a regression surface: recursive partitioning. The left panel of Figure 8.6 illustrates the method for a hypothetical situation involving two predictor variables, X_1 and X_2 (e.g., r and m in the galaxy example). At the top of the tree, the sample population of N cases has been split into two groups: those with X_1 equal to or less than value t_1 go to the left, those with $X_1 > t_1$ to the right. The leftward group is itself then divided into two groups depending on whether or not $X_2 \leq t_2$. The division stops there, leaving two *terminal nodes* R_1 and R_2. On the tree's right side, two other splits give terminal nodes R_3, R_4, and R_5.

A prediction value $\hat{y}_{R_j}$ is attached to each terminal node R_j. The prediction $\hat{y}$ applying to a new observation $x = (x_1, x_2)$ is calculated by starting x at the top of the tree and following the splits downward until a terminal node, and its attached prediction $\hat{y}_{R_j}$, is reached. The corresponding regression surface $\hat{S}$ is shown in the right panel of Figure 8.6 (here the $\hat{y}_{R_j}$ happen to be in ascending order).

Various algorithmic rules are used to decide which variable to split and which splitting value t to take at each step of the tree's construction. Here is the most common method: suppose at step k of the algorithm, group$_k$ of N_k cases remains to be split, those cases having mean and sum of squares

$$m_k = \sum_{i \in \text{group}_k} y_i / N_k \quad \text{and} \quad s_k^2 = \sum_{i \in \text{group}_k} (y_i - m_k)^2. \tag{8.48}$$

Dividing group$_k$ into group$_{k,\text{left}}$ and group$_{k,\text{right}}$ produces means $m_{k,\text{left}}$ and $m_{k,\text{right}}$, and corresponding sums of squares $s^2_{k,\text{left}}$ and $s^2_{k,\text{right}}$. The algorithm proceeds by choosing the splitting variable X_k and the threshold t_k to minimize

$$s^2_{k,\text{left}} + s^2_{k,\text{right}}. \tag{8.49}$$

In other words, it splits group$_k$ into two groups that are as different from
†7 each other as possible.†

Cross-validation estimates of prediction error, Chapter 12, are used to decide when the splitting process should stop. If group$_k$ is not to be further divided, it becomes terminal node R_k, with prediction value $\hat{y}_{R_k} = m_k$. None of this would be feasible without electronic computation, but even quite large prediction problems can be short work for modern computers.

Figure 8.7 shows a regression tree analysis[14] of the `spam` data, Table 8.3. There are seven terminal nodes, labeled 0 or 1 for decision `ham` or `spam`. The leftmost node, say R_1, is a 0, and contains 2462 `ham` cases and 275 `spam` (compared with 2788 and 1813 in the full data set). Starting at the top of the tree, R_1 is reached if it has a low proportion of $ symbols

[14] Using the R program `rpart`, in classification mode, employing a different splitting rule than the version based on (8.49).

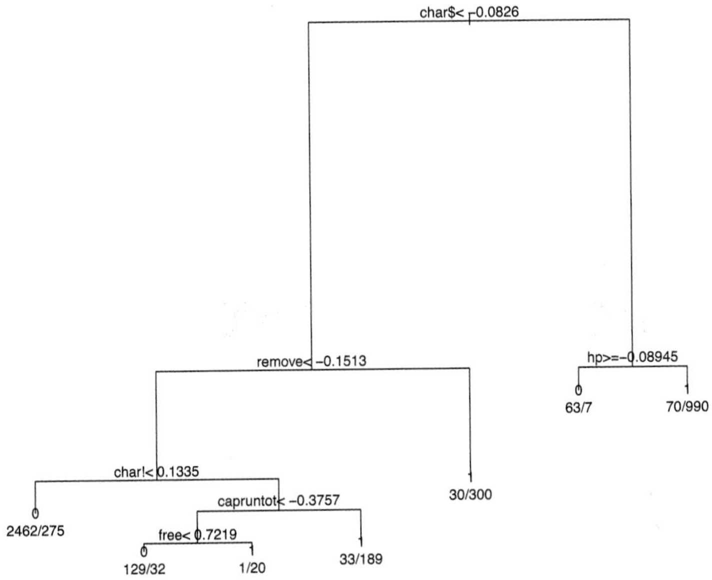

Figure 8.7 Regression tree on the **spam** data; 0 = **ham**, 1 = **spam**. Error rates: **ham** 5.2%, **spam** 17.4%. Captions indicate leftward (ham) moves.

char$, a low proportion of the word **remove**, and a low proportion of exclamation marks **char!**.

Regression trees are easy to interpret ("Too many dollar signs means spam!") seemingly suiting them for use 3, explanation. Unfortunately, they are also easy to overinterpret, with a reputation for being unstable in practice. Discontinuous regression surfaces $\hat{S}$, as in Figure 8.6, disqualify them for use 2, estimation. Their principal use in what follows will be as key parts of prediction algorithms, use 1. The tree in Figure 8.6 has apparent error rates (8.46) of 5.2% and 17.4%. This can be much improved upon by "bagging" (bootstrap aggregation), Chapters 17 and 20, and by other computer-intensive techniques.

Compared with generalized linear models, regression trees represent a break from classical methodology that is more stark. First of all, they are totally nonparametric; bigger but less structured data sets have promoted nonparametrics in twenty-first-century statistics. Regression trees are more computer-intensive and less efficient than GLMs but, as will be seen in Part III, the availability massive data sets and modern computational equipment has diminished the appeal of efficiency in favor of easy assumption-free application.

8.5 Notes and Details

Computer-age algorithms depend for their utility on statistical computing languages. After a period of evolution, the language **S** (Becker *et al.*, 1988) and its open-source successor **R** (R Core Team, 2015), have come to dominate applied practice.[15] Generalized linear models are available from a single **R** command, e.g.,

```
glm(y~X,family=binomial)
```

for logistic regression (Chambers and Hastie, 1993), and similarly for regression trees and hundreds of other applications.

The classic version of bioassay, *probit analysis*, assumes that each test animal has its own lethal dose level X, and that the population distribution of X is normal,

$$\Pr\{X \leq x\} = \Phi(\alpha_0 + \alpha_1 x) \tag{8.50}$$

for unknown parameters (α_0, α_1) and standard normal cdf Φ. Then the number of animals dying at dose x is binomial $\text{Bi}(n_x, \pi_x)$ as in (8.3), with $\pi_x = \Phi(\alpha_0 + \alpha_1 x)$, or

$$\Phi^{-1}(\pi_x) = \alpha_0 + \alpha_1 x. \tag{8.51}$$

Replacing the standard normal cdf $\Phi(z)$ with the logistic cdf $1/(1+e^{-z})$ (which resembles Φ), changes (8.51) into logistic regression (8.5). The usual goal of bioassay was to estimate "LD50," the dose lethal to 50% of the test population; it is indicated by the open circle in Figure 8.2.

Cox (1970), the classic text on logistic regression, lists Berkson (1944) as an early practitioner. Wedderburn (1974) is credited with generalized linear models in McCullagh and Nelder's influential text of that name, first edition 1983; Birch (1964) developed an important and suggestive special case of GLM theory.

The twenty-first century has seen an efflorescence of computer-based regression techniques, as described extensively in Hastie *et al.* (2009). The discussion of regression trees here is taken from their Section 9.2, including our Figure 8.6. They use the **spam** data as a central example; it is publicly

[15] Previous computer packages such as SAS and SPSS continue to play a major role in application areas such as the social sciences, biomedical statistics, and the pharmaceutical industry.

available at `ftp.ics.uci.edu`. Breiman *et al.* (1984) propelled regression trees into wide use with their CART algorithm.

†1 [p. 112] *Sufficiency as in* (8.13). The Fisher–Neyman criterion says that if $f_\alpha(\boldsymbol{x}) = h_\alpha(S(\boldsymbol{x}))g(\boldsymbol{x})$, when $g(\cdot)$ does not depend on α, then $S(\boldsymbol{x})$ is sufficient for α.

†2 [p. 118] *Equation* (8.28). From (8.24)–(8.25) we have the log likelihood function

$$l_\alpha(\boldsymbol{y}) = \alpha' z - \psi(\alpha) \tag{8.52}$$

with sufficient statistic $z = \boldsymbol{X}'\boldsymbol{y}$ and $\psi(\alpha) = \sum_{i=1}^N \gamma(x_i'\alpha)$. Differentiating with respect to α,

$$\dot{l}(\boldsymbol{y}) = z - \dot{\psi}(\alpha) = \boldsymbol{X}'\boldsymbol{y} - \boldsymbol{X}'\boldsymbol{\mu}(\alpha), \tag{8.53}$$

where we have used $d\gamma/d\lambda = \mu_\lambda$ (5.55), so $\dot{\gamma}(x_i'\alpha) = x_i'\mu_i(\alpha)$. But (8.53) says $\dot{l}_\alpha(\boldsymbol{y}) = \boldsymbol{X}'(\boldsymbol{y} - \boldsymbol{\mu}(\alpha))$, verifying the MLE equation (8.28).

†3 [p. 118] *Concavity of the log likelihood.* From (8.53), the second derivative matrix $\ddot{l}_\alpha(\boldsymbol{y})$ with respect to α is

$$-\ddot{\psi}(\alpha) = -\operatorname{cov}_\alpha(z), \tag{8.54}$$

(5.57)–(5.59). But $z = \boldsymbol{X}'\boldsymbol{y}$ has

$$\operatorname{cov}_\alpha(z) = \boldsymbol{X}'\boldsymbol{\Sigma}(\alpha)\boldsymbol{X}, \tag{8.55}$$

a positive definite $p \times p$ matrix, verifying the concavity of $l_\alpha(\boldsymbol{y})$ (which in fact applies to any exponential family, not only GLMs).

†4 [p. 118] *Formula* (8.30). The sufficient statistic z has mean vector and covariance matrix

$$z \sim (\beta, V_\alpha), \tag{8.56}$$

with $\beta = E_\alpha\{z\}$ (5.58) and $V_\alpha = \boldsymbol{X}'\boldsymbol{\Sigma}(\alpha)\boldsymbol{X}$ (8.55). Using (5.60), the first-order Taylor series for $\hat{\alpha}$ as a function of z is

$$\hat{\alpha} \doteq \alpha + V_\alpha^{-1}(z - \beta). \tag{8.57}$$

Taken literally, (8.57) gives (8.30).

†5 [p. 118] *Formula* (8.33). This formula, attributed to Hoeffding (1965), is a key result in the interpretation of GLM fitting. Applying definition (8.31) to family (8.32) gives

$$\begin{aligned}\frac{1}{2}D(\lambda_1, \lambda_2) &= E_{\lambda_1}\{(\lambda_1 - \lambda_2)y - [\gamma(\lambda_1) - \gamma(\lambda_2)]\} \\ &= (\lambda_1 - \lambda_2)\mu_1 - [\gamma(\lambda_1) - \gamma(\lambda_2)]\,.\end{aligned} \tag{8.58}$$

If λ_1 is the MLE $\hat{\lambda}$ then $\mu_1 = y$ (from the maximum likelihood equation $0 = d[\log f_\lambda(y)]/d\lambda = y - \dot{\gamma}(\lambda) = y - \mu_\lambda$), giving[16]

$$\frac{1}{2}D\left(\hat{\lambda},\lambda\right) = \left(\hat{\lambda}-\lambda\right)y - \left[\gamma\left(\hat{\lambda}\right) - \gamma(\lambda)\right] \tag{8.59}$$

for any choice of λ. But the right-hand side of (8.59) is $f_\lambda(y)/f_y(y)$, verifying (8.33).

†6 [p. 120] *Table 8.5.* The galaxy counts are from Loh and Spillar's 1988 redshift survey, as discussed in Efron and Petrosian (1992).

†7 [p. 126] *Criteria* (8.49). Abbreviating "left" and "right" by l and r, we have

$$s_k^2 = s_{kl}^2 + s_{kr}^2 + \frac{N_{kl}N_{kr}}{N_k}(m_{kl} - m_{kr})^2, \tag{8.60}$$

with N_{kl} and N_{kr} the subgroup sizes, showing that minimizing (8.49) is the same as maximizing the last term in (8.60). Intuitively, a *good* split is one that makes the left and right groups as different as possible, the ideal being all 0s on the left and all 1s on the right, making the terminal nodes "pure."

[16] In some cases $\hat{\lambda}$ is undefined; for example, when $y = 0$ for a Poisson response, $\hat{\lambda} = \log(y)$ which is undefined. But, in (8.59), we assume that $\hat{\lambda}y = 0$. Similarly for binary y and the binomial family.

9

Survival Analysis and the EM Algorithm

Survival analysis had its roots in governmental and actuarial statistics, spanning centuries of use in assessing life expectancies, insurance rates, and annuities. In the 20 years between 1955 and 1975, survival analysis was adapted by statisticians for application to biomedical studies. Three of the most popular post-war statistical methodologies emerged during this period: the Kaplan–Meier estimate, the log-rank test,[1] and Cox's proportional hazards model, the succession showing increased computational demands along with increasingly sophisticated inferential justification. A connection with one of Fisher's ideas on maximum likelihood estimation leads in the last section of this chapter to another statistical method that has "gone platinum," the EM algorithm.

9.1 Life Tables and Hazard Rates

An insurance company's *life table* appears in Table 9.1, showing its number of clients (that is, life insurance policy holders) by age, and the number of deaths during the past year in each age group,[2] for example five deaths among the 312 clients aged 59. The column labeled $\hat{S}$ is of great interest to the company's actuaries, who have to set rates for new policy holders. It is an estimate of survival probability: probability 0.893 of a person aged 30 (the beginning of the table) surviving past age 59, etc. $\hat{S}$ is calculated according to an ancient but ingenious algorithm.

Let X represent a typical lifetime, so

$$f_i = \Pr\{X = i\} \tag{9.1}$$

[1] Also known as the Mantel–Haenszel or Cochran–Mantel–Haenszel test.

[2] The insurance company is fictitious but the deaths y are based on the true 2010 rates for US men, per Social Security Administration data.

Table 9.1 *Insurance company life table; at each age, n = number of policy holders, y = number of deaths, $\hat{h}$ = hazard rate y/n, $\hat{S}$ = survival probability estimate* (9.6).

Age	n	y	$\hat{h}$	$\hat{S}$	Age	n	y	$\hat{h}$	$\hat{S}$
30	116	0	.000	1.000	60	231	1	.004	.889
31	44	0	.000	1.000	61	245	5	.020	.871
32	95	0	.000	1.000	62	196	5	.026	.849
33	97	0	.000	1.000	63	180	4	.022	.830
34	120	0	.000	1.000	64	170	2	.012	.820
35	71	1	.014	.986	65	114	0	.000	.820
36	125	0	.000	.986	66	185	5	.027	.798
37	122	0	.000	.986	67	127	2	.016	.785
38	82	0	.000	.986	68	127	5	.039	.755
39	113	0	.000	.986	69	158	2	.013	.745
40	79	0	.000	.986	70	100	3	.030	.723
41	90	0	.000	.986	71	155	4	.026	.704
42	154	0	.000	.986	72	92	1	.011	.696
43	103	0	.000	.986	73	90	1	.011	.689
44	144	0	.000	.986	74	110	2	.018	.676
45	192	2	.010	.976	75	122	5	.041	.648
46	153	1	.007	.969	76	138	8	.058	.611
47	179	1	.006	.964	77	46	0	.000	.611
48	210	0	.000	.964	78	75	4	.053	.578
49	259	2	.008	.956	79	69	6	.087	.528
50	225	2	.009	.948	80	95	4	.042	.506
51	346	1	.003	.945	81	124	6	.048	.481
52	370	2	.005	.940	82	67	7	.104	.431
53	568	4	.007	.933	83	112	12	.107	.385
54	1081	8	.007	.927	84	113	8	.071	.358
55	1042	2	.002	.925	85	116	12	.103	.321
56	1094	10	.009	.916	86	124	17	.137	.277
57	597	4	.007	.910	87	110	21	.191	.224
58	359	1	.003	.908	88	63	9	.143	.192
59	312	5	.016	.893	89	79	10	.127	.168

is the probability of dying at age i, and

$$S_i = \sum_{j \geq i} f_j = \Pr\{X \geq i\} \tag{9.2}$$

is the probability of surviving past age $i - 1$. The *hazard rate* at age i is by definition

$$h_i = f_i/S_i = \Pr\{X = i|X \geq i\}, \tag{9.3}$$

the probability of dying at age i given survival past age $i - 1$.

A crucial observation is that the probability S_{ij} of surviving past age j given survival past age $i - 1$ is the product of surviving each intermediate year,

$$S_{ij} = \prod_{k=i}^{j}(1 - h_k) = \Pr\{X > j|X \geq i\}; \tag{9.4}$$

first you have to survive year i, probability $1 - h_i$; then year $i + 1$, probability $1 - h_{i+1}$, etc., up to year j, probability $1 - h_j$. Notice that S_i (9.2) equals $S_{1,i-1}$.

$\hat{S}$ in Table 9.1 is an estimate of S_{ij} for $i = 30$. First, each h_i was estimated as the binomial proportion of the number of deaths y_i among the n_i clients,

$$\hat{h}_i = y_i/n_i, \tag{9.5}$$

and then we set

$$\hat{S}_{30,j} = \prod_{k=30}^{j}\left(1 - \hat{h}_k\right). \tag{9.6}$$

The insurance company doesn't have to wait 50 years to learn the probability of a 30-year-old living past 80 (estimated to be 0.506 in the table). One year's data suffices.[3]

Hazard rates are more often described in terms of a *continuous* positive random variable T (often called "time"), having density function $f(t)$ and "reverse cdf," or survival function,

$$S(t) = \int_t^\infty f(x)\,dx = \Pr\{T \geq t\}. \tag{9.7}$$

The hazard rate

$$h(t) = f(t)/S(t) \tag{9.8}$$

satisfies

$$h(t)dt \doteq \Pr\{T \in (t, t + dt)|T \geq t\} \tag{9.9}$$

for $dt \to 0$, in analogy with (9.3). The analog of (9.4) is† †1

[3] Of course the estimates can go badly wrong if the hazard rates change over time.

$$\Pr\{T \geq t_1 | T \geq t_0\} = \exp\left\{-\int_{t_0}^{t_1} h(x)\,dx\right\} \tag{9.10}$$

so in particular the reverse cdf (9.7) is given by

$$S(t) = \exp\left\{-\int_0^t h(x)\,dx\right\}. \tag{9.11}$$

A one-sided exponential density

$$f(t) = (1/c)e^{-t/c} \qquad \text{for } t \geq 0 \tag{9.12}$$

has $S(t) = \exp\{-t/c\}$ and constant hazard rate

$$h(t) = 1/c. \tag{9.13}$$

The name "memoryless" is quite appropriate for density (9.12): having survived to any time t, the probability of surviving dt units more is always the same, about $1 - dt/c$, no matter what t is. If human lifetimes were exponential there wouldn't be old or young people, only lucky or unlucky ones.

9.2 Censored Data and the Kaplan–Meier Estimate

Table 9.2 reports the survival data from a randomized clinical trial run by **NCOG** (the Northern California Oncology Group) comparing two treatments for head and neck cancer: **Arm_A**, chemotherapy, versus **Arm_B**, chemotherapy plus radiation. The response for each patient is survival time in months. The + sign following some entries indicates *censored data*, that is, survival times known only to exceed the reported value. These are patients "lost to followup," mostly because the **NCOG** experiment ended with some of the patients still alive.

This is what the experimenters hoped to see of course, but it complicates the comparison. Notice that there is more censoring in **Arm_B**. In the absence of censoring we could run a simple two-sample test, maybe Wilcoxon's test, to see whether the more aggressive treatment of **Arm_B** was increasing the survival times. *Kaplan–Meier* curves provide a graphical comparison that takes proper account of censoring. (The next section describes an appropriate censored data two-sample test.) Kaplan–Meier curves have become familiar friends to medical researchers, a *lingua franca* for reporting clinical trial results.

Life table methods are appropriate for censored data. Table 9.3 puts the **Arm_A** results into the same form as the insurance study of Table 9.1, now

Table 9.2 *Censored survival times in days, from two arms of the* NCOG *study of head/neck cancer.*

Arm_A: Chemotherapy

7	34	42	63	64	74+	83	84	91
108	112	129	133	133	139	140	140	146
149	154	157	160	160	165	173	176	185+
218	225	241	248	273	277	279+	297	319+
405	417	420	440	523	523+	583	594	1101
1116+	1146	1226+	1349+	1412+	1417			

Arm_B: Chemotherapy+Radiation

37	84	92	94	110	112	119	127	130
133	140	146	155	159	169+	173	179	194
195	209	249	281	319	339	432	469	519
528+	547+	613+	633	725	759+	817	1092+	1245+
1331+	1557	1642+	1771+	1776	1897+	2023+	2146+	2297+

with the time unit being months. Of the 51 patients enrolled[4] in **Arm_A**, $y_1 = 1$ was observed to die in the first month after treatment; this left 50 at risk, $y_2 = 2$ of whom died in the second month; $y_3 = 5$ of the remaining 48 died in their third month after treatment, and one was lost to followup, this being noted in the l column of the table, leaving $n_4 = 40$ patients "at risk" at the beginning of month 5, etc.

$\hat{S}$ here is calculated as in (9.6) except starting at time 1 instead of 30. There is nothing wrong with this estimate, but binning the NCOG survival data by months is arbitrary. Why not go down to days, as the data was originally presented in Table 9.2? A Kaplan–Meier survival curve is the limit of life table survival estimates as the time unit goes to zero.

Observations z_i for censored data problems are of the form

$$z_i = (t_i, d_i), \tag{9.14}$$

where t_i equals the observed survival time while d_i indicates whether or not there was censoring,

$$d_i = \begin{cases} 1 & \text{if death observed} \\ 0 & \text{if death not observed} \end{cases} \tag{9.15}$$

[4] The patients were enrolled at different calendar times, as they entered the study, but for each patient "time zero" in the table is set at the beginning of his or her treatment.

Table 9.3 **Arm_A** *of the* **NCOG** *head/neck cancer study, binned by month; n = number at risk, y = number of deaths, l = lost to followup, h = hazard rate y/n; $\hat{S}$ = life table survival estimate.*

Month	n	y	l	h	$\hat{S}$	Month	n	y	l	h	$\hat{S}$
1	51	1	0	.020	.980	25	7	0	0	.000	.184
2	50	2	0	.040	.941	26	7	0	0	.000	.184
3	48	5	1	.104	.843	27	7	0	0	.000	.184
4	42	2	0	.048	.803	28	7	0	0	.000	.184
5	40	8	0	.200	.642	29	7	0	0	.000	.184
6	32	7	0	.219	.502	30	7	0	0	.000	.184
7	25	0	1	.000	.502	31	7	0	0	.000	.184
8	24	3	0	.125	.439	32	7	0	0	.000	.184
9	21	2	0	.095	.397	33	7	0	0	.000	.184
10	19	2	1	.105	.355	34	7	0	0	.000	.184
11	16	0	1	.000	.355	35	7	0	0	.000	.184
12	15	0	0	.000	.355	36	7	0	0	.000	.184
13	15	0	0	.000	.355	37	7	1	1	.143	.158
14	15	3	0	.200	.284	38	5	1	0	.200	.126
15	12	1	0	.083	.261	39	4	0	0	.000	.126
16	11	0	0	.000	.261	40	4	0	0	.000	.126
17	11	0	0	.000	.261	41	4	0	1	.000	.126
18	11	1	1	.091	.237	42	3	0	0	.000	.126
19	9	0	0	.000	.237	43	3	0	0	.000	.126
20	9	2	0	.222	.184	44	3	0	0	.000	.126
21	7	0	0	.000	.184	45	3	0	1	.000	.126
22	7	0	0	.000	.184	46	2	0	0	.000	.126
23	7	0	0	.000	.184	47	2	1	1	.500	.063
24	7	0	0	.000	.184						

(so $d_i = 0$ corresponds to a + in Table 9.2). Let

$$t_{(1)} < t_{(2)} < t_{(3)} < \ldots < t_{(n)} \tag{9.16}$$

denote the *ordered* survival times,[5] censored or not, with corresponding indicator $d_{(k)}$ for $t_{(k)}$. The *Kaplan–Meier estimate* for survival probability
†2 $S_{(j)} = \Pr\{X > t_{(j)}\}$ is then† the life table estimate

$$\hat{S}_{(j)} = \prod_{k \le j} \left(\frac{n-k}{n-k+1} \right)^{d_{(k)}}. \tag{9.17}$$

[5] Assuming no ties among the survival times, which is convenient but not crucial for what follows.

$\hat{S}$ jumps downward at death times t_j, and is constant between observed deaths.

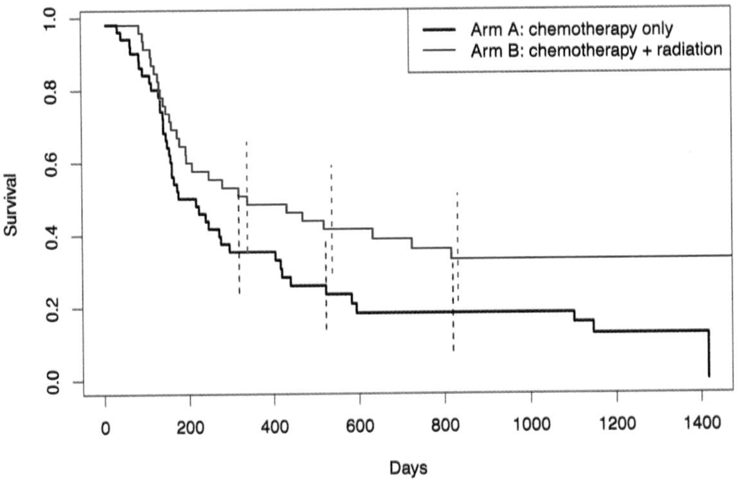

Figure 9.1 NCOG Kaplan–Meier survival curves; lower Arm_A (chemotherapy only); upper Arm_B (chemotherapy+radiation). Vertical lines indicate approximate 95% confidence intervals.

The Kaplan–Meier curves for both arms of the NCOG study are shown in Figure 9.1. Arm_B, the more aggressive treatment, looks better: its 50% survival estimate occurs at 324 days, compared with 182 days for Arm_A. The answer to the inferential question—is B really better than A or is this just random variability?—is less clear-cut.

The accuracy of $\hat{S}_{(j)}$ can be estimated from Greenwood's formula† for †3
its standard deviation (now back in life table notation),

$$\text{sd}\left(\hat{S}_{(j)}\right) = \hat{S}_{(j)} \left[\sum_{k \le j} \frac{y_k}{n_k(n_k - y_k)} \right]^{1/2}. \tag{9.18}$$

The vertical bars in Figure 9.1 are approximate 95% confidence limits for the two curves based on Greenwood's formula. They overlap enough to cast doubt on the superiority of Arm_B at any one choice of "days," but the two-sample test of the next section, which compares survival at all timepoints, will provide more definitive evidence.

Life tables and the Kaplan–Meier estimate seem like a textbook example of frequentist inference as described in Chapter 2: a useful probabilistic result is derived (9.4), and then implemented by the plug-in principle (9.6). There is more to the story though, as discussed below.

†4 Life table curves are nonparametric, in the sense that no particular relationship is assumed between the hazard rates h_i. A parametric approach can greatly improve the curves' accuracy.† Reverting to the life table form of Table 9.3, we assume that the death counts y_k are independent binomials,

$$y_k \overset{\text{ind}}{\sim} \text{Bi}(n_k, h_k), \tag{9.19}$$

and that the logits $\lambda_k = \log\{h_k/(1-h_k)\}$ satisfy some sort of regression equation

$$\boldsymbol{\lambda} = \boldsymbol{X}\alpha, \tag{9.20}$$

as in (8.22). A cubic regression for instance would set $x_k = (1, k, k^2, k^3)'$ for the kth row of $\boldsymbol{X}$, with $\boldsymbol{X}$ 47 × 4 for Table 9.3.

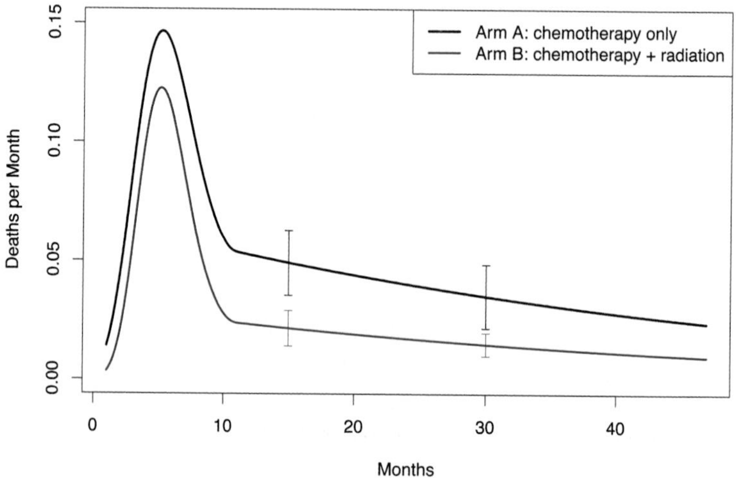

Figure 9.2 Parametric hazard rate estimates for the **NCOG** study. **Arm_A**, black curve, has about 2.5 times higher hazard than **Arm_B** for all times more than a year after treatment. Standard errors shown at 15 and 30 months.

The parametric hazard-rate estimates in Figure 9.2 were instead based on a "cubic-linear spline,"

$$x_k = \left(1, k, (k-11)_-^2, (k-11)_-^3\right)', \tag{9.21}$$

where $(k-11)_-$ equals $k-11$ for $k \le 11$, and 0 for $k \ge 11$. The vector

$\boldsymbol{\lambda} = X\alpha$ describes a curve that is cubic for $k \leq 11$, linear for $k \geq 11$, and joined smoothly at 11. The logistic regression maximum likelihood estimate $\hat{\alpha}$ produced hazard rate curves

$$\hat{h}_k = 1 \Big/ \left(1 + e^{-x_k'\hat{\alpha}}\right) \tag{9.22}$$

as in (8.8). The black curve in Figure 9.2 traces $\hat{h}_k$ for Arm_A, while the red curve is that for Arm_B, fit separately.

Comparison in terms of hazard rates is more informative than the survival curves of Figure 9.1. Both arms show high initial hazards, peaking at five months, and then a long slow decline.[6] Arm_B hazard is always below Arm_A, in a ratio of about 2.5 to 1 after the first year. Approximate 95% confidence limits, obtained as in (8.30), don't overlap, indicating superiority of Arm_B at 15 and 30 months after treatment.

In addition to its frequentist justification, survival analysis takes us into the Fisherian realm of conditional inference, Section 4.3. The y_k's in model (9.19) are considered *conditionally* on the n_k's, effectively treating the n_k values in Table 9.3 as *ancillaries*, that is as fixed constants, by themselves containing no statistical information about the unknown hazard rates. We will examine this tactic more carefully in the next two sections.

9.3 The Log-Rank Test

A randomized clinical trial, interpreted by a two-sample test, remains the gold standard of medical experimentation. Interpretation usually involves Student's two-sample t-test or its nonparametric cousin Wilcoxon's test, but neither of these is suitable for censored data. The *log-rank test* † †5
employs an ingenious extension of life tables for the nonparametric two-sample comparison of censored survival data.

Table 9.4 compares the results of the NCOG study for the first six months[7] after treatment. At the beginning[8] of month 1 there were 45 patients "at risk" in Arm_B, none of whom died, compared with 51 at risk and 1 death in Arm_A. This left 45 at risk in Arm_B at the beginning of month 2, and 50 in Arm_A, with 1 and 2 deaths during the month respectively. (Losses

[6] The cubic–linear spline (9.21) is designed to show more detail in the early months, where there is more available patient data and where hazard rates usually change more quickly.

[7] A month is defined here as 365/12=30.4 days.

[8] The "beginning of month 1" is each patient's initial treatment time, at which all 45 patients ever enrolled in Arm_B were at risk, that is, available for observation.

Table 9.4 *Life table comparison for the first six months of the* **NCOG** *study. For example, at the beginning of the sixth month after treatment, there were 33 remaining* **Arm_B** *patients, of whom 4 died during the month, compared with 32 at risk and 7 dying in* **Arm_A**. *The conditional expected number of deaths in* **Arm_A**, *assuming the null hypothesis of equal hazard rates in both arms, was 5.42, using expression* (9.24).

Month	Arm_B		Arm_A		Expected number Arm_A deaths
	At risk	Died	At risk	Died	
1	45	0	51	1	.53
2	45	1	50	2	1.56
3	44	1	48	5	3.13
4	43	5	42	2	3.46
5	38	5	40	8	6.67
6	33	4	32	7	5.42

to followup were assumed to occur at the *end* of each month; there was 1 such at the end of month 3, reducing the number at risk in **Arm_A** to 42 for month 4.)

The month 6 data is displayed in two-by-two tabular form in Table 9.5, showing the notation used in what follows: n_A for the number at risk in **Arm_A**, n_d for the number of deaths, etc.; y indicates the number of **Arm_A** deaths. If the marginal totals n_A, n_B, n_d, and n_s are given, then y determines the other three table entries by subtraction, so we are not losing any information by focusing on y.

Table 9.5 *Two-by-two display of month-6 data for the* **NCOG** *study. E is the expected number of* **Arm_A** *deaths assuming the null hypothesis of equal hazard rates (last column of Table 9.4).*

	Died	Survived	
Arm_A	$y = 7$ $E = 5.42$	25	$n_A = 32$
Arm_B	4	29	$n_B = 33$
	$n_d = 11$	$n_s = 54$	$n = 65$

Consider the null hypothesis that the hazard rates (9.3) for month 6 are the same in **Arm_A** and **Arm_B**,

$$H_0(6) : h_{A6} = h_{B6}. \tag{9.23}$$

Under $H_0(6)$, y has mean E and variance V,

$$\begin{aligned} E &= n_A n_d / n \\ V &= n_A n_B n_d n_s \big/ \left[n^2 (n-1) \right], \end{aligned} \tag{9.24}$$

as calculated according to the *hypergeometric distribution.*† $E = 5.42$ and †6
$V = 2.28$ in Table 9.5.

We can form a two-by-two table for each of the $N = 47$ months of the **NCOG** study, calculating y_i, E_i, and V_i for month i. The log-rank statistic Z is then defined to be† †7

$$Z = \sum_{i=1}^{N} (y_i - E_i) \Bigg/ \left(\sum_{i=1}^{N} V_i \right)^{1/2}. \tag{9.25}$$

The idea here is simple but clever. Each month we test the null hypothesis of equal hazard rates

$$H_0(i) : h_{Ai} = h_{Bi}. \tag{9.26}$$

The numerator $y_i - E_i$ has expectation 0 under $H_0(i)$, but, if h_{Ai} is greater than h_{Bi}, that is, if treatment B is superior, then the numerator has a positive expectation. Adding up the numerators gives us power to detect a general superiority of treatment B over A, against the null hypothesis of equal hazard rates, $h_{Ai} = h_{Bi}$ for all i.

For the **NCOG** study, binned by months,

$$\sum_1^N y_i = 42, \quad \sum_1^N E_i = 32.9, \quad \sum_1^N V_i = 16.0, \tag{9.27}$$

giving log-rank test statistic

$$Z = 2.27. \tag{9.28}$$

Asymptotic calculations based on the central limit theorem suggest

$$Z \dot{\sim} \mathcal{N}(0, 1) \tag{9.29}$$

under the null hypothesis that the two treatments are equally effective, i.e., that $h_{Ai} = h_{Bi}$ for $i = 1, 2, \ldots, N$. In the usual interpretation, $Z = 2.27$ is significant at the one-sided 0.012 level, providing moderately strong evidence in favor of treatment B.

An impressive amount of inferential guile goes into the log-rank test.

1 Working with hazard rates instead of densities or cdfs is essential for survival data.
2 Conditioning at each period on the numbers at risk, n_A and n_B in Table 9.5, finesses the difficulties of censored data; censoring only changes the at-risk numbers in future periods.
3 Also conditioning on the number of deaths and survivals, n_d and n_s in Table 9.5, leaves only the *univariate* statistic y to interpret at each period, which is easily done through the null hypothesis of equal hazard rates (9.26).
4 Adding the discrepancies $y_i - E_i$ in the numerator of (9.25) (rather than say, adding the individual Z values $Z_i = (y_i - E_i)/V_i^{1/2}$, or adding the Z_i^2 values) accrues power for the natural alternative hypothesis "$h_{Ai} > h_{Bi}$ for all i," while avoiding destabilization from small values of V_i.

Each of the four tactics had been used separately in classical applications. Putting them together into the log-rank test was a major inferential accomplishment, foreshadowing a still bigger step forward, the *proportional hazards model*, our subject in the next section.

Conditional inference takes on an aggressive form in the log-rank test. Let $\boldsymbol{D}_i$ indicate all the data except y_i available at the end of the ith period. For month 6 in the **NCOG** study, $\boldsymbol{D}_6$ includes all data for months 1–5 in Table 9.4, and the marginals n_A, n_B, n_d, and n_s in Table 9.5, but not the y value for month 6. The key assumption is that, under the null hypothesis of equal hazard rates (9.26),

$$y_i | \boldsymbol{D}_i \overset{\text{ind}}{\sim} (E_i, V_i), \tag{9.30}$$

"ind" here meaning that the y_i's can be treated as independent quantities with means and variances (9.24). In particular, we can add the variances V_i to get the denominator of (9.25). (A "partial likelihood" argument, described in the endnotes, justifies adding the variances.)

The purpose of all this Fisherian conditioning is to simplify the inference: the conditional distribution $y_i | \boldsymbol{D}_i$ depends only on the hazard rates h_{Ai} and h_{Bi}; "nuisance parameters," relating to the survival times and censoring mechanism of the data in Table 9.2, are hidden away. There is a price to pay in testing power, though usually a small one. The lost-to-followup values l in Table 9.3 have been ignored, even though they might contain useful information, say if all the early losses occurred in one arm.

9.4 The Proportional Hazards Model

The Kaplan–Meier estimator is a one-sample device, dealing with data coming from a single distribution. The log-rank test makes two-sample comparisons. *Proportional hazards* ups the ante to allow for a full regression analysis of censored data. Now the individual data points z_i are of the form

$$z_i = (c_i, t_i, d_i), \tag{9.31}$$

where t_i and d_i are observed survival time and censoring indicator, as in (9.14)–(9.15), and c_i is a known $1 \times p$ vector of covariates whose effect on survival we wish to assess. Both of the previous methods are included here: for the log-rank test, c_i indicates treatment, say c_i equals 0 or 1 for `Arm_A` or `Arm_B`, while c_i is absent for Kaplan–Meier.

Table 9.6 *Pediatric cancer data, first 20 of 1620 children.* `Sex` *1 = male, 2 = female;* `race` *1 = white, 2 = nonwhite;* `age` *in years;* `entry` *= calendar date of entry in days since July 1, 2001;* `far` *= home distance from treatment center in miles;* `t` *= survival time in days;* `d` *= 1 if death observed, 0 if not.*

sex	race	age	entry	far	t	d
1	1	2.50	710	108	325	0
2	1	10.00	1866	38	1451	0
2	2	18.17	2531	100	221	0
2	1	3.92	2210	100	2158	0
1	1	11.83	875	78	760	0
2	1	11.17	1419	0	168	0
2	1	5.17	1264	28	2976	0
2	1	10.58	670	120	1833	0
1	1	1.17	1518	73	131	0
2	1	6.83	2101	104	2405	0
1	1	13.92	1239	0	969	0
1	1	5.17	518	117	1894	0
1	1	2.50	1849	99	193	1
1	1	.83	2758	38	1756	0
2	1	15.50	2004	12	682	0
1	1	17.83	986	65	1835	0
2	1	3.25	1443	58	2993	0
1	1	10.75	2807	42	1616	0
1	2	18.08	1229	23	1302	0
2	2	5.83	2727	23	174	1

Medical studies regularly produce data of form (9.31). An example, the *pediatric cancer* data, is partially listed in Table 9.6. The first 20 of $n = 1620$ cases are shown. There are five explanatory covariates (defined in the table's caption): **sex**, **race**, **age** at entry, calendar date of **entry** into the study, and **far**, the distance of the child's home from the treatment center. The response variable t is survival in days from time of treatment until death. Happily, only 160 of the children were observed to die ($d = 1$). Some left the study for various reasons, but most of the $d = 0$ cases were those children still alive at the end of the study period. Of particular interest was the effect of **far** on survival. We wish to carry out a regression analysis of this heavily censored data set.

The proportional hazards model assumes that the hazard rate $h_i(t)$ for the ith individual (9.8) is

$$h_i(t) = h_0(t)e^{c_i'\beta}. \tag{9.32}$$

Here $h_0(t)$ is a baseline hazard (which we need not specify) and β is an unknown p-parameter vector we want to estimate. For concise notation, let

$$\theta_i = e^{c_i'\beta}; \tag{9.33}$$

model (9.32) says that individual i's hazard is a constant nonnegative factor θ_i times the baseline hazard. Equivalently, from (9.11), the ith survival function $S_i(t)$ is a power of the baseline survival function $S_0(t)$,

$$S_i(t) = S_0(t)^{\theta_i}. \tag{9.34}$$

Larger values of θ_i lead to more quickly declining survival curves, i.e., to worse survival (as in (9.11)).

Let J be the number of observed deaths, $J = 160$ here, occurring at times

$$T_{(1)} < T_{(2)} < \ldots < T_{(J)}, \tag{9.35}$$

again for convenience assuming no ties.[9] Just before time $T_{(j)}$ there is a *risk set* of individuals still under observation, whose indices we denote by $\mathcal{R}_j$,

$$\mathcal{R}_j = \{i : t_i \geq T_{(j)}\}. \tag{9.36}$$

Let i_j be the index of the individual observed to die at time $T_{(j)}$. The key to proportional hazards regression is the following result.

[9] More precisely, assuming only one event, a death, occurred at $T_{(j)}$, with none of the other individuals being lost to followup at exact time $T_{(j)}$.

Lemma † *Under the proportional hazards model* (9.32), *the conditional* †8
probability, given the risk set $\mathcal{R}_j$, *that individual* i *in* $\mathcal{R}_j$ *is the one observed to die at time* $T_{(j)}$ *is*

$$\Pr\{i_j = i | \mathcal{R}_j\} = e^{c_i'\beta} \Big/ \sum_{k\in\mathcal{R}_j} e^{c_k'\beta}. \tag{9.37}$$

To put it in words, given that one person dies at time $T_{(j)}$, *the probability it is individual* i *is proportional to* $\exp(c_i'\beta)$, *among the set of individuals at risk.*

For the purpose of estimating the parameter vector β in model (9.32), we multiply factors (9.37) to form the *partial likelihood*

$$L(\beta) = \prod_{j=1}^{J} \left(e^{c_{i_j}'\beta} \Big/ \sum_{k\in\mathcal{R}_j} e^{c_k'\beta} \right). \tag{9.38}$$

$L(\beta)$ is then treated as an ordinary likelihood function, yielding an approximately unbiased MLE-like estimate

$$\hat{\beta} = \arg\max_{\beta} \{L(\beta)\}, \tag{9.39}$$

with an approximate covariance obtained from the second-derivative matrix of $l(\beta) = \log L(\beta)$,† as in Section 4.3, †9

$$\hat{\beta} \,\dot{\sim}\, \left(\beta, \left[-\ddot{l}\left(\hat{\beta}\right)\right]^{-1}\right). \tag{9.40}$$

Table 9.7 shows the proportional hazards analysis of the pediatric cancer data, with the covariates `age`, `entry`, and `far` standardized to have mean 0 and standard deviation 1 for the 1620 cases.[10] Neither `sex` nor `race` seems to make much difference. We see that `age` is a mildly significant factor, with older children doing better (i.e., the estimated regression coefficient is negative). However, the dramatic effects are date of `entry` and `far`. Individuals who entered the study later survived longer—perhaps the treatment protocol was being improved—while children living farther away from the treatment center did worse.

Justification of the partial likelihood calculations is similar to that for the log-rank test, but there are some important differences, too: the proportional hazards model is semiparametric ("semi" because we don't have to specify $h_0(t)$ in (9.31)), rather than nonparametric as before; and the

[10] Table 9.7 was obtained using the `R` program `coxph`.

Table 9.7 *Proportional hazards analysis of pediatric cancer data (*`age`*,* `entry` *and* `far` *standardized).* `Age` *significantly negative, older children doing better;* `entry` *very significantly negative, showing hazard rate declining with calendar date of entry;* `far` *very significantly positive, indicating worse results for children living farther away from the treatment center. Last two columns show limits of approximate 95% confidence intervals for* $\exp(\beta)$.

	β	sd	z-value	p-value	$\exp(\beta)$	Lower	Upper
`sex`	−.023	.160	−.142	.887	.98	.71	1.34
`race`	.282	.169	1.669	.095	1.33	.95	1.85
`age`	−.235	.088	−2.664	.008	.79	.67	.94
`entry`	−.460	.079	−5.855	.000	.63	.54	.74
`far`	.296	.072	4.117	.000	1.34	1.17	1.55

emphasis on likelihood has increased the Fisherian nature of the inference, moving it further away from pure frequentism. Still more Fisherian is the emphasis on likelihood inference in (9.38)–(9.40), rather than the direct frequentist calculations of (9.24)–(9.25).

The conditioning argument here is less obvious than that for the Kaplan–Meier estimate of the log-rank test. Has its convenience possibly come at too high a price? In fact it can be shown that inference based on the partial likelihood is highly efficient, assuming of course the correctness of the proportional hazards model (9.31).

9.5 Missing Data and the EM Algorithm

Censored data, the motivating factor for survival analysis, can be thought of as a special case of a more general statistical topic, *missing data*. What's missing, in Table 9.2 for example, are the actual survival times for the + cases, which are known only to exceed the tabled values. If the data were *not* missing, we could use standard statistical methods, for instance Wilcoxon's test, to compare the two arms of the **NCOG** study. The EM algorithm is an iterative technique for solving missing-data inferential problems using only standard methods.

A missing-data situation is shown in Figure 9.3: $n = 40$ points have been independently sampled from a bivariate normal distribution (5.12),

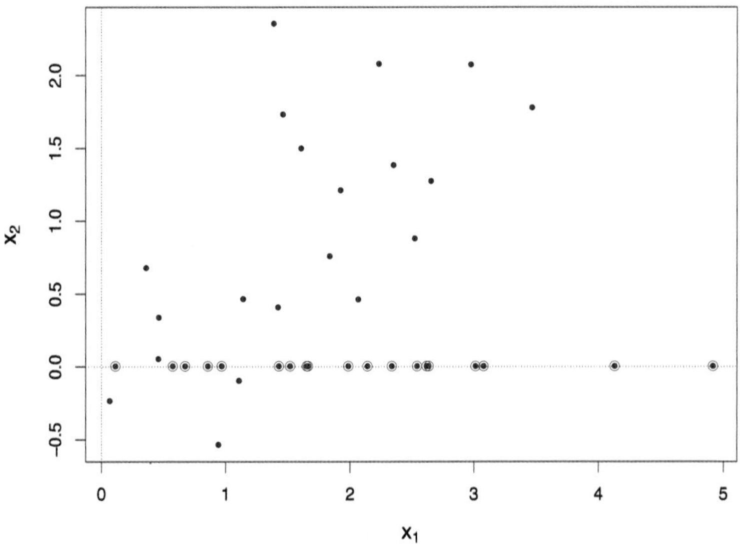

Figure 9.3 Forty points from a bivariate normal distribution, the last 20 with x_2 missing (circled).

means (μ_1, μ_2), variances (σ_1^2, σ_2^2), and correlation ρ,

$$\begin{pmatrix} x_{1i} \\ x_{2i} \end{pmatrix} \overset{\text{ind}}{\sim} \mathcal{N}_2\left(\begin{pmatrix} \mu_1 \\ \mu_2 \end{pmatrix}, \begin{pmatrix} \sigma_1^2 & \sigma_1\sigma_2\rho \\ \sigma_1\sigma_2\rho & \sigma_2^2 \end{pmatrix} \right). \tag{9.41}$$

However, the second coordinates of the last 20 points have been lost. These are represented by the circled points in Figure 9.3, with their x_2 values arbitrarily set to 0.

We wish to find the maximum likelihood estimate of the parameter vector $\theta = (\mu_1, \mu_2, \sigma_1, \sigma_2, \rho)$. The standard maximum likelihood estimates

$$\hat{\mu}_1 = \sum_1^{40} x_{1i}/40, \quad \hat{\mu}_2 = \sum_1^{40} x_{2i}/40,$$

$$\hat{\sigma}_1 = \left[\sum_1^{40} (x_{1i} - \hat{\mu}_1)^2 / 40 \right]^{1/2}, \quad \hat{\sigma}_2 = \left[\sum_1^{40} (x_{2i} - \hat{\mu}_2)^2 / 40 \right]^{1/2},$$

$$\hat{\rho} = \left[\sum_1^{40} (x_{1i} - \hat{\mu}_1)(x_{2i} - \hat{\mu}_2) / 40 \right] \Big/ (\hat{\sigma}_1 \hat{\sigma}_2), \tag{9.42}$$

are unavailable for μ_2, σ_2, and ρ because of the missing data.

The EM algorithm begins by filling in the missing data in some way, say by setting $x_{2i} = 0$ for the 20 missing values, giving an artificially complete data set $\textit{data}^{(0)}$. Then it proceeds as follows.

- The standard method (9.42) is applied to the filled-in $\textit{data}^{(0)}$ to produce $\hat{\theta}^{(0)} = (\hat{\mu}_1^{(0)}, \hat{\mu}_2^{(0)}, \hat{\sigma}_1^{(0)}, \hat{\sigma}_2^{(0)}, \hat{\rho}^{(0)})$; this is the M ("maximizing") step.[11]
- Each of the missing values is replaced by its conditional expectation (assuming $\theta = \hat{\theta}^{(0)}$) given the nonmissing data; this is the E ("expectation") step. In our case the missing values x_{2i} are replaced by

$$\hat{\mu}_2^{(0)} + \hat{\rho}^{(0)} \frac{\hat{\sigma}_2^{(0)}}{\hat{\sigma}_1^{(0)}} \left(x_{1i} - \hat{\mu}_1^{(0)} \right). \tag{9.43}$$

- The E and M steps are repeated, at the jth stage giving a new artificially complete data set $\textit{data}^{(j)}$ and an updated estimate $\hat{\theta}^{(j)}$. The iteration stops when $\|\hat{\theta}^{(j)+1} - \hat{\theta}^{(j)}\|$ is suitably small.

Table 9.8 shows the EM algorithm at work on the bivariate normal example of Figure 9.3. In exponential families the algorithm is guaranteed to converge to the MLE $\hat{\theta}$ based on just the observed data $\boldsymbol{o}$; moreover, the likelihood $f_{\hat{\theta}^{(j)}}(\boldsymbol{o})$ increases with every step j. (The convergence can be sluggish, as it is here for $\hat{\sigma}_2$ and $\hat{\rho}$.)

The EM algorithm ultimately derives from the *fake-data principle*, a property of maximum likelihood estimation going back to Fisher that can †10 only briefly be summarized here.† Let $\boldsymbol{x} = (\boldsymbol{o}, \boldsymbol{u})$ represent the "complete data," of which $\boldsymbol{o}$ is observed while $\boldsymbol{u}$ is unobserved or missing. Write the density for $\boldsymbol{x}$ as

$$f_\theta(\boldsymbol{x}) = f_\theta(\boldsymbol{o}) f_\theta(\boldsymbol{u}|\boldsymbol{o}), \tag{9.44}$$

and let $\hat{\theta}(\boldsymbol{o})$ be the MLE of θ based just on $\boldsymbol{o}$.

Suppose we now generate simulations of $\boldsymbol{u}$ by sampling from the conditional distribution $f_{\hat{\theta}(\boldsymbol{o})}(\boldsymbol{u}|\boldsymbol{o})$,

$$\boldsymbol{u}^{*k} \sim f_{\hat{\theta}(\boldsymbol{o})}(\boldsymbol{u}|\boldsymbol{o}) \qquad \text{for } k = 1, 2, \ldots, K \tag{9.45}$$

(the stars indicating creation by the statistician and not by observation), giving fake complete-data values $\boldsymbol{x}^{*k} = (\boldsymbol{o}, \boldsymbol{u}^{*k})$. Let

$$\textit{data}^* = \{\boldsymbol{x}^{*1}, \boldsymbol{x}^{*2}, \ldots, \boldsymbol{x}^{*K}\}, \tag{9.46}$$

[11] In this example, $\hat{\mu}_1^{(0)}$ and $\hat{\sigma}_1^{(0)}$ are available as the complete-data estimates in (9.42), and, as in Table 9.8, stay the same in subsequent steps of the algorithm.

Table 9.8 *EM algorithm for estimating means, standard deviations, and the correlation of the bivariate normal distribution that gave the data in Figure 9.3.*

Step	μ_1	μ_2	σ_1	σ_2	ρ
1	1.86	.463	1.08	.738	.162
2	1.86	.707	1.08	.622	.394
3	1.86	.843	1.08	.611	.574
4	1.86	.923	1.08	.636	.679
5	1.86	.971	1.08	.667	.736
6	1.86	1.002	1.08	.694	.769
7	1.86	1.023	1.08	.716	.789
8	1.86	1.036	1.08	.731	.801
9	1.86	1.045	1.08	.743	.808
10	1.86	1.051	1.08	.751	.813
11	1.86	1.055	1.08	.756	.816
12	1.86	1.058	1.08	.760	.819
13	1.86	1.060	1.08	.763	.820
14	1.86	1.061	1.08	.765	.821
15	1.86	1.062	1.08	.766	.822
16	1.86	1.063	1.08	.767	.822
17	1.86	1.064	1.08	.768	.823
18	1.86	1.064	1.08	.768	.823
19	1.86	1.064	1.08	.769	.823
20	1.86	1.064	1.08	.769	.823

whose notional likelihood $\prod_1^K f_\theta(\boldsymbol{x}^{*k})$ yields MLE $\hat{\theta}^*$. It then turns out that $\hat{\theta}^*$ goes to $\hat{\theta}(\boldsymbol{o})$ as K goes to infinity. In other words, maximum likelihood estimation is *self-consistent*: generating artificial data from the MLE density $f_{\hat{\theta}(\boldsymbol{o})}(\boldsymbol{u}|\boldsymbol{o})$ doesn't change the MLE. Moreover, any value $\hat{\theta}^{(0)}$ not equal to the MLE $\hat{\theta}(\boldsymbol{o})$ cannot be self-consistent: carrying through (9.45)–(9.46) using $f_{\hat{\theta}^{(0)}}(\boldsymbol{u}|\boldsymbol{o})$ leads to hypothetical MLE $\hat{\theta}^{(1)}$ having $f_{\hat{\theta}^{(1)}}(\boldsymbol{o}) > f_{\hat{\theta}^{(0)}}(\boldsymbol{o})$, etc., a more general version of the EM algorithm.[12]

Modern technology allows social scientists to collect huge data sets, perhaps hundreds of responses for each of thousands or even millions of individuals. Inevitably, some entries of the individual responses will be missing. *Imputation* amounts to employing some version of the fake-data principle to fill in the missing values. Imputation's goal goes beyond find-

[12] Simulation (9.45) is unnecessary in exponential families, where at each stage $data^*$ can be replaced by $(\boldsymbol{o}, E^{(j)}(\boldsymbol{u}|\boldsymbol{o}))$, with $E^{(j)}$ indicating expectation with respect to $\hat{\theta}^{(j)}$, as in (9.43).

ing the MLE, to the creation of graphs, confidence intervals, histograms, and more, using only convenient, standard complete-data methods.

Finally, returning to survival analysis, the Kaplan–Meier estimate (9.17) is itself self-consistent.† Consider the `Arm_A` censored observation 74+ in Table 9.2. We know that that patient's survival time exceeded 74. Suppose we distribute his probability mass (1/51 of the `Arm_A` sample) to the right, in accordance with the conditional distribution for $x > 74$ defined by the `Arm_A` Kaplan–Meier survival curve. It turns out that redistributing all the censored cases does not change the original Kaplan–Meier survival curve; Kaplan–Meier is self-consistent, leading to its identification as the "nonparametric MLE" of a survival function.

9.6 Notes and Details

The progression from life tables, Kaplan–Meier curves, and the log-rank test to proportional hazards regression was modest in its computational demands, until the final step. Kaplan–Meier curves lie within the capabilities of mechanical calculators. Not so for proportional hazards, which is emphatically a child of the computer age. As the algorithms grew more intricate, their inferential justification deepened in scope and sophistication. This is a pattern we also saw in Chapter 8, in the progression from bioassay to logistic regression to generalized linear models, and will reappear as we move from the jackknife to the bootstrap in Chapter 10.

Censoring is not the same as truncation. For the truncated galaxy data of Section 8.3, we learn of the existence of a galaxy only if it falls into the observation region (8.38). The censored individuals in Table 9.2 are known to exist, but with imperfect knowledge of their lifetimes. There is a version of the Kaplan–Meier curve applying to truncated data, which was developed in the astronomy literature by Lynden-Bell (1971).

The methods of this chapter apply to data that is left-truncated as well as right-censored. In a survival time study of a new HIV drug, for instance, subject i might not enter the study until some time τ_i after his or her initial diagnosis, in which case t_i would be left-truncated at τ_i, as well as possibly later right-censored. This only modifies the composition of the various risk sets. However, other missing-data situations, e.g., left- *and* right-censoring, require more elaborate, less elegant, treatments.

†1 [p. 133] *Formula* (9.10). Let the interval $[t_0, t_1]$ be partitioned into a large number of subintervals of length dt, with t_k the midpoint of subinterval k.

As in (9.4), using (9.9),

$$\begin{aligned}\Pr\{T \geq t_1 | T \geq t_0\} &\doteq \prod (1 - h(t_i)\,dt) \\ &= \exp\left\{\sum \log(1 - h(t_i)\,dt)\right\} \\ &\doteq \exp\left\{-\sum h(t_i)\,dt\right\},\end{aligned} \tag{9.47}$$

which, as $dt \to 0$, goes to (9.10).

†2 [p. 136] *Kaplan–Meier estimate.* In the life table formula (9.6) (with $k = 1$), let the time unit be small enough to make each bin contain at most one value $t_{(k)}$ (9.16). Then at $t_{(k)}$,

$$\hat{h}_{(k)} = \frac{d_{(k)}}{n - k + 1}, \tag{9.48}$$

giving expression (9.17).

†3 [p. 137] *Greenwood's formula* (9.18). In the life table formulation of Section 9.1, (9.6) gives

$$\log \hat{S}_j = \sum_1^j \log\left(1 - \hat{h}_k\right). \tag{9.49}$$

From $\hat{h}_k \overset{\text{ind}}{\sim} \text{Bi}(n_k, h_k)$ we get

$$\begin{aligned}\text{var}\left\{\log \hat{S}_j\right\} &= \sum_1^j \text{var}\left\{\log\left(1 - \hat{h}_k\right)\right\} \doteq \sum_1^j \frac{\text{var}\,\hat{h}_k}{(1 - h_k)^2} \\ &= \sum_1^j \frac{h_k}{1 - h_k}\frac{1}{n_k},\end{aligned} \tag{9.50}$$

where we have used the delta-method approximation $\text{var}\{\log X\} \doteq \text{var}\{X\}/E\{X\}^2$. Plugging in $\hat{h}_k = y_k/n_k$ yields

$$\text{var}\left\{\log \hat{S}_j\right\} \doteq \sum_1^j \frac{y_k}{n_k(n_k - y_k)}. \tag{9.51}$$

Then the inverse approximation $\text{var}\{X\} = E\{X\}^2\,\text{var}\{\log X\}$ gives Greenwood's formula (9.18).

The censored data situation of Section 9.2 does not enjoy independence between the $\hat{h}_k$ values. However, successive conditional independence, given the n_k values, is enough to verify the result, as in the partial likelihood calculations below. *Note*: the confidence intervals in Figure 9.1 were obtained by exponentiating the intervals,

$$\log \hat{S}_j \pm 1.96 \left[\text{var} \left\{ \log \hat{S}_j \right\} \right]^{1/2}. \tag{9.52}$$

†4 [p. 138] *Parametric life tables analysis.* Figure 9.2 and the analysis behind it is developed in Efron (1988), where it is called "partial logistic regression" in analogy with partial likelihood.

†5 [p. 139] *The log-rank test.* This chapter featured an all-star cast, including four of the most referenced papers of the post-war era: Kaplan and Meier (1958), Cox (1972) on proportional hazards, Dempster *et al.* (1977) codifying and naming the EM algorithm, and Mantel and Haenszel (1959) on the log-rank test. (Cox (1958) gives a careful, and early, analysis of the Mantel–Haenszel idea.) The not very helpful name "log-rank" does at least remind us that the test depends only on the ranks of the survival times, and will give the same result if all the observed survival times t_i are monotonically transformed, say to $\exp(t_i)$ or $t_i^{1/2}$. It is often referred to as the Mantel–Haenszel or Cochran–Mantel–Haenszel test in older literature. Kaplan–Meier and proportional hazards are also rank-based procedures.

†6 [p. 141] *Hypergeometric distribution.* Hypergeometric calculations, as for Table 9.5, are often stated as follows: n marbles are placed in an urn, n_A labeled A and n_B labeled B; n_d marbles are drawn out at random; y is the number of these labeled A. Elementary (but not simple) calculations then produce the conditional distribution of y given the table's marginals n_A, n_B, n, n_d, and n_s,

$$\Pr\{y|\text{marginals}\} = \binom{n_A}{y}\binom{n_B}{n_d - y} \Big/ \binom{n}{n_d} \tag{9.53}$$

for

$$\max(n_A - n_s, 0) \le y \le \min(n_d, n_A),$$

and expressions (9.24) for the mean and variance. If n_A and n_B go to infinity such that $n_A/n \to p_A$ and $n_B/n \to 1 - p_A$, then $V \to n_d p_A(1 - p_A)$, the variance of $y \sim \text{Bi}(n_d, p_A)$.

†7 [p. 141] *Log-rank statistic Z (9.25).* Why is $(\sum_1^N V_i)^{1/2}$ the correct denominator for Z? Let $u_i = y_i - E_i$ in (9.30), so Z's numerator is $\sum_1^N u_i$, with

$$u_i | \boldsymbol{D}_i \sim (0, V_i) \tag{9.54}$$

under the null hypothesis of equal hazard rates. This implies that, unconditionally, $E\{u_i\} = 0$. For $j < i$, u_j is a function of $\boldsymbol{D}_i$ (since y_j and

E_j are), so $E\{u_j u_i | \boldsymbol{D}_i\} = 0$, and, again unconditionally, $E\{u_j u_i\} = 0$. Therefore, assuming equal hazard rates,

$$E\left(\sum_1^N u_i\right)^2 = E\left(\sum_1^N u_i^2\right) = \sum_1^N \text{var}\{u_i\} \doteq \sum_1^N V_i. \tag{9.55}$$

The last approximation, replacing unconditional variances var$\{u_i\}$ with conditional variances V_i, is justified in Crowley (1974), as is the asymptotic normality (9.29).

†8 [p. 145] *Lemma* (9.37). For $i \in \mathcal{R}_j$, the probability p_i that death occurs in the infinitesimal interval $(T_{(j)}, T_{(j)} + dT)$ is $h_i(T_{(j)})\, dT$, so

$$p_i = h_0(T_{(j)}) e^{c_i' \beta} dT, \tag{9.56}$$

and the probability of event A_i that individual i dies while the others don't is

$$P_i = p_i \prod_{k \in \mathcal{R}_j - i} (1 - p_k). \tag{9.57}$$

But the A_i are disjoint events, so, given that $\cup A_i$ has occurred, the probability that it is individual i who died is

$$P_i \Big/ \sum_{\mathcal{R}_j} P_j \doteq e^{c_i \beta} \Big/ \sum_{k \in \mathcal{R}_j} e^{c_k \beta}, \tag{9.58}$$

this becoming exactly (9.37) as $dT \to 0$.

†9 [p. 145] *Partial likelihood* (9.40). Cox (1975) introduced partial likelihood as inferential justification for the proportional hazards model, which had been questioned in the literature. Let $\boldsymbol{D}_j$ indicate all the observable information available just before time $T_{(j)}$(9.35), including all the death or loss times for individuals having $t_i < T_{(j)}$. (Notice that $\boldsymbol{D}_j$ determines the risk set $\mathcal{R}_j$.) By successive conditioning we write the full likelihood $f_\theta(\text{data})$ as

$$\begin{aligned} f_\theta(\text{data}) &= f_\theta(\boldsymbol{D}_1) f_\theta(i_1|\mathcal{R}_1) f_\theta(\boldsymbol{D}_2|\boldsymbol{D}_1) f_\theta(i_2|\mathcal{R}_2)\dots \\ &= \prod_{j=1}^J f_\theta(\boldsymbol{D}_j|\boldsymbol{D}_{j-1}) \prod_{j=1}^J f_\theta(i_j|\mathcal{R}_j). \end{aligned} \tag{9.59}$$

Letting $\theta = (\alpha, \beta)$, where α is a nuisance parameter vector having to do with the occurrence and timing of events between observed deaths,

$$f_{\alpha,\beta}(\text{data}) = \left[\prod_{j=1}^{J} f_{\alpha,\beta}(\boldsymbol{D}_j | \boldsymbol{D}_{j-1}) \right] L(\beta), \tag{9.60}$$

where $L(\beta)$ is the partial likelihood (9.38).

The proportional hazards model simply ignores the bracketed factor in (9.60); $l(\beta) = \log L(\beta)$ is treated as a genuine likelihood, maximized to give $\hat{\beta}$, and assigned covariance matrix $(-\ddot{l}(\hat{\beta}))^{-1}$ as in Section 4.3. Efron (1977) shows this tactic is highly efficient for the estimation of β.

†10 [p. 148] *Fake-data principle.* For any two values of the parameters θ_1 and θ_2 define

$$l_{\theta_1}(\theta_2) = \int [\log f_{\theta_2}(\boldsymbol{o}, \boldsymbol{u})] \, f_{\theta_1}(\boldsymbol{u}|\boldsymbol{o}) \, d\boldsymbol{u}, \tag{9.61}$$

this being the limit as $K \to \infty$ of

$$l_{\theta_1}(\theta_2) = \lim_{K \to \infty} \frac{1}{K} \sum_{k=1}^{K} \log f_{\theta_2}(\boldsymbol{o}, \boldsymbol{u}^{*k}), \tag{9.62}$$

the fake-data log likelihood (9.46) under θ_2, if θ_1 were the true value of θ. Using $f_\theta(\boldsymbol{o}, \boldsymbol{u}) = f_\theta(\boldsymbol{o}) f_\theta(\boldsymbol{u}|\boldsymbol{o})$, definition (9.61) gives

$$\begin{aligned} l_{\theta_1}(\theta_2) - l_{\theta_1}(\theta_1) &= \log\left(\frac{f_{\theta_2}(\boldsymbol{o})}{f_{\theta_1}(\boldsymbol{o})}\right) - \int \log\left(\frac{f_{\theta_2}(\boldsymbol{u}|\boldsymbol{o})}{f_{\theta_1}(\boldsymbol{u}|\boldsymbol{o})}\right) f_{\theta_1}(\boldsymbol{u}|\boldsymbol{o}) \\ &= \log\left(\frac{f_{\theta_2}(\boldsymbol{o})}{f_{\theta_1}(\boldsymbol{o})}\right) - \frac{1}{2} D\left(f_{\theta_1}(\boldsymbol{u}|\boldsymbol{o}), f_{\theta_2}(\boldsymbol{u}|\boldsymbol{o})\right), \end{aligned} \tag{9.63}$$

with D the deviance (8.31), which is always positive unless $\boldsymbol{u}|\boldsymbol{o}$ has the same distribution under θ_1 and θ_2, which we will assume doesn't happen.

Suppose we begin the EM algorithm at $\theta = \theta_1$ and find the value θ_2 maximizing $l_{\theta_1}(\theta)$. Then $l_{\theta_1}(\theta_2) > l_{\theta_1}(\theta_1)$ and $D > 0$ implies $f_{\theta_2}(\boldsymbol{o}) > f_{\theta_1}(\boldsymbol{o})$ in (9.63); that is, we have increased the likelihood of the observed data. Now take $\theta_1 = \hat{\theta} = \arg\max_\theta f_\theta(\boldsymbol{o})$. Then the right side of (9.63) is negative, implying $l_{\hat{\theta}}(\hat{\theta}) > l_{\hat{\theta}}(\theta_2)$ for any θ_2 not equaling $\theta_1 = \hat{\theta}$. Putting this together,[13] successively computing $\theta_1, \theta_2, \theta_3, \ldots$ by fake-data MLE calculations increases $f_\theta(\boldsymbol{o})$ at every step, and the only stable point of the algorithm is at $\theta = \hat{\theta}(\boldsymbol{o})$.

†11 [p. 150] *Kaplan–Meier self-consistency.* This property was verified in Efron (1967), where the name was coined.

[13] Generating the fake data is equivalent to the E step of the algorithm, the M step being the maximization of $l_{\theta_j}(\theta)$.

10

The Jackknife and the Bootstrap

A central element of frequentist inference is the *standard error*. An algorithm has produced an estimate of a parameter of interest, for instance the mean $\bar{x} = 0.752$ for the 47 **ALL** scores in the top panel of Figure 1.4. How accurate is the estimate? In this case, formula (1.2) for the standard deviation[1] of a sample mean gives estimated standard error

$$\widehat{\text{se}} = 0.040, \tag{10.1}$$

so one can't take the third digit of $\bar{x} = 0.752$ very seriously, and even the 5 is dubious.

Direct standard error formulas like (1.2) exist for various forms of averaging, such as linear regression (7.34), and for hardly anything else. Taylor series approximations ("device 2" of Section 2.1) extend the formulas to smooth functions of averages, as in (8.30). Before computers, applied statisticians needed to be Taylor series experts in laboriously pursuing the accuracy of even moderately complicated statistics.

The jackknife (1957) was a first step toward a computation-based, nonformulaic approach to standard errors. The bootstrap (1979) went further toward automating a wide variety of inferential calculations, including standard errors. Besides sparing statisticians the exhaustion of tedious routine calculations the jackknife and bootstrap opened the door for more complicated estimation algorithms, which could be pursued with the assurance that their accuracy would be easily assessed. This chapter focuses on standard errors, with more adventurous bootstrap ideas deferred to Chapter 11. We end with a brief discussion of accuracy estimation for robust statistics.

[1] We will use the terms "standard error" and "standard deviation" interchangeably.

10.1 The Jackknife Estimate of Standard Error

The basic applications of the jackknife apply to *one-sample problems*, where the statistician has observed an independent and identically distributed (iid) sample $\boldsymbol{x} = (x_1, x_2, \ldots, x_n)'$ from an unknown probability distribution F on some space $\mathcal{X}$,

$$x_i \overset{\text{iid}}{\sim} F \qquad \text{for } i = 1, 2, \ldots, n. \tag{10.2}$$

$\mathcal{X}$ can be anything: the real line, the plane, a function space.[2] A *real-valued* statistic $\hat{\theta}$ has been computed by applying some algorithm $s(\cdot)$ to $\boldsymbol{x}$,

$$\hat{\theta} = s(\boldsymbol{x}), \tag{10.3}$$

and we wish to assign a standard error to $\hat{\theta}$. That is, we wish to estimate the standard deviation of $\hat{\theta} = s(\boldsymbol{x})$ under sampling model (10.2).

Let $\boldsymbol{x}_{(i)}$ be the sample with x_i removed,

$$\boldsymbol{x}_{(i)} = (x_1, x_2, \ldots, x_{i-1}, x_{i+1}, \ldots, x_n)', \tag{10.4}$$

and denote the corresponding value of the statistic of interest as

$$\hat{\theta}_{(i)} = s(\boldsymbol{x}_{(i)}). \tag{10.5}$$

Then the *jackknife estimate of standard error* for $\hat{\theta}$ is

$$\widehat{\text{se}}_{\text{jack}} = \left[\frac{n-1}{n} \sum_1^n \left(\hat{\theta}_{(i)} - \hat{\theta}_{(\cdot)} \right)^2 \right]^{1/2}, \quad \text{with } \hat{\theta}_{(\cdot)} = \sum_1^n \hat{\theta}_{(i)} / n. \tag{10.6}$$

In the case where $\hat{\theta}$ is the mean $\bar{x}$ of real values $x_1, x_2, \ldots, x_n$ (i.e., $\mathcal{X}$ is an interval of the real line), $\hat{\theta}_{(i)}$ is their average excluding x_i, which can be expressed as

$$\hat{\theta}_{(i)} = (n\bar{x} - x_i)/(n-1). \tag{10.7}$$

Equation (10.7) gives $\hat{\theta}_{(\cdot)} = \bar{x}$, $\hat{\theta}_{(i)} - \hat{\theta}_{(\cdot)} = (\bar{x} - x_i)/(n-1)$, and

$$\widehat{\text{se}}_{\text{jack}} = \left[\sum_{i=1}^n (x_i - \bar{x})^2 / \left(n(n-1)\right) \right]^{1/2}, \tag{10.8}$$

exactly the same as the classic formula (1.2). This is no coincidence. The fudge factor $(n-1)/n$ in definition (10.6) was inserted to make $\widehat{\text{se}}_{\text{jack}}$ agree with (1.2) when $\hat{\theta}$ is $\bar{x}$.

[2] If $\mathcal{X}$ is an interval of the real line we might take F to be the usual cumulative distribution function, but here we will just think of F as any full description of the probability distribution for an x_i on $\mathcal{X}$.

The advantage of $\widehat{\text{se}}_{\text{jack}}$ is that definition (10.6) can be applied in an automatic way to *any* statistic $\hat{\theta} = s(\boldsymbol{x})$. All that is needed is an algorithm that computes $s(\cdot)$ for the deleted data sets $\boldsymbol{x}_{(i)}$. Computer power is being substituted for theoretical Taylor series calculations. Later we will see that the underlying inferential ideas—plug-in estimation of frequentist standard errors—haven't changed, only their implementation.

As an example, consider the kidney function data set of Section 1.1. Here the data consists of $n = 157$ points (x_i, y_i), with x = age and y = tot in Figure 1.1. (So the generic x_i in (10.2) now represents the pair (x_i, y_i), and F describes a distribution in the plane.) Suppose we are interested in the correlation between age and tot, estimated by the usual sample correlation $\hat{\theta} = s(\boldsymbol{x})$,

$$s(\boldsymbol{x}) = \sum_{i=1}^{n}(x_i - \bar{x})(y_i - \bar{y}) \Bigg/ \left[\sum_{1}^{n}(x_i - \bar{x})^2 \sum_{1}^{n}(y_i - \bar{y})^2\right]^{1/2}, \tag{10.9}$$

computed to be $\hat{\theta} = -0.572$ for the kidney data.

Applying (10.6) gave $\widehat{\text{se}}_{\text{jack}} = 0.058$ for the accuracy of $\hat{\theta}$. Nonparametric bootstrap computations, Section 10.2, also gave estimated standard error 0.058. The classic Taylor series formula looks quite formidable in this case,

$$\widehat{\text{se}}_{\text{taylor}} = \left\{\frac{\hat{\theta}^2}{4n}\left[\frac{\hat{\mu}_{40}}{\hat{\mu}_{20}^2} + \frac{\hat{\mu}_{04}}{\hat{\mu}_{02}^2} + \frac{2\hat{\mu}_{22}}{\hat{\mu}_{20}\hat{\mu}_{02}} + \frac{4\hat{\mu}_{22}}{\hat{\mu}_{11}^2} - \frac{4\hat{\mu}_{31}}{\hat{\mu}_{11}\hat{\mu}_{20}} - \frac{4\hat{\mu}_{13}}{\hat{\mu}_{11}\hat{\mu}_{02}}\right]\right\}^{1/2} \tag{10.10}$$

where

$$\hat{\mu}_{hk} = \sum_{i=1}^{n}(x_i - \bar{x})^h (y_i - \bar{y})^k / n. \tag{10.11}$$

It gave $\widehat{\text{se}} = 0.057$.

It is worth emphasizing some features of the jackknife formula (10.6).

- It is nonparametric; no special form of the underlying distribution F need be assumed.
- It is completely automatic: a single master algorithm can be written that inputs the data set $\boldsymbol{x}$ and the function $s(\boldsymbol{x})$, and outputs $\widehat{\text{se}}_{\text{jack}}$.
- The algorithm works with data sets of size $n-1$, not n. There is a hidden assumption of smooth behavior across sample sizes. This can be worrisome for statistics like the sample median that have a different definition for odd and even sample size.

- The jackknife standard error is upwardly biased as an estimate of the
†1 true standard error.†
- The connection of the jackknife formula (10.6) with Taylor series methods is closer than it appears. We can write

$$\widehat{\text{se}}_{\text{jack}} = \left[\frac{\sum_1^n D_i^2}{n^2}\right]^{1/2}, \qquad \text{where } D_i = \frac{\hat{\theta}_{(i)} - \hat{\theta}_{(\cdot)}}{1/\sqrt{n(n-1)}}. \tag{10.12}$$

As discussed in Section 10.3, the D_i are approximate *directional derivatives*, measures of how fast the statistic $s(\boldsymbol{x})$ is changing as we decrease the weight on data point x_i. So $\text{se}^2_{\text{jack}}$ is proportional to the sum of squared derivatives of $s(\boldsymbol{x})$ in the n component directions. Taylor series expressions such as (10.10) amount to doing the derivatives by formula rather than numerically.

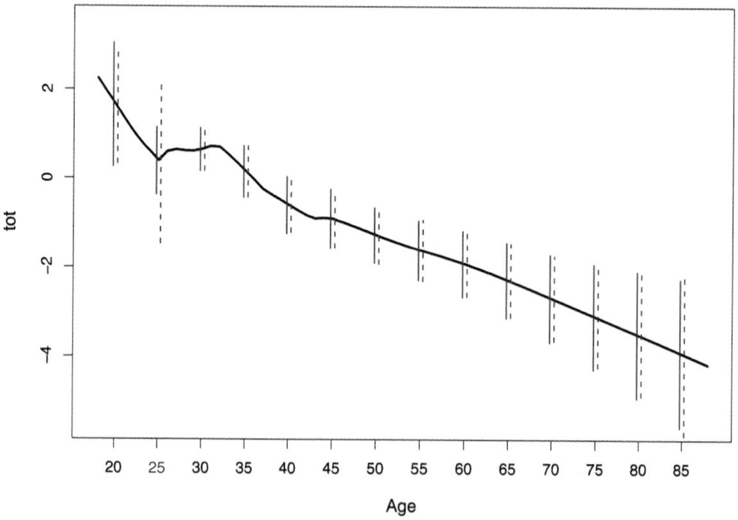

Figure 10.1 The `lowess` curve for the kidney data of Figure 1.2. Vertical bars indicate ±2 standard errors: *jackknife* (10.6) blue dashed; *bootstrap* (10.16) red solid. The jackknife greatly overestimates variability at age 25.

The principal weakness of the jackknife is its dependence on local derivatives. Unsmooth statistics $s(\boldsymbol{x})$, such as the kidney data `lowess` curve in Figure 1.2, can result in erratic behavior for $\widehat{\text{se}}_{\text{jack}}$. Figure 10.1 illustrates the point. The dashed blue vertical bars indicate ±2 jackknife standard errors for the `lowess` curve evaluated at ages $20, 25, \ldots, 85$. For the most part these agree with the dependable bootstrap standard errors, solid red bars, described in Section 10.2. But things go awry at age 25, where the local derivatives greatly overstate the sensitivity of the `lowess` curve to global changes in the sample $\boldsymbol{x}$.

10.2 The Nonparametric Bootstrap

From the point of view of the bootstrap, the jackknife was a halfway house between classical methodology and a full-throated use of electronic computation. (The term "computer-intensive statistics" was coined to describe the bootstrap.) The frequentist standard error of an estimate $\hat{\theta} = s(\boldsymbol{x})$ is, ideally, the standard deviation we would observe by repeatedly sampling new versions of $\boldsymbol{x}$ from F. This is impossible since F is unknown. Instead, the bootstrap ("ingenious device" number 4 in Section 2.1) substitutes an estimate $\hat{F}$ for F and then estimates the frequentist standard by direct simulation, a feasible tactic only since the advent of electronic computation.

The bootstrap estimate of standard error for a statistic $\hat{\theta} = s(\boldsymbol{x})$ computed from a random sample $\boldsymbol{x} = (x_1, x_2, \ldots, x_n)$ (10.2) begins with the notion of a *bootstrap sample*

$$\boldsymbol{x}^* = (x_1^*, x_2^*, \ldots, x_n^*), \tag{10.13}$$

where each x_i^* is drawn randomly with equal probability and with replacement from $\{x_1, x_2, \ldots, x_n\}$. Each bootstrap sample provides a *bootstrap replication* of the statistic of interest,[3]

$$\hat{\theta}^* = s(\boldsymbol{x}^*). \tag{10.14}$$

Some large number B of bootstrap samples are independently drawn ($B = 500$ in Figure 10.1). The corresponding bootstrap replications are calculated, say

$$\hat{\theta}^{*b} = s(\boldsymbol{x}^{*b}) \qquad \text{for } b = 1, 2, \ldots, B. \tag{10.15}$$

The resulting bootstrap estimate of standard error for $\hat{\theta}$ is the empirical standard deviation of the $\hat{\theta}^{*b}$ values,

$$\widehat{\text{se}}_{\text{boot}} = \left[\sum_{b=1}^{B} \left(\hat{\theta}^{*b} - \hat{\theta}^{*\cdot}\right)^2 \Big/ (B-1)\right]^{1/2}, \quad \text{with } \hat{\theta}^{*\cdot} = \sum_{1}^{B} \hat{\theta}^{*b} \big/ B. \tag{10.16}$$

[3] The star notation $\boldsymbol{x}^*$ is intended to avoid confusion with the original data $\boldsymbol{x}$, which stays fixed in bootstrap computations, and likewise $\hat{\theta}^*$ vis-a-vis $\hat{\theta}$.

Motivation for $\widehat{\text{se}}_{\text{boot}}$ begins by noting that $\hat{\theta}$ is obtained in two steps: first $\boldsymbol{x}$ is generated by iid sampling from probability distribution F, and then $\hat{\theta}$ is calculated from $\boldsymbol{x}$ according to algorithm $s(\cdot)$,

$$F \xrightarrow{\text{iid}} \boldsymbol{x} \xrightarrow{s} \hat{\theta}. \tag{10.17}$$

We don't know F, but we can estimate it by the *empirical probability distribution* $\hat{F}$ that puts probability $1/n$ on each point x_i (e.g., weight $1/157$ on each point (x_i, y_i) in Figure 1.2). Notice that a bootstrap sample $\boldsymbol{x}^*$ (10.13) *is an iid sample drawn from* $\hat{F}$, since then each $\boldsymbol{x}^*$ independently has equal probability of being any member of $\{x_1, x_2, \ldots, x_n\}$. It can be shown that $\hat{F}$ maximizes the probability of obtaining the observed sample $\boldsymbol{x}$ under all possible choices of F in (10.2), i.e., it is the *nonparametric MLE* of F.

Bootstrap replications $\hat{\theta}^*$ are obtained by a process analogous to (10.17),

$$\hat{F} \xrightarrow{\text{iid}} \boldsymbol{x}^* \xrightarrow{s} \hat{\theta}^*. \tag{10.18}$$

In the real world (10.17) we only get to see the single value $\hat{\theta}$, but the bootstrap world (10.18) is more generous: we can generate as many bootstrap replications $\hat{\theta}^{*b}$ as we want, or have time for, and directly estimate their variability as in (10.16). The fact that $\hat{F}$ approaches F as n grows large suggests, correctly in most cases, that $\widehat{\text{se}}_{\text{boot}}$ approaches the true standard error of $\hat{\theta}$.

The true standard deviation of $\hat{\theta}$, i.e., its standard error, can be thought of as a function of the probability distribution F that generates the data, say $\text{Sd}(F)$. Hypothetically, $\text{Sd}(F)$ inputs F and outputs the standard deviation of $\hat{\theta}$, which we can imagine being evaluated by independently running (10.17) some enormous number of times N, and then computing the empirical standard deviation of the resulting $\hat{\theta}$ values,

$$\text{Sd}(F) = \left[\sum_{j=1}^{N} \left(\hat{\theta}^{(j)} - \hat{\theta}^{(\cdot)} \right)^2 \Big/ (N-1) \right]^{1/2}, \quad \text{with } \hat{\theta}^{(\cdot)} = \sum_{1}^{N} \hat{\theta}^{(j)} / N. \tag{10.19}$$

The bootstrap standard error of $\hat{\theta}$ is the plug-in estimate

$$\widehat{\text{se}}_{\text{boot}} = \text{Sd}(\hat{F}). \tag{10.20}$$

More exactly, $\text{Sd}(\hat{F})$ is the *ideal bootstrap estimate* of standard error, what we would get by letting the number of bootstrap replications B go to infinity. In practice we have to stop at some finite value of B, as discussed in what follows.

As with the jackknife, there are several important points worth emphasizing about $\widehat{\text{se}}_{\text{boot}}$.

- It is completely automatic. Once again, a master algorithm can be written that inputs the data $\boldsymbol{x}$ and the function $s(\cdot)$, and outputs $\widehat{\text{se}}_{\text{boot}}$.
- We have described the *one-sample nonparametric bootstrap*. Parametric and multisample versions will be taken up later.
- Bootstrapping "shakes" the original data more violently than jackknifing, producing nonlocal deviations of $\boldsymbol{x}^*$ from $\boldsymbol{x}$. The bootstrap is more dependable than the jackknife for unsmooth statistics since it doesn't depend on local derivatives.
- $B = 200$ is usually sufficient† for evaluating $\widehat{\text{se}}_{\text{boot}}$. Larger values, 1000 or 2000, will be required for the bootstrap confidence intervals of Chapter 11. †2
- There is nothing special about standard errors. We could just as well use the bootstrap replications to estimate the expected absolute error $E\{|\hat{\theta} - \theta|\}$, or any other accuracy measure.
- Fisher's MLE formula (4.27) is applied in practice via

$$\widehat{\text{se}}_{\text{fisher}} = (n\mathcal{I}_{\hat{\theta}})^{-1/2}, \tag{10.21}$$

that is, by plugging in $\hat{\theta}$ for θ after a theoretical calculation of se. The bootstrap operates in the same way at (10.20), though the plugging in is done before rather than after the calculation. The connection with Fisherian theory is more obvious for the parametric bootstrap of Section 10.4.

The jackknife is a completely frequentist device, both in its assumptions and in its applications (standard errors and biases). The bootstrap is also basically frequentist, but with a touch of the Fisherian as in the relation with (10.21). Its versatility has led to applications in a variety of estimation and prediction problems, with even some Bayesian connections.† †3 Unusual applications can also pop up for the jackknife; see the jackknife-after-bootstrap comment in the chapter endnotes.† †4

From a classical point of view, the bootstrap is an incredible computational spendthrift. Classical statistics was fashioned to minimize the hard labor of mechanical computation. The bootstrap seems to go out of its way to multiply it, by factors of $B = 200$ or 2000 or more. It is nice to report that all this computational largesse can have surprising data analytic payoffs.

The 22 students of Table 3.1 actually each took five tests, `mechanics`, `vectors`, `algebra`, `analytics`, and `statistics`. Table 10.1 shows

Table 10.1 *Correlation matrix for the student score data. The eigenvalues are 3.463, 0.660, 0.447, 0.234, and 0.197. The eigenratio statistic* $\hat{\theta} = 0.693$, *and its bootstrap standard error estimate is 0.075* ($B = 2000$).

	mechanics	vectors	algebra	analytics	statistics
mechanics	1.00	.50	.76	.65	.54
vectors	.50	1.00	.59	.51	.38
algebra	.76	.59	1.00	.76	.67
analysis	.65	.51	.76	1.00	.74
statistics	.54	.38	.67	.74	1.00

the sample correlation matrix and also its eigenvalues. The "eigenratio" statistic,

$$\hat{\theta} = \text{largest eigenvalue/sum eigenvalues}, \tag{10.22}$$

measures how closely the five scores can be predicted by a single linear combination, essentially an IQ score for each student: $\hat{\theta} = 0.693$ here, indicating strong predictive power for the IQ score. How accurate is 0.693?

$B = 2000$ bootstrap replications (10.15) yielded bootstrap standard error estimate (10.16) $\widehat{\text{se}}_{\text{boot}} = 0.075$. (This was 10 times more bootstraps than necessary for $\widehat{\text{se}}_{\text{boot}}$, but will be needed for Chapter 11's bootstrap confidence interval calculations.) The jackknife (10.6) gave a bigger estimate, $\widehat{\text{se}}_{\text{jack}} = 0.083$.

Standard errors are usually used to suggest approximate confidence intervals, often $\hat{\theta} \pm 1.96\widehat{\text{se}}$ for 95% coverage. These are based on an assumption of normality for $\hat{\theta}$. The histogram of the 2000 bootstrap replications of $\hat{\theta}$, as seen in Figure 10.2, disabuses belief in even approximate normality. Compared with classical methods, a massive amount of computation has gone into the histogram, but this will pay off in Chapter 11 with more accurate confidence limits. We can claim a double reward here for bootstrap methods: much wider applicability and improved inferences. The bootstrap histogram—invisible to classical statisticians—nicely illustrates the advantages of computer-age statistical inference.

10.3 Resampling Plans

There is a second way to think about the jackknife and the bootstrap: as algorithms that reweight, or *resample*, the original data vector $\boldsymbol{x} =$

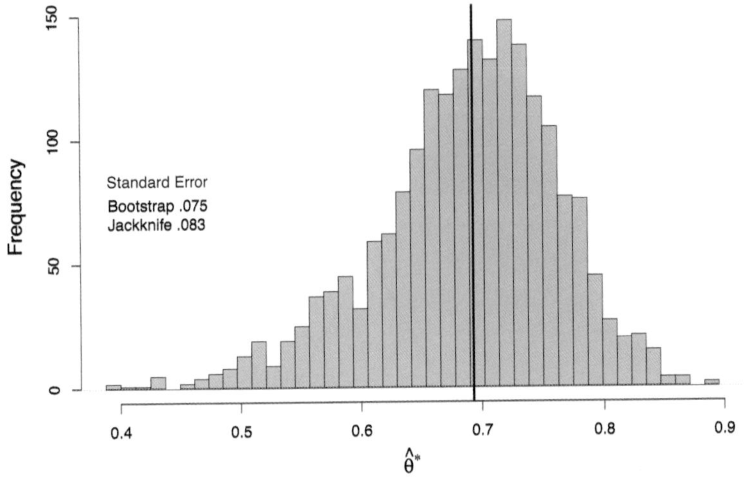

Figure 10.2 Histogram of $B = 2000$ bootstrap replications $\hat{\theta}^*$ for the eigenratio statistic (10.22) for the student score data. The vertical black line is at $\hat{\theta} = .693$. The long left tail shows that normality is a dangerous assumption in this case.

$(x_1, x_2, \ldots, x_n)'$. At the price of a little more abstraction, resampling connects the two algorithms and suggests a class of other possibilities.

A *resampling vector* $\boldsymbol{P} = (P_1, P_2, \ldots, P_n)'$ is by definition a vector of nonnegative weights summing to 1,

$$\boldsymbol{P} = (P_1, P_2, \ldots, P_n)' \qquad \text{with } P_i \geq 0 \text{ and } \sum_{i=1}^{n} P_i = 1. \qquad (10.23)$$

That is, $\boldsymbol{P}$ is a member of the simplex $\mathcal{S}_n$ (5.39). Resampling plans operate by holding the original data set $\boldsymbol{x}$ fixed, and seeing how the statistic of interest $\hat{\theta}$ changes as the weight vector $\boldsymbol{P}$ varies across $\mathcal{S}_n$.

We denote the value of $\hat{\theta}$ for a vector putting weight P_i on x_i as

$$\hat{\theta}^* = S(\boldsymbol{P}), \qquad (10.24)$$

the star notation now indicating any reweighting, not necessarily from bootstrapping; $\hat{\theta} = s(\boldsymbol{x})$ describes the behavior of $\hat{\theta}$ in the real world (10.17), while $\hat{\theta}^* = S(\boldsymbol{P})$ describes it in the resampling world. For the sample mean $s(\boldsymbol{x}) = \bar{x}$, we have $S(\boldsymbol{P}) = \sum_1^n P_i x_i$. The unbiased estimate of variance $s(\boldsymbol{x}) = \sum_i^n (x_i - \bar{x})^2/(n-1)$ can be seen to have

$$S(\boldsymbol{P}) = \frac{n}{n-1}\left[\sum_{i=1}^{n} P_i x_i^2 - \left(\sum_{i=1}^{n} P_i x_i\right)^2\right]. \tag{10.25}$$

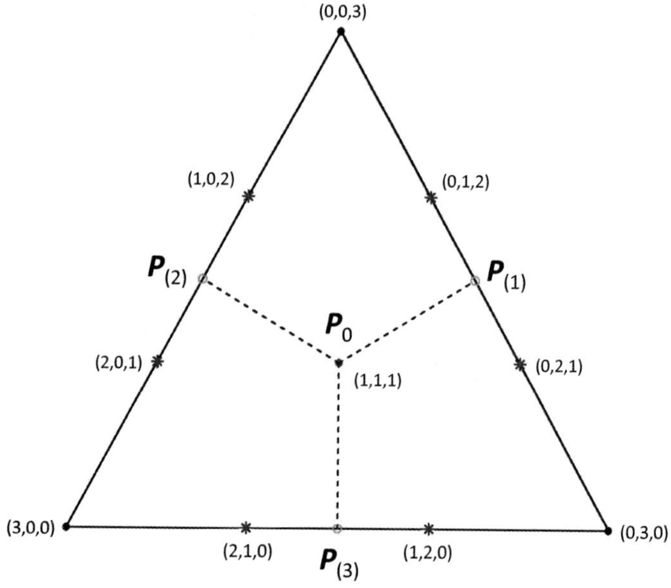

Figure 10.3 Resampling simplex for sample size $n = 3$. The center point is $\boldsymbol{P}_0$ (10.26); the green circles are the jackknife points $\boldsymbol{P}_{(i)}$ (10.28); triples indicate bootstrap resampling numbers (N_1, N_2, N_3) (10.29). The *bootstrap probabilities* are $6/27$ for $\boldsymbol{P}_0$, $1/27$ for each corner point, and $3/27$ for each of the six starred points.

Letting

$$\boldsymbol{P}_0 = (1, 1, \ldots, 1)'/n, \tag{10.26}$$

the resampling vector putting equal weight on each value x_i, we require in the definition of $S(\cdot)$ that

$$S(\boldsymbol{P}_0) = s(\boldsymbol{x}) = \hat{\theta}, \tag{10.27}$$

the original estimate. The ith jackknife value $\hat{\theta}_{(i)}$ (10.5) corresponds to resampling vector

$$P_{(i)} = (1, 1, \ldots, 1, 0, 1, \ldots, 1)'/(n-1), \tag{10.28}$$

with 0 in the ith place. Figure 10.3 illustrates the resampling simplex $\mathcal{S}_3$ applying to sample size $n = 3$, with the center point being $\boldsymbol{P}_0$ and the open circles the three possible jackknife vectors $\boldsymbol{P}_{(i)}$.

With $n = 3$ sample points $\{x_1, x_2, x_3\}$ there are only 10 distinct bootstrap vectors (10.13), also shown in Figure 10.3. Let

$$N_i = \#\{x_j^* = x_i\}, \tag{10.29}$$

the number of bootstrap draws in $\boldsymbol{x}^*$ equaling x_i. The triples in the figure are (N_1, N_2, N_3), for example $(1, 0, 2)$ for $\boldsymbol{x}^*$ having x_1 once and x_3 twice.[4] The bootstrap resampling vectors are of the form

$$\boldsymbol{P}^* = (N_1, N_2, \ldots, N_n)'/n, \tag{10.30}$$

where the N_i are nonnegative integers summing to n. According to definition (10.13) of bootstrap sampling, the vector $\boldsymbol{N} = (N_1, N_2, \ldots, N_n)'$ follows a multinomial distribution (5.38) with n draws on n equally likely categories,

$$\boldsymbol{N} \sim \text{Mult}_n(n, \boldsymbol{P}_0). \tag{10.31}$$

This gives bootstrap probability (5.37)

$$\left(\frac{n!}{N_1!N_2!\ldots N_n!}\right)\frac{1}{n^n} \tag{10.32}$$

on $\boldsymbol{P}^*$ (10.30).

Figure 10.3 is misleading in that the jackknife vectors $\boldsymbol{P}_{(i)}$ appear only slightly closer to $\boldsymbol{P}_0$ than are the bootstrap vectors $\boldsymbol{P}^*$. As n grows large they are, in fact, an order of magnitude closer. Subtracting (10.26) from (10.28) gives Euclidean distance

$$\|\boldsymbol{P}_{(i)} - \boldsymbol{P}_0\| = 1\Big/\sqrt{n(n-1)}. \tag{10.33}$$

For the bootstrap, notice that N_i in (10.29) has a binomial distribution,

$$N_i \sim \text{Bi}\left(n, \frac{1}{n}\right), \tag{10.34}$$

[4] A hidden assumption of definition (10.24) is that $\hat{\theta} = s(\boldsymbol{x})$ has the same value for any permutation of $\boldsymbol{x}$, so for instance $s(x_1, x_3, x_3) = s(x_3, x_1, x_3) = S(1/3, 0, 2/3)$.

with mean 1 and variance $(n-1)/n$. Then $P_i^* = N_i/n$ has mean and variance $(1/n, (n-1)/n^3)$. Adding over the n coordinates gives the expected root mean square distance for bootstrap vector $\boldsymbol{P}^*$,

$$\left(E\|\boldsymbol{P}^* - \boldsymbol{P}_0\|^2\right)^{1/2} = \sqrt{(n-1)/n^2}, \tag{10.35}$$

an order of magnitude $\sqrt{n}$ times further than (10.33).

The function $S(\boldsymbol{P})$ has approximate directional derivative

$$D_i = \frac{S(\boldsymbol{P}_{(i)}) - S(\boldsymbol{P}_0)}{\|\boldsymbol{P}_{(i)} - \boldsymbol{P}_0\|} \tag{10.36}$$

in the direction from $\boldsymbol{P}_0$ toward $\boldsymbol{P}_{(i)}$ (measured along the dashed lines in Figure 10.3). D_i measures the slope of function $S(\boldsymbol{P})$ at $\boldsymbol{P}_0$, in the direction of $\boldsymbol{P}_{(i)}$. Formula (10.12) shows $\widehat{\text{se}}_{\text{jack}}$ as proportional to the root mean square of the slopes.

If $S(\boldsymbol{P})$ is a *linear* function of $\boldsymbol{P}$, as it is for the sample mean, it turns out that $\widehat{\text{se}}_{\text{jack}}$ equals $\widehat{\text{se}}_{\text{boot}}$ (except for the fudge factor $(n-1)/n$ in (10.6)). Most statistics are not linear, and then the local jackknife resamples may provide a poor approximation to the full resampling behavior of $S(\boldsymbol{P})$. This was the case at one point in Figure 10.1.

With only 10 possible resampling points $\boldsymbol{P}^*$, we can easily evaluate the *ideal* bootstrap standard error estimate

$$\widehat{\text{se}}_{\text{boot}} = \left[\sum_{k=1}^{10} p_k \left(\hat{\theta}^{*k} - \hat{\theta}^{*\cdot}\right)^2\right]^{1/2}, \quad \hat{\theta}^{*\cdot} = \sum_{k=1}^{10} p_k \hat{\theta}^{*k}, \tag{10.37}$$

with $\hat{\theta}^{*k} = S(\boldsymbol{P}^k)$ and p_k the probability from (10.32) (listed in Figure 10.3). This rapidly becomes impractical. The number of distinct bootstrap samples for n points turns out to be

$$\binom{2n-1}{n}. \tag{10.38}$$

For $n = 10$ this is already 92,378, while $n = 20$ gives 6.9×10^{10} distinct possible resamples. Choosing B vectors $\boldsymbol{P}^*$ at random, which is what algorithm (10.13)–(10.15) effectively is doing, makes the un-ideal bootstrap standard error estimate (10.16) almost as accurate as (10.37) for B as small as 200 or even less.

The luxury of examining the resampling surface provides a major advantage to modern statisticians, both in inference and methodology. A variety of other resampling schemes have been proposed, a few of which follow.

The Infinitesimal Jackknife

Looking at Figure 10.3 again, the vector

$$\boldsymbol{P}_i(\epsilon) = (1-\epsilon)\boldsymbol{P}_0 + \epsilon\boldsymbol{P}_{(i)} = \boldsymbol{P}_0 + \epsilon(\boldsymbol{P}_{(i)} - \boldsymbol{P}_0) \tag{10.39}$$

lies proportion ϵ of the way from $\boldsymbol{P}_0$ to $\boldsymbol{P}_{(i)}$. Then

$$\tilde{D}_i = \lim_{\epsilon\to 0} \frac{S(\boldsymbol{P}_i(\epsilon)) - S(\boldsymbol{P}_0)}{\epsilon\|\boldsymbol{P}_{(i)} - \boldsymbol{P}_0\|} \tag{10.40}$$

exactly defines the direction derivative at $\boldsymbol{P}_0$ in the direction of $\boldsymbol{P}_{(i)}$. The infinitesimal jackknife estimate of standard error is

$$\widehat{\text{se}}_{\text{IJ}} = \left(\sum_{i=1}^{n} \tilde{D}_i^2/n^2\right)^{1/2}, \tag{10.41}$$

usually evaluated numerically by setting ϵ to some small value in (10.40)–(10.41) (rather than $\epsilon = 1$ in (10.12)). We will meet the infinitesimal jackknife again in Chapters 17 and 20.

Multisample Bootstrap

The median difference between the **AML** and the **ALL** scores in Figure 1.4 is

$$\texttt{mediff} = 0.968 - 0.733 = 0.235. \tag{10.42}$$

How accurate is 0.235? An appropriate form of bootstrapping draws 25 times with replacement from the 25 **AML** patients, 47 times with replacement from the 47 **ALL** patients, and computes $\texttt{mediff}^*$ as the difference between the medians of the two bootstrap samples. (Drawing one bootstrap sample of size 72 from all the patients would result in random sample sizes for the $\texttt{AML}^*/\texttt{ALL}^*$ groups, adding inappropriate variability to the frequentist standard error estimate.)

A histogram of $B = 500$ $\texttt{mediff}^*$ values appears in Figure 10.4. They give $\widehat{\text{se}}_{\text{boot}} = 0.074$. The estimate (10.42) is 3.18 $\widehat{\text{se}}$ units above zero, agreeing surprisingly well with the usual two-sample t-statistic 3.13 (based on *mean* differences), and its permutation histogram Figure 4.3. Permutation testing can be considered another form of resampling.

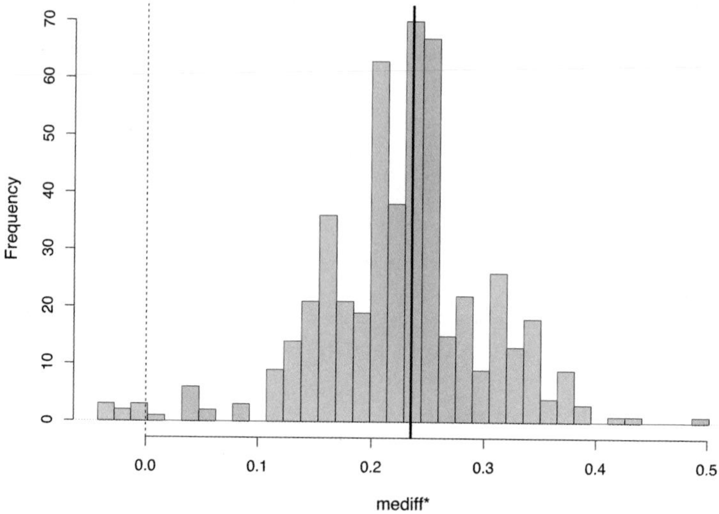

Figure 10.4 $B = 500$ bootstrap replications for the median difference between the **AML** and **ALL** scores in Figure 1.4, giving $\widehat{\text{se}}_{\text{boot}} = 0.074$. The observed value **mediff** $= 0.235$ (vertical black line) is more than 3 standard errors above zero.

Moving Blocks Bootstrap

Suppose $\boldsymbol{x} = (x_1, x_2, \ldots, x_n)$, instead of being an iid sample (10.2), is a time series. That is, the x values occur in a meaningful order, perhaps with nearby observations highly correlated with each other. Let $\mathcal{B}_m$ be the set of contiguous blocks of length m, for example

$$\mathcal{B}_3 = \{(x_1, x_2, x_3), (x_2, x_3, x_4), \ldots, (x_{n-2}, x_{n-1}, x_n)\}. \tag{10.43}$$

Presumably, m is chosen large enough that correlations between x_i and x_j, $|j - i| > m$, are neglible. The moving block bootstrap first selects n/m blocks from $\mathcal{B}_m$, and assembles them in random order to construct a bootstrap sample $\boldsymbol{x}^*$. Having constructed B such samples, $\widehat{\text{se}}_{\text{boot}}$ is calculated as in (10.15)–(10.16).

The Bayesian Bootstrap

Let $G_1, G_2, \ldots, G_n$ be independent one-sided exponential variates (denoted Gam(1,1) in Table 5.1), each having density $\exp(-x)$ for $x > 0$.

The Bayesian bootstrap uses resampling vectors

$$\boldsymbol{P}^* = (G_1, G_2, \ldots, G_n) \Big/ \sum_1^n G_i. \tag{10.44}$$

It can be shown that $\boldsymbol{P}^*$ is then uniformly distributed over the resampling simplex $\mathcal{S}_n$; for $n = 3$, uniformly distributed over the triangle in Figure 10.3. Prescription (10.44) is motivated by assuming a Jeffreys-style uninformative prior distribution (Section 3.2) on the unknown distribution F (10.2).

Distribution (10.44) for $\boldsymbol{P}^*$ has mean vector and covariance matrix

$$\boldsymbol{P}^* \sim \left[\boldsymbol{P}_0, \frac{1}{n+1}\left(\text{diag}(\boldsymbol{P}_0) - \boldsymbol{P}_0\boldsymbol{P}_0'\right)\right]. \tag{10.45}$$

This is almost identical to the mean and covariance of bootstrap resamples $\boldsymbol{P}^* \sim \text{Mult}_n(n, \boldsymbol{P}_0)/n$,

$$\boldsymbol{P}^* \sim \left[\boldsymbol{P}_0, \frac{1}{n}\left(\text{diag}(\boldsymbol{P}_0) - \boldsymbol{P}_0\boldsymbol{P}_0'\right)\right], \tag{10.46}$$

(5.40). The Bayesian bootstrap and the ordinary bootstrap tend to agree, at least for smoothly defined statistics $\hat{\theta}^* = S(\boldsymbol{P}^*)$.

There was some Bayesian disparagement of the bootstrap when it first appeared because of its blatantly frequentist take on estimation accuracy. And yet connections like (10.45)–(10.46) have continued to pop up, as we will see in Chapter 13.

10.4 The Parametric Bootstrap

In our description (10.18) of bootstrap resampling,

$$\hat{F} \xrightarrow{\text{iid}} \boldsymbol{x}^* \longrightarrow \hat{\theta}^*, \tag{10.47}$$

there is no need to insist that $\hat{F}$ be the nonparametric MLE of F. Suppose we are willing to assume that the observed data vector $\boldsymbol{x}$ comes from a *parametric family* $\mathcal{F}$ as in (5.1),

$$\mathcal{F} = \left\{f_\mu(\boldsymbol{x}),\ \mu \in \Omega\right\}. \tag{10.48}$$

Let $\hat{\mu}$ be the MLE of μ. The *bootstrap parametric* resamples from $f_{\hat{\mu}}(\cdot)$,

$$f_{\hat{\mu}} \longrightarrow \boldsymbol{x}^* \longrightarrow \hat{\theta}^*, \tag{10.49}$$

and proceeds as in (10.14)–(10.16) to calculate $\widehat{\text{se}}_{\text{boot}}$.

As an example, suppose that $\boldsymbol{x} = (x_1, x_2, \ldots, x_n)$ is an iid sample of size n from a normal distribution,

$$x_i \overset{\text{iid}}{\sim} \mathcal{N}(\mu, 1), \qquad i = 1, 2, \ldots, n. \tag{10.50}$$

Then $\hat{\mu} = \bar{x}$, and a parametric bootstrap sample is $\boldsymbol{x}^* = (x_1^*, x_2^*, \ldots, x_n^*)$, where

$$x_i^* \overset{\text{iid}}{\sim} \mathcal{N}(\bar{x}, 1), \qquad i = 1, 2, \ldots, n. \tag{10.51}$$

More adventurously, if $\mathcal{F}$ were a family of time series models for $\boldsymbol{x}$, algorithm (10.49) would still apply (now without any iid structure): $\boldsymbol{x}^*$ would be a time series sampled from model $f_{\hat{\mu}}(\cdot)$, and $\hat{\theta}^* = s(\boldsymbol{x}^*)$ the resampled statistic of interest. B independent realizations $\boldsymbol{x}^{*b}$ would give $\hat{\theta}^{*b}$, $b = 1, 2, \ldots, B$, and $\widehat{\text{se}}_{\text{boot}}$ from (10.16).

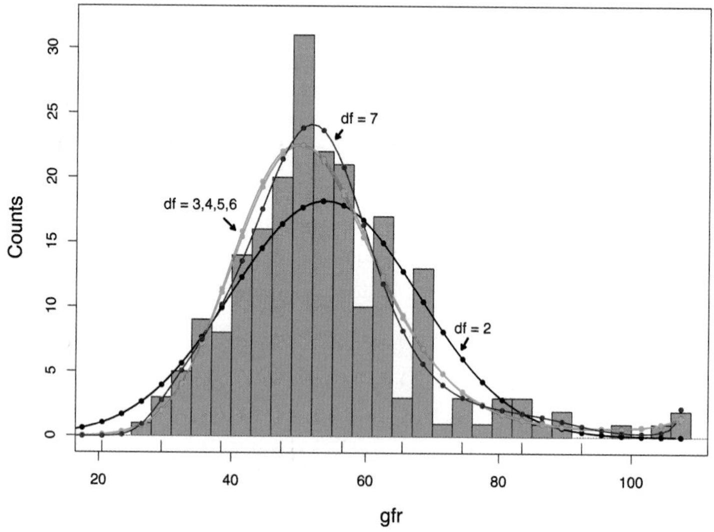

Figure 10.5 The **gfr** data of Figure 5.7 (histogram). Curves show the MLE fits from polynomial Poisson models, for degrees of freedom df $= 2, 3, \ldots, 7$. The points on the curves show the fits computed at the centers $x_{(j)}$ of the bins, with the responses being the counts in the bins. The dashes at the base of the plot show the nine **gfr** values appearing in Table 10.2.

As an example of parametric bootstrapping, Figure 10.5 expands the **gfr** investigation of Figure 5.7. In addition to the seventh-degree polynomial fit (5.62), we now show lower-degree polynomial fits for 2, 3, 4, 5, and 6 degrees of freedom; df = 2 obviously gives a poor fit; df = 3, 4, 5, 6 give nearly identical curves; df = 7 gives only a slightly better fit to the raw data.

The plotted curves were obtained from the Poisson regression method used in Section 8.3.[5]

- The x-axis was partitioned into $K = 32$ bins, with endpoints 13, 16, 19, ..., 109, and centerpoints, say,

$$\boldsymbol{x}_{(\,)} = (x_{(1)}, x_{(2)}, \ldots, x_{(K)}), \tag{10.52}$$

  $x_{(1)} = 14.5$, $x_{(2)} = 17.5$, etc.
- Count vector $\boldsymbol{y} = (y_1, y_2, \ldots, y_K)$ was computed

$$y_k = \#\{x_i \text{ in bin}_k\} \tag{10.53}$$

  (so $\boldsymbol{y}$ gives the heights of the bars in Figure 10.5).
- An independent Poisson model was assumed for the counts,

$$y_k \overset{\text{ind}}{\sim} \text{Poi}(\mu_k) \qquad \text{for } k = 1, 2, \ldots, K. \tag{10.54}$$

- The parametric model of degree "df" assumed that the μ_k values were described by an exponential polynomial of degree df in the $x_{(k)}$ values,

$$\log(\mu_k) = \sum_{j=0}^{\text{df}} \beta_j x_{(k)}^j. \tag{10.55}$$

- The MLE $\hat{\boldsymbol{\beta}} = (\hat{\beta}_0, \hat{\beta}_1, \ldots, \hat{\beta}_{\text{df}})$ in model (10.54)–(10.55) was found.[6]
- The plotted curves in Figure 10.5 trace the MLE values $\hat{\mu}_k$,

$$\log(\hat{\mu}_k) = \sum_{j=0}^{\text{df}} \hat{\beta}_j x_{(k)}^j. \tag{10.56}$$

How accurate are the curves? Parametric bootstraps were used to assess their standard errors. That is, Poisson resamples were generated according to

$$y_k^* \overset{\text{ind}}{\sim} \text{Poi}(\hat{\mu}_k) \qquad \text{for } k = 1, 2, \ldots, K, \tag{10.57}$$

and bootstrap MLE values $\hat{\mu}_k^*$ calculated as above, but now based on count vector $\boldsymbol{y}^*$ rather than $\boldsymbol{y}$. All of this was done $B = 200$ times, yielding bootstrap standard errors (10.16).

[5] "Lindsey's method," discussed further in Chapter 15.

[6] A single R command, `glm(y~poly(x,df),family=poisson)` accomplishes this.

Table 10.2 *Bootstrap estimates of standard error for the* `gfr` *density. Poisson regression models* (10.54)–(10.55), df $= 2, 3, \ldots, 7$*, as in Figure 10.5; each* $B = 200$ *bootstrap replications; nonparametric standard errors based on binomial bin counts.*

	Degrees of freedom						Nonparametric standard error
`gfr`	2	3	4	5	6	7	
20.5	.28	.07	.13	.13	.12	.05	.00
29.5	.65	.57	.57	.66	.74	1.11	1.72
38.5	1.05	1.39	1.33	1.52	1.72	1.73	2.77
47.5	1.47	1.91	2.12	1.93	2.15	2.39	4.25
56.5	1.57	1.60	1.79	1.93	1.87	2.28	4.35
65.5	1.15	1.10	1.07	1.31	1.34	1.27	1.72
74.5	.76	.61	.62	.68	.81	.71	1.72
83.5	.40	.30	.40	.38	.49	.68	1.72
92.5	.13	.20	.29	.29	.34	.46	.00

The results appear in Table 10.2, showing $\widehat{\text{se}}_{\text{boot}}$ for df $= 2, 3, \ldots, 7$ degrees of freedom evaluated at nine values of `gfr`. Variability generally increases with increasing df, as expected. Choosing a "best" model is a compromise between standard error and possible definitional bias as suggested by Figure 10.5, with perhaps df $= 3$ or 4, the winner.

If we kept increasing the degrees of freedom, eventually (at df $= 32$) we would exactly match the bar heights y_k in the histogram. At this point the parametric bootstrap would merge into the nonparametric bootstrap. "Nonparametric" is another name for "very highly parameterized." The huge sample sizes associated with modern applications have encouraged nonparametric methods, on the sometimes mistaken ground that estimation efficiency is no longer of concern. It is costly here, as the "nonparametric" column of Table 10.2 shows.[7]

Figure 10.6 returns to the student score eigenratio calculations of Figure 10.2. The solid histogram shows 2000 parametric bootstrap replications (10.49), with $f_{\hat{\mu}}$ the five-dimensional bivariate normal distribution $\mathcal{N}_5(\bar{x}, \hat{\Sigma})$. Here $\bar{x}$ and $\hat{\Sigma}$ are the usual MLE estimates for the expectation vector and covariance matrix based on the 22 five-component student score vectors. It is narrower than the corresponding nonparametric bootstrap histogram, with $\widehat{\text{se}}_{\text{boot}} = 0.070$ compared with the nonparametric estimate

[7] These are the binomial standard errors $[y_k(1 - y_k)/n]^{1/2}$, $n = 211$. The nonparametric results look much more competitive when estimating cdf's rather than densities.

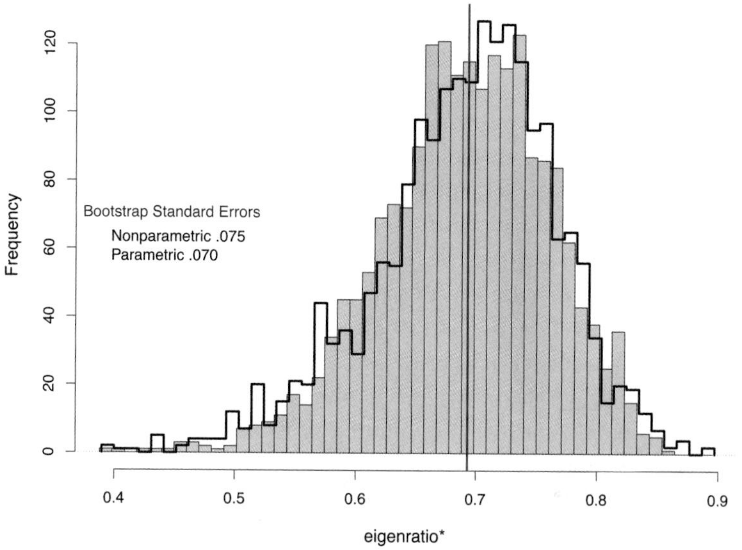

Figure 10.6 Eigenratio example, student score data. *Solid histogram* $B = 2000$ parametric bootstrap replications $\hat{\theta}^*$ from the five-dimensional normal MLE; *line histogram* the 2000 nonparametric replications of Figure 10.2. MLE $\hat{\theta} = .693$ is vertical red line.

0.075. (Note the different histogram bin limits from Figure 10.2, changing the details of the nonparametric histogram.)

Parametric families act as *regularizers*, smoothing out the raw data and de-emphasizing outliers. In fact the student score data is not a good candidate for normal modeling, having at least one notable outlier,[8] casting doubt on the smaller estimate of standard error.

The classical statistician could only imagine a mathematical device that given any statistic $\hat{\theta} = s(\boldsymbol{x})$ would produce a formula for its standard error, as formula (1.2) does for $\bar{x}$. The electronic computer *is* such a device. As harnessed by the bootstrap, it automatically produces a numerical estimate of standard error (though not a formula), with no further cleverness required. Chapter 11 discusses a more ambitious substitution of computer power for mathematical analysis: the bootstrap computation of confidence intervals.

[8] As revealed by examining scatterplots of the five variates taken two at a time. Fast and painless plotting is another advantage for twenty-first-century data analysts.

10.5 Influence Functions and Robust Estimation

The sample mean played a dominant role in classical statistics for reasons heavily weighted toward mathematical tractibility. Beginning in the 1960s, an important counter-movement, *robust estimation*, aimed to improve upon the statistical properties of the mean. A central element of that theory, the *influence function*, is closely related to the jackknife and infinitesimal jackknife estimates of standard error.

We will only consider the case where $\mathcal{X}$, the sample space, is an interval of the real line. The unknown probability distribution F yielding the iid sample $\boldsymbol{x} = (x_1, x_2, \ldots, x_n)$ in (10.2) is now the cdf of a density function $f(x)$ on $\mathcal{X}$. A parameter of interest, i.e., a function of F, is to be estimated by the plug-in principle, $\hat{\theta} = T(\hat{F})$, where, as in Section 10.2, $\hat{F}$ is the empirical probability distribution putting probability $1/n$ on each sample point x_i. For the mean,

$$\theta = T(F) = \int_{\mathcal{X}} x f(x)\, dx \quad \text{and} \quad \hat{\theta} = T\left(\hat{F}\right) = \frac{1}{n}\sum_{i=1}^{n} x_i. \tag{10.58}$$

(In Riemann–Stieltjes notation, $\theta = \int x dF(x)$ and $\hat{\theta} = \int x d\hat{F}(x)$.)

The influence function of $T(F)$, evaluated at point x in $\mathcal{X}$, is defined to be

$$\text{IF}(x) = \lim_{\epsilon \to 0} \frac{T\left((1-\epsilon)F + \epsilon\delta_x\right) - T(F)}{\epsilon}, \tag{10.59}$$

where δ_x is the "one-point probability distribution" putting probability 1 on x. In words, IF(x) measures the differential effect of modifying F by putting additional probability on x. For the mean $\theta = \int x f(x) dx$ we calculate that

$$\text{IF}(x) = x - \theta. \tag{10.60}$$

†5 A fundamental theorem† says that $\hat{\theta} = T(\hat{F})$ is approximately

$$\hat{\theta} \doteq \theta + \frac{1}{n}\sum_{i=1}^{n} \text{IF}(x_i), \tag{10.61}$$

with the approximation becoming exact as n goes to infinity. This implies that $\hat{\theta} - \theta$ is, approximately, the mean of the n iid variates IF(x_i), and that the variance of $\hat{\theta}$ is approximately

$$\text{var}\left\{\hat{\theta}\right\} \doteq \frac{1}{n}\,\text{var}\left\{\text{IF}(x)\right\}, \tag{10.62}$$

var{IF(x)} being the variance of IF(x) for any one draw of x from F. For the sample mean, using (10.60) in (10.62) gives the familiar equality

$$\text{var}\{\bar{x}\} = \frac{1}{n}\,\text{var}\{x\}. \tag{10.63}$$

The sample mean suffers from an *unbounded* influence function (10.60), which grows ever larger as x moves farther from θ. This makes $\bar{x}$ unstable against heavy-tailed densities such as the Cauchy (4.39). Robust estimation theory seeks estimators $\hat{\theta}$ of bounded influence, that do well against heavy-tailed densities without giving up too much efficiency against light-tailed densities such as the normal. Of particular interest have been the trimmed mean and its close cousin the winsorized mean.

Let $x^{(\alpha)}$ denote the 100αth percentile of distribution F, satisfying $F(x^{(\alpha)}) = \alpha$ or equivalently

$$\alpha = \int_{-\infty}^{x^{(\alpha)}} f(x)\,dx. \tag{10.64}$$

The *αth trimmed mean of F*, $\theta_{\text{trim}}(\alpha)$, is defined as

$$\theta_{\text{trim}}(\alpha) = \frac{1}{1-2\alpha}\int_{x^{(\alpha)}}^{x^{(1-\alpha)}} x f(x)\,dx, \tag{10.65}$$

the mean of the central $1-2\alpha$ portion of F, trimming off the lower and upper α portions. This is not the same as the *αth winsorized mean* $\theta_{\text{wins}}(\alpha)$,

$$\theta_{\text{wins}}(\alpha) = \int_{\mathcal{X}} W(x) f(x)\,dx, \tag{10.66}$$

where

$$W(x) = \begin{cases} x^{(\alpha)} & \text{if } x \le x^{(\alpha)} \\ x & \text{if } x^{(\alpha)} \le x \le x^{(1-\alpha)} \\ x^{(1-\alpha)} & \text{if } x \ge x^{(1-\alpha)}; \end{cases} \tag{10.67}$$

$\theta_{\text{trim}}(\alpha)$ removes the outer portions of F, while $\theta_{\text{wins}}(\alpha)$ moves them into $x^{(\alpha)}$ or $x^{(1-\alpha)}$. In practice, empirical versions $\hat{\theta}_{\text{trim}}(\alpha)$ and $\hat{\theta}_{\text{wins}}(\alpha)$ are used, substituting the empirical density $\hat{f}$, with probability $1/n$ at each x_i, for f.

There turns out to be an interesting relationship between the two: the influence function of $\theta_{\text{trim}}(\alpha)$ is a function of $\theta_{\text{wins}}(\alpha)$,

$$\text{IF}_{\alpha}(x) = \frac{W(x) - \theta_{\text{wins}}(\alpha)}{1-2\alpha}. \tag{10.68}$$

This is pictured in Figure 10.7, where we have plotted empirical influence

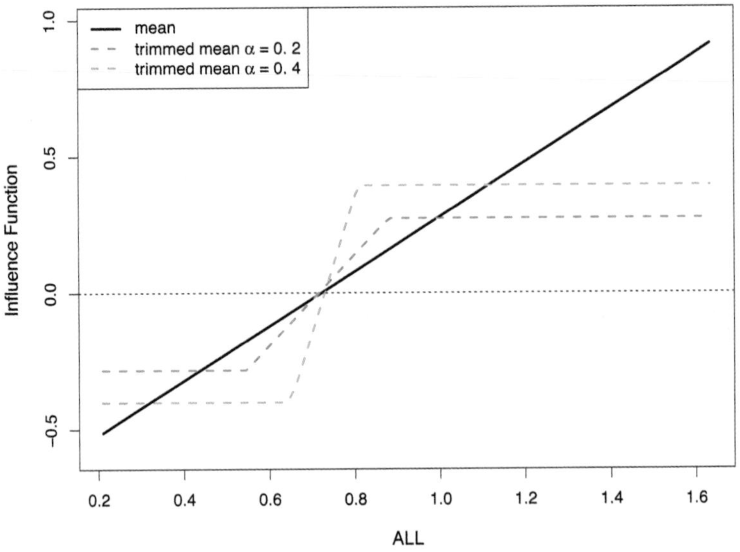

Figure 10.7 Empirical influence functions for the 47 leukemia **ALL** scores of Figure 1.4. The two dashed curves are $\text{IF}_\alpha(x)$ for the trimmed means (10.68), for $\alpha = 0.2$ and $\alpha = 0.4$. The solid curve is $\text{IF}(x)$ for the sample mean $\bar{x}$ (10.60).

functions (plugging in $\hat{F}$ for F in definition (10.59)) relating to the 47 leukemia **ALL** scores of Figure 1.4: $\text{IF}_{0.2}(x)$ and $\text{IF}_{0.4}(x)$ are plotted, along with $\text{IF}_0(x)$ (10.60), that is, for the mean.

Table 10.3 *Trimmed means and their bootstrap standard deviations for the 47 leukemia* **ALL** *scores of Figure 1.4;* $B = 1000$ *bootstrap replications for each trim value. The last column gives empirical influence function estimates of the standard error, which are also the infinitesimal jackknife estimates* (10.41). *These fail for the median.*

	Trim	Trimmed mean	Bootstrap sd	(IFse)
Mean	.0	.752	.040	(.040)
	.1	.729	.038	(.034)
	.2	.720	.035	(.034)
	.3	.725	.044	(.044)
	.4	.734	.047	(.054)
Median	.5	.733	.053	

The upper panel of Figure 1.4 shows a moderately heavy right tail for the **ALL** distribution. Would it be more efficient to estimate the center of the distribution with a trimmed mean rather than $\bar{x}$? The bootstrap provides an answer: $\widehat{\text{se}}_{\text{boot}}$ (10.16) was calculated for $\bar{x}$ and $\hat{\theta}_{\text{trim}}(\alpha)$, $\alpha = 0.1, 0.2, 0.3, 0.4$, and 0.5, the last being the sample median. It appears that $\hat{\theta}_{\text{trim}}(0.2)$ is moderately better than $\bar{x}$. This brings up an important question discussed in Chapter 20: if we use something like Table 10.3 to select an estimator, how does the selection process affect the accuracy of the resulting estimate?

We might also use the square root of formula (10.62) to estimate the standard errors of the various estimators, plugging in the empirical influence function for IF(x). This turns out to be the same as using the infinitesimal jackknife (10.41). These appear in the last column of Table 10.3. Predictably, this approach fails for the sample median, whose influence function is a square wave, sharply discontinuous at the median θ,

$$\text{IF}(x) = \pm 1 \big/ \left(2 f(\theta)\right). \tag{10.69}$$

Robust estimation offers a nice illustration of statistical progress in the computer age. Trimmed means go far back into the classical era. Influence functions are an insightful inferential tool for understanding the tradeoffs in trimmed mean estimation. And finally the bootstrap allows easy assessment of the accuracy of robust estimation, including some more elaborate ones not discussed here.

10.6 Notes and Details

Quenouille (1956) introduced what is now called the jackknife estimate of bias. Tukey (1958) realized that Quenouille-type calculations could be repurposed for nonparametric standard-error estimation, inventing formula (10.6) and naming it "the jackknife," as a rough and ready tool. Miller's important 1964 paper, "A trustworthy jackknife," asked when formula (10.6) could be trusted. (Not for the median.)

The bootstrap (Efron, 1979) began as an attempt to better understand the jackknife's successes and failures. Its name celebrates Baron Munchausen's success in pulling himself up by his own bootstraps from the bottom of a lake. Burgeoning computer power soon overcame the bootstrap's main drawback, prodigous amounts of calculation, propelling it into general use. Meanwhile, 1000+ theoretical papers were published asking when the bootstrap itself could be trusted. (Most but not all of the time in common practice).

A main reference for the chapter is Efron's 1982 monograph *The Jackknife, the Bootstrap and Other Resampling Plans*. Its Chapter 6 shows the equality of three nonparametric standard error estimates: Jaeckel's (1972) infinitesimal jackknife (10.41); the empirical influence function estimate, based on (10.62); and what is known as the nonparametric delta method.

Bootstrap Packages

Various bootstrap packages in **R** are available on the CRAN contributed-packages web site, **bootstrap** being an ambitious one. Algorithm 10.1 shows a simple **R** program for nonparametric bootstrapping. Aside from bookkeeping, it's only a few lines long.

Algorithm 10.1 *An R program for the nonparametric bootstrap.*

```
Boot <- function (x, B, func, ...){
 # x is data vector or matrix (with each row a case)
 # B is number of bootstrap replications
 # func is R function that inputs a data vector or
 # matrix and returns a numeric number or vector
 # ... other arguments for func
    x <- as.matrix(x)
    n <- nrow(x)
    f0=func(x,...) # get size of output
    fmat <- matrix(0,length(f0),B)
    for (b in 1:B) {
         i=sample(1:n, n, replace = TRUE)
         fmat[,b] <- func(x[i, ],...)
     }
    drop(fmat)
}
```

†1 [p. 158] *The jackknife standard error.* The 1982 monograph also contains Efron and Stein's (1981) result on the bias of the jackknife variance estimate, the square of formula (10.6): modulo certain sample size considerations, the expectation of the jackknife variance estimate is biased upward for the true variance.

For the sample mean $\bar{x}$, the jackknife yields exactly the usual variance estimate (1.2), $\sum_i (x_i - \bar{x})^2/(n(n-1))$, while the ideal bootstrap estimate $(B \to \infty)$ gives

$$\sum_{i=1}^{n} (x_i - \bar{x})^2/n^2. \tag{10.70}$$

As with the jackknife, we could append a fudge factor to get perfect agreement with (1.2), but there is no real gain in doing so.

†2 [p. 161] *Bootstrap sample sizes.* Let $\widehat{\text{se}}_B$ indicate the bootstrap standard error estimate (10.16) based on B replications, and $\widehat{\text{se}}_\infty$ the "ideal bootstrap," $B \to \infty$. In any actual application, there are diminishing returns from increasing B past a certain point, because $\widehat{\text{se}}_\infty$ is itself a statistic whose value varies with the observed sample $\boldsymbol{x}$ (as in (10.70)), leaving an irreducible remainder of randomness in any standard error estimate. Section 6.4 of Efron and Tibshirani (1993) shows that $B = 200$ will almost always be plenty (for standard errors, but not for bootstrap confidence intervals, Chapter 11). Smaller numbers, 25 or even less, can still be quite useful in complicated situations where resampling is expensive. An early complaint, "Bootstrap estimates are random," is less often heard in an era of frequent and massive simulations.

†3 [p. 161] *The Bayesian bootstrap.* Rubin (1981) suggested the Bayesian bootstrap (10.44). Section 10.6 of Efron (1982) used (10.45)–(10.46) as an objective Bayes justification for what we will call the percentile-method bootstrap confidence intervals in Chapter 12.

†4 [p. 161] *Jackknife-after-bootstrap.* For the eigenratio example displayed in Figure 10.2, $B = 2000$ nonparametric bootstrap replications gave $\widehat{\text{se}}_{\text{boot}} = 0.075$. How accurate is this value? Bootstrapping the bootstrap seems like too much work, perhaps 200 times 2000 resamples. It turns out, though, that we can use the jackknife to estimate the variability of $\widehat{\text{se}}_{\text{boot}}$ based on just the original 2000 replications.

Now the deleted sample estimate in (10.6) is $\widehat{\text{se}}_{\text{boot}(i)}$. The key idea is to consider those bootstrap samples $\boldsymbol{x}^*$ (10.13), among the original 2000, that *do not include the point* x_i. About 37% of the original B samples will be in this subset. Section 19.4 of Efron and Tibshirani (1993) shows that applying definition (10.16) to this subset gives $\widehat{\text{se}}_{\text{boot}(i)}$. For the estimate of Figure 10.2, the jackknife-after-bootstrap calculations gave $\widehat{\text{se}}_{\text{jack}} = 0.022$ for $\widehat{\text{se}}_{\text{boot}} = 0.075$. In other words, 0.075 isn't very accurate, which is to be expected for the standard error of a complicated statistic estimated from only $n = 22$ observations. An infinitesimal jackknife version of this technique will play a major role in Chapter 20.

†5 [p. 174] *A fundamental theorem.* Tukey can justly be considered the founding father of robust statistics, his 1960 paper being especially influential. Huber's celebrated 1964 paper brought the subject into the realm of high-concept mathematical statistics. *Robust Statistics: The Approach Based on Influence Functions*, the 1986 book by Hampel *et al.*, conveys the breadth of a subject only lightly scratched in our Section 10.5. Hampel (1974)

introduced the influence function as a statistical tool. Boos and Serfling (1980) verified expression (10.62). Qualitative notions of robustness, more than specific theoretical results, have had a continuing influence on modern data analysis.

11

Bootstrap Confidence Intervals

The jackknife and the bootstrap represent a different use of modern computer power: rather than extending classical methodology—from ordinary least squares to generalized linear models, for example—they extend the reach of classical inference.

Chapter 10 focused on standard errors. Here we will take up a more ambitious inferential goal, the bootstrap automation of confidence intervals. The familiar *standard intervals*

$$\hat{\theta} \pm 1.96\,\widehat{\text{se}}, \tag{11.1}$$

for approximate 95% coverage, are immensely useful in practice but often not very accurate. If we observe $\hat{\theta} = 10$ from a Poisson model $\hat{\theta} \sim \text{Poi}(\theta)$, the standard 95% interval (3.8, 16.2) (using $\widehat{\text{se}} = \hat{\theta}^{1/2}$) is a mediocre approximation to the exact interval[1]

$$(5.1, 17.8). \tag{11.2}$$

Standard intervals (11.1) are symmetric around $\hat{\theta}$, this being their main weakness. Poisson distributions grow more variable as θ increases, which is why interval (11.2) extends farther to the right of $\hat{\theta} = 10$ than to the left. Correctly capturing such effects in an automatic way is the goal of bootstrap confidence interval theory.

11.1 Neyman's Construction for One-Parameter Problems

The student score data of Table 3.1 comprised $n = 22$ pairs,

$$x_i = (m_i, v_i), \qquad i = 1, 2, \ldots, 22, \tag{11.3}$$

[1] Using the Neyman construction of Section 11.1, as explained there; see also Table 11.2 in Section 11.4.

where m_i and v_i were student i's scores on the "mechanics" and "vectors" tests. The sample correlation coefficient $\hat{\theta}$ between m_i and v_i was computed to be

$$\hat{\theta} = 0.498. \tag{11.4}$$

Question: What can we infer about the true correlation θ between m and v? Figure 3.2 displayed three possible Bayesian answers. Confidence intervals provide the frequentist solution, by far the most popular in applied practice.

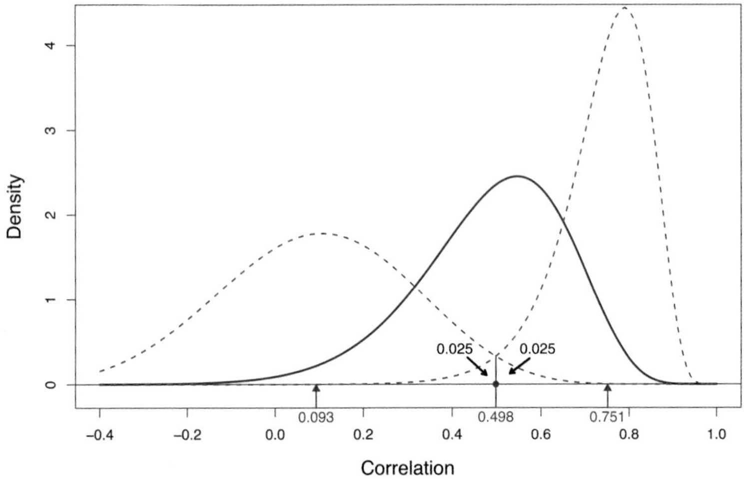

Figure 11.1 The solid curve is the normal correlation coefficient density $f_{\hat{\theta}}(r)$ (3.11) for $\hat{\theta} = 0.498$, the MLE estimate for the student score data; $\hat{\theta}(\text{lo}) = 0.093$ and $\hat{\theta}(\text{up}) = 0.751$ are the endpoints of the 95% confidence interval for θ, with corresponding densities shown by dashed curves. These yield tail areas 0.025 at $\hat{\theta}$ (11.6).

Suppose, first, that we assume a bivariate normal model (5.12) for the pairs (m_i, v_i). In that case the probability density $f_\theta(\hat{\theta})$ for sample correlation $\hat{\theta}$ given true correlation θ has known form (3.11). The solid curve in Figure 11.1 graphs f for $\theta = 0.498$, that is, for θ set equal to the observed value $\hat{\theta}$. In more careful notation, the curve graphs $f_{\hat{\theta}}(r)$ as a function of the dummy variable[2] r taking values in $[-1, 1]$.

[2] This is an example of a parametric bootstrap distribution (10.49), here with $\hat{\mu}$ being $\hat{\theta}$.

Two other curves $f_\theta(r)$ appear in Figure 11.1: for θ equaling

$$\hat\theta(\text{lo}) = 0.093 \quad \text{and} \quad \hat\theta(\text{up}) = 0.751. \tag{11.5}$$

These were numerically calculated as the solutions to

$$\int_{\hat\theta}^{1} f_{\hat\theta(\text{lo})}(r)\,dr = 0.025 \quad \text{and} \quad \int_{-1}^{\hat\theta} f_{\hat\theta(\text{up})}(r)\,dr = 0.025. \tag{11.6}$$

In words, $\hat\theta(\text{lo})$ is the smallest value of θ putting probability at least 0.025 above $\hat\theta = 0.498$, while $\hat\theta(\text{up})$ is the largest value with probability at least 0.025 below $\hat\theta$;

$$\theta \in \left[\hat\theta(\text{lo}), \hat\theta(\text{up})\right] \tag{11.7}$$

is a 95% confidence interval for the true correlation, statement (11.7) holding true with probability 0.95, for every possible value of θ.

We have just described *Neyman's construction* of confidence intervals for one-parameter problems $f_\theta(\hat\theta)$. (Later we will consider the more difficult situation where there are "nuisance parameters" in addition to the parameter of interest θ.) One of the jewels of classical frequentist inference, it depends on a *pivotal argument*—"ingenious device" number 5 of Section 2.1—to show that it produces genuine confidence intervals, i.e., ones that contain the true parameter value θ at the claimed probability level, 0.95 in Figure 11.1. The argument appears in the chapter endnotes.† †1

For the Poisson calculation (11.2) it was necessary to define exactly what "the smallest value of θ putting probability at least 0.025 above $\hat\theta$" meant. This was done assuming that, for any θ, half of the probability $f_\theta(\hat\theta)$ at $\hat\theta = 10$ counted as "above," and similarly for calculating the upper limit.

Transformation Invariance

Confidence intervals enjoy the important and useful property of transformation invariance. In the Poisson example (11.2), suppose our interest shifts from parameter θ to parameter $\phi = \log\theta$. The 95% exact interval (11.2) for θ then transforms to the exact 95% interval for ϕ simply by taking logs of the endpoints,

$$(\log(5.1), \log(17.8)) = (1.63, 2.88). \tag{11.8}$$

To state things generally, suppose we observe $\hat\theta$ from a family of densities $f_\theta(\hat\theta)$ and construct a confidence interval $\mathcal{C}(\hat\theta)$ for θ of coverage level

α ($\alpha = 0.95$ in our examples). Now let parameter ϕ be a monotonic increasing function of θ, say

$$\phi = m(\theta) \tag{11.9}$$

($m(\theta) = \log\theta$ in (11.8)), and likewise $\hat{\phi} = m(\hat{\theta})$ for the point estimate. Then $\mathcal{C}(\hat{\theta})$ maps point by point into $\mathcal{C}^\phi(\hat{\phi})$, a level-$\alpha$ confidence interval for ϕ,

$$\mathcal{C}^\phi(\hat{\phi}) = \left\{\phi = m(\theta) \text{ for } \theta \in \mathcal{C}\left(\hat{\theta}\right)\right\}. \tag{11.10}$$

This just says that the event $\{\theta \in \mathcal{C}(\hat{\theta})\}$ is the same as the event $\{\phi \in \mathcal{C}^\phi(\hat{\phi})\}$, so if the former always occurs with probability α then so must the latter.

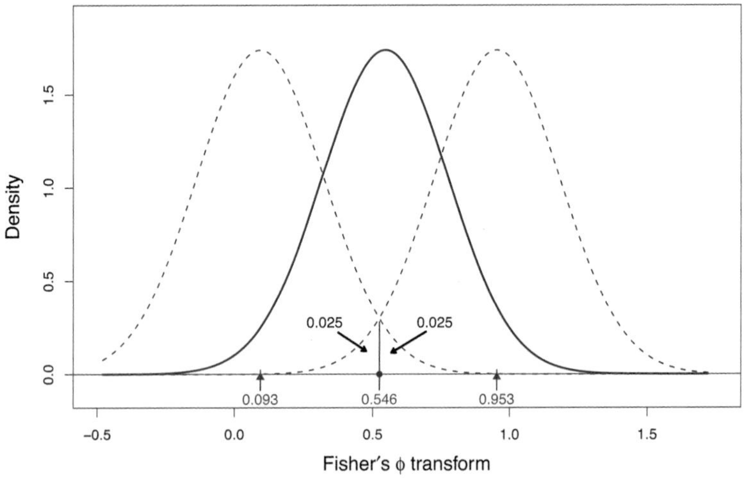

Figure 11.2 The situation in Figure 11.1 after transformation to $\phi = m(\theta)$ according to (11.11). The curves are nearly $N(\phi, \sigma^2)$ with standard deviation $\sigma = 1/\sqrt{19} = 0.229$.

Transformation invariance has an historical resonance with the normal correlation coefficient. Fisher's derivation of $f_\theta(\hat{\theta})$ (3.11) in 1915 was a mathematical triumph, but a difficult one to exploit in an era of mechanical computation. Most ingeniously, Fisher suggested instead working with the transformed parameter $\phi = m(\theta)$ where

$$\phi = m(\theta) = \frac{1}{2}\log\left(\frac{1+\theta}{1-\theta}\right), \tag{11.11}$$

and likewise with statistic $\hat{\phi} = m(\hat{\theta})$. Then, to a surprisingly good approximation,

$$\hat{\phi} \dot{\sim} \mathcal{N}\left(\phi, \frac{1}{n-3}\right). \tag{11.12}$$

See Figure 11.2, which shows Neyman's construction on the ϕ scale.

In other words, we are back in Fisher's favored situation (4.31), the simple normal translation problem, where

$$\mathcal{C}^{\phi}\left(\hat{\phi}\right) = \hat{\phi} \pm 1.96\frac{1}{\sqrt{n-3}} \tag{11.13}$$

is the "obviously correct" 95% confidence interval[3] for ϕ, closely approximating Neyman's construction. The endpoints of (11.13) are then transformed back to the θ scale according to the inverse transformation

$$\theta = \frac{e^{2\phi} - 1}{e^{2\phi} + 1}, \tag{11.14}$$

giving (almost) the interval $\mathcal{C}(\hat{\theta})$ seen in Figure 11.1, but without the involved computations.

Bayesian confidence statements are inherently transformation invariant. The fact that the Neyman intervals are also invariant, unlike the standard intervals (11.1), has made them more palatable to Bayesian statisticians. Transformation invariance will play a major role in justifying the bootstrap confidence intervals introduced next.

11.2 The Percentile Method

Our goal is to automate the calculation of confidence intervals: given the bootstrap distribution of a statistical estimator $\hat{\theta}$, we want to automatically produce an appropriate confidence interval for the unseen parameter θ. To this end, a series of four increasingly accurate bootstrap confidence interval algorithms will be described.

The first and simplest method is to use the standard interval (11.1), $\hat{\theta} \pm 1.96\widehat{\text{se}}$ for 95% coverage, with $\widehat{\text{se}}$ taken to be the bootstrap standard error $\widehat{\text{se}}_{\text{boot}}$ (10.16). The limitations of this approach become obvious in Figure 11.3, where the histogram shows $B = 2000$ nonparametric bootstrap replications $\hat{\theta}^*$ of the sample correlation coefficient for the student

[3] This is an anachronism. Fisher hated the term "confidence interval" after it was later coined by Neyman for his comprehensive theory. He thought of (11.13) as an example of the *logic of inductive inference*.

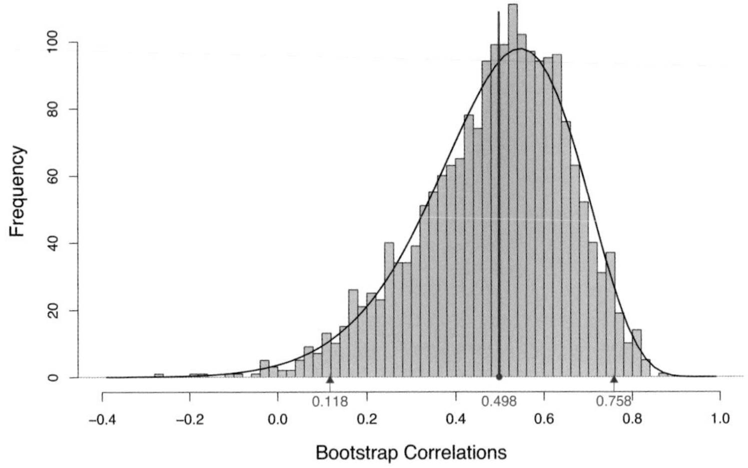

Figure 11.3 Histogram of $B = 2000$ nonparametric bootstrap replications $\hat{\theta}^*$ for the student score sample correlation; the solid curve is the ideal parametric bootstrap distribution $f_{\hat{\theta}}(r)$ as in Figure 11.1. Observed correlation $\hat{\theta} = 0.498$. Small triangles show histogram's 0.025 and 0.975 quantiles.

score data, obtained as in Section 10.2. The standard intervals are justified by taking literally the asymptotic normality of $\hat{\theta}$,

$$\hat{\theta} \dot{\sim} \mathcal{N}(\theta, \sigma^2), \tag{11.15}$$

σ the true standard error.

Relation (11.15) will generally hold for large enough sample size n, but we can see that for the student score data asymptotic normality has *not* yet set in, with the histogram being notably long-tailed to the left. We can't expect good performance from the standard method in this case. (The parametric bootstrap distribution is just as nonnormal, as shown by the smooth curve.)

The *percentile method* uses the shape of the bootstrap distribution to improve upon the standard intervals (11.1). Having generated B bootstrap replications $\hat{\theta}^{*1}, \hat{\theta}^{*2}, \ldots, \hat{\theta}^{*B}$, either nonparametrically as in Section 10.2 or parametrically as in Section 10.4, we use the obvious percentiles of their distribution to define the percentile confidence limits. The histogram in Figure 11.3 has its 0.025 and 0.975 percentiles equal to 0.118 and 0.758, and these are the endpoints of the central 95% nonparametric percentile interval.

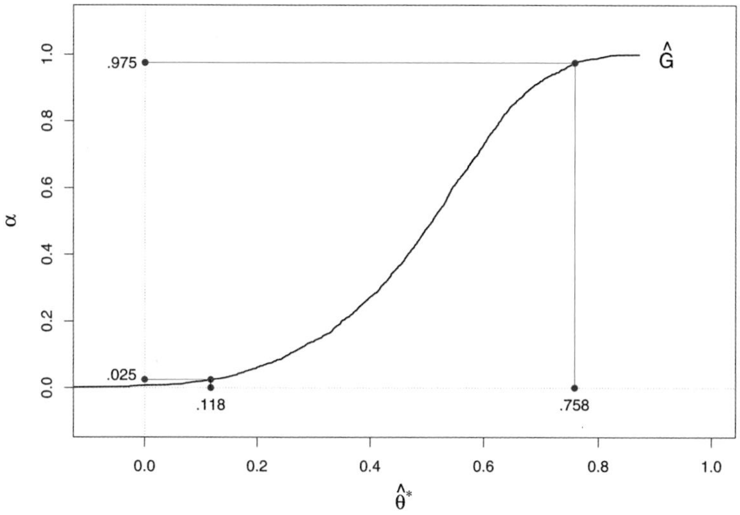

Figure 11.4 A 95% central confidence interval via the percentile method, based on the 2000 nonparametric replications $\hat{\theta}^*$ of Figure 11.3.

We can state things more precisely in terms of the *bootstrap cdf* $\hat{G}(t)$, the proportion of bootstrap samples less than t,

$$\hat{G}(t) = \#\left\{\hat{\theta}^{*b} \leq t\right\} \Big/ B. \tag{11.16}$$

The αth percentile point $\hat{\theta}^{*(\alpha)}$ of the bootstrap distribution is given by the inverse function of $\hat{G}$,

$$\hat{\theta}^{*(\alpha)} = \hat{G}^{-1}(\alpha); \tag{11.17}$$

$\hat{\theta}^{*(\alpha)}$ is the value putting proportion α of the bootstrap sample to its left. The level-α upper endpoint of the percentile interval, say $\hat{\theta}_{\%\mathrm{ile}}[\alpha]$, is by definition

$$\hat{\theta}_{\%\mathrm{ile}}[\alpha] = \hat{\theta}^{*(\alpha)} = \hat{G}^{-1}(\alpha). \tag{11.18}$$

In this notation, the 95% central percentile interval is

$$\left(\hat{\theta}_{\%\mathrm{ile}}[.025], \hat{\theta}_{\%\mathrm{ile}}[.975]\right). \tag{11.19}$$

The construction is illustrated in Figure 11.4.

The percentile intervals are transformation invariant. Let $\phi = m(\theta)$ as in (11.9), and likewise $\hat{\phi} = m(\hat{\theta})$ ($m(\cdot)$ monotonically increasing), with bootstrap replications $\hat{\phi}^{*b} = m(\hat{\theta}^{*b})$ for $b = 1, 2, \ldots, B$. The bootstrap percentiles transform in the same way,

$$\hat{\phi}^{*(\alpha)} = m\left(\hat{\theta}^{*(\alpha)}\right), \tag{11.20}$$

so that, as in (11.18),

$$\hat{\phi}_{\%\text{ile}}[\alpha] = m\left(\hat{\theta}_{\%\text{ile}}[\alpha]\right), \tag{11.21}$$

verifying transformation invariance.

In what sense does the percentile method improve upon the standard intervals? One answer involves transformation invariance. Suppose there exists a monotone transformation $\phi = m(\theta)$ and $\hat{\phi} = m(\hat{\theta})$ such that

$$\hat{\phi} \sim \mathcal{N}(\phi, \sigma^2) \tag{11.22}$$

for every θ, with σ^2 constant. Fisher's transformation (11.11)–(11.12) almost accomplishes this for the normal correlation coefficient.

It would then be true that parametric bootstrap replications would also follow (11.22),

$$\hat{\phi}^* \sim \mathcal{N}\left(\hat{\phi}, \sigma^2\right). \tag{11.23}$$

That is, the bootstrap cdf $\hat{G}^{\phi}$ would be normal with mean $\hat{\phi}$ and variance σ^2. The αth percentile of $\hat{G}^{\phi}$ would equal

$$\hat{\phi}_{\%\text{ile}}[\alpha] = \hat{\phi}^{*(\alpha)} = \hat{\phi} + z^{(\alpha)}\sigma, \tag{11.24}$$

where $z^{(\alpha)}$ denotes the αth percentile of a standard normal distribution,

$$z^{(\alpha)} = \Phi^{-1}(\alpha) \tag{11.25}$$

($z^{(.975)} = 1.96$, $z^{(.025)} = -1.96$, etc.).

In other words, the percentile method would provide Fisher's "obviously correct" intervals for ϕ,

$$\hat{\phi} \pm 1.96\sigma \tag{11.26}$$

for 95% coverage for example. But, because of transformation invariance, the percentile intervals for our original parameter θ would also be exactly correct.

Some comments concerning the percentile method are pertinent.

- The method does not require actually knowing the transformation to normality $\hat{\phi} = m(\hat{\theta})$, it only assumes its existence.
- If a transformation to form (11.22) exists, then the percentile intervals are not only accurate, but also *correct* in the Fisherian sense of giving the logically appropriate inference.† †2
- The justifying assumption for the standard intervals (11.15), $\hat{\theta} \dot{\sim} \mathcal{N}(\theta, \sigma^2)$, becomes more accurate as the sample size n increases (usually with σ decreasing as $1/\sqrt{n}$), but the convergence can be slow in cases like that of the normal correlation coefficient. The broader assumption (11.22), that $m(\hat{\theta}) \dot{\sim} \mathcal{N}(m(\theta), \sigma^2)$ for *some* transformation $m(\cdot)$, speeds up convergence, irrespective of whether or not it holds exactly. Section 11.4 makes this point explicit, in terms of asymptotic rates of convergence.
- The standard method works fine once it is applied on an appropriate scale, as in Figure 11.2. The trouble is that the method is *not* transformation invariant, leaving the statistician the job of finding the correct scale. The percentile method can be thought of as a transformation-invariant version of the standard intervals, an "automatic Fisher" that substitutes massive computations for mathematical ingenuity.
- The method requires bootstrap sample sizes† on the order of $B = 2000$. †3
- The percentile method is not the last word in bootstrap confidence intervals. Two improvements, the "BC" and "BCa" methods, will be discussed in the next section. Table 11.1 compares the various intervals as applied to the student score correlation, $\hat{\theta} = 0.498$.

Table 11.1 *Bootstrap confidence limits for student score correlation, $\hat{\theta} = 0.498$, $n = 22$. Parametric exact limits from Neyman's construction as in Figure 11.1. The BC and BCa methods are discussed in the next two sections; (z_0, a), two constants required for BCa, are $(-0.055, 0.005)$ parametric, and $(0.000, 0.006)$ nonparametric.*

	Parametric		Nonparametric	
	.025	.975	.025	.975
1. Standard	.17	.83	.18	.82
2. Percentile	.11	.77	.13	.76
3. BC	.08	.75	.13	.76
4. BCa	.08	.75	.12	.76
Exact	**.09**	**.75**		

The label "computer-intensive inference" seems especially apt as applied to bootstrap confidence intervals. Neyman and Fisher's constructions are expanded from a few special theoretically tractable cases to almost any situation where the statistician has a repeatable algorithm. Automation, the replacement of mathematical formulas with wide-ranging computer algorithms, will be a major theme of succeeding chapters.

11.3 Bias-Corrected Confidence Intervals

The ideal form (11.22) for the percentile method, $\hat{\phi}^* \sim \mathcal{N}(\hat{\phi}, \sigma^2)$, says that the transformation $\hat{\phi} = m(\hat{\theta})$ yields an unbiased estimator of constant variance. The improved methods of this section and the next take into account the possibility of bias and changing variance. We begin with bias.

If $\hat{\phi} \sim \mathcal{N}(\phi, \sigma^2)$ for all $\phi = m(\theta)$, as hypothesized in (11.22), then $\hat{\phi}^* \sim \mathcal{N}(\hat{\phi}, \sigma^2)$ and

$$\Pr{}_* \left\{\hat{\phi}^* \le \hat{\phi}\right\} = 0.50 \tag{11.27}$$

($\Pr_*$ indicating bootstrap probability), in which case the monotonicity of $m(\cdot)$ gives

$$\Pr{}_* \left\{\hat{\theta}^* \le \hat{\theta}\right\} = 0.50. \tag{11.28}$$

That is, $\hat{\theta}^*$ is *median unbiased*[4] for $\hat{\theta}$, and likewise $\hat{\theta}$ for θ.

We can check that. For a parametric family of densities $f_\theta(\hat{\theta})$, (11.28) implies

$$\int_{-\infty}^{\hat{\theta}} f_{\hat{\theta}}\left(\hat{\theta}^*\right)\, d\hat{\theta}^* = 0.50. \tag{11.29}$$

For the normal correlation coefficient density (3.11), $n = 22$, numerical integration gives

$$\int_{-1}^{.498} f_{.498}\left(\hat{\theta}^*\right)\, d\hat{\theta}^* = 0.478, \tag{11.30}$$

which is not far removed from 0.50, but far enough to have a small impact on proper inference. It suggests that $\hat{\theta}^*$ is biased *upward* relative to $\hat{\theta}$—that's why *less* than half of the bootstrap probability lies below $\hat{\theta}$—and by implication that $\hat{\theta}$ is upwardly biased for estimating θ. Accordingly, confidence intervals should be adjusted a little bit downward. The *bias-corrected percentile method* (BC for short) is a data-based algorithm for making such adjustments.

[4] Median unbiasedness, unlike the usual mean unbiasedness definition, has the advantage of being transformation invariant.

Having simulated B bootstrap replications $\hat{\theta}^{*1}, \hat{\theta}^{*2}, \ldots, \hat{\theta}^{*B}$, parametric or nonparametric, let p_0 be the proportion of replications less than $\hat{\theta}$,

$$p_0 = \#\left\{\hat{\theta}^{*b} \leq \hat{\theta}\right\} \Big/ B \tag{11.31}$$

(an estimate of (11.29)), and define the *bias-correction value*

$$z_0 = \Phi^{-1}(p_0), \tag{11.32}$$

where Φ^{-1} is the inverse function of the standard normal cdf. The BC level-α confidence interval endpoint is defined to be

$$\hat{\theta}_{\text{BC}}[\alpha] = \hat{G}^{-1}\left[\Phi\left(2z_0 + z^{(\alpha)}\right)\right], \tag{11.33}$$

where $\hat{G}$ is the bootstrap cdf (11.16) and $z^{(\alpha)} = \Phi^{-1}(\alpha)$ (11.25).

If $p_0 = 0.50$, the median unbiased situation, then $z_0 = 0$ and

$$\hat{\theta}_{\text{BC}}[\alpha] = \hat{G}^{-1}\left[\Phi\left(z^{(\alpha)}\right)\right] = \hat{G}^{-1}(\alpha) = \hat{\theta}_{\%\text{ile}}[\alpha], \tag{11.34}$$

the percentile limit (11.18). Otherwise, a bias correction is made. Taking $p_0 = 0.478$ for the normal correlation example (the value we would get from an infinite number of parametric bootstrap replications) gives bias correction value -0.055. Notice that the BC limits are indeed shifted downward from the parametric percentile limits in Table 11.1. Nonparametric bootstrapping gave p_0 about 0.50 in this case, making the BC limits nearly the same as the percentile limits.

A more general transformation argument motivates the BC definition (11.33). Suppose there exists a monotone transformation $\phi = m(\theta)$ and $\hat{\phi} = m(\hat{\theta})$ such that for any θ

$$\hat{\phi} \sim \mathcal{N}(\phi - z_0\sigma, \sigma^2), \tag{11.35}$$

with z_0 and σ fixed constants. Then the BC endpoints are accurate, i.e., have the claimed coverage probabilities, and are also "obviously correct" in the Fisherian sense. See the chapter endnotes† for proof and discussion. †4

As before, the statistican does not need to know the transformation $m(\cdot)$ that leads to $\hat{\phi} \sim \mathcal{N}(\phi - z_0\sigma, \sigma^2)$, only that it exists. It is a broader target than $\hat{\phi} \sim \mathcal{N}(\phi, \sigma^2)$ (11.22), making the BC method better justified than the percentile method, irrespective of whether or not such a transformation exists. There is no extra computational burden: the bootstrap replications $\{\hat{\theta}^{*b},\ b = 1, 2, \ldots, B\}$, parametric or nonparametric, provide $\hat{G}$ (11.16) and z_0 (11.31)–(11.32), giving $\hat{\theta}_{\text{BC}}[\alpha]$ from (11.33).

11.4 Second-Order Accuracy

Coverage errors of the standard confidence intervals typically decrease at order $O(1/\sqrt{n})$ in the sample size n: having calculated $\hat{\theta}_{\text{stan}}[\alpha] = \hat{\theta} + z^{(\alpha)}\hat{\sigma}$ for an iid sample $\boldsymbol{x} = (x_1, x_2, \dots, x_n)$, we can expect the actual coverage probability to be

$$\Pr_\theta\left\{\theta \le \hat{\theta}_{\text{stan}}[\alpha]\right\} \doteq \alpha + c_1/\sqrt{n}, \tag{11.36}$$

where c_1 depends on the problem at hand; (11.36) defines "first-order accuracy." It can connote painfully slow convergence to the nominal coverage level α, requiring sample size $4n$ to cut the error in half.

A *second-order accurate* method, say $\hat{\theta}_{\text{2nd}}[\alpha]$, makes errors of order only $O(1/n)$,

$$\Pr_\theta\left\{\theta \le \hat{\theta}_{\text{2nd}}[\alpha]\right\} \doteq \alpha + c_2/n. \tag{11.37}$$

The improvement is more than theoretical. In practical problems like that of Table 11.1, second-order accurate methods—BCa, defined in the following, is one such—often provide nearly the claimed coverage probabilities, even in small-size samples.

Neither the percentile method nor the BC method is second-order accurate (although, as in Table 11.1, they tend to be more accurate than the standard intervals). The difficulty for $\hat{\theta}_{\text{BC}}[\alpha]$ lies in the ideal form (11.35), $\hat{\phi} \sim \mathcal{N}(\phi - z_0\sigma, \sigma^2)$, where it is assumed $\hat{\phi} = m(\hat{\theta})$ has *constant* standard error σ. Instead, we now postulate the existence of a monotone transformation $\phi = m(\theta)$ and $\hat{\phi} = m(\hat{\theta})$ less restrictive than (11.35),

$$\hat{\phi} \sim \mathcal{N}(\phi - z_0\sigma_\phi, \sigma_\phi^2), \qquad \sigma_\phi = 1 + a\phi. \tag{11.38}$$

†5 Here the "acceleration"† a is a small constant describing how the standard deviation of $\hat{\phi}$ varies with ϕ. If $a = 0$ we are back in situation (11.34)[5], but if not, an amendment to the BC formula (11.33) is required.

The *BCa method* ("bias-corrected and accelerated") takes its level-α confidence limit to be

$$\hat{\theta}_{\text{BCa}}[\alpha] = \hat{G}^{-1}\left[\Phi\left(z_0 + \frac{z_0 + z^{(\alpha)}}{1 - a(z_0 + z^{(\alpha)})}\right)\right]. \tag{11.39}$$

A still more elaborate transformation argument shows that, if there exists a monotone transformation $\phi = m(\theta)$ and constants z_0 and a yielding

[5] This assumes $\sigma_0 = 1$ on the right side of (11.38), which can always be achieved by further transforming ϕ to ϕ/σ.

(11.38), then the BCa limits have their claimed coverage probabilities and, moreover, are correct in the Fisherian sense.

BCa makes three corrections to the standard intervals (11.1): for nonnormality of $\hat{\theta}$ (through using the bootstrap percentiles rather than just the bootstrap standard error); for bias (through the bias correction value z_0); and for nonconstant standard error (through a). Notice that if $a = 0$ then BCa (11.39) reduces to BC (11.33). If $z_0 = 0$ then BC reduces to the percentile method (11.18); and if $\hat{G}$, the bootstrap histogram, is normal, then (11.18) reduces to the standard interval (11.1). All three of the corrections, for nonnormality, bias, and acceleration, can have substantial effects in practice and are necessary to achieve second-order accuracy. A great deal of theoretical effort was devoted to verifying the second-order accuracy and BCa intervals under reasonably general assumptions.[6]

Table 11.2 *Nominal 95% central confidence intervals for Poisson parameter θ having observed $\hat{\theta} = 10$; actual tail areas above and below $\hat{\theta} = 10$ defined as in Figure 11.1 (atom of probability split at 10). For instance, lower standard limit 3.80 actually puts probability 0.004 above 10, rather than nominal value 0.025. Bias correction value z_0 (11.32) and acceleration a (11.38) both equal 0.050.*

	Nominal limits		Tail areas	
	.025	.975	Above	Below
1. Standard	3.80	16.20	.004	.055
2. %ile	4.18	16.73	.007	.042
3. BC	4.41	17.10	.010	.036
4. BCa	5.02	17.96	.023	.023
Exact	**5.08**	**17.82**	**.025**	**.025**

The advantages of increased accuracy are not limited to large sample sizes. Table 11.2 returns to our original example of observing $\hat{\theta} = 10$ from Poisson model $\hat{\theta} \sim \text{Poi}(\theta)$. According to Neyman's construction, the 0.95 exact limits give tail areas 0.025 in both the above and below directions, as in Figure 11.1, and this is nearly matched by the BCa limits. However the standard limits are much too conservative at the left end and anti-conservative at the right.

[6] The mathematical side of statistics has also been affected by electronic computation, where it is called upon to establish the properties of general-purpose computer algorithms such as the bootstrap. Asymptotic analysis in particular has been aggressively developed, the verification of second-order accuracy being a nice success story.

Table 11.3 *95% nominal confidence intervals for the parametric and nonparametric eigenratio examples of Figures 10.2 and 10.6.*

	Parametric		Nonparametric	
	.025	.975	.025	.975
1. Standard	.556	.829	.545	.840
2. %ile	.542	.815	.517	.818
3. BC	.523	.828	.507	.813
4. BCa	.555	.820	.523	.828
	($z_0 = -.029$, $a = .058$)		($z_0 = -.049$, $a = .051$)	

Bootstrap confidence limits continue to provide better inferences in the vast majority of situations too complicated for exact analysis. One such situation is examined in Table 11.3. It relates to the eigenratio example illustrated in Figures 10.2–10.6. In this case the nonnormality and bias corrections stretch the bootstrap intervals to the left, but the acceleration effect pulls right, partially canceling out the net change from the standard intervals.

The percentile and BC methods are completely automatic, and can be applied whenever a sufficiently large number of bootstrap replications are available. The same cannot be said of BCa. A drawback of the BCa method is that the acceleration a is not a function of the bootstrap distribution and must be computed separately. Often this is straightforward:

- For one-parameter exponential families such as the Poisson, a equals z_0.
- In one-sample nonparametric problems, a can be estimated from the jackknife resamples $\hat{\theta}_{(i)}$ (10.5),

$$\hat{a} = \frac{1}{6} \frac{\sum_{i=1}^{n} \left(\hat{\theta}_{(i)} - \hat{\theta}_{(\cdot)}\right)^3}{\left[\sum_{i=1}^{n} \left(\hat{\theta}_i - \hat{\theta}_{(\cdot)}\right)^2\right]^{1.5}}. \tag{11.40}$$

- The *abc method* computes a in multiparameter exponential families (5.54), as does the resampling-based **R** algorithm `accel`.

Confidence intervals require the number of bootstrap replications B to be on the order of 2000, rather than the 200 or fewer needed for standard errors; the corrections made to the standard intervals are more delicate than standard errors and require greater accuracy.

There is one more cautionary note to sound concerning nuisance parameters: biases can easily get out of hand when the parameter vector μ is high-dimensional. Suppose we observe

$$x_i \overset{\text{ind}}{\sim} \mathcal{N}(\mu_i, 1) \qquad \text{for } i = 1, 2, \ldots, n, \tag{11.41}$$

and wish to set a confidence interval for $\theta = \sum_1^n \mu_i^2$. The MLE $\hat{\theta} = \sum_1^n x_i^2$ will be sharply biased upward if n is at all large. To be specific, if $n = 10$ and $\hat{\theta} = 20$, we compute† †6

$$z_0 = \Phi^{-1}(0.156) = -1.01. \tag{11.42}$$

This makes[7] $\hat{\theta}_{\text{BC}}[.025]$ (11.33) equal a ludicrously small bootstrap percentile,

$$\hat{G}^{-1}(0.000034), \tag{11.43}$$

a warning sign against the BC or BCa intervals, which work most dependably for $|z_0|$ and $|a|$ small, say ≤ 0.2.

A more general warning would be against blind trust in maximum likelihood estimates in high dimensions. Computing z_0 is a wise precaution even if it is not used for BC or BCa purposes, in case it alerts one to dangerous biases.

Confidence intervals for classical applications were most often based on the standard method (11.1) (with $\widehat{\text{se}}$ estimated by the delta method) except in a few especially simple situations such as the Poisson. Second-order accurate intervals are very much a computer-age development, with both the algorithms and the inferential theory presupposing high-speed electronic computation.

11.5 Bootstrap-*t* Intervals

The initial breakthrough on exact confidence intervals came in the form of Student's t distribution in 1908. Suppose we independently observe data from two possibly different normal distributions, $\boldsymbol{x} = (x_1, x_2, \ldots, x_{n_x})$ and $\boldsymbol{y} = (y_1, y_2, \ldots, y_{n_y})$,

$$x_i \overset{\text{iid}}{\sim} \mathcal{N}(\mu_x, \sigma^2) \quad \text{and} \quad y_i \overset{\text{iid}}{\sim} \mathcal{N}(\mu_y, \sigma^2), \tag{11.44}$$

and wish to form a 0.95 central confidence interval for

$$\theta = \mu_y - \mu_x. \tag{11.45}$$

The obvious estimate is

$$\hat{\theta} = \bar{y} - \bar{x}, \tag{11.46}$$

[7] Also $\hat{\theta}_{\text{BCa}}[.025]$, a is zero in this model.

but its distribution depends on the nuisance parameter σ^2.

Student's masterstroke was to base inference about θ on the *pivotal quantity*

$$t = \frac{\hat{\theta} - \theta}{\widehat{\text{se}}} \tag{11.47}$$

where $\widehat{\text{se}}^2$ is an unbiased estimate of σ^2,

$$\widehat{\text{se}}^2 = \left(\frac{1}{n_x} + \frac{1}{n_y}\right) \frac{\sum_1^{n_x}(x_i - \bar{x})^2 + \sum_1^{n_y}(y_i - \bar{y})^2}{n_x + n_y - 2}; \tag{11.48}$$

t then has the "Student's t distribution" with df $= n_x + n_y - 2$ degrees of freedom if $\mu_x = \mu_y$, no matter what σ^2 may be.

Letting $t_{\text{df}}^{(\alpha)}$ represent the 100αth percentile of a t_{df} distribution yields

$$\hat{\theta}_t[\alpha] = \hat{\theta} - \widehat{\text{se}} \cdot t_{\text{df}}^{(1-\alpha)} \tag{11.49}$$

as the upper level-α interval of a Student's t confidence limit. Applied to the difference between the **AML** and **ALL** scores in Figure 1.4, the central 0.95 Student's t interval for $\theta = E\{\textbf{AML}\} - E\{\textbf{ALL}\}$ was calculated to be

$$\left(\hat{\theta}_t[.025], \hat{\theta}_t[.975]\right) = (.062, .314). \tag{11.50}$$

Here $n_x = 47$, $n_y = 25$, and df $= 70$.

Student's theory depends on the normality assumptions of (11.44). The *bootstrap-t* approach is to accept (or pretend) that t in (11.47) is pivotal, but to estimate its distribution via bootstrap resampling. Nonparametric bootstrap samples are drawn separately from $\boldsymbol{x}$ and $\boldsymbol{y}$,

$$\boldsymbol{x}^* = (x_1^*, x_2^*, \ldots, x_{n_x}^*) \quad \text{and} \quad \boldsymbol{y}^* = (y_1^*, y_2^*, \ldots, y_{n_y}^*), \tag{11.51}$$

from which we calculate $\hat{\theta}^*$ and $\widehat{\text{se}}^*$, (11.46) and (11.48), giving

$$t^* = \frac{\hat{\theta}^* - \hat{\theta}}{\widehat{\text{se}}^*}, \tag{11.52}$$

with $\hat{\theta}$ playing the role of θ, as appropriate in the bootstrap world. Replications $\{t^{*b}, b = 1, 2, \ldots, B\}$ provide estimated percentiles $t^{*(\alpha)}$ and corresponding confidence limits

$$\hat{\theta}_t^*[\alpha] = \hat{\theta} - \widehat{\text{se}} \cdot t^{*(1-\alpha)}. \tag{11.53}$$

For the **AML–ALL** example, the t^* distribution differed only slightly from a t_{70} distribution; the resulting 0.95 interval was $(0.072, 0.323)$, nearly the same as (11.50), lending credence to the original normality assumptions.

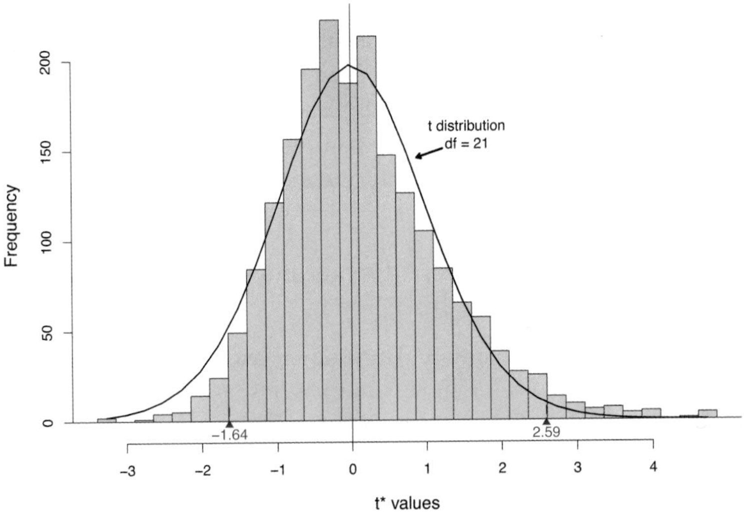

Figure 11.5 $B = 2000$ nonparametric replications of bootstrap-t statistic for the student score correlation; small triangles show 0.025 and 0.975 percentile points. The histogram is sharply skewed to the right; the solid curve is Student's t density for 21 degrees of freedom.

Returning to the student score correlation example of Table 11.1, we can apply bootstrap-t methods by still taking $t = (\hat{\theta} - \theta)/\widehat{\text{se}}$ to be notionally pivotal, but now with θ the true correlation, $\hat{\theta}$ the sample correlation, and $\widehat{\text{se}}$ the approximate standard error $(1 - \hat{\theta}^2)/\sqrt{19}$. Figure 11.5 shows the histogram of $B = 2000$ nonparametric bootstrap replications $t^* = (\hat{\theta}^* - \hat{\theta})/\widehat{\text{se}}^*$. These gave bootstrap percentiles

$$\left(t^{*(.025)}, t^{*(.975)}\right) = (-1.64, 2.59) \tag{11.54}$$

(which might be compared with $(-2.08, 2.08)$ for a standard t_{21} distribution), and 0.95 interval $(0.051, 0.781)$ from (11.53), somewhat out of place compared with the other entries in the right panel of Table 11.1.

Bootstrap-t intervals are *not* transformation invariant. This means they can perform poorly or well depending on the scale of application. If performed on Fisher's scale (11.11) they agree well with exact intervals for the correlation coefficient. A practical difficulty is the requirement of a formula for $\widehat{\text{se}}$.

Nevertheless, the idea of estimating the actual distribution of a proposed pivotal quantity has great appeal to the modern statistical spirit. Calculating the percentiles of the original Student t distribution was a multi-year project in the early twentieth century. Now we can afford to calculate our own special "t table" for each new application. Spending such computational wealth wisely, while not losing one's inferential footing, is the central task and goal of twenty-first-century statisticians.

11.6 Objective Bayes Intervals and the Confidence Distribution

Interval estimates are ubiquitous. They play a major role in the scientific discourse of a hundred disciplines, from physics, astronomy, and biology to medicine and the social sciences. Neyman-style frequentist confidence intervals dominate the literature, but there have been influential Bayesian and Fisherian developments as well, as discussed next.

Given a one-parameter family of densities $f_\theta(\hat{\theta})$ and a prior density $g(\theta)$, Bayes' rule (3.5) produces the posterior density of θ,

$$g(\theta|\hat{\theta}) = g(\theta) f_\theta(\hat{\theta}) / f(\hat{\theta}), \tag{11.55}$$

where $f(\hat{\theta})$ is the marginal density $\int f_\theta(\hat{\theta}) g(\theta) d\theta$. The Bayes 0.95 *credible interval* $C(\theta|\hat{\theta})$ spans the central 0.95 region of $g(\theta|\hat{\theta})$, say

$$C(\theta|\hat{\theta}) = (a(\hat{\theta}), b(\hat{\theta})), \tag{11.56}$$

with

$$\int_{a(\hat{\theta})}^{b(\hat{\theta})} g(\theta|\hat{\theta})\, d\theta = 0.95, \tag{11.57}$$

and with posterior probability 0.025 in each tail region.

Confidence intervals, of course, require no prior information, making them eminently useful in day-to-day applied practice. The Bayesian equivalents are credible intervals based on uninformative priors, Section 3.2. "Matching priors," those whose credible intervals nearly match Neyman confidence intervals, have been of particular interest. Jeffreys' prior (3.17),

$$\begin{aligned} g(\theta) &= \mathcal{I}_\theta^{1/2}, \\ \mathcal{I}_\theta &= \int \left[\frac{\partial}{\partial\theta} \log f_\theta(\hat{\theta}) \right]^2 f_\theta(\hat{\theta})\, d\hat{\theta}, \end{aligned} \tag{11.58}$$

provides a generally accurate matching prior for one-parameter problems. Figure 3.2 illustrates this for the student score correlation, where the credible interval $(0.093, 0.750)$ is a near-exact match to the Neyman 0.95 interval of Figure 11.1.

Difficulties begin with multiparameter families $f_\mu(x)$ (5.1): we wish to construct an interval estimate for a one-dimensional function $\theta = t(\mu)$ of the p-dimensional parameter vector μ, and must somehow remove the effects of the $p-1$ "nuisance parameters." In a few rare situations, including the normal theory correlation coefficient, this can be done exactly. Pivotal methods do the job for Student's t construction. Bootstrap confidence intervals greatly extend the reach of such methods, at a cost of greatly increased computation.

Bayesians get rid of nuisance parameters by integrating them out of the posterior density $g(\mu|\boldsymbol{x}) = g(\mu) f_\mu(\boldsymbol{x})/f(\boldsymbol{x})$ (3.6) ($\boldsymbol{x}$ now representing all the data, "$\boldsymbol{x}$" equaling $(\boldsymbol{x}, \boldsymbol{y})$ for the Student t setup (11.44)). That is, we calculate[8] the marginal density of $\theta = t(\mu)$ given $\boldsymbol{x}$, and call it $h(\theta|\boldsymbol{x})$. A credible interval for θ, $\mathcal{C}(\theta|\boldsymbol{x})$, is then constructed as in (11.56)–(11.57), with $h(\theta|\boldsymbol{x})$ playing the role of $g(\theta|\hat{\theta})$. This leaves us the knotty problem of choosing an uninformative multidimensional prior $g(\mu)$. We will return to the question after first discussing fiducial methods, a uniquely Fisherian device.

Fiducial constructions begin with what seems like an obviously incorrect interpretation of pivotality. We rewrite the Student t pivotal $t = (\hat{\theta} - \theta)/\widehat{\text{se}}$ (11.47) as

$$\theta = \hat{\theta} - \widehat{\text{se}} \cdot t, \tag{11.59}$$

where t has a Student's t distribution with df degrees of freedom, $t \sim t_{\text{df}}$. Having observed the data $(\boldsymbol{x}, \boldsymbol{y})$ (11.44), fiducial theory assigns θ the distribution implied by (11.59), as if $\hat{\theta}$ and $\widehat{\text{se}}$ were *fixed* at their calculated values while t was distributed as t_{df}. Then $\hat{\theta}_t[\alpha]$ (11.49), the Student t level-α confidence limit, is the 100αth percentile of θ's fiducial distribution.

We seem to have achieved a Bayesian posterior conclusion without any prior assumptions.[9] The historical development here is confused by Fisher's refusal to accept Neyman's confidence interval theory, as well as his disparagement of Bayesian ideas. As events worked out, all of Fisher's immense prestige was not enough to save fiducial theory from the scrapheap of failed statistical methods.

[8] Often a difficult calculation, as discussed in Chapter 13.

[9] "Enjoying the Bayesian omelette without breaking the Bayesian eggs," in L. J. Savage's words.

And yet, in Arthur Koestler's words, "The history of ideas is filled with barren truths and fertile errors." Fisher's underlying rationale went something like this: $\hat{\theta}$ and $\widehat{\text{se}}$ exhaust the information about θ available from the data, after which there remains an irreducible component of randomness described by t. This is an idea of substantial inferential appeal, and one that can be rephrased in more general terms discussed next that bear on the question of uninformative priors.

By definition, an upper confidence limit $\hat{\theta}_{\boldsymbol{x}}[\alpha]$ satisfies

$$\Pr\left\{\theta \le \hat{\theta}_{\boldsymbol{x}}[\alpha]\right\} = \alpha \tag{11.60}$$

(where now we have indicated the observed data $\boldsymbol{x}$ in the notation), and so

$$\Pr\left\{\hat{\theta}_{\boldsymbol{x}}[\alpha] \le \theta \le \hat{\theta}_{\boldsymbol{x}}[\alpha + \epsilon]\right\} = \epsilon. \tag{11.61}$$

We can consider $\hat{\theta}_{\boldsymbol{x}}[\alpha]$ as a one-to-one function between α in $(0, 1)$ and θ a point in its parameter space Θ (assuming that $\hat{\theta}_{\boldsymbol{x}}[\alpha]$ is smoothly increasing in α). Letting ϵ go to zero in (11.61) determines the *confidence density* of θ, say $\tilde{g}_{\boldsymbol{x}}(\theta)$,

$$\tilde{g}_{\boldsymbol{x}}(\theta) = d\alpha/d\theta, \tag{11.62}$$

the local derivative of probability at location θ for the unknown parameter, the derivative being taken at $\theta = \hat{\theta}_{\boldsymbol{x}}[\alpha]$.

Integrating $\tilde{g}_{\boldsymbol{x}}(\theta)$ recovers α as a function of θ. Let $\theta_1 = \hat{\theta}_{\boldsymbol{x}}[\alpha_1]$ and $\theta_2 = \hat{\theta}_{\boldsymbol{x}}[\alpha_2]$ for any two values $\alpha_1 < \alpha_2$ in $(0, 1)$. Then

$$\begin{aligned}\int_{\theta_1}^{\theta_2} \tilde{g}_{\boldsymbol{x}}(\theta)\, d\theta &= \int_{\theta_1}^{\theta_2} \frac{d\alpha}{d\theta}\, d\theta = \alpha_2 - \alpha_1 \\ &= \Pr\{\theta_1 \le \theta \le \theta_2\},\end{aligned} \tag{11.63}$$

as in (11.60). There is nothing controversial about (11.63) as long as we remember that the random quantity in $\Pr\{\theta_1 \le \theta \le \theta_2\}$ is not θ but rather the interval (θ_1, θ_2), which varies as a function of $\boldsymbol{x}$. Forgetting this leads to the textbook error of attributing Bayesian properties to frequentist results: "There is 0.95 probability that θ is in its 0.95 confidence interval," etc.

This is exactly what the fiducial argument does.[10] Whether or not one accepts (11.63), there is an immediate connection with *matching priors*.

[10] Fiducial and confidence densities agree, as can be seen in the Student t situation (11.59), at least in the somewhat limited catalog of cases Fisher thought appropriate for fiducial calculations.

Suppose prior $g(\mu)$ gives a perfect match to the confidence interval system $\hat{\theta}_{\boldsymbol{x}}[\alpha]$. Then, by definition, its posterior density $h(\theta|\boldsymbol{x})$ must satisfy

$$\int_{-\infty}^{\hat{\theta}_x[\alpha]} h(\theta|\boldsymbol{x})\, d\theta = \alpha = \int_{-\infty}^{\hat{\theta}_x[\alpha]} \tilde{g}_{\boldsymbol{x}}(\theta)\, d\theta \tag{11.64}$$

for $0 < \alpha < 1$. But this implies $h(\theta|\boldsymbol{x})$ equals $\tilde{g}_{\boldsymbol{x}}(\theta)$ for all θ. That is, *the confidence density $\tilde{g}_{\boldsymbol{x}}(\theta)$ is the posterior density of θ given x for any matching prior.*

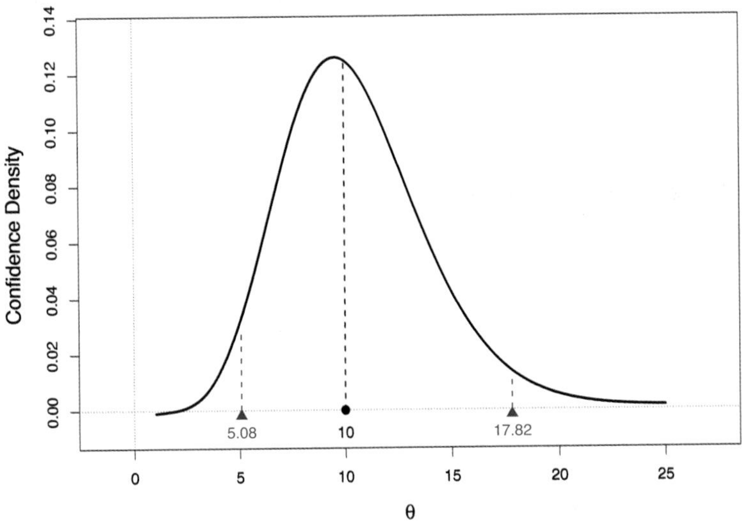

Figure 11.6 Confidence density (11.62) for Poisson parameter θ having observed $\hat{\theta} = 10$. There is area 0.95 under the curve between 5.08 and 17.82, as in Table 11.2, and areas 0.025 in each tail.

Figure 11.6 graphs the confidence density for $\hat{\theta} \sim \text{Poi}(\theta)$ having observed $\hat{\theta} = 10$. This was obtained by numerically differentiating α as a function of θ (11.62),

$$\alpha = \Pr\{10 \le \text{Poi}(\theta)\}, \tag{11.65}$$

"$\le$" including splitting the atom of probability at 10. According to Table 11.2, $\tilde{g}_{10}(\theta)$ has area 0.95 between 5.08 and 17.82, and area 0.025 in each tail. Whatever its provenance, the graph delivers a striking picture of the uncertainty in the unknown value of θ.

Bootstrap confidence intervals provide easily computable confidence densities. Let $\hat{G}(\theta)$ be the bootstrap cdf and $\hat{g}(\theta)$ its density function (obtained by differentiating a smoothed version of $\hat{G}(\theta)$ when $\hat{G}$ is based on B bootstrap replications). The percentile confidence limits $\hat{\theta}[\alpha] = \hat{G}^{-1}(\alpha)$ (11.16) have $\alpha = \hat{G}(\theta)$, giving

$$\tilde{g}_{\boldsymbol{x}}(\theta) = \hat{g}(\theta). \tag{11.66}$$

(It is helpful to picture this in Figure 11.4.) For the percentile method, the bootstrap density *is* the confidence density.

For the BCa intervals (11.39), the confidence density is obtained by reweighting $\hat{g}(\theta)$,

$$\tilde{g}_{\boldsymbol{x}}(\theta) = cw(\theta)\hat{g}(\theta), \tag{11.67}$$

†7 where†

$$w(\theta) = \frac{\varphi\left[z_\theta/(1+az_\theta) - z_0\right]}{(1+az_\theta)^2\varphi(z_\theta + z_0)}, \qquad \text{with } z_\theta = \Phi^{-1}\hat{G}(\theta) - z_0. \tag{11.68}$$

Here φ is the standard normal density, Φ its cdf, and c the constant that makes $\tilde{g}_{\boldsymbol{x}}(\theta)$ integrate to 1. In the usual case where the bootstrap cdf is estimated from replications $\hat{\theta}^{*b}$, $b = 1, 2, \ldots, B$ (either parametric or nonparametric), the BCa confidence density is a reweighted version of $\hat{g}(\theta)$. Define

$$W_b = w\left(\hat{\theta}^{*b}\right) \Big/ \sum_{i=1}^{B} w\left(\hat{\theta}^{*i}\right). \tag{11.69}$$

Then the BCa confidence density is the discrete density putting weight W_b on $\hat{\theta}^{*b}$.

Figure 11.7 returns to the student score data, $n = 22$ students, five scores each, modeled normally as in Figure 10.6,

$$x_i \overset{\text{iid}}{\sim} \mathcal{N}_5(\lambda, \boldsymbol{\Sigma}) \qquad \text{for } i = 1, 2, \ldots, 22. \tag{11.70}$$

This is a $p = 20$-dimensional parametric family: 5 expectations, 5 variances, 10 covariances. The parameter of interest was taken to be

$$\theta = \text{maximum eigenvalue of } \boldsymbol{\Sigma}. \tag{11.71}$$

It had MLE $\hat{\theta} = 683$, this being the maximum eigenvalue of the MLE sample covariance matrix $\hat{\Sigma}$ (dividing each sum of squares by 22 rather than 21).

$B = 8000$ parametric bootstrap replications[11] $\hat{\theta}^{*b}$ gave percentile and

[11] $B = 2000$ would have been enough for most purposes, but $B = 8000$ gave a sharper picture of the different curves.

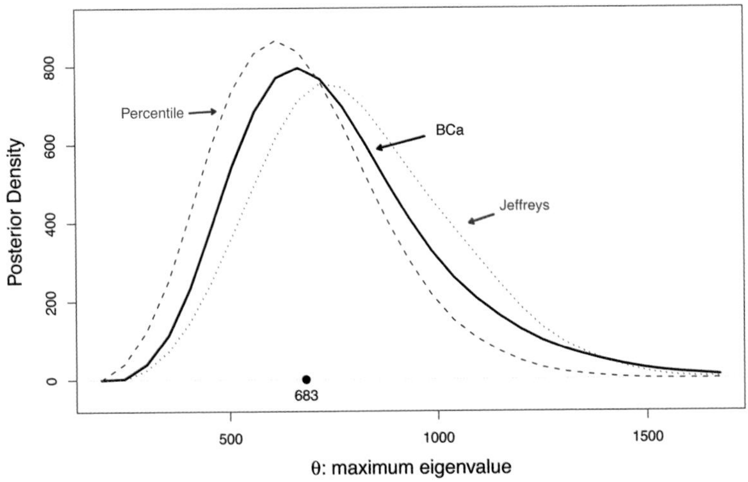

Figure 11.7 Confidence densities for the maximum eigenvalue parameter (11.71), using a multivariate normal model (11.70) for the student score data. The dashed red curve is the percentile method, solid black the BCa (with $(z_0, a) = (0.178, 0.093)$). The dotted blue curve is the Bayes posterior density for θ, using Jeffreys' prior (11.72).

BCa confidence densities as shown. In this case the weights W_b (11.69) increased with $\hat{\theta}^{*b}$, pushing the BCa density to the right. Also shown is the Bayes posterior density† for θ starting from Jeffreys' multiparameter prior †8
density

$$g^{\text{Jeff}}(\mu) = |\mathcal{I}_\mu|^{1/2}, \tag{11.72}$$

where $\mathcal{I}_\mu$ is the Fisher information matrix (5.26). It isn't truly uninformative here, moving its credible limits upward from the second-order accurate BCa confidence limits. Formula (11.72) is discussed further in Chapter 13.

Bayesian data analysis has the attractive property that, after examining the data, we can express our remaining uncertainty in the language of probability. Fiducial and confidence densities provide something similar for confidence intervals, at least partially freeing the frequentist from the interpretive limitations of Neyman's intervals.

11.7 Notes and Details

Fisher's theory of fiducial inference (1930) preceded Neyman's approach, formalized in (1937), which was presented as an attempt to put interval estimation on a firm probabilistic basis, as opposed to the mysteries of fiducialism. The result was an elegant theory of exact and optimal intervals, phrased in hard-edged frequentistic terms. Readers familiar with the theory will know that Neyman's construction—a favorite name in the physics literature—as pictured in Figure 11.1, requires some conditions on the family of densities $f_\theta(\hat{\theta})$ to yield optimal intervals, a sufficient condition being monotone likelihood ratios.

Bootstrap confidence intervals, Efron (1979, 1987), are neither exact nor optimal, but aim instead for wide applicability combined with near-exact accuracy. Second-order acuracy of BCa intervals was established by Hall (1988). BCa is emphatically a child of the computer age, routinely requiring $B = 2000$ or more bootstrap replications per use. Shortcut methods are available. The "abc method" (DiCiccio and Efron, 1992) needs only 1% as much computation, at the expense of requiring smoothness properties for $\theta = t(\mu)$, and a less automatic coding of the exponential family setting for individual situations. In other words, it is less convenient.

†1 [p. 183] *Neyman's construction.* For any given value of θ, let $(\theta^{(.025)}, \theta^{(.975)})$ denote the central 95% interval of density $f_\theta(\hat{\theta})$, satisfying

$$\int_{-\infty}^{\theta^{(.025)}} f_\theta\left(\hat{\theta}\right)\, d\hat{\theta} = 0.025 \quad \text{and} \quad \int_{-\infty}^{\theta^{(.975)}} f_\theta\left(\hat{\theta}\right)\, d\hat{\theta} = 0.975; \tag{11.73}$$

and let $I_\theta(\hat{\theta})$ be the indicator function for $\hat{\theta} \in (\theta^{(.025)}, \theta^{(.975)})$,

$$I_\theta\left(\hat{\theta}\right) = \begin{cases} 1 & \text{if } \theta^{(.025)} < \hat{\theta} < \theta^{(.975)} \\ 0 & \text{otherwise.} \end{cases} \tag{11.74}$$

By definition, $I_\theta(\hat{\theta})$ has a two-point probability distribution,

$$I_\theta\left(\hat{\theta}\right) = \begin{cases} 1 & \text{probability } 0.95 \\ 0 & \text{probability } 0.05. \end{cases} \tag{11.75}$$

This makes $I_\theta(\hat{\theta})$ a pivotal statistic, one whose distribution does not depend upon θ.

Neyman's construction takes the confidence interval $\mathcal{C}(\hat{\theta})$ corresponding to observed value $\hat{\theta}$ to be

$$\mathcal{C}\left(\hat{\theta}\right) = \left\{\theta : I_\theta\left(\hat{\theta}\right) = 1\right\}. \tag{11.76}$$

Then $\mathcal{C}(\hat{\theta})$ has the desired coverage property

$$\Pr_{\theta_0}\left\{\theta \in \mathcal{C}\left(\hat{\theta}\right)\right\} = \Pr_{\theta_0}\left\{I_\theta\left(\hat{\theta}\right) = 1\right\} = 0.95 \tag{11.77}$$

for any choice of the true parameter θ_0. (For the normal theory correlation density of $f_\theta(\hat{\theta})$, $\hat{\theta}(.025)$ and $\hat{\theta}(.975)$ are increasing functions of θ. This makes our previous construction (11.6) agree with (11.76).) The construction applies quite generally, as long as we are able to define acceptance regions of the sample space having the desired target probability content for every choice of θ. This can be challenging in multiparameter families.

†2 [p. 189] *Fisherian correctness.* Fisher, arguing against the Neyman paradigm, pointed out that confidence intervals could be accurate without being correct: having observed $x_i \overset{\text{iid}}{\sim} \mathcal{N}(\theta, 1)$ for $i = 1, 2, \ldots, 20$, the standard 0.95 interval based on just the first 10 observations would provide exact 0.95 coverage while giving obviously incorrect inferences for θ. If we can reduce the situation to form (11.22), the percentile method intervals satisfy Fisher's "logic of inductive inference" for correctness, as at (4.31).

†3 [p. 189] *Bootstrap sample sizes.* Why we need bootstrap sample sizes on the order of $B = 2000$ for confidence interval construction can be seen in the estimation of the bias correction value z_0 (11.32). The delta-method standard error of $z_0 = \Phi^{-1}(p_0)$ is calculated to be

$$\frac{1}{\varphi(z_0)}\left[\frac{p_0(1-p_0)}{B}\right]^{1/2}, \tag{11.78}$$

with $\varphi(z)$ the standard normal density. With $p_0 \doteq 0.5$ and $z_0 \doteq 0$ this is about $1.25/B^{1/2}$, equaling 0.028 at $B = 2000$, a none-too-small error for use in the BC formula (11.33) or the BCa formula (11.39).

†4 [p. 191] *The acceleration a.* This a appears in (11.38) as $d\sigma_\phi/d\phi$, the rate of change of $\hat{\phi}$'s standard deviation as a function of its expectation. In one-parameter exponential families it turns out that this is one-third of $d\sigma_\theta/d\theta$; that is, the transformation to normality $\phi = m(\theta)$ also decreases the instability of the standard deviation, though not to zero.

The variance of the score function $\dot{l}_x(\theta)$ determines the standard deviation of the MLE $\hat{\theta}$ (4.17)–(4.18). In one-parameter exponential families, one-sixth the *skewness* of $\dot{l}_x(\theta)$ gives a. The skewness connection can be seen at work in estimate (11.40). In multivariate exponential families (5.50), the skewness must be evaluated in the "least favorable" direction, discussed further in Chapter 13. The R algorithm `accel` (book web site) uses B parametric bootstrap replications $(\hat{\beta}^{*b}, \hat{\theta}^{*b})$ to estimate a. The percentile and BC intervals require only the replications $\hat{\theta}^{*b}$, while BCa also requires knowledge of the underlying exponential family. See Sections 4, 6, and 7 of Efron (1987).

†5 [p. 192] *BCa accuracy and correctness.* The BCa confidence limit $\hat{\theta}_{\text{BCa}}[\alpha]$ (11.39) is transformation invariant. Define

$$z[\alpha] = z_0 + \frac{z_0 + z^{(\alpha)}}{1 - a(z_0 + z^{(\alpha)})}, \tag{11.79}$$

so $\hat{\theta}_{\text{BCa}}[\alpha] = \hat{G}^{-1}\{\Phi[z[\alpha]]\}$. For a monotone increasing transformation $\phi = m(\theta)$, $\hat{\phi} = m(\hat{\theta})$, and $\hat{\phi}^* = m(r)$, the bootstrap cdf $\hat{H}$ of $\hat{\phi}^*$ satisfies $\hat{H}^{-1}(\alpha) = m[\hat{G}^{-1}(\alpha)]$ since $\hat{\phi}^{*(\alpha)} = m(\hat{\theta}^{*(\alpha)})$ for the bootstrap percentiles. Therefore

$$\hat{\phi}_{\text{BCa}}[\alpha] = \hat{H}^{-1}\left\{\Phi\left(z[\alpha]\right)\right\} = m\left(\hat{G}^{-1}\left\{\Phi\left(z[\alpha]\right)\right\}\right) = m\left(\hat{\theta}_{\text{BCa}}[\alpha]\right), \tag{11.80}$$

verifying transformation invariance. (Notice that $z_0 = \Phi^{-1}[\hat{G}(\hat{\theta})]$ equals $\Phi^{-1}[\hat{H}(\hat{\phi})]$ and is also transformation invariant, as is a, as discussed previously.)

Exact confidence intervals are transformation invariant, adding considerably to their inferential appeal. For approximate intervals, transformation invariance means that if we can demonstrate good behavior on any one scale then it remains good on all scales. The model (11.38) to the ϕ scale can be re-expressed as

$$\left\{1 + a\hat{\phi}\right\} = \{1 + a\phi\}\{1 + a(Z - z_0)\}, \tag{11.81}$$

where Z is a standard normal variate, $Z \sim \mathcal{N}(0, 1)$.

Taking logarithms,

$$\hat{\gamma} = \gamma + U, \tag{11.82}$$

where $\hat{\gamma} = \log\{1 + a\hat{\phi}\}$, $\gamma = \log\{1 + a\phi\}$, and U is the random variable $\log\{1 + a(Z - z_0)\}$; (11.82) represents the simplest kind of translation model, where the unknown value of γ rigidly shifts the distribution of U. The obvious confidence limit for γ,

$$\hat{\gamma}[\alpha] = \hat{\gamma} - U^{(1-\alpha)}, \tag{11.83}$$

where $U^{(1-\alpha)}$ is the $100(1-\alpha)$th percentile of U, is then accurate, and also "correct," according to Fisher's (admittedly vague) logic of inductive inference. It is an algebraic exercise, given in Section 3 of Efron (1987), to reverse the transformations $\theta \to \phi \to \gamma$ and recover $\hat{\theta}_{\text{BCa}}[\alpha]$ (11.39). Setting $a = 0$ shows the accuracy and correctness of $\hat{\theta}_{\text{BC}}[\alpha]$ (11.33).

†6 [p. 195] *Equation* (11.42). Model (11.41) makes $\hat{\theta} = \sum x_i^2$ a *noncentral chi-square variable* with noncentrality parameter $\theta = \sum \mu_i^2$ and n degrees of freedom, written as $\hat{\theta} \sim \chi^2_{\theta,n}$. With $\hat{\theta} = 20$ and $n = 10$, the parametric bootstrap distribution is $r \sim \chi^2_{20,10}$. Numerical evaluation gives $\Pr\{\chi^2_{20,10} \leq 20\} = 0.156$, leading to (11.42).

Efron (1985) concerns confidence intervals for parameters $\theta = t(\boldsymbol{\mu})$ in model (11.41), where third-order accurate confidence intervals can be calculated. The acceleration a equals zero for such problems, making the BC intervals second-order accurate. In practice, the BC intervals usually perform well, and are a reasonable choice if the accleration a is unavailable.

†7 [p. 202] *BCa confidence density* (11.68). Define

$$z_\theta = \Phi^{-1}\left[\hat{G}(\theta)\right] - z_0 = \frac{z_0 + z^{(\alpha)}}{1 - a\left(z_0 + z^{(\alpha)}\right)}, \tag{11.84}$$

so that

$$z^{(\alpha)} = \frac{z_\theta}{1 + az_\theta} - z_0 \quad \text{and} \quad \alpha = \Phi\left(\frac{z_\theta}{1 + az_\theta} - z_0\right). \tag{11.85}$$

Here we are thinking of α and θ as functionally related by $\theta = \hat{\theta}_{\text{BCa}}[\alpha]$. Differentiation yields

$$\begin{aligned} \frac{d\alpha}{dz_\theta} &= \frac{\varphi\left(\frac{z_\theta}{1+az_\theta} - z_0\right)}{(1 + az_\theta)^2}, \\ \frac{dz_\theta}{d\theta} &= \frac{\varphi\left(\frac{z_\theta}{1+az_\theta} - z_0\right)}{(1 + az_\theta)^2 \varphi(z_\theta + z_0)} \hat{g}(\theta), \end{aligned} \tag{11.86}$$

which together give $d\alpha/d\theta$, verifying (11.68).

The name "confidence density" seems to appear first in Efron (1993), though the idea is familiar in the fiducial literature. An ambitious frequentist theory of confidence distributions is developed in Xie and Singh (2013).

†8 [p. 203] *Jeffreys' prior.* Formula (11.72) is discussed further in Chapter 13, in the more general context of uninformative prior distributions. The theory of matching priors was initiated by Welch and Peers (1963), another important reference being Tibshirani (1989).

12

Cross-Validation and C_p Estimates of Prediction Error

Prediction has become a major branch of twenty-first-century commerce. Questions of prediction arise naturally: how credit-worthy is a loan applicant? Is a new email message `spam`? How healthy is the kidney of a potential donor? Two problems present themselves: how to construct an effective prediction rule, and how to estimate the accuracy of its predictions. In the language of Chapter 1, the first problem is more algorithmic, the second more inferential. Chapters 16–19, on *machine learning*, concern prediction rule construction. Here we will focus on the second question: having chosen a particular rule, how do we estimate its predictive accuracy?

Two quite distinct approaches to prediction error assessment developed in the 1970s. The first, depending on the classical technique of cross-validation, was fully general and nonparametric. A narrower (but more efficient) model-based approach was the second, emerging in the form of Mallows' C_p estimate and the Akaike information criterion (AIC). Both theories will be discussed here, beginning with cross-validation, after a brief overview of prediction rules.

12.1 Prediction Rules

Prediction problems typically begin with a *training set* $\boldsymbol{d}$ consisting of N pairs (x_i, y_i),

$$\boldsymbol{d} = \{(x_i, y_i),\ i = 1, 2, \dots, N\}, \tag{12.1}$$

where x_i is a vector of p *predictors* and y_i a real-valued *response*. On the basis of the training set, a *prediction rule* $r_{\boldsymbol{d}}(x)$ is constructed such that a prediction $\hat{y}$ is produced for any point x in the predictor's sample space $\mathcal{X}$,

$$\hat{y} = r_{\boldsymbol{d}}(x) \qquad \text{for } x \in \mathcal{X}. \tag{12.2}$$

The inferential task is to assess the accuracy of the rule's predictions. (In practice there are usually several competing rules under consideration and the main question is determining which is best.)

In the **spam** data of Section 8.1, x_i comprised $p = 57$ keyword counts, while y_i (8.18) indicated whether or not message i was **spam**. The rule $r_{\boldsymbol{d}}(x)$ in Table 8.3 was an MLE logistic regression fit. Given a new message's count vector, say x_0, $r_{\boldsymbol{d}}(x_0)$ provided an estimated probability $\hat{\pi}_0$ of it being **spam**, which could be converted into a prediction $\hat{y}_0$ according to

$$\hat{y}_0 = \begin{cases} 1 & \text{if } \hat{\pi}_0 \geq 0.5 \\ 0 & \text{if } \hat{\pi}_0 < 0.5. \end{cases} \tag{12.3}$$

The diabetes data of Table 7.2, Section 7.3, involved the $p = 10$ predictors $x = (\texttt{age}, \texttt{sex}, \ldots, \texttt{glu})$, obtained at baseline, and a response y measuring disease progression one year later. Given a new patient's baseline measurements x_0, we would like to predict his or her progression y_0. Table 7.3 suggests two possible prediction rules, ordinary least squares and ridge regression using ridge parameter $\lambda = 0.1$, either of which will produce a prediction $\hat{y}_0$. In this case we might assess prediction error in terms of squared error, $(y_0 - \hat{y}_0)^2$.

In both of these examples, $r_{\boldsymbol{d}}(x)$ was a regression estimator suggested by a probability model. One of the charms of prediction is that the rule $r_{\boldsymbol{d}}(x)$ need not be based on an explicit model. Regression trees, as pictured in Figure 8.7, are widely used[1] prediction algorithms that do not require model specifications. Prediction, perhaps because of its model-free nature, is an area where algorithmic developments have run far ahead of their inferential justification.

Quantifying the prediction error of a rule $r_{\boldsymbol{d}}(x)$ requires specification of the discrepancy $D(y, \hat{y})$ between a prediction $\hat{y}$ and the actual response y. The two most common choices are *squared error*

$$D(y, \hat{y}) = (y - \hat{y})^2, \tag{12.4}$$

and *classification error*

$$D(y, \hat{y}) = \begin{cases} 1 & \text{if } y \neq \hat{y} \\ 0 & \text{if } y = \hat{y}, \end{cases} \tag{12.5}$$

when, as with the **spam** data, the response y is dichotomous. (Prediction of a dichotomous response is often called "classification.")

1 *Random forests*, one of the most popular machine learning prediction algorithms, is an elaboration of regression trees. See Chapter 17.

For the purpose of error estimation, we suppose that the pairs (x_i, y_i) in the training set $\boldsymbol{d}$ of (12.1) have been obtained by random sampling from some probability distribution F on $(p+1)$-dimensional space $\mathcal{R}^{p+1}$,

$$(x_i, y_i) \overset{\text{iid}}{\sim} F \qquad \text{for } i = 1, 2, \ldots, N. \tag{12.6}$$

The *true error rate* $\text{Err}_{\boldsymbol{d}}$ of rule $r_{\boldsymbol{d}}(x)$ is the expected discrepancy of $\hat{y}_0 = r_{\boldsymbol{d}}(x_0)$ from y_0 given a new pair (x_0, y_0) drawn from F independently of $\boldsymbol{d}$,

$$\text{Err}_{\boldsymbol{d}} = E_F\{D(y_0, \hat{y}_0)\}; \tag{12.7}$$

$\boldsymbol{d}$ (and $r_{\boldsymbol{d}}(\cdot)$) is held fixed in expectation (12.7), only (x_0, y_0) varying.

Figure 12.1 concerns the *supernova data*, an example we will return to in
†1 the next section.† Absolute magnitudes y_i have been measured for $N = 39$ relatively nearby Type Ia supernovas, with the data scaled such that

$$y_i \overset{\text{ind}}{\sim} \mathcal{N}(\mu_i, 1), \qquad i = 1, 2, \ldots, 39, \tag{12.8}$$

is a reasonable model. For each supernova, a vector x_i of $p = 10$ spectral energies has been observed,

$$x_i = (x_{i1}, x_{i2}, \ldots, x_{i10}), \qquad i = 1, 2, \ldots, 39. \tag{12.9}$$

Table 12.1 shows (x_i, y_i) for $i = 1, 2, \ldots, 5$. (The frequency measurements have been standardized to have mean 0 and variance 1, while y has been adjusted to have mean 0.)

On the basis of the training set $\boldsymbol{d} = \{(x_i, y_i),\ i = 1, 2, \ldots, 39\}$, we wish to construct a rule $r_{\boldsymbol{d}}(x)$ that, given the frequency vector x_0 for a newly observed Type Ia supernova, accurately predicts[2] its absolute magnitude y_0. To this end, a *lasso* estimate $\tilde{\beta}(\lambda)$ was fit, with y in (7.42) the vector $(y_1, y_2, \ldots, y_{39})$ and x the 39×10 matrix having ith row x_i; λ was selected to minimize a C_p estimate of prediction error, Section 12.3, yielding prediction rule

$$\hat{y}_0 = x_0' \tilde{\beta}(\lambda). \tag{12.10}$$

(So in this case constructing $r_{\boldsymbol{d}}(x)$ itself involves error rate estimation.)

[2] Type Ia supernovas were used as "standard candles" in the discovery of dark energy and the cosmological expansion of the Universe, on the grounds that they have constant absolute magnitude. This isn't exactly true. Our training set is unusual in that the 39 supernovas are close enough to Earth to have y ascertained directly. This allows the construction of a prediction rule based on the frequency vector x, which *is* observable for distant supernovas, leading to improved calibration of the cosmological expansion.

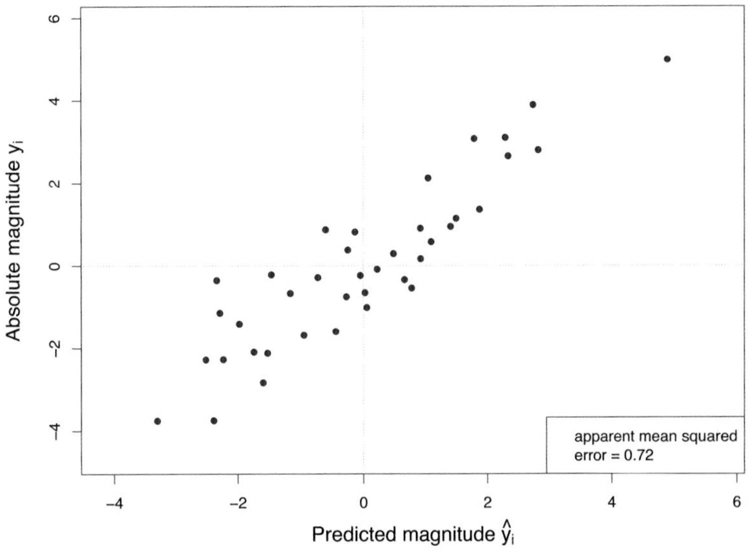

Figure 12.1 The supernova data; observed absolute magnitudes y_i (on log scale) plotted versus predictions $\hat{y}_i$ obtained from lasso rule (12.10), for $N = 39$ nearby Type Ia supernovas. Predictions based on 10 spectral power measurements, 7 of which had nonzero coefficients in $\tilde{\beta}(\lambda)$.

The plotted points in Figure 12.1 are $(\hat{y}_i, y_i)$ for $i = 1, 2, \ldots, N = 39$. These gave *apparent error*

$$\text{err} = \frac{1}{N}\sum_{i=1}^{N}(y_i - \hat{y}_i)^2 = 0.720. \tag{12.11}$$

Comparing this with $\sum(y_i - \bar{y})^2/N = 3.91$ yields an impressive-looking "R squared" value

$$R^2 = 1 - 0.720/3.91 = 0.816. \tag{12.12}$$

Things aren't really that good (see (12.23)). Cross-validation and C_p methods allow us to correct apparent errors for the fact that $r_{\boldsymbol{d}}(x)$ was chosen to make the predictions $\hat{y}_i$ fit the data y_i.

Prediction and estimation are close cousins but they are not twins. As discussed earlier, prediction is less model-dependent, which partly accounts for the distinctions made in Section 8.4. The prediction criterion Err (12.7)

Table 12.1 *Supernova data; 10 frequency measurements and response variable "absolute magnitude" for the first 5 of $N = 39$ Type Ia supernovas. In terms of notation* (12.1), *frequency measurements are x and magnitude y.*

	SN1	SN2	SN3	SN4	SN5
x_1	−.84	−1.89	.26	−.08	.41
x_2	−.93	−.46	−.80	1.02	−.81
x_3	.32	2.41	1.14	−.21	−.13
x_4	.18	.77	−.86	−1.12	1.31
x_5	−.68	−.94	.68	−.86	−.65
x_6	−1.27	−1.53	−.35	.72	.30
x_7	.34	.09	−1.04	.62	−.82
x_8	−.43	.26	−1.10	.56	−1.53
x_9	−.02	.18	−1.32	.62	−1.49
x_{10}	−.3	−.54	−1.70	−.49	−1.09
mag	**−.54**	**2.12**	**−.22**	**.95**	**−3.75**

is an expectation over the (x, y) space. This emphasizes good overall performance, without much concern for behavior at individual points x in $\mathcal{X}$.

Shrinkage usually improves prediction. Consider a Bayesian model like that of Section 7.1,

$$\mu_i \sim \mathcal{N}(0, A) \quad \text{and} \quad x_i|\mu_i \sim \mathcal{N}(\mu_i, 1) \qquad \text{for } i = 1, 2, \dots, N. \tag{12.13}$$

The Bayes shrinkage estimator, which is ideal for estimation,

$$\hat{\mu}_i = Bx_i, \qquad B = A/(A+1), \tag{12.14}$$

is also ideal for prediction. Suppose that in addition to the observations x_i there are independent unobserved replicates, one for each of the N x_i values,

$$y_i \sim \mathcal{N}(\mu_i, 1) \qquad \text{for } i = 1, 2, \dots, N, \tag{12.15}$$

that we wish to predict. The Bayes predictor

$$\hat{y}_i = Bx_i \tag{12.16}$$

has overall Bayes prediction error

$$E\left\{\frac{1}{N}\sum_{i=1}^{N}(y_i - \hat{y}_i)^2\right\} = B + 1, \tag{12.17}$$

which cannot be improved upon. The MLE rule $\hat{y}_i = x_i$ has Bayes prediction error 2, which is always worse than (12.17).

As far as prediction is concerned it pays to overshrink, as illustrated in Figure 7.1 for the James–Stein version of situation (12.13). This is fine for prediction, but less fine for estimation if we are concerned about extreme cases; see Table 7.4. Prediction rules sacrifice the extremes for the sake of the middle, a particularly effective tactic in dichotomous situations (12.5), where the cost of individual errors is bounded. The most successful machine learning prediction algorithms, discussed in Chapters 16–19, carry out a version of local Bayesian shrinkage in selected regions of $\mathcal{X}$.

12.2 Cross-Validation

Having constructed a prediction rule $r_{\boldsymbol{d}}(x)$ on the basis of training set $\boldsymbol{d}$, we wish to know its prediction error $\text{Err} = E_F\{D(y_0, \hat{y}_0)\}$ (12.7) for a new case obtained independently of $\boldsymbol{d}$. A first guess is the apparent error

$$\text{err} = \frac{1}{N}\sum_{i=1}^{N} D(y_i, \hat{y}_i), \tag{12.18}$$

the average discrepancy in the training set between y_i and its prediction $\hat{y}_i = r_{\boldsymbol{d}}(x_i)$; err usually underestimates Err since $r_{\boldsymbol{d}}(x)$ has been adjusted[3] to fit the observed responses y_i.

The ideal remedy, discussed in Section 12.4, would be to have an independent *validation set* (or *test* set) $\boldsymbol{d}_{\text{val}}$ of N_{val} additional cases,

$$\boldsymbol{d}_{\text{val}} = \left\{(x_{0j}, y_{0j}),\ j = 1, 2, \dots, N_{\text{val}}\right\}. \tag{12.19}$$

This would provide as unbiased estimate of Err,

$$\widehat{\text{Err}}_{\text{val}} = \frac{1}{N_{\text{val}}}\sum_{j=1}^{N_{\text{val}}} D(y_{0j}, \hat{y}_{0j}), \qquad \hat{y}_{0j} = r_{\boldsymbol{d}}(x_{0j}). \tag{12.20}$$

Cross-validation attempts to mimic $\widehat{\text{Err}}_{\text{val}}$ without the need for a validation set. Define $\boldsymbol{d}(i)$ to be the reduced training set in which pair (x_i, y_i) has been omitted, and let $r_{\boldsymbol{d}(i)}(\cdot)$ indicate the rule constructed on the basis

[3] Linear regression using ordinary least squares fitting provides a classical illustration: $\text{err} = \sum_i (y_i - \hat{y}_i)^2/N$ must be increased to $\sum_i (y_i - \hat{y}_i)^2/(N - p)$, where p is the degrees of freedom, to obtain an unbiased estimate of the noise variance σ^2.

of $\boldsymbol{d}(i)$. The *cross-validation estimate* of prediction error is

$$\widehat{\text{Err}}_{\text{cv}} = \frac{1}{N}\sum_{i=1}^{N} D(y_i, \hat{y}_{(i)}), \qquad \hat{y}_{(i)} = r_{\boldsymbol{d}(i)}(x_i). \tag{12.21}$$

Now (x_i, y_i) is *not* involved in the construction of the prediction rule for y_i.

$\widehat{\text{Err}}_{\text{cv}}$ (12.21) is the "leave one out" version of cross-validation. A more common tactic is to leave out several pairs at a time: $\boldsymbol{d}$ is randomly partitioned into J groups of size about N/J each; $\boldsymbol{d}(j)$, the training set with group j omitted, provides rule $r_{\boldsymbol{d}(j)}(x)$, which is used to provide predictions for the y_i in group j. Then $\widehat{\text{Err}}_{\text{cv}}$ is evaluated as in (12.21). Besides reducing the number of rule constructions necessary, from N to J, grouping induces larger changes among the J training sets, improving the predictive performance on rules $r_{\boldsymbol{d}}(x)$ that include discontinuities. (The argument here is similar to that for the jackknife, Section 10.1.)

Cross-validation was applied to the supernova data pictured in Figure 12.1. The 39 cases were split, randomly, into $J = 13$ groups of three cases each. This gave

$$\widehat{\text{Err}}_{\text{cv}} = 1.17, \tag{12.22}$$

(12.21), 62% larger than $\text{err} = 0.72$ (12.9). The R^2 calculation (12.12) now yields the smaller value

$$R^2 = 1 - 1.17/3.91 = 0.701. \tag{12.23}$$

We can apply cross-validation to the `spam` data of Section 8.1, having $N = 4061$ cases, $p = 57$ predictors, and dichotomous response y. For this example, each of the 57 predictors was itself dichotomized to be either 0 or 1 depending on whether the original value x_{ij} equaled zero or not. A logistic regression, Section 8.1, regressing y_i on the 57 dichotomized predictors, gave apparent classification error (12.5)

$$\text{err} = 0.064, \tag{12.24}$$

i.e., 295 wrong predictions among the 4061 cases. Cross-validation, with $J = 10$ groups of size 460 or 461 each, increased this to

$$\widehat{\text{Err}}_{\text{cv}} = 0.069, \tag{12.25}$$

an increase of 8%.

`Glmnet` is an automatic model building program that, among other things, constructs a lasso sequence of logistic regression models, adding

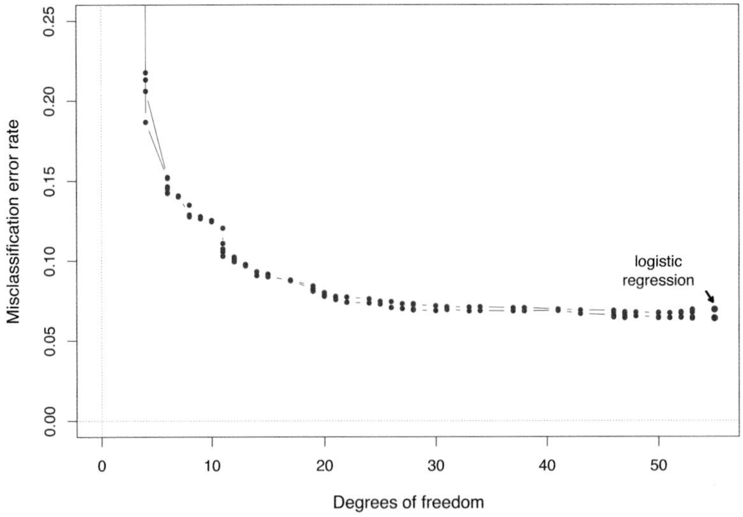

Figure 12.2 Spam data. Apparent error rate err (blue) and cross-validated estimate (red) for a sequence of prediction rules generated by `glmnet`. The degrees of freedom are the number of nonzero regression coefficients: df = 57 corresponds to ordinary logistic regression, which gave apparent err 0.064, cross-validated rate 0.069. The minimum cross-validated error rate is 0.067.

variables one at a time in their order of apparent predictive power; see Chapter 16. The solid curve in Figure 12.2 tracks the apparent error err (12.18) as a function of the number of predictors employed. Aside from numerical artifacts, err is monotonically decreasing, declining to err $= 0.064$ for the full model that employs all 57 predictors, i.e., for the usual logistic regression model, as in (12.24).

`Glmnet` produced prediction error estimates $\widehat{\mathrm{Err}}_{\mathrm{cv}}$ for each of the successive models, shown by the dashed curve. These are a little noisy themselves, but settle down between 4% and 8% above the corresponding err estimates. The minimum value

$$\widehat{\mathrm{Err}}_{\mathrm{cv}} = 0.067 \tag{12.26}$$

occurred for the model using 47 predictors.

The difference between (12.26) and (12.25) is too small to take seriously given the noise in the $\widehat{\mathrm{Err}}_{\mathrm{cv}}$ estimates. There is a more subtle objection: the choice of "best" prediction rule based on comparative $\widehat{\mathrm{Err}}_{\mathrm{cv}}$ estimates is not itself cross-validated. Each case (x_i, y_i) is involved in choosing its own best prediction, so $\widehat{\text{Err}}_{\text{cv}}$ at the apparently optimum choice cannot be taken entirely at face value.

Nevertheless, perhaps the principal use of cross-validation lies in choosing among competing prediction rules. Whether or not this is fully justified, it is often the only game in town. That being said, minimum predictive error, no matter how effectuated, is a notably weaker selection principle than minimum variance of estimation.

As an example, consider an iid normal sample

$$x_i \overset{\text{iid}}{\sim} \mathcal{N}(\mu, 1), \qquad i = 1, 2, \ldots, 25, \tag{12.27}$$

having mean $\bar{x}$ and median $\breve{x}$. Both are unbiased for estimating μ, but $\bar{x}$ is much more efficient,

$$\text{var}(\breve{x}) / \text{var}(\bar{x}) \doteq 1.57. \tag{12.28}$$

Suppose we wish to predict a future observation x_0 independently selected from the same $\mathcal{N}(\mu, 1)$ distribution. In this case there is very little advantage to $\bar{x}$,

$$E\left\{(x_0 - \breve{x})^2\right\} / E\left\{(x_0 - \bar{x})^2\right\} = 1.02. \tag{12.29}$$

The noise in $x_0 \sim \mathcal{N}(\mu, 1)$ dominates its prediction error. Perhaps the proliferation of prediction algorithms to be seen in Part III reflects how weakly changes in strategy affect prediction error.

Table 12.2 *Ratio of predictive errors $E\{(\bar{x}_0 - \breve{x})^2\}/E\{(\bar{x}_0 - \bar{x})^2\}$ for $\bar{x}_0$ the mean of an independent sample of size N_0 from $\mathcal{N}(\mu, 1)$; $\bar{x}$ and $\breve{x}$ are the mean and median from $x_i \sim \mathcal{N}(\mu, 1)$ for $i = 1, 2, \ldots, 25$.*

N_0	1	10	100	1000	∞
Ratio	1.02	1.16	1.46	1.56	1.57

In this last example, suppose that our task was to predict the *average* $\bar{x}_0$ of N_0 further draws from the $\mathcal{N}(\mu, 1)$ distribution. Table 12.2 shows the ratio of predictive errors as a function of N_0. The superiority of the mean compared to the median reveals itself as N_0 gets larger. In this super-simplified example, the difference between prediction and estimation lies in predicting the average of *one* versus an *infinite number* of future observations.

Does $\widehat{\text{Err}}_{\text{cv}}$ actually estimate $\text{Err}_{\boldsymbol{d}}$ as defined in (12.7)? It seems like the answer must be yes, but there is some doubt expressed in the literature, for reasons demonstrated in the following simulation: we take the true distribution F in (12.6) to be the discrete distribution $\hat{F}$ that puts weight 1/39 on each of the 39 (x_i, y_i) pairs of the supernova data.[4] A random sample with replacement of size 39 from $\hat{F}$ gives simulated data set $\boldsymbol{d}^*$ and prediction rule $r_{\boldsymbol{d}^*}(\cdot)$ based on the lasso/C_p recipe used originally. The same cross-validation procedure as before, applied to $\boldsymbol{d}^*$, gives $\widehat{\text{Err}}^*_{\text{cv}}$. Because this is a simulation, we can also compute the actual mean-squared error rate of rule $r_{\boldsymbol{d}^*}(\cdot)$ applied to the true distribution $\hat{F}$,

$$\text{Err}^* = \frac{1}{39} \sum_{i=1}^{39} D\left(y_i, r_{\boldsymbol{d}^*}(x_i)\right). \tag{12.30}$$

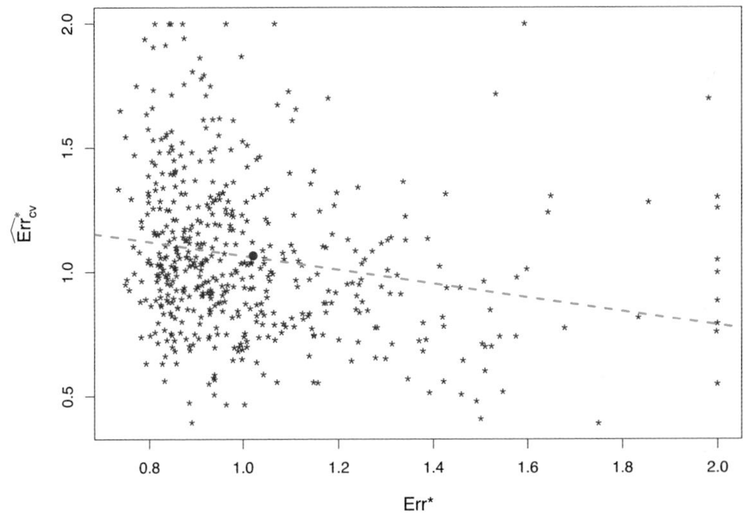

Figure 12.3 Simulation experiment comparing true error Err with cross-validation estimate $\widehat{\text{Err}}^*_{\text{cv}}$; 500 simulations based on the supernova data. $\widehat{\text{Err}}^*_{\text{cv}}$ and Err are negatively correlated.

Figure 12.3 plots $(\text{Err}^*, \widehat{\text{Err}}^*_{\text{cv}})$ for 500 simulations, using squared error discrepancy $D(y, \hat{y}) = (y - \hat{y})^2$. Summary statistics are given in Table 12.3. $\widehat{\text{Err}}^*_{\text{cv}}$ has performed well overall, averaging 1.07, quite near the true Err 1.02, both estimates being 80% greater than the average apparent error 0.57. However, the figure shows something unsettling: there is a

4 Simulation based on $\hat{F}$ is the same as nonparametric bootstrap analysis, Chapter 10.

Table 12.3 *True error* Err^*, *cross-validated error* $\widehat{\text{Err}}^*_{\text{cv}}$, *and apparent error* err^*; *500 simulations based on supernova data. Correlation* -0.175 *between* Err^* *and* $\widehat{\text{Err}}^*_{\text{cv}}$.

	Err^*	$\widehat{\text{Err}}^*_{\text{cv}}$	err^*
Mean	1.02	1.07	.57
St dev	.27	.34	.16

negative correlation between $\widehat{\text{Err}}^*_{\text{cv}}$ and Err^*. Large values of $\widehat{\text{Err}}^*_{\text{cv}}$ go with smaller values of the true prediction error, and vice versa.

Our original definition of Err,

$$\text{Err}_{\boldsymbol{d}} = E_F\left\{D(y_0, r_{\boldsymbol{d}}(x_0))\right\}, \tag{12.31}$$

took $r_{\boldsymbol{d}}(\cdot)$ fixed as constructed from $\boldsymbol{d}$, only $(x_0, y_0) \sim F$ random. In other words, $\text{Err}_{\boldsymbol{d}}$ was the expected prediction error for the specific rule $r_{\boldsymbol{d}}(\cdot)$, as is Err^* for $r_{\boldsymbol{d}^*}(\cdot)$. If $\widehat{\text{Err}}^*_{\text{cv}}$ is tracking Err^* we would expect to see a positive correlation in Figure 12.3.

As it is, all we can say is that $\widehat{\text{Err}}^*_{\text{cv}}$ is estimating the *expected* predictive error, where $\boldsymbol{d}$ as well as (x_0, y_0) is random in definition (12.31). This makes cross-validation a *strongly* frequentist device: $\widehat{\text{Err}}_{\text{cv}}$ is estimating the average prediction error of the algorithm producing $r_{\boldsymbol{d}}(\cdot)$, not of $r_{\boldsymbol{d}}(\cdot)$ itself.

12.3 Covariance Penalties

Cross-validation does its work nonparametrically and without the need for probabilistic modeling. Covariance penalty procedures require probability models, but within their ambit they provide less noisy estimates of prediction error. Some of the most prominent covariance penalty techniques will be examined here, including *Mallows'* C_p, *Akaike's information criterion* (AIC), and *Stein's unbiased risk estimate* (SURE).

The covariance penalty approach treats prediction error estimation in a regression framework: the predictor vectors x_i in the training set $\boldsymbol{d} = \{(x_i, y_i)\}, i = 1, 2, \ldots, N\}$ (12.1) are considered *fixed* at their observed values, not random as in (12.6). An unknown vector $\boldsymbol{\mu}$ of expectations $\mu_i = E\{y_i\}$ has yielded the observed vector of responses $\boldsymbol{y}$ according to some given probability model, which to begin with we assume to have the simple form

$$\boldsymbol{y} \sim (\boldsymbol{\mu}, \sigma^2 \boldsymbol{I}); \tag{12.32}$$

that is, the y_i are uncorrelated, with y_i having unknown mean μ_i and variance σ^2. We take σ^2 as known, though in practice it must usually be estimated.

A regression rule $r(\cdot)$ has been used to produce an estimate of vector $\boldsymbol{\mu}$,

$$\hat{\boldsymbol{\mu}} = r(\boldsymbol{y}). \tag{12.33}$$

(Only $\boldsymbol{y}$ is included in the notation since the predictors x_i are considered fixed and known.) For instance we might take

$$\hat{\boldsymbol{\mu}} = r(\boldsymbol{y}) = \boldsymbol{X}(\boldsymbol{X}'\boldsymbol{X})^{-1}\boldsymbol{X}'\boldsymbol{y}, \tag{12.34}$$

where $\boldsymbol{X}$ is the $N \times p$ matrix having x_i as the ith row, as suggested by the linear regression model $\boldsymbol{\mu} = \boldsymbol{X}\beta$.

In covariance penalty calculations, the estimator $\hat{\boldsymbol{\mu}}$ also functions as a predictor. We wonder how accurate $\hat{\boldsymbol{\mu}} = r(\boldsymbol{y})$ will be in predicting a new vector of observations $\boldsymbol{y}_0$ from model (12.32),

$$\boldsymbol{y}_0 \sim (\boldsymbol{\mu}, \sigma^2 \boldsymbol{I}), \qquad \text{independent of } \boldsymbol{y}. \tag{12.35}$$

To begin with, prediction error will be assessed in terms of squared discrepancy,

$$\text{Err}_i = E_0\left\{(y_{0i} - \hat{\mu}_i)^2\right\} \tag{12.36}$$

for component i, where E_0 indicates expectation with y_{0i} random but $\hat{\mu}_i$ held fixed. Overall prediction error is the average[5]

$$\text{Err.} = \frac{1}{N} \sum_{i=1}^{N} \text{Err}_i \,. \tag{12.37}$$

The *apparent error* for component i is

$$\text{err}_i = (y_i - \hat{\mu}_i)^2. \tag{12.38}$$

A simple but powerful lemma underlies the theory of covariance penalties.

Lemma *Let E indicate expectation over both $\boldsymbol{y}$ in* (12.32) *and $\boldsymbol{y}_0$ in* (12.35). *Then*

$$E\{\text{Err}_i\} = E\{\text{err}_i\} + 2\,\text{cov}(\hat{\mu}_i, y_i), \tag{12.39}$$

[5] Err. is sometimes called "insample error," as opposed to "outsample error" Err (12.7), though in practice the two tend to behave similarly.

where the last term is the covariance between the ith components of $\hat{\boldsymbol{\mu}}$ and $\boldsymbol{y}$,

$$\operatorname{cov}(\hat{\mu}_i, y_i) = E\left\{(\hat{\mu}_i - \mu_i)(y_i - \mu_i)\right\}. \tag{12.40}$$

(Note: (12.40) does not require $E\{\hat{\mu}_i\} = \mu_i$.)

Proof Letting $\epsilon_i = y_i - \mu_i$ and $\delta_i = (\hat{\mu}_i - \mu_i)$, the elementary equality $(\epsilon_i - \delta_i)^2 = \epsilon_i^2 - 2\epsilon_i\delta_i + \delta_i^2$ becomes

$$(y_i - \hat{\mu}_i)^2 = (y_i - \mu_i)^2 - 2(\hat{\mu}_i - \mu_i)(y_i - \mu_i) + (\hat{\mu}_i - \mu_i)^2, \tag{12.41}$$

and likewise

$$(y_{0i} - \hat{\mu}_i)^2 = (y_{0i} - \mu_i)^2 - 2(\hat{\mu}_i - \mu_i)(y_{0i} - \mu_i) + (\hat{\mu}_i - \mu_i)^2. \tag{12.42}$$

Taking expectations, (12.41) gives

$$E\{\text{err}_i\} = \sigma^2 - 2\operatorname{cov}(\hat{\mu}_i, y_i) + E(\hat{\mu}_i - \mu_i)^2, \tag{12.43}$$

while (12.42) gives

$$E\{\text{Err}_i\} = \sigma^2 + E(\hat{\mu}_i - \mu_i)^2, \tag{12.44}$$

the middle term on the right side of (12.42) equaling zero because of the independence of y_{0i} and $\hat{\mu}_i$. Taking the difference between (12.44) and (12.43) verifies the lemma. ■

Note: The lemma remains valid if σ^2 varies with i.

The lemma says that, on average, the apparent error err_i understimates the true prediction error Err_i by the *covariance penalty* $2\operatorname{cov}(\hat{\mu}_i, y_i)$. (This makes intuitive sense since $\operatorname{cov}(\mu_i, y_i)$ measures the amount by which y_i influences its own prediction $\hat{\mu}_i$.) Covariance penalty estimates of prediction error take the form

$$\widehat{\text{Err}}_i = \text{err}_i + 2\widehat{\operatorname{cov}}(\hat{\mu}_i, y_i), \tag{12.45}$$

where $\widehat{\operatorname{cov}}(\hat{\mu}_i, y_i)$ approximates $\operatorname{cov}(\mu_i, y_i)$; overall prediction error (12.37) is estimated by

$$\widehat{\text{Err}}_{\cdot} = \text{err} + \frac{2}{N}\sum_{i=1}^{N}\widehat{\operatorname{cov}}(\hat{\mu}_i, y_i), \tag{12.46}$$

where $\text{err} = \sum \text{err}_i / N$ as before.

The form of $\widehat{\operatorname{cov}}(\hat{\mu}_i, y_i)$ in (12.45) depends on the context assumed for the prediction problem.

(1) Suppose that $\hat{\boldsymbol{\mu}} = r(\boldsymbol{y})$ in (12.32)–(12.33) is *linear*,

$$\hat{\boldsymbol{\mu}} = \boldsymbol{c} + \boldsymbol{M}\boldsymbol{y}, \tag{12.47}$$

where $\boldsymbol{c}$ is a known N-vector and $\boldsymbol{M}$ a known $N \times N$ matrix. Then the covariance matrix between $\hat{\boldsymbol{\mu}}$ and $\boldsymbol{y}$ is

$$\text{cov}(\hat{\boldsymbol{\mu}}, \boldsymbol{y}) = \sigma^2 \boldsymbol{M}, \tag{12.48}$$

giving $\text{cov}(\hat{\mu}_i, y_i) = \sigma^2 M_{ii}$, M_{ii} the ith diagonal element of $\boldsymbol{M}$,

$$\widehat{\text{Err}}_i = \text{err}_i + 2\sigma^2 M_{ii}, \tag{12.49}$$

and, since $\text{err} = \sum_i (y_i - \hat{\mu}_i)^2/N$,

$$\widehat{\text{Err}}. = \frac{1}{N}\sum_{i=1}^{N}(y_i - \mu_i)^2 + \frac{2\sigma^2}{N}\text{tr}(\boldsymbol{M}). \tag{12.50}$$

Formula (12.50) is *Mallows' C_p* estimate of prediction error. For OLS estimation (12.34), $\boldsymbol{M} = \boldsymbol{X}(\boldsymbol{X}'\boldsymbol{X})^{-1}\boldsymbol{X}'$ has $\text{tr}(\boldsymbol{M}) = p$, the number of predictors, so

$$\widehat{\text{Err}}. = \frac{1}{N}\sum_{i=1}^{N}(y_i - \hat{\mu}_i)^2 + \frac{2}{N}\sigma^2 p. \tag{12.51}$$

For the supernova data (12.8)–(12.9), the OLS predictor $\hat{\boldsymbol{\mu}} = \boldsymbol{X}(\boldsymbol{X}'\boldsymbol{X})^{-1}\boldsymbol{X}'\boldsymbol{y}$ yielded $\text{err} = \sum(y_i - \hat{\mu}_i)^2/39 = 0.719$. The covariance penalty, with $N = 39$, $\sigma^2 = 1$, and[6] $p = 10$, was 0.513, giving C_p estimate of prediction error

$$\widehat{\text{Err}}. = 0.719 + 0.513 = 1.23. \tag{12.52}$$

For OLS regression, the *degrees of freedom* p, the rank of matrix $\boldsymbol{X}$ in (12.34), determines the covariance penalty $(2/N)\sigma^2 p$ in (12.51). Comparing this with (12.46) leads to a general definition of degrees of freedom df for a regression rule $\hat{\boldsymbol{\mu}} = r(\boldsymbol{y})$,

$$\text{df} = (1/\sigma^2)\sum_{i=1}^{N}\widehat{\text{cov}}(\hat{\mu}_i, y_i). \tag{12.53}$$

This definition provides common ground for comparing different types of regression rules. Rules with larger df are more flexible and tend toward better apparent fits to the data, but require bigger covariance penalties for fair comparison.

[6] We are not counting the intercept as an 11th predictor since $\boldsymbol{y}$ and all the x_i were standardized to have mean 0, all our models assuming zero intercept.

(2) For lasso estimation (7.42) and (12.10), it can be shown that formula (12.51), with p equaling the number of nonzero regression coeffi-
†2 cients, holds to a good approximation.† The lasso rule used in Figure 12.1 for the supernova data had $p = 7$; err was 0.720 for this rule, almost the same as for the OLS rule above, but the C_p penalty is less, $2\cdot 7/39 = 0.359$, giving

$$\widehat{\text{Err}}. = 0.720 + 0.359 = 1.08, \tag{12.54}$$

compared with 1.23 for OLS. This estimate does not account for the data-based selection of the choice $p = 7$, see item **(4)** below.

(3) If we are willing to add multivariate normality to model (12.32),

$$\boldsymbol{y} \sim \mathcal{N}_p(\boldsymbol{\mu}, \sigma^2 \boldsymbol{I}), \tag{12.55}$$

we can drop the assumption of linearity (12.47). In this case it can be shown that, for any differentiable estimator $\hat{\boldsymbol{\mu}} = r(\boldsymbol{y})$, the covariance in
†3 formula (12.51) is given by†

$$\text{cov}(\hat{\mu}_i, y_i) = \sigma^2 E\{\partial\hat{\mu}_i/\partial y_i\}, \tag{12.56}$$

σ^2 times the partial derivative of $\hat{\mu}_i$ with respect to y_i. (Another measure of y_i's influence on its own prediction.) The SURE formula (Stein's unbiased risk estimator) is

$$\widehat{\text{Err}}_i = \text{err}_i + 2\sigma^2 \frac{\partial\hat{\mu}_i}{\partial y_i}, \tag{12.57}$$

with corresponding estimate for overall prediction error

$$\widehat{\text{Err}}. = \text{err} + \frac{2\sigma^2}{N} \sum_{i=1}^{N} \frac{\partial\hat{\mu}_i}{\partial y_i}. \tag{12.58}$$

SURE was applied to the rule $\hat{\boldsymbol{\mu}} =$ `lowess(x,y,1/3)` for the kidney fitness data of Figure 1.2. The open circles in Figure 12.4 plot the component-wise degrees of freedom estimates[7]

$$\frac{\partial\hat{\mu}_i}{\partial y_i}, \qquad i = 1, 2, \ldots, N = 157, \tag{12.59}$$

(obtained by numerical differentiation) versus age_i. Their sum

$$\sum_{i=1}^{N} \frac{\partial\hat{\mu}_i}{\partial y_i} = 6.67 \tag{12.60}$$

[7] Notice that the factor σ^2 in (12.56) cancels out in (12.53).

estimates the total degrees of freedom, as in (12.53), implying that `lowess(x,y,1/3)` is about as flexible as a sixth-degree polynomial fit, with df = 7.

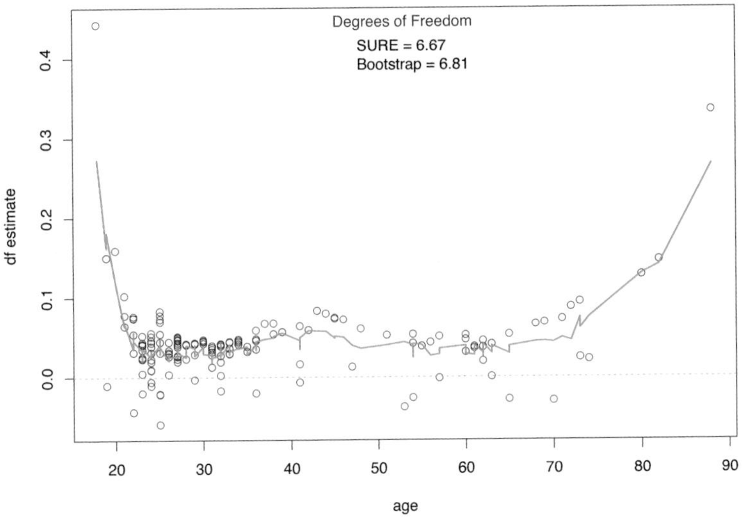

Figure 12.4 Analysis of the `lowess(x,y,1/3)` fit to kidney data of Figure 1.2. Open circles are SURE coordinate-wise df estimates $\partial\hat{\mu}_i/\partial y_i$, plotted versus age$_i$, giving total degrees of freedom 6.67. The solid curve tracks bootstrap coordinate-wise estimates (12.65), with their sum giving total df = 6.81.

(4) The *parametric bootstrap*[8] of Section 10.4 can be used to estimate the covariances $\text{cov}(\hat{\mu}_i, y_i)$ in the lemma (12.39). The data vector $\boldsymbol{y}$ is assumed to be generated from a member $f_{\boldsymbol{\mu}}(\boldsymbol{y})$ of a given parametric family

$$\mathcal{F} = \left\{f_{\boldsymbol{\mu}}(\boldsymbol{y}),\ \boldsymbol{\mu} \in \Omega\right\}, \tag{12.61}$$

yielding $\hat{\boldsymbol{\mu}} = r(\boldsymbol{y})$,

$$f_{\boldsymbol{\mu}} \rightarrow \boldsymbol{y} \rightarrow \hat{\boldsymbol{\mu}} = r(\boldsymbol{y}). \tag{12.62}$$

Parametric bootstrap replications of $\boldsymbol{y}$ and $\hat{\boldsymbol{\mu}}$ are obtained by analogy with

[8] There is also a *nonparametric* bootstrap competitor to cross-validation, the ".632 estimate;" see the chapter endnote †4.

(12.62),[9]

$$f_{\hat{\boldsymbol{\mu}}} \rightarrow \boldsymbol{y}^* \rightarrow \hat{\boldsymbol{\mu}}^* = r(\boldsymbol{y}^*). \tag{12.63}$$

A large number B of replications then yield bootstrap estimates

$$\widehat{\text{cov}}(\hat{\mu}_i, y_i) = \frac{1}{B}\sum_{j=1}^{B}(\hat{\mu}_i^{*b} - \hat{\mu}_i^{*\cdot})(y_i^{*b} - y_i^{*\cdot}), \tag{12.64}$$

the dot notation indicating averages over the B replications.

$B = 1000$ parametric bootstrap replications $(\hat{\boldsymbol{\mu}}^*, \boldsymbol{y}^*)$ were obtained from the normal model (12.55), taking $\hat{\boldsymbol{\mu}}$ in (12.63) to be the estimate from `lowess(x,y,1/3)` as in Figure 1.2. A standard linear regression, of y as a 12th-degree polynomial function of age, gave $\hat{\sigma}^2 = 3.28$. Covariances were computed as in (12.64), yielding coordinate-wise degrees of freedom estimates (12.53),

$$\text{df}_i = \widehat{\text{cov}}(\hat{\mu}_i, y_i)/\hat{\sigma}^2. \tag{12.65}$$

The solid curve in Figure 12.4 plots df_i as a function of age_i. These are seen to be similar to but less noisy than the SURE estimates. They totaled 6.81, nearly the same as (12.60). The overall covariance penalty term in (12.46) equaled 0.284, increasing $\widehat{\text{Err}}$ by about 9% over $\text{err} = 3.15$.

The advantage of parametric bootstrap estimates (12.64) of covariance penalties is their applicability to *any* prediction rule $\hat{\boldsymbol{\mu}} = r(\boldsymbol{y})$ no matter how exotic. Applied to the lasso estimates for the supernova data, $B = 1000$ replications yielded total $\text{df} = 6.85$ for the rule that always used $p = 7$ predictors, compared with the theoretical approximation $\text{df} = 7$. Another 1000 replications, now letting $\hat{\boldsymbol{\mu}}^* = r(\boldsymbol{y}^*)$ choose the apparently best p^* each time, increased the df estimate to 7.48, so the adaptive choice of p cost about 0.6 extra degrees of freedom. These calculations exemplify modern computer-intensive inference, carrying through error estimation for complicated adaptive prediction rules on a totally automatic basis.

(5) Covariance penalties can apply to measures of prediction error other than squared error $D(y_i, \hat{\mu}_i) = (y_i - \hat{\mu}_i)^2$. We will discuss two examples of a general theory. First consider *classification*, where y_i equals 0 or 1 and

[9] It isn't necessary for the $\hat{\boldsymbol{\mu}}$ in (12.63) to equal $\hat{\boldsymbol{\mu}} = r(\boldsymbol{y})$. The calculation (12.64) was rerun taking $\hat{\boldsymbol{\mu}}$ in (12.63) from `lowess(x,y,1/6)` (but with $r(\boldsymbol{y})$ still from `lowess(x,y,1/3)`) with almost identical results. In general, one might take $\hat{\boldsymbol{\mu}}$ in (12.63) to be from a more flexible, less biased, estimator than $r(\boldsymbol{y})$.

similarly the predictor $\hat{\mu}_i$, with dichotomous error

$$D(y_i, \hat{\mu}_i) = \begin{cases} 1 & \text{if } y_i \neq \hat{\mu}_i \\ 0 & \text{if } y_i = \hat{\mu}_i, \end{cases} \tag{12.66}$$

as in (12.5).[10] In this situation, the apparent error is the observed proportion of prediction mistakes in the training set (12.1),

$$\text{err} = \#\{y_i \neq \hat{\mu}_i\}/N. \tag{12.67}$$

Now the true prediction error for case i is

$$\text{Err}_i = \text{Pr}_0\{y_{0i} \neq \hat{\mu}_i\}, \tag{12.68}$$

the conditional probability given $\hat{\mu}_i$ that an independent replicate y_{0i} of y_i will be incorrectly predicted. The lemma holds as stated in (12.39), leading to the prediction error estimate

$$\widehat{\text{Err}}. = \frac{\#\{y_i \neq \hat{\mu}_i\}}{N} + \frac{2}{N} \sum_{i=1}^{N} \text{cov}(\hat{\mu}_i, y_i). \tag{12.69}$$

Some algebra yields

$$\text{cov}(\hat{\mu}_i, y_i) = \mu_i(1 - \mu_i) \left(\Pr\{\hat{\mu}_i = 1 | y_i = 1\} - \Pr\{\hat{\mu}_1 = 1 | y_i = 0\}\right), \tag{12.70}$$

with $\mu_i = \Pr\{y_i = 1\}$, showing again the covariance penalty measuring the self-influence of y_i on its own prediction.

As a second example, suppose that the observations y_i are obtained from different members of a one-parameter exponential family $f_\mu(y) = \exp\{\lambda y - \gamma(\lambda)\} f_0(y)$ (8.32),

$$y_i \sim f_{\mu_i}(y_i) \qquad \text{for } i = 1, 2, \ldots, N, \tag{12.71}$$

and that error is measured by the deviance (8.31),

$$D(y, \hat{\mu}) = 2 \int_{\mathcal{Y}} f_y(Y) \log \left(\frac{f_y(Y)}{f_{\hat{\mu}}(Y)} \right) dY. \tag{12.72}$$

According to (8.33), the apparent error $\sum D(y_i, \hat{\mu}_i)$ is then

$$\text{err} = \frac{2}{N} \sum_{i=1}^{N} \log \left(\frac{f_{y_i}(y_i)}{f_{\hat{\mu}_i}(y_i)} \right) = \frac{2}{N} \left\{ \log\left(f_{\boldsymbol{y}}(\boldsymbol{y})\right) - \log\left(f_{\hat{\boldsymbol{\mu}}}(\boldsymbol{y})\right) \right\}. \tag{12.73}$$

[10] More generally, $\hat{\pi}_i$ is some predictor of $\Pr\{y_i = 1\}$, and $\hat{\mu}_i$ is the indicator function $I(\hat{\pi}_i \geq 0.5)$.

In this case the general theory gives overall covariance penalty

$$\text{penalty} = \frac{2}{N}\sum_{i=1}^{N}\text{cov}\left(\hat{\lambda}_i, \hat{\mu}_i\right), \tag{12.74}$$

where $\hat{\lambda}_i$ is the natural parameter in family (8.32) corresponding to $\hat{\mu}_\lambda$ (e.g., $\hat{\lambda}_i = \log\hat{\mu}_i$ for Poisson observations). Moreover, if $\hat{\boldsymbol{\mu}}$ is obtained as the MLE of $\boldsymbol{\mu}$ in a generalized linear model with p degrees of freedom (8.22),

$$\text{penalty} \doteq \frac{2p}{N} \tag{12.75}$$

to a good approximation. The corresponding version of $\widehat{\text{Err}}.$ (12.46) can then be written as

$$\widehat{\text{Err}}. \doteq -\frac{2}{N}\left\{\log\left(f_{\hat{\boldsymbol{\mu}}}(\boldsymbol{y})\right) - p\right\} + \text{ constant}, \tag{12.76}$$

the constant $(2/N)\log(f_{\boldsymbol{y}}(\boldsymbol{y}))$ not depending on $\hat{\boldsymbol{\mu}}$.

The term in brackets is the *Akaike information criterion* (AIC): if the statistician is comparing possible prediction rules $r^{(j)}(\boldsymbol{y})$ for a given data set $\boldsymbol{y}$, the AIC says to select the rule maximizing the *penalized* maximum likelihood

$$\log\left(f_{\hat{\boldsymbol{\mu}}^{(j)}}(\boldsymbol{y})\right) - p^{(j)}, \tag{12.77}$$

where $\hat{\boldsymbol{\mu}}^{(j)}$ is rule j's MLE and $p^{(j)}$ its degrees of freedom. Comparison with (12.76) shows that for GLMs, the AIC amounts to selecting the rule with the smallest value of $\widehat{\text{Err}}.^{(j)}$.

Cross-validation does not require a probability model, but if such a model is available then the error estimate $\widehat{\text{Err}}_{\text{cv}}$ can be improved by *bootstrap smoothing*.[11] With the predictor vectors x_i considered fixed as observed, a parametric model generates the data set $\boldsymbol{d} = \{(x_i, y_i), i = 1, \ldots, N\}$ as in (12.62), from which we calculate the prediction rule $r_{\boldsymbol{d}}(\cdot)$ and the error estimate $\widehat{\text{Err}}_{\text{cv}}$ (12.21),

$$f_{\boldsymbol{\mu}} \to \boldsymbol{d} \to r_{\boldsymbol{d}}(\cdot) \to \widehat{\text{Err}}_{\text{cv}}. \tag{12.78}$$

Substituting the estimated density $f_{\hat{\boldsymbol{\mu}}}$ for f_μ, as in (12.63), provides

[11] Perhaps better known as "bagging;" see Chapter 17.

parametric bootstrap replicates of $\widehat{\text{Err}}_{\text{cv}}$,

$$f_{\hat{\mu}} \to \boldsymbol{d}^* \to r_{\boldsymbol{d}^*}(\cdot) \to \widehat{\text{Err}}_{\text{cv}}^{*}. \tag{12.79}$$

Some large number B of replications can then be averaged to give the smoothed estimate

$$\overline{\text{Err}} = \frac{1}{B}\sum_{b=1}^{B} \widehat{\text{Err}}_{\text{cv}}^{*b}. \tag{12.80}$$

$\overline{\text{Err}}$ averages out the considerable noise in $\widehat{\text{Err}}_{\text{cv}}$, often significantly reducing its variability.[12]

A surprising result, referenced in the endnotes, shows that $\overline{\text{Err}}$ approximates the covariance penalty estimate Err.. Speaking broadly, Err. is what's left after excess randomness is squeezed out of $\widehat{\text{Err}}_{\text{cv}}$ (an example of "Rao–Blackwellization," to use classical terminology). Improvements can be quite substantial.† Covariance penalty estimates, when believable parametric models are available, should be preferred to cross-validation. †4

12.4 Training, Validation, and Ephemeral Predictors

Good Practice suggests splitting the full set of observed predictor–response pairs (x, y) into a training set $\boldsymbol{d}$ of size N (12.1), and a validation set $\boldsymbol{d}_{\text{val}}$, of size N_{val} (12.16). The validation set is put into a vault while the training set is used to develop an effective prediction rule $r_{\boldsymbol{d}}(x)$. Finally, $\boldsymbol{d}_{\text{val}}$ is removed from the vault and used to calculate $\widehat{\text{Err}}_{\text{val}}$ (12.20), an honest estimate of the predictive error rate of $r_{\boldsymbol{d}}$.

This *is* a good idea, and seems foolproof, at least if one has enough data to afford setting aside a substantial portion for a validation set during the training process. Nevertheless, there remains some peril of underestimating the true error rate, arising from *ephemeral* predictors, those whose predictive powers fade away over time. A contrived, but not completely fanciful, example illustrates the danger.

The example takes the form of an imaginary microarray study involving 360 subjects, 180 patients and 180 healthy controls, coded

$$y_i = \begin{cases} 1 & \text{patient} \\ 0 & \text{control,} \end{cases} \qquad i = 1, 2, \ldots, 360. \tag{12.81}$$

[12] A related tactic pertaining to grouped cross-validation is to repeat calculation (12.21) for several different randomly selected splits into J groups, and then average the resulting $\widehat{\text{Err}}_{\text{cv}}$ estimates.

Each subject is assessed on a microarray measuring the genetic activity of $p = 100$ genes, these being the predictors

$$x_i = (x_{i1}, x_{i2}, x_{i3}, \dots, x_{i100})'. \tag{12.82}$$

One subject per day is assessed, alternating patients and controls.

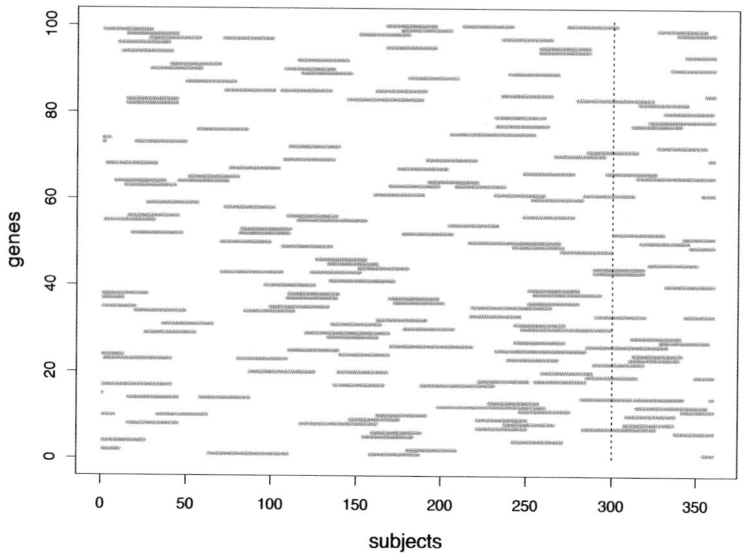

Figure 12.5 Orange bars indicate transient episodes, (12.84) and the reverse, for imaginary medical study (12.81)–(12.82).

The measurements x_{ij} are independent of each other and of the y_i's,

$$x_{ij} \overset{\text{ind}}{\sim} \mathcal{N}(\mu_{ij}, 1) \qquad \text{for } i = 1, 2, \dots, 360 \quad \text{and} \quad j = 1, 2, \dots, 100. \tag{12.83}$$

Most of the μ_{ij} equal zero, but each gene's measurements can experience "transient episodes" of two possible types: in type 1,

$$\mu_{ij} = \begin{cases} 2 & \text{if } y_i = 1 \\ -2 & \text{if } y_i = 0, \end{cases} \tag{12.84}$$

while type 2 reverses signs. The episodes are about 30 days long, randomly and independently located between days 1 and 360, with an average of two episodes per gene. The orange bars in Figure 12.5 indicate the episodes.

For the purpose of future diagnoses we wish to construct a prediction rule $\hat{y} = r_{\boldsymbol{d}}(x)$. To this end we *randomly* divide the 360 subjects into a training set $\boldsymbol{d}$ of size $N = 300$ and a validation set $\boldsymbol{d}_{\text{val}}$ of size $N_{\text{val}} = 60$. The popular "machine learning" prediction program *Random Forests*, Chapter 17, is applied. Random Forests forms $r_{\boldsymbol{d}}(x)$ by averaging the predictions of a large number of randomly subsampled regression trees (Section 8.4).

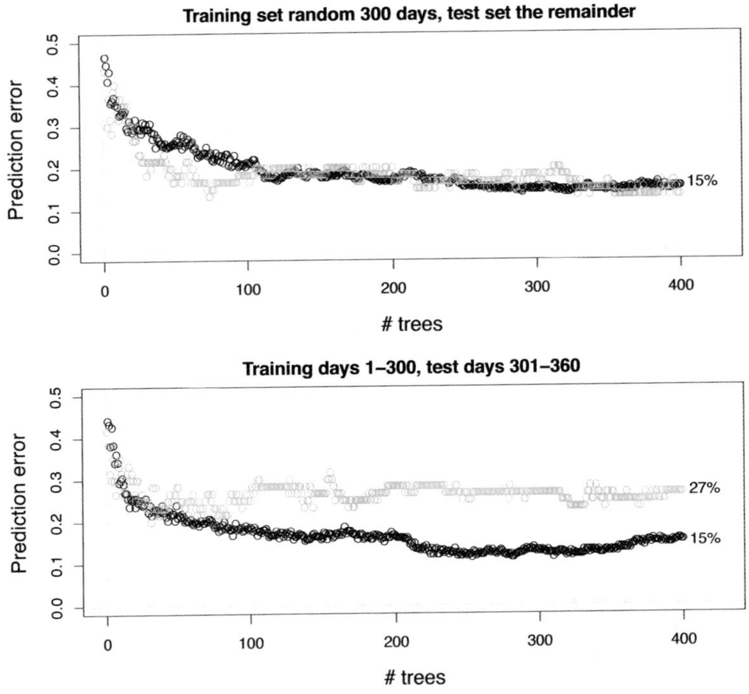

Figure 12.6 Test error (blue) and cross-validated training error (black), for Random Forest prediction rules using the imaginary medical study (12.81)–(12.82). *Top panel*: training set randomly selected 300 days, test set the remaining 60 days. *Bottom panel*: training set the first 300 days, test set the last 60 days.

The top panel of Figure 12.6 shows the results, with blue points indicating test-set error and black the (cross-validated) training-set error. Both converge to 15% as the number of Random Forest trees grows large. This seems to confirm an 85% success rate for prediction rule $r_{\boldsymbol{d}}(x)$.

One change has been made for the bottom panel: now the training set is the data for days 1 through 300, and the test set days 301 through 360.

The cross-validated training-set prediction error still converges to 15%, but $\widehat{\text{Err}}_{\text{val}}$ is now 27%, nearly double.

The reason isn't hard to see. Any predictive power must come from the transient episodes, which lose efficacy outside of their limited span. In the first example the test days are located among the training days, and inherit their predictive accuracy from them. This mostly fails in the second setup, where the test days are farther removed from the training days. (Only the orange bars crossing the 300-day line can help lower $\widehat{\text{Err}}_{\text{val}}$ in this situation.)

An obvious, but often ignored, dictum is that $\widehat{\text{Err}}_{\text{val}}$ is more believable if the test set is further separated from the training set. "Further" has a clear meaning in studies with a time or location factor, but not necessarily in general. For J-fold cross-validation, separation is improved by removing contiguous blocks of N/J cases for each group, rather than by random selection, but the amount of separation is still limited, making $\widehat{\text{Err}}_{\text{cv}}$ less believable than a suitably constructed $\widehat{\text{Err}}_{\text{val}}$.

The distinction between transient, ephemeral predictors and dependable ones is sometimes phrased as the difference between correlation and causation. For prediction purposes, if not for scientific exegesis, we may be happy to settle for correlations as long as they are persistent enough for our purposes. We return to this question in Chapter 15 in the discussion of large-scale hypothesis testing.

A notorious cautionary tale of fading correlations concerns *Google Flu*
†5 *Trends*,† a machine-learning algorithm for predicting influenza outbreaks. Introduced in 2008, the algorithm, based on counts of internet search terms, outperformed traditional medical surveys in terms of speed and predictive accuracy. Four years later, however, the algorithm failed, badly overestimating what turned out to be a nonexistent flu epidemic. Perhaps one lesson here is that the Google algorithmists needed a validation set years—not weeks or months—removed from the training data.

Error rate estimation is mainly frequentist in nature, but the very large data sets available from the internet have encouraged a disregard for inferential justification of any type. This can be dangerous. The heterogeneous nature of "found" data makes statistical principles of analysis more, not less, relevant.

12.5 Notes and Details

The evolution of prediction algorithms and their error estimates nicely illustrates the influence of electronic computation on statistical theory and practice. The classical recipe for cross-validation recommended splitting the full data set in two, doing variable selection, model choice, and data fitting on the first half, and then testing the resulting procedure on the second half. Interest revived in 1974 with the independent publication of papers by Geisser and by Stone, featuring leave-one-out cross-validation of predictive error rates.

A question of bias versus variance arises here. A rule based on only $N/2$ cases is less accurate than the actual rule based on all N. Leave-one-out cross-validation minimizes this type of bias, at the expense of increased variability of error rate estimates for "jumpy" rules of a discontinuous nature. Current best practice is described in Section 7.10 of Hastie *et al.* (2009), where J-fold cross-validation with J perhaps 10 is recommended, possibly averaged over several random data splits.

Nineteen seventy-three was another good year for error estimation, featuring Mallows' C_p estimator and Akaike's information criterion. Efron (1986) extended C_p methods to a general class of situations (see below), established the connection with AIC, and suggested bootstrapping methods for covariance penalties. The connection between cross-validation and covariance penalties was examined in Efron (2004), where the Rao–Blackwell-type relationship mentioned at the end of Section 12.3 was demonstrated. The SURE criterion appeared in Charles Stein's 1981 paper. Ye (1998) suggested the general degrees of freedom definition (12.53).

†1 [p. 210] *Standard candles and dark energy.* Adam Riess, Saul Perlmutter, and Brian Schmidt won the 2011 Nobel Prize in physics for discovering increasing rates of expansion of the Universe, attributed to an Einsteinian concept of *dark energy*. They measured cosmic distances using Type Ia supernovas as "standard candles." The type of analysis suggested by Figure 12.1 is intended to improve the cosmological distance scale.

†2 [p. 222] *Data-based choice of a lasso estimate.* The regularization parameter λ for a lasso estimator (7.42) controls the number of nonzero coefficients of $\tilde{\beta}(\lambda)$, with larger λ yielding fewer nonzeros. Efron *et al.* (2004) and Zou *et al.* (2007) showed that a good approximation for the degrees of freedom df (12.53) of a lasso estimate is the number of its nonzero coefficients. Substituting this for p in (12.51) provides a quick version of $\widehat{\text{Err}}$.. This was minimized at $\text{df} = 7$ for the supernova example in Figure 12.1 (12.54).

†3 [p. 222] *Stein's unbiased risk estimate.* The covariance formula (12.56) is obtained directly from integration by parts. The computation is clear from

the one-dimensional version of (12.55), $N = 1$:

$$\begin{aligned}\operatorname{cov}(\hat{\mu}, y) &= \int_{-\infty}^{\infty} \left[\frac{1}{\sqrt{2\pi\sigma^2}} e^{-\frac{1}{2}\frac{(y-\mu)^2}{\sigma^2}} (y-\mu) \right] \hat{\mu}(y)\, dy \\ &= \sigma^2 \int_{-\infty}^{\infty} \left[\frac{1}{\sqrt{2\pi\sigma^2}} e^{-\frac{1}{2}\frac{(y-\mu)^2}{\sigma^2}} \right] \frac{\partial \hat{\mu}(y)}{\partial y}\, dy \\ &= \sigma^2 E\left\{ \frac{\partial \hat{\mu}(y)}{\partial y} \right\}. \end{aligned} \tag{12.85}$$

Broad regularity conditions for SURE are given in Stein (1981).

†4 [p. 227] *The .632 rule.* Bootstrap competitors to cross-validation are discussed in Efron (1983) and Efron and Tibshirani (1997). The most successful of these, the ".632 rule" is generally less variable than leave-one-out cross-validation. We suppose that nonparametric bootstrap data sets $\boldsymbol{d}^{*b}$, $b = 1, 2, \ldots, B$, have been formed, each by sampling with replacement N times from the original N members of $\boldsymbol{d}$ (12.1). Data set $\boldsymbol{d}^{*b}$ produces rule

$$r^{*b}(x) = r_{\boldsymbol{d}^{*b}}(x), \tag{12.86}$$

giving predictions

$$y_i^{*b} = r^{*b}(x_i). \tag{12.87}$$

Let $I_i^b = 1$ if pair (x_i, y_i) is *not* in $\boldsymbol{d}^{*b}$, and 0 if it is. (About $e^{-1} = 0.368$ of the $N \cdot B$ I_i^b will equal 1, the remaining 0.632 equaling 0.) The "out of bootstrap" estimate of prediction error is

$$\widehat{\text{Err}}_{\text{out}} = \sum_{i=1}^{N} \sum_{j=1}^{B} I_i^b D\left(y_i, \hat{y}_i^{*b}\right) \Big/ \sum_{i=1}^{N} \sum_{j=1}^{B} I_i^b, \tag{12.88}$$

the average discrepancy in the omitted cases.

$\widehat{\text{Err}}_{\text{out}}$ is similar to a grouped cross-validation estimate that omits about 37% of the cases each time. The .632 rule compensates for the upward bias in $\widehat{\text{Err}}_{\text{out}}$ by incorporating the downwardly biased apparent error (12.18),

$$\widehat{\text{Err}}_{.632} = 0.632\, \widehat{\text{Err}}_{\text{out}} + 0.368\, \text{err}\,. \tag{12.89}$$

$\widehat{\text{Err}}_{\text{out}}$ has resurfaced in the popular Random Forests prediction algorithm, Chapter 17, where a closely related procedure gives the "out of bag" estimate of Err.

†5 [p. 230] *Google Flu Trends.* Harford's 2014 article, "Big data: A big mistake?," concerns the enormous "found" data sets available in the internet age, and the dangers of forgetting the principles of statistical inference in their analysis. Google Flu Trends is his primary cautionary example.

13

Objective Bayes Inference and Markov Chain Monte Carlo

From its very beginnings, Bayesian inference exerted a powerful influence on statistical thinking. The notion of a single coherent methodology employing only the rules of probability to go from assumption to conclusion was and is immensely attractive. For 200 years, however, two impediments stood between Bayesian theory's philosophical attraction and its practical application.

1 In the absence of relevant past experience, the choice of a prior distribution introduces an unwanted subjective element into scientific inference.
2 Bayes' rule (3.5) looks simple enough, but carrying out the numerical calculation of a posterior distribution often involves intricate higher-dimensional integrals.

The two impediments fit neatly into the dichotomy of Chapter 1, the first being inferential and the second algorithmic.[1]

A renewed cycle of Bayesian enthusiasm took hold in the 1960s, at first concerned mainly with coherent inference. Building on work by Bruno de Finetti and L. J. Savage, a principled theory of *subjective probability* was constructed: the Bayesian statistician, by the careful elicitation of prior knowledge, utility, and belief, arrives at the correct *subjective prior distribution* for the problem at hand. Subjective Bayesianism is particularly appropriate for individual decision making, say for the business executive trying to choose the best investment in the face of uncertain information.

It is less appropriate for scientific inference, where the sometimes skeptical world of science puts a premium on objectivity. An answer came from the school of *objective Bayes inference*. Following the approach of Laplace and Jeffreys, as discussed in Section 3.2, their goal was to fashion objective, or "uninformative," prior distributions that in some sense were unbiased in their effects upon the data analysis.

[1] The exponential family material in this chapter provides technical support, but is not required in detail for a general understanding of the main ideas.

In what came as a surprise to the Bayes community, the objective school has been the most successful in bringing Bayesian ideas to bear on scientific data analysis. Of the 24 articles in the December 2014 issue of the *Annals of Applied Statistics*, 8 employed Bayesian analysis, predominantly based on objective priors.

This is where electronic computation enters the story. Commencing in the 1980s, dramatic steps forward were made in the numerical calculation of high-dimensional Bayes posterior distributions. *Markov chain Monte Carlo* (MCMC) is the generic name for modern posterior computation algorithms. These proved particularly well suited for certain forms of objective Bayes prior distributions.

Taken together, objective priors and MCMC computations provide an attractive package for the statistician faced with a complicated data analysis situation. Statistical inference becomes almost automatic, at least compared with the rigors of frequentist analysis. This chapter discusses both parts of the package, the choice of prior and the subsequent computational methods. Criticisms arise, both from the frequentist viewpoint and that of informative Bayesian analysis, which are brought up here and also in Chapter 21.

13.1 Objective Prior Distributions

A *flat*, or uniform, distribution over the space of possible parameter values seems like the obvious choice for an uninformative prior distribution, and has been so ever since Laplace's advocacy in the late eighteenth century. For a finite parameter space Ω, say

$$\Omega = \{\mu_{(1)}, \mu_{(2)}, \dots, \mu_{(K)}\}, \tag{13.1}$$

"flat" has the obvious meaning

$$g^{\text{flat}}(\mu) = \frac{1}{K} \qquad \text{for all } \mu \in \Omega. \tag{13.2}$$

If K is infinite, or if Ω is continuous, we can still take

$$g^{\text{flat}}(\mu) = \text{constant}. \tag{13.3}$$

Bayes' rule (3.5) gives the same posterior distribution for any choice of the constant,

$$\begin{aligned} g^{\text{flat}}(\mu|x) &= g^{\text{flat}}(\mu) f_\mu(x)/f(x), \quad \text{with} \\ f(x) &= \int_\Omega f_\mu(x) g^{\text{flat}}(\mu)\, d\mu. \end{aligned} \tag{13.4}$$

Notice that $g^{\text{flat}}(\mu)$ cancels out of $g^{\text{flat}}(\mu|x)$. The fact that $g^{\text{flat}}(\mu)$ is "improper," that is, it integrates to infinity, doesn't affect the formal use of Bayes' rule in (13.4) as long as $f(x)$ is finite.

Notice also that $g^{\text{flat}}(\mu|x)$ amounts to taking the posterior density of μ to be proportional to the likelihood function $L_x(\mu) = f_\mu(x)$ (with x fixed and μ varying over Ω). This brings us close to Fisherian inference, with its emphasis on the direct interpretation of likelihoods, but Fisher was adamant in his insistance that likelihood was not probability.

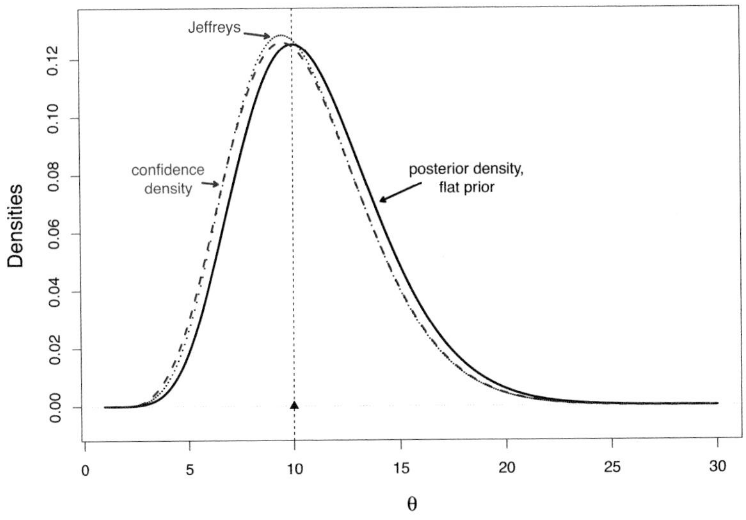

Figure 13.1 The solid curve is flat-prior posterior density (13.4) having observed $x = 10$ from Poisson model $x \sim \text{Poi}(\mu)$; it is shifted about 0.5 units right from the confidence density (dashed) of Figure 11.6. Jeffreys' prior gives a posterior density (dotted) nearly the same as the confidence density.

The solid curve in Figure 13.1 shows $g^{\text{flat}}(\mu|x)$ for the Poisson situation of Table 11.2,

$$x \sim \text{Poi}(\mu), \tag{13.5}$$

with $x = 10$ observed; $g^{\text{flat}}(\mu|x)$ is shifted almost exactly 0.5 units right of the confidence density from Figure 11.6. ("θ" is μ itself in this case.)[2]

Fisher's withering criticism of flat-prior Bayes inference focused on its

[2] The reader may wish to review Chapter 11, particularly Section 11.6, for these constructions.

lack of transformation invariance. If we were interested in $\theta = \log(\mu)$ rather than μ, $g^{\text{flat}}(\theta|x)$ would not be the transformation to the log scale of $g^{\text{flat}}(\mu|x)$. Jeffreys' prior, (3.17) or (11.72), which *does* transform correctly, is

$$g^{\text{Jeff}}(\mu) = 1/\sqrt{\mu} \tag{13.6}$$

for $x \sim \text{Poi}(\mu)$; $g^{\text{Jeff}}(\mu|x = 10)$ is then a close match to the confidence density in Figure 13.1.

Coverage Matching Priors

A variety of improvements and variations on Jeffreys' prior have been suggested for use as general-purpose uninformative prior distributions, as
†1 briefly discussed in the chapter endnotes.† All share the drawback seen in Figure 11.7: the posterior distribution $g(\mu|x)$ can have unintended effects on the resulting inferences for a real-valued parameter of interest $\theta = t(\mu)$. This is unavoidable; it is mathematically impossible for any single prior to be uninformative for every choice of $\theta = t(\mu)$.

The label "uninformative" for a prior sometimes means "gives Bayes posterior intervals that closely match confidence intervals." Perhaps surprisingly, this definition has considerable resonance in the Bayes community. Such priors can be constructed for any given scalar parameter of interest $\theta = t(\mu)$, for instance the maximum eigenvalue parameter of Fig-
†2 ure 11.7. In brief, the construction proceeds as follows.†

- The p-dimensional parameter vector μ is transformed to a form that makes θ the first coordinate, say

$$\mu \to (\theta, \nu), \tag{13.7}$$

  where ν is a $(p-1)$-dimensioned nuisance parameter.
- The transformation is chosen so that the Fisher information matrix (11.72) for (θ, ν) has the "diagonal" form

$$\begin{pmatrix} \mathcal{I}_{\theta\theta} & 0 \\ 0' & \mathcal{I}_{\nu\nu} \end{pmatrix}. \tag{13.8}$$

  (This is always possible.)
- Finally, the prior for (θ, ν) is taken proportional to

$$g(\theta, \nu) = \mathcal{I}_{\theta\theta}^{1/2} h(\nu), \tag{13.9}$$

  where $h(\nu)$ is an arbitrary $(p-1)$-dimensional density. In other words,

$g(\theta, \nu)$ combines the one-dimensional Jeffreys' prior (3.16) for θ with an arbitrary independent prior for the orthogonal nuisance parameter vector ν.

The main thing to notice about (13.9) is that $g(\theta, \nu)$ represents different priors on the original parameter vector μ for different functions $\theta = t(\mu)$. No single prior $g(\mu)$ can be uninformative for all choices of the parameter of interest θ.

Calculating $g(\theta, \nu)$ can be difficult. One alternative is to go directly to the BCa confidence density (11.68)–(11.69), which can be interpreted as the posterior distribution from an uninformative prior (because its integrals agree closely with confidence interval endpoints).

Coverage matching priors are not much used in practice, and in fact none of the eight *Annals of Applied Statistics* objective Bayes papers mentioned earlier were of type (13.9). A form of "almost uninformative" priors, the conjugates, is more popular, mainly because of the simpler computation of their posterior distributions.

13.2 Conjugate Prior Distributions

A mathematically convenient class of prior distributions, the *conjugate priors*, applies to samples from an exponential family,[3] Section 5.5,

$$f_\mu(x) = e^{\alpha x - \psi(\alpha)} f_0(x). \tag{13.10}$$

Here we have indexed the family with the expectation parameter

$$\mu = E_f\{x\}, \tag{13.11}$$

rather than the canonical parameter α. On the right-hand side of (13.10), α can be thought of as a one-to-one function of μ (the so-called "link function"), e.g., $\alpha = \log(\mu)$ for the Poisson family. The observed data is a random sample $\boldsymbol{x} = (x_1, x_2, \ldots, x_n)$ from f_μ,

$$x_1, x_2, \ldots, x_n \overset{\text{iid}}{\sim} f_\mu, \tag{13.12}$$

having density function

$$f_\mu(\boldsymbol{x}) = e^{n[\alpha\bar{x} - \psi(\alpha)]} f_0(\boldsymbol{x}), \tag{13.13}$$

the average $\bar{x} = \sum x_i/n$ being sufficient.

[3] We will concentrate on one-parameter families, though the theory extends to the multiparameter case. Figure 13.2 relates to a two-parameter situation.

The family of conjugate priors for μ, $g_{n_0,x_0}(\mu)$, allows the statistician to choose two parameters, n_0 and x_0,

$$g_{n_0,x_0}(\mu) = c e^{n_0[x_0\alpha - \psi(\alpha)]} \big/ V(\mu), \tag{13.14}$$

$V(\mu)$ the variance of an x from f_μ,

$$V(\mu) = \text{var}_f\{x\}; \tag{13.15}$$

c is the constant that makes $g_{n_0,x_0}(\mu)$ integrate to 1 with respect to Lebesgue measure on the interval of possible μ values. The interpretation is that x_0 represents the average of n_0 hypothetical prior observations from f_μ.

The utility of conjugate priors is seen in the following theorem.

†3 **Theorem 13.1** † *Define*

$$n_+ = n_0 + n \quad \textit{and} \quad \bar{x}_+ = \frac{n_0}{n_+}x_0 + \frac{n}{n_+}\bar{x}. \tag{13.16}$$

Then the posterior density of μ *given* $\boldsymbol{x} = (x_1, x_2, \ldots, x_n)$ *is*

$$g(\mu|\boldsymbol{x}) = g_{n_+,\bar{x}_+}(\mu); \tag{13.17}$$

moreover, the posterior expectation of μ *given* $\boldsymbol{x}$ *is*

$$E\{\mu|\boldsymbol{x}\} = \frac{n_0}{n_+}x_0 + \frac{n}{n_+}\bar{x}. \tag{13.18}$$

The intuitive interpretation is quite satisfying: we begin with a hypothetical prior sample of size n_0, sufficient statistic x_0; observe $\boldsymbol{x}$, a sample of size n; and update our prior distribution $g_{n_0,x_0}(\mu)$ to a distribution $g_{n_+,\bar{x}_+}(\mu)$ of the same form. Moreover, $E\{\mu|\boldsymbol{x}\}$ equals the average of a hypothetical sample with n_0 copies of x_0,

$$(x_0, x_0, \ldots, x_0, x_1, x_2, \ldots, x_n). \tag{13.19}$$

As an example, suppose $x_i \overset{\text{iid}}{\sim} \text{Poi}(\mu)$, that is we have n i.i.d. observations from a Poisson distribution, Table 5.1. Formula (13.14) gives conjugate prior †
†4

$$g_{n_0,x_0}(\mu) = c\mu^{n_0x_0-1}e^{-n_0\mu}, \tag{13.20}$$

c not depending on μ. So in the notation of Table 5.1, $g_{n_0,x_0}(\mu)$ is a gamma distribution, $\text{Gam}(n_0x_0, 1/n_0)$. The posterior distribution is

$$\begin{aligned} g(\mu|\boldsymbol{x}) = g_{n_+,\bar{x}_+}(\mu) &\sim \text{Gam}(n_+\bar{x}_+, 1/n_+) \\ &\sim \frac{1}{n_+}G_{n_+\bar{x}_+}, \end{aligned} \tag{13.21}$$

†5 where G_ν indicates a standard gamma distribution,†

$$G_\nu = \text{Gam}(\nu, 1). \tag{13.22}$$

Table 13.1 *Conjugate priors* (13.14)–(13.16) *for four familiar one-parameter exponential families, using notation in Table 5.1; the last column shows the posterior distribution of μ given n observations x_i, starting from prior $g_{n_0,x_0}(\mu)$. In line 4, G_ν is the standard gamma distribution* Gam$(\nu, 1)$, *with μ the same as gamma parameter σ in Table 5.1. The chapter endnotes give the density of the inverse gamma distribution $1/G_\nu$, and corresponding results for chi-squared variates.*

	Name	x_i distribution	$g_{n_0,x_0}(\mu)$	$g(\mu \mid \boldsymbol{x})$
1.	Normal	$\mathcal{N}(\mu, \sigma_1^2)$ (σ_1^2 known)	$\mathcal{N}(x_0, \sigma_1^2/n_0)$	$\mathcal{N}(\bar{x}_+, \sigma_1^2/n_+)$
2.	Poisson	Poi(μ)	Gam$(n_0 x_0, 1/n_0)$	Gam$(n_+ \bar{x}_+, 1/n_+)$
3.	Binomial	Bi$(1, \mu)$	Be$(n_0 x_0, n_0(1 - x_0))$	Be$(n_+ \bar{x}_+, n_+(1 - \bar{x}_+))$
4.	Gamma	$\mu G_\nu/\nu$ (ν known)	$n_0 x_0 \nu / G_{n_0 \nu + 1}$	$n_+ \bar{x}_+ \nu / G_{n_+ \nu + 1}$

Table 13.1 describes the conjugate prior and posterior distributions for four familiar one-parameter families. The binomial case, where μ is the "success probability" π in Table 5.1, is particularly evocative: independent coin flips $x_1, x_2, \dots, x_n$ give, say, $s = \sum_i x_i = n\bar{x}$ successes. Prior $g_{n_0,x_0}(\pi)$ amounts to assuming proportion $x_0 = s_0/n_0$ prior successes in n_0 flips. Formula (13.18) becomes

$$E\{\pi \mid \boldsymbol{x}\} = \frac{s_0 + s}{n_0 + n} \tag{13.23}$$

for the posterior expectation of π. The choice $(n_0, x_0) = (2, 1/2)$ for instance gives Bayesian estimate $(s+1)/(n+2)$ for π, pulling the MLE s/n a little bit toward 1/2.

The size of n_0, the number of hypothetical prior observations, determines how informative or uninformative the prior $g_{n_0,x_0}(\mu)$ is. Recent objective Bayes literature has favored choosing n_0 small, $n_0 = 1$ being popular. The hope here is to employ a *proper* prior (one that has a finite integral), while still not injecting much unwarranted information into the analysis. The choice of x_0 is also by convention. One possibility is to set

$x_0 = \bar{x}$, in which case the posterior expectation $E\{\mu|\boldsymbol{x}\}$ (13.18) equals the MLE $\bar{x}$. Another possibility is choosing x_0 equal to a "null" value, for instance $x_0 = 0$ for effect size estimation in (3.28).

Table 13.2 *Vasoconstriction data; volume of air inspired in 39 cases, 19 without vasoconstriction ($y = 0$) and 20 with vasoconstriction ($y = 1$).*

$y = 0$		$y = 1$	
60	98	85	115
74	98	88	120
78	104	88	126
78	104	90	126
78	113	90	128
88	118	93	136
90	120	104	143
95	123	108	151
95	137	110	154
98		111	157

As a miniature example of objective Bayes inference, we consider the
†6 *vasoconstriction data*† of Table 13.2: $n = 39$ measurements of lung volume have been obtained, 19 without vasoconstriction ($y = 0$) and 20 with ($y = 1$). Here we will think of the y_i as binomial variates,

$$y_i \overset{\text{ind}}{\sim} \text{Bi}(1, \pi_i), \qquad i = 1, 2, \dots, 39, \tag{13.24}$$

following logistic regression model (8.5),

$$\log\left(\frac{\pi_i}{1 - \pi_i}\right) = \alpha_0 + \alpha_1 x_i, \tag{13.25}$$

with the x_i as fixed covariates.

Letting $X_i = (1, x_i)'$, (13.24)–(13.25) results in a two-parameter exponential family (8.24),

$$f_\alpha(y) = e^{n\left[\alpha'\hat{\beta} - \psi(\alpha)\right]} f_0(y), \tag{13.26}$$

having

$$\hat{\beta} = \frac{1}{n}\left(\sum_{i=1}^{n} y_i, \sum_{i=1}^{n} x_i y_i\right)' \quad \text{and} \quad \psi(\alpha) = \frac{1}{n}\sum_{i=1}^{n}\log(1 + e^{\alpha' X_i}).$$

The MLE $\hat{\alpha}$ has approximate 2×2 covariance matrix $\hat{V}$ as given in (8.30).

In Figure 13.2, the posterior distributions are graphed in terms of

$$\gamma = \hat{V}^{-1/2}(\alpha - \hat{\alpha}) \tag{13.27}$$

rather than α or μ, making the contours of equal density roughly circular and centered at zero.

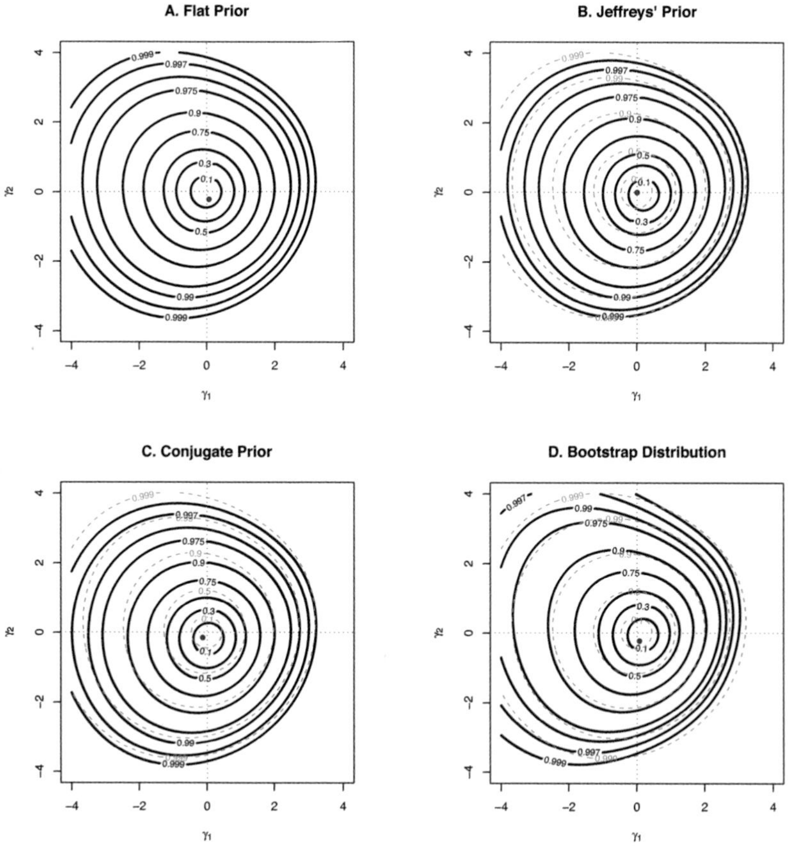

Figure 13.2 Vasoconstriction data; contours of equal posterior density of γ (13.27) from four uninformative priors, as described in the text. Numbers indicate probability content within contours; light dashed contours from Panel A, flat prior.

Panel A of Figure 13.2 illustrates the flat prior posterior density of γ given the data $\boldsymbol{y}$ in model (13.24)–(13.25). The heavy lines are contours of equal density, with the one labeled "0.9" containing 90% of the posterior probability, etc. Panel B shows the corresponding posterior density contours obtained from Jeffreys' multiparameter prior (11.72), in this case

$$g^{\text{Jeff}}(\alpha) = |V_\alpha|^{1/2}, \tag{13.28}$$

V_α the covariance matrix of $\hat{\alpha}$, as calculated from (8.30). For comparison purposes the light dashed curves show some of the flat prior contours from panel A. The effect of $g^{\text{Jeff}}(\alpha)$ is to reduce the flat prior bulge toward the upper left corner.

Panel C relates to the conjugate prior[4] $g_{1,0}(\alpha)$. Besides reducing the flat prior bulge, $g_{1,0}(\alpha)$ pulls the contours slightly downward.

Panel D shows the parametric bootstrap distribution: model (13.24)–(13.25), with $\hat{\alpha}$ replacing α, gave resamples $\boldsymbol{y}^*$ and MLE replications $\hat{\alpha}^*$. The contours of $\hat{\gamma}^* = \hat{V}^{-1/2}(\hat{\alpha}^* - \hat{\alpha})$ considerably accentuate the bulge toward the left.

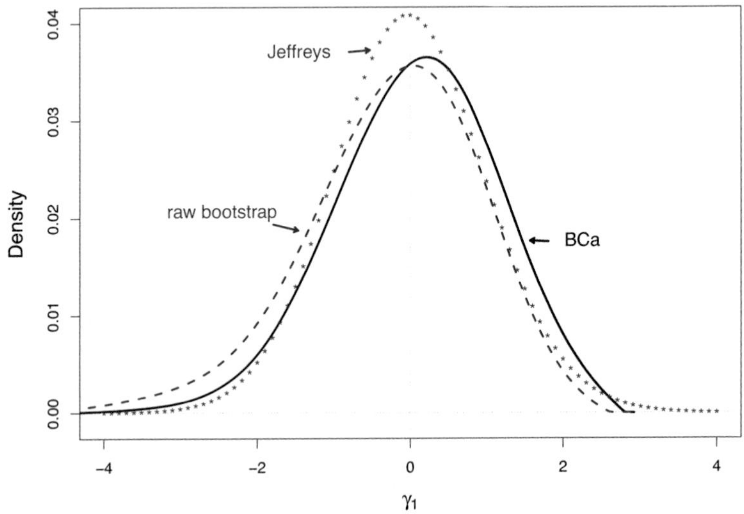

Figure 13.3 Posterior densities for γ_1, first coordinate of γ in (13.27), for the vasoconstriction data. Dashed red curve: raw (unweighted) distribution of $B = 8000$ parametric replications from model (13.24)–(13.25); solid black curve: BCa density (11.68) ($z_0 = 0.123$, $a = 0.053$); dotted blue curve: posterior density using Jeffreys multiparameter prior (11.72).

[4] The role of $\bar{x}$ in (13.13) is taken by $\hat{\beta}$ in (13.26), so $g_{1,0}$ has $\hat{\beta} = 0$, $n_0 = 1$. This makes $g_{1,0}(\alpha) = \exp\{-\psi(\alpha)\}$. The factor $V(\mu)$ in (13.14) is absent in the conjugate prior for α (as opposed to μ).

This doesn't necessarily imply that a bootstrap analysis would give much different answers than the three (quite similar) objective Bayes results. For any particular real-valued parameter of interest θ, the raw bootstrap distribution (equal weight on each replication) would be reweighted according to the BCa formula (11.68) in order to produce accurate confidence intervals. Figure 13.3 compares the raw bootstrap distribution, the BCa confidence density, and the posterior density obtained from Jeffreys' prior, for θ equal to γ_1, the first coordinate of γ in (13.27). The BCa density is shifted to the right of Jeffreys'.

Critique of Objective Bayes Inference

Despite its simplicity, or perhaps because of it, objective Bayes procedures are vulnerable to criticism from both ends of the statistical spectrum. From the subjectivist point of view, objective Bayes is only partially Bayesian: it employs Bayes' theorem but without doing the hard work of determining a convincing prior distribution. This introduces frequentist elements into its practice—clearly so in the case of Jeffreys' prior—along with frequentist incoherencies.

For the frequentist, objective Bayes analysis can seem dangerously untethered from the usual standards of accuracy, having only tenuous large-sample claims to legitimacy. This is more than a theoretical objection. The practical advantages claimed for Bayesian methods depend crucially on the fine structure of the prior. Can we safely ignore stopping rules or selective inference (e.g., choosing the largest of many estimated parameters for special attention) for a prior not based on some form of genuine experience?

In an era of large, complicated, and difficult data-analytic problems, objective Bayes methods are answering a felt need for relatively straightforward paths to solution. Granting their usefulness, it is still reasonable to hope for better justification,[5] or at least for more careful comparisons with competing methods as in Figure 13.3.

13.3 Model Selection and the Bayesian Information Criterion

Data-based model selection has become a major theme of modern statistical inference. In the problem's simplest form, the statistician observes data $\boldsymbol{x}$ and wishes to choose between a smaller model $\mathcal{M}_0$ and a larger model

[5] Chapter 20 discusses the frequentist assessment of Bayes and objective Bayes estimates.

$\mathcal{M}_1$. The classic textbook example takes $\boldsymbol{x} = (x_1, x_2, \ldots, x_n)'$ as an independent normal sample,

$$x_i \overset{\text{iid}}{\sim} \mathcal{N}(\mu, 1) \qquad \text{for } i = 1, 2, \ldots, n, \tag{13.29}$$

with $\mathcal{M}_0$ the null hypothesis $\mu = 0$ and $\mathcal{M}_1$ the general two-sided alternative,

$$\mathcal{M}_0 : \mu = 0, \qquad \mathcal{M}_1 : \mu \neq 0. \tag{13.30}$$

(We can include $\mu = 0$ in $\mathcal{M}_1$ with no effect on what follows.) From a frequentist viewpoint, choosing between $\mathcal{M}_0$ and $\mathcal{M}_1$ in (13.29)–(13.30) amounts to running a hypothesis test of $H_0 : \mu = 0$, perhaps augmented with a confidence interval for μ.

Bayesian model selection aims for more: an evaluation of the posterior probabilities of $\mathcal{M}_0$ and $\mathcal{M}_1$ given $\boldsymbol{x}$. A full Bayesian specification requires prior probabilities for the two models,

$$\pi_0 = \Pr\{\mathcal{M}_0\} \quad \text{and} \quad \pi_1 = 1 - \pi_0 = \Pr\{\mathcal{M}_1\}, \tag{13.31}$$

and conditional prior densities for μ within each model,

$$g_0(\mu) = g(\mu|\mathcal{M}_0) \quad \text{and} \quad g_1(\mu) = g(\mu|\mathcal{M}_1). \tag{13.32}$$

Let $f_\mu(\boldsymbol{x})$ be the density of $\boldsymbol{x}$ given μ. Each model induces a marginal density for $\boldsymbol{x}$, say

$$f_0(\boldsymbol{x}) = \int_{\mathcal{M}_0} f_\mu(\boldsymbol{x}) g_0(\mu)\, d\mu \quad \text{and} \quad f_1(\boldsymbol{x}) = \int_{\mathcal{M}_1} f_\mu(\boldsymbol{x}) g_1(\mu)\, d\mu. \tag{13.33}$$

Bayes' theorem, in its ratio form (3.8), then gives posterior probabilities

$$\pi_0(\boldsymbol{x}) = \Pr\{\mathcal{M}_0|\boldsymbol{x}\} \quad \text{and} \quad \pi_1(\boldsymbol{x}) = \Pr\{\mathcal{M}_1|\boldsymbol{x}\} \tag{13.34}$$

satisfying

$$\frac{\pi_1(\boldsymbol{x})}{\pi_0(\boldsymbol{x})} = \frac{\pi_1}{\pi_0} B(\boldsymbol{x}), \tag{13.35}$$

where $B(\boldsymbol{x})$ is the *Bayes factor*

$$B(\boldsymbol{x}) = \frac{f_1(\boldsymbol{x})}{f_0(\boldsymbol{x})}, \tag{13.36}$$

leading to the elegant statement that the posterior odds ratio is the prior odds ratio times the Bayes factor.

All of this is of more theoretical than applied use. Prior specifications (13.31)–(13.32) are usually unavailable in practical settings (which is why standard hypothesis testing is so popular). The objective Bayes school has concentrated on estimating the Bayes factor $B(\boldsymbol{x})$, with the understanding that the prior odds ratio π_1/π_0 in (13.35) would be roughly evaluated depending on the specific circumstances—perhaps set to the Laplace choice $\pi_1/\pi_0 = 1$.

Table 13.3 *Jeffreys' scale of evidence for the interpretation of Bayes factors.*

Bayes factor	Evidence for M_1
< 1	negative
1–3	barely worthwhile
3–20	positive
20–50	strong
> 150	very strong

Jeffreys suggested a scale of evidence for interpreting Bayes factors, reproduced in Table 13.3;[†] $B(\boldsymbol{x}) = 10$ for instance constitutes *positive* but not *strong* evidence in favor of the bigger model. Jeffreys' scale is a Bayesian version of Fisher's interpretive scale for the outcome of a hypothetic test, with coverage value (one minus the significance level) 0.95 famously constituting "significant" evidence against the null hypothesis. Table 13.4 shows Fisher's scale, as commonly interpreted in the biomedical and social sciences. †7

Table 13.4 *Fisher's scale of evidence against null hypothesis $\mathcal{M}_0$ and in favor of $\mathcal{M}_1$, as a function of coverage level (1 minus the p-value).*

Coverage	(p-value)	Evidence for M_1
.80	(.20)	null
.90	(.10)	borderline
.95	(.05)	moderate
.975	(.025)	substantial
.99	(.01)	strong
.995	(.005)	very strong
.999	(.001)	overwhelming

Even if we accept the reduction of model selection to assessing the Bayes factor $B(\boldsymbol{x})$ in (13.35), and even if we accept Jeffreys' scale of interpretation, this still leaves a crucial question: how to compute $B(\boldsymbol{x})$ in practice, without requiring informative choices of the priors g_0 and g_1 in (13.32).

A popular objective Bayes answer is provided by the *Bayesian information criterion*† (BIC). For a given model $\mathcal{M}$ we define

†8

$$\mathrm{BIC}(\mathcal{M}) = \log\left\{f_{\hat{\mu}}(\boldsymbol{x})\right\} - \frac{p}{2}\log(n), \tag{13.37}$$

where $\hat{\mu}$ is the MLE, p the degrees of freedom (number of free parameters) in $\mathcal{M}$, and n the sample size. Then the BIC approximation to Bayes factor $B(\boldsymbol{x})$ (13.36) is

$$\begin{aligned}\log B_{\mathrm{BIC}}(\boldsymbol{x}) &= \mathrm{BIC}(\mathcal{M}_1) - \mathrm{BIC}(\mathcal{M}_0)\\ &= \log\left\{f_{\hat{\mu}_1}(\boldsymbol{x})/f_{\hat{\mu}_0}(\boldsymbol{x})\right\} - \frac{p_1 - p_0}{2}\log(n),\end{aligned} \tag{13.38}$$

the subscripts indexing the MLEs and degrees of freedom in $\mathcal{M}_1$ and $\mathcal{M}_0$.

This can be restated in somewhat more familiar terms. Letting $W(\boldsymbol{x})$ be *Wilks' likelihood ratio statistic*,

$$W(\boldsymbol{x}) = 2\log\left\{f_{\hat{\mu}_1}(\boldsymbol{x})/f_{\hat{\mu}_0}(\boldsymbol{x})\right\}, \tag{13.39}$$

we have

$$\log B_{\mathrm{BIC}}(\boldsymbol{x}) = \frac{1}{2}\left\{W(\boldsymbol{x}) - d\log(n)\right\}, \tag{13.40}$$

with $d = p_1 - p_0$. $W(\boldsymbol{x})$ approximately follows a χ^2_d distribution under model $\mathcal{M}_0$, $E_0\{W(\boldsymbol{x})\} \doteq d$, implying $B_{\mathrm{BIC}}(\boldsymbol{x})$ will tend to be less than one, favoring $\mathcal{M}_0$ if it is true, ever more strongly as n increases.

We can apply BIC selection to the vasoconstriction data of Table 13.2, taking $\mathcal{M}_1$ to be model (13.24)–(13.25), and $\mathcal{M}_0$ to be the submodel having $\alpha_1 = 0$. In this case $d = 1$ in (13.40). Direct calculation gives $W = 7.07$ and

$$B_{\mathrm{BIC}} = 5.49, \tag{13.41}$$

positive but not *strong* evidence against $\mathcal{M}_0$ according to Jeffreys' scale. By comparison, the usual frequentist z-value for testing $\alpha_1 = 0$ is 2.36, coverage level 0.982, between *substantial* and *strong* evidence against $\mathcal{M}_0$ on Fisher's scale.

The BIC was named in reference to Akaike's information criterion (AIC),

$$\mathrm{AIC}(\mathcal{M}) = \log\left\{f_{\hat{\mu}}(\boldsymbol{x})\right\} - p, \tag{13.42}$$

which suggests, as in (12.73), basing model selection on the sign of

$$\mathrm{AIC}(\mathcal{M}_1) - \mathrm{AIC}(\mathcal{M}_0) = \frac{1}{2}\left\{W(\boldsymbol{x}) - 2d\right\}. \tag{13.43}$$

The BIC penalty $d\log(n)$ in (13.40) grows more severe than the AIC penalty $2d$ as n gets larger, increasingly favoring selection of $\mathcal{M}_0$ rather than $\mathcal{M}_1$. The distinction is rooted in Bayesian notions of coherent behavior, as discussed in what follows.

Where does the BIC penalty term $d\log(n)$ in (13.40) come from? A first answer uses the simple normal model $x_i \sim \mathcal{N}(\mu, 1)$, (13.29)–(13.30). $\mathcal{M}_0$ has prior $g_0(\mu) = g(\mu|\mathcal{M}_0)$ equal a delta function at zero. Suppose we take $g_1(\mu) = g(\mu|\mathcal{M}_1)$ in (13.32) to be the Gaussian conjugate prior

$$g_1(\mu) \sim \mathcal{N}(M, A). \tag{13.44}$$

The discussion following (13.23) in Section 13.2 suggests setting $M = 0$ and $A = 1$, corresponding to prior information equivalent to one of the n actual observations. In this case we can calculate the actual Bayes factor $B(\boldsymbol{x})$,

$$\log B(\boldsymbol{x}) = \frac{1}{2}\left\{\frac{n}{n+1}W(\boldsymbol{x}) - \log(n+1)\right\}, \tag{13.45}$$

nearly equaling $\log B_{\text{BIC}}(\boldsymbol{x})$ ($d = 1$), for large n. Justifications of the BIC formula as an approximate Bayes factor follow generalizations of this kind of argument, as discussed in the chapter endnotes.

The difference between BIC and frequentist hypothesis testing grows more drastic for large n. Suppose $\mathcal{M}_0$ is a regression model and $\mathcal{M}_1$ is $\mathcal{M}_0$ augmented with one additional covariate (so $d = 1$). Let z be a standard z-value for testing the hypothesis that $\mathcal{M}_1$ is no improvement over $\mathcal{M}_0$,

$$z \mathrel{\dot\sim} \mathcal{N}(0, 1) \text{ under } \mathcal{M}_0. \tag{13.46}$$

Table 13.5 shows $B_{\text{BIC}}(\boldsymbol{x})$ as a function of z and n. At $n = 15$ Fisher's and Jeffreys' scales give roughly similar assessments of the evidence against $\mathcal{M}_0$ (though Jeffreys' nomenclature is more conservative). At the other end of the table, at $n = 10{,}000$, the inferences are contradictory: $z = 3.29$, with p-value 0.001 and coverage level 0.999, is *overwhelming* evidence for $\mathcal{M}_1$ on Fisher's scale, but *barely worthwhile* for Jeffreys'. Bayesian coherency, the axiom that inferences should be consistent over related situations, lies behind the contradiction.

Suppose $n = 1$ in the simple normal model (13.29)–(13.30). That is, we observe only the single variable

$$x \sim \mathcal{N}(\mu, 1), \tag{13.47}$$

and wish to decide between $\mathcal{M}_0 : \mu = 0$ and $\mathcal{M}_1 : \mu \neq 0$. Let $g_1^{(1)}(\mu)$ denote our $\mathcal{M}_1$ prior density (13.32) for this situation.

Table 13.5 *BIC Bayes factors corresponding to z-values for testing one additional covariate; coverage value (1 minus the significance level) of a two-sided hypothesis test as interpreted by Fisher's scale of evidence, right. Jeffreys' scale of evidence, Table 13.3, is in rough agreement with Fisher for $n = 15$, but favors the null much more strongly for larger sample sizes.*

		n							
Cover	z-value	15	50	250	1000	2500	5000	10000	Fisher
.80	1.28	.59	.32	.14	.07	.05	.03	.02	null
.90	1.64	1.00	.55	.24	.12	.08	.05	.04	borderline
.95	1.96	1.76	.97	.43	.22	.14	.10	.07	moderate
.975	2.24	3.18	1.74	.78	.39	.25	.17	.12	substantial
.99	2.58	7.12	3.90	1.74	.87	.55	.39	.28	strong
.995	2.81	13.27	7.27	3.25	1.63	1.03	.73	.51	very strong
.999	3.29	57.96	31.75	14.20	7.10	4.49	3.17	2.24	overwhelming

The case $n > 1$ in (13.29) is logically identical to (13.47). Letting $x^{(n)} = \sqrt{n}(\sum x_i/n)$ and $\mu^{(n)} = \sqrt{n}\mu$ gives

$$x^{(n)} \sim \mathcal{N}\left(\mu^{(n)}, 1\right), \tag{13.48}$$

with (13.30) becoming $\mathcal{M}_0 : \mu^{(n)} = 0$ and $\mathcal{M}_1 : \mu^{(n)} \neq 0$. Coherency requires that $\mu^{(n)}$ in (13.48) have the same $\mathcal{M}_1$ prior as μ in (13.47). Since $\mu = \mu^{(n)}/\sqrt{n}$, this implies that $g_1^{(n)}(\mu)$, the $\mathcal{M}_1$ prior for sample size n, satisfies

$$g_1^{(n)}(\mu) = g_1^{(1)}\left(\mu/\sqrt{n}\right)/\sqrt{n}, \tag{13.49}$$

this being "sample size coherency."

The effect of (13.49) is to spread the $\mathcal{M}_1$ prior density $g_1^{(n)}(\mu)$ farther away from the null value $\mu = 0$ at rate $\sqrt{n}$, while the $\mathcal{M}_0$ prior $g_0^{(n)}(\mu)$ stays fixed. For any fixed value of the sufficient statistic $x^{(n)}$ ($x^{(n)}$ being "z" in Table 13.5), this results in the Bayes factor $B(x^{(n)})$ decreasing at rate $1/\sqrt{n}$; the frequentist/Bayesian contradiction seen in Table 13.5 goes beyond the specifics of the BIC algorithm.

A *general information criterion* takes the form

$$\text{GIC}(\mathcal{M}) = \log f_{\hat{\mu}}(\boldsymbol{x}) - p\, c_n, \tag{13.50}$$

where c_n is any sequence of positive numbers; $c_n = \log(n)/2$ for BIC (13.37) and $c_n = 1$ for AIC (13.42). The difference

$$\Delta \equiv \text{GIC}(\mathcal{M}_1) - \text{GIC}(\mathcal{M}_0) = \frac{1}{2}\left(W(\boldsymbol{x}) - 2c_n d\right), \tag{13.51}$$

$d = p_1 - p_0$, will be positive if $W(\boldsymbol{x}) > 2c_n d$. For $d = 1$, as in Table 13.5, Δ will favor $\mathcal{M}_1$ if $W(\boldsymbol{x}) \geq 2c_n$, with approximate probability, if $\mathcal{M}_0$ is actually true,

$$\Pr\{\chi_1^2 \geq 2c_n\}. \tag{13.52}$$

This equals 0.157 for the AIC choice $c_n = 1$; for BIC, $n = 10{,}000$, it equals 0.0024. The choice

$$c_n = 1.92 \tag{13.53}$$

makes $\Pr\{\Delta > 0|\mathcal{M}_0\} \doteq 0.05$, agreeing with the usual frequentist 0.05 rejection level.

The BIC is *consistent*: $\Pr\{\Delta > 0\}$ goes to zero as $n \to \infty$ if $\mathcal{M}_0$ is true. This isn't true of (13.53) for instance, where we will have $\Pr\{\Delta > 0\} \doteq 0.05$ no matter how large n may be, but consistency is seldom compelling as a practical argument.

Confidence intervals help compensate for possible frequentist overfitting. With $z = 3.29$ and $n = 10{,}000$, the 95% confidence interval for μ in model $\mathcal{M}_1$ (13.30) is $(0.013, 0.053)$. Whether or not such a small effect is interesting depends on the scientific context. The fact that BIC says "not interesting" speaks to its inherent small-model bias.

The prostate cancer study data of Section 3.3 provides a more challenging model selection problem. Figure 3.4 shows the histogram of $N = 6033$ observations x_i, each measuring the effects of one gene. The histogram has 49 bins, each of width 0.2, with centers c_j ranging from -4.4 to 5.2; y_j, the height of the histogram at c_j, is the number of x_i in bin j,

$$y_j = \#\{x_i \in \text{bin } j\} \qquad \text{for } j = 1, 2, \ldots, 49. \tag{13.54}$$

We assume that the y_j follow a Poisson regression model as in Section 8.3,

$$y_j \overset{\text{ind}}{\sim} \text{Poi}(\nu_j), \qquad j = 1, 2, \ldots, 49, \tag{13.55}$$

and wish to fit a log polynomial GLM model to the ν_j. The model selection question is "What degree polynomial?" Degree 2 corresponds to normal densities, but the long tails seen in Figure 3.4 suggest otherwise.

Models of degree 2 through 8 are assessed in Figure 13.4. Four model selection measures are compared: AIC (13.42); BIC (13.37) with $n = 49$,

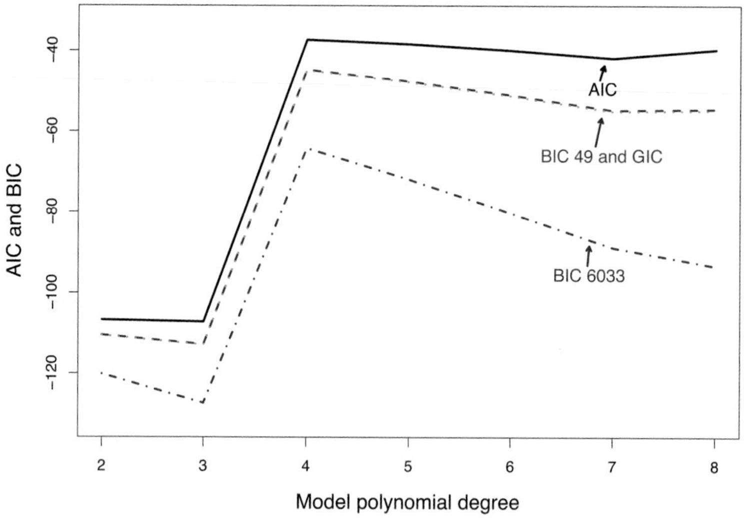

Figure 13.4 Log polynomial models of degree 2 through 8 applied to the prostate study histogram of Figure 3.4. Model selection criteria: AIC (13.42); BIC (13.37) with $n = 49$, number of bins, or 6033, number of genes; GIC (13.50) using classic Fisher hypothesis choice $c_n = 1.92$. All four selected the fourth-degree model as best.

the number of y_j values (bins), and also $n = 6033$, the number of genes; and GIC (13.50), with $c_n = 1.92$ (13.53), the choice based on classic Fisherian hypothesis testing. (This is almost the same as BIC $n = 49$, since $\log(49)/2 = 1.95$.) A fourth-degree polynomial model was the winner under all four criteria.

The "untethered" criticism made against objective Bayes methods in general is particularly applicable to BIC. The concept of "sample size" is not well defined, as the prostate study example shows. Sample size coherency (13.49), the rationale for BIC's strong bias toward smaller models, is less convincing in the absence of priors based on genuine experience (especially if there is no prospect of the sample size changing). Whatever its vulnerabilities, BIC model selection has nevertheless become a mainstay of objective Bayes model selection, not least because of its freedom from the choice of Bayesian priors.

13.4 Gibbs Sampling and MCMC

Miraculously blessed with visions of the future, a Bayesian statistician of the 1970s would certainly be pleased with the prevalence of Bayes methodology in twenty-first-century applications. But his pleasure might be tinged with surprise that the applications were mostly of the objective, "uninformative" type, rather than taken from the elegant de Finetti–Savage school of subjective inference.

The increase in Bayesian applications, and the change in emphasis from subjective to objective, had more to do with computation than philosophy. Better computers and algorithms facilitated the calculation of formerly intractable Bayes posterior distributions. Technology determines practice, and the powerful new algorithms encouraged Bayesian analyses of large and complicated models where subjective priors (or those based on actual past experience) were hard to come by. Add in the fact that the algorithms worked most easily with simple "convenience" priors like the conjugates of Section 13.2, and the stage was set for an objective Bayes renaissance.

At first glance it's hard to see why Bayesian computations should be daunting. From parameter vector $\boldsymbol{\theta}$, data $\boldsymbol{x}$, density function $f_{\boldsymbol{\theta}}(\boldsymbol{x})$, and prior density $g(\boldsymbol{\theta})$, Bayes' rule (3.5)–(3.6) directly produces the posterior density

$$g(\boldsymbol{\theta}|\boldsymbol{x}) = g(\boldsymbol{\theta}) f_{\boldsymbol{\theta}}(\boldsymbol{x}) / f(\boldsymbol{x}), \tag{13.56}$$

where $f(\boldsymbol{x})$ is the marginal density

$$f(\boldsymbol{x}) = \int_{\Omega} g(\boldsymbol{\theta}) f_{\boldsymbol{\theta}}(\boldsymbol{x})\, d\boldsymbol{\theta}. \tag{13.57}$$

The posterior probability of any set A in the parameter space Ω is then

$$P\{A|\boldsymbol{x}\} = \int_{A} g(\boldsymbol{\theta}) f_{\boldsymbol{\theta}}(\boldsymbol{x})\, d\boldsymbol{\theta} \bigg/ \int_{\Omega} g(\boldsymbol{\theta}) f_{\boldsymbol{\theta}}(\boldsymbol{x})\, d\boldsymbol{\theta}. \tag{13.58}$$

This is easy to write down but usually difficult to evaluate if $\boldsymbol{\theta}$ is multidimensional.

Modern Bayes methods attack the problem through the application of computer power. Even if we can't integrate $g(\boldsymbol{\theta}|\boldsymbol{x})$, perhaps we can *sample* from it. If so, a sufficiently large sample, say

$$\boldsymbol{\theta}^{(1)}, \boldsymbol{\theta}^{(2)}, \dots, \boldsymbol{\theta}^{(B)} \sim g(\boldsymbol{\theta}|\boldsymbol{x}) \tag{13.59}$$

would provide estimates

$$\hat{P}\{A|\boldsymbol{x}\} = \#\left\{\boldsymbol{\theta}^{(j)} \in A\right\} \Big/ B, \tag{13.60}$$

and similarly for posterior moments, correlations, etc. We would in this way be employing the same general tactic as the bootstrap, applied now for Bayesian rather than frequentist purposes—toward the same goal as the bootstrap, of freeing practical applications from the constraints of mathematical tractability.

The two most popular computational methods,[6] *Gibbs sampling* and *Markov chain Monte Carlo* (MCMC), are based on Markov chain algorithms; that is, the posterior samples $\boldsymbol{\theta}^{(b)}$ are produced in sequence, each one depending only on $\boldsymbol{\theta}^{(b-1)}$ and not on its more distant predecessors. We begin with Gibbs sampling.

The central idea of Gibbs sampling is to reduce the generation of multidimensional vectors $\boldsymbol{\theta} = (\theta_1, \theta_2, \dots, \theta_K)$ to a series of univariate calculations. Let $\boldsymbol{\theta}_{(k)}$ denote $\boldsymbol{\theta}$ with component k removed, and $g_{(k)}$ the conditional density of θ_k given $\boldsymbol{\theta}_{(k)}$ and the data $\boldsymbol{x}$,

$$\theta_k | \boldsymbol{\theta}_{(k)}, \boldsymbol{x} \sim g_{(k)}\left(\theta_k | \boldsymbol{\theta}_{(k)}, \boldsymbol{x}\right). \tag{13.61}$$

The algorithm begins at some arbitrary initial value $\boldsymbol{\theta}^{(0)}$. Having computed $\boldsymbol{\theta}^{(1)}, \boldsymbol{\theta}^{(2)}, \dots, \boldsymbol{\theta}^{(b-1)}$, the components of $\boldsymbol{\theta}^{(b)}$ are generated according to conditional distributions (13.61),

$$\theta_k^{(b)} \sim g_{(k)}\left(\theta_k \middle| \boldsymbol{\theta}_{(k)}^{(b-1)}, \boldsymbol{x}\right) \qquad \text{for } k = 1, 2, \dots, K. \tag{13.62}$$

As an example, we take $\boldsymbol{x}$ to be the $n = 20$ observations for $y = 1$ in the vasoconstriction data of Table 13.2, and assume that these are a normal sample,

$$x_i \overset{\text{iid}}{\sim} \mathcal{N}(\mu, \tau), \qquad i = 1, 2, \dots, n = 20. \tag{13.63}$$

The sufficient statistics for estimating the bivariate parameter $\theta = (\mu, \tau)$ are the sample mean and variance

$$\bar{x} = \sum_1^n x_i/n \quad \text{and} \quad T = \sum_1^n (x_i - \bar{x})^2/(n-1), \tag{13.64}$$

having independent normal and gamma distributions,

$$\bar{x} \sim \mathcal{N}(\mu, \tau/n) \quad \text{and} \quad T \sim \tau G_\nu/\nu, \tag{13.65}$$

with $\nu = \frac{n-1}{2}$, the latter being $\text{Gam}(\nu, \tau/\nu)$ in the notation of Table 5.1.

[6] The two methods are often referred to collectively as MCMC because of mathematical connections, with "Metropolis-Hasting algorithm" referring to the second type of procedure.

For our Bayes prior distribution we take the conjugates

$$\tau \sim k_1\tau_1/G_{k_1+1} \quad \text{and} \quad \mu|\tau \sim \mathcal{N}(\mu_0, \tau/n_0). \tag{13.66}$$

In terms of Table 13.1, $(x_0, n_0\nu) = (\tau_1, k_1)$ for the gamma, while $(x_0, \sigma_1^2) = (\mu_0, \tau)$ for the normal. (A simple specification would take $\mu \sim \mathcal{N}(\mu_0, \tau_1/n_0)$.)

Multiplying the normal and gamma functional forms in Table 5.1 yields density function

$$f_{\mu,\tau}(\bar{x}, T) = c\tau^{-(\nu+\frac{1}{2})} \exp\left\{-\frac{1}{\tau}\left[\nu T + \frac{n}{2}(\bar{x}-\mu)^2\right]\right\} \tag{13.67}$$

and prior density

$$g(\mu, \tau) = c\tau^{-(k_1+2.5)} \exp\left\{-\frac{1}{\tau}\left[k_1\tau_1 + \frac{n_0}{2}(\mu-\mu_0)^2\right]\right\}, \tag{13.68}$$

c indicating positive constants that do not affect the posterior computations. The posterior density $cg(\mu,\tau)f_{\mu,\tau}(\bar{x}, T)$ is then calculated to be

$$\begin{aligned} g(\mu,\tau|\bar{x},T) &= c\tau^{-(\nu+k_1+3)}\exp\{-Q/\tau\}, \\ \text{where } Q &= (k_1\tau_1 + T) + \frac{n_+}{2}(\mu-\bar{\mu}_+)^2 + \frac{n_0 n}{2n_+}(\mu_0-\bar{x})^2. \end{aligned} \tag{13.69}$$

Here $n_+ = n_0 + n$ and $\bar{\mu}_+ = (n_0\mu_0 + n\bar{x})/n_+$.

In order to make use of Gibbs sampling we need to know the *full conditional* distributions $g(\mu|\tau,\bar{x},T)$ and $g(\tau|\mu,\bar{x},T)$, as in (13.62). (In this case, $k = 2$, $\theta_1 = \mu$, and $\theta_2 = \tau$.) This is where the conjugate expressions in Table 13.1 come into play. Inspection of density (13.69) shows that

$$\mu|\tau,\bar{x},T \sim \mathcal{N}\left(\bar{\mu}_+, \frac{\tau}{n_+}\right) \quad \text{and} \quad \tau|\mu,\bar{x},T \sim \frac{Q}{G_{\nu+k_1+2}}. \tag{13.70}$$

$B = 10{,}000$ Gibbs samples $\boldsymbol{\theta}^{(b)} = (\mu^{(b)}, \tau^{(b)})$ were generated starting from $\boldsymbol{\theta}^{(0)} = (\bar{x}, T) = (116, 554)$. The prior specifications were chosen to be (presumably) uninformative or mildly informative,

$$n_0 = 1, \quad \mu_0 = \bar{x}, \quad k_1 = 1 \text{ or } 9.5, \quad \text{and} \quad \tau_1 = T. \tag{13.71}$$

(In which case $\bar{\mu}_+ = \bar{x}$ and $Q = (\nu + k_1)T + n_+(\mu - \bar{x})^2$. From $\nu = (n-1)/2$, we see that k_1 corresponds to about $2k_1$ hypothetical prior observations.) The resulting posterior distributions for τ are shown by the histograms in Figure 13.5.

As a point of frequentist comparison, $B = 10{,}000$ parametric bootstrap replications (which involve no prior assumptions),

$$\hat{\tau}^* \sim \hat{\tau}G_\nu/\nu, \qquad \hat{\tau} = T, \tag{13.72}$$

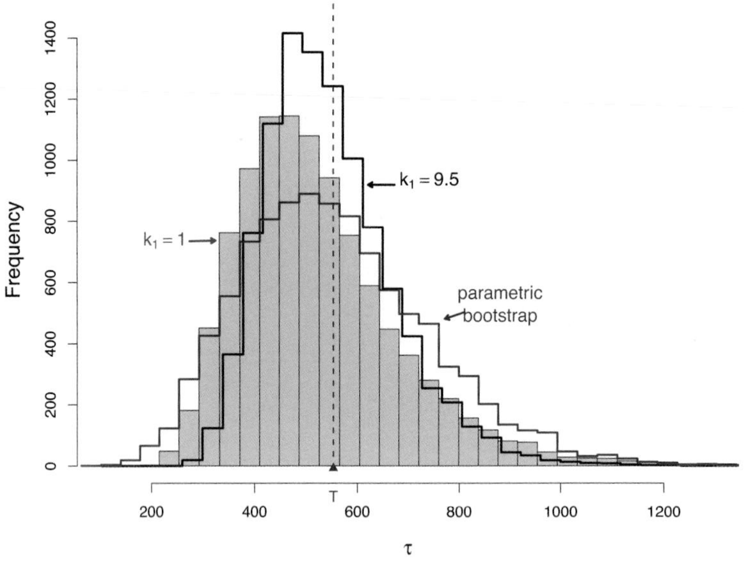

Figure 13.5 Posterior distributions for variance parameter τ, model (13.63)–(13.65), volume of air inspired for vasoconstriction group $y = 1$ from Table 13.2. Solid teal histogram: $B = 10,000$ Gibbs samples with $k_1 = 1$; black line histogram: $B = 10,000$ samples with $k_1 = 9.5$; red line histogram: 10,000 parametric bootstrap samples (13.72) suggests even the $k_1 = 1$ prior has substantial posterior effect.

are seen to be noticeably more dispersed than even the $k_1 = 1$ Bayes posterior distribution, the likely choice for an objective Bayes analysis. Bayes techniques, even objective ones, have regularization effects that may or may not be appropriate.

A similar, independent Gibbs sample of size 10,000 was obtained for the 19 $y = 0$ vasoconstriction measurements in Table 13.2, with specifications as in (13.71), $k = 1$. Let

$$\delta^{(b)} = \frac{\mu_1^{(b)} - \mu_0^{(b)}}{\left(\tau_1^{(b)} - \tau_0^{(b)}\right)^{1/2}}, \tag{13.73}$$

where $(\mu_1^{(b)}, \tau_1^{(b)})$ and $(\mu_0^{(b)}, \tau_0^{(b)})$ denote the bth Gibbs samples from the $y = 1$ and $y = 0$ runs.

Figure 13.6 shows the posterior distribution of δ. Twenty-eight of the

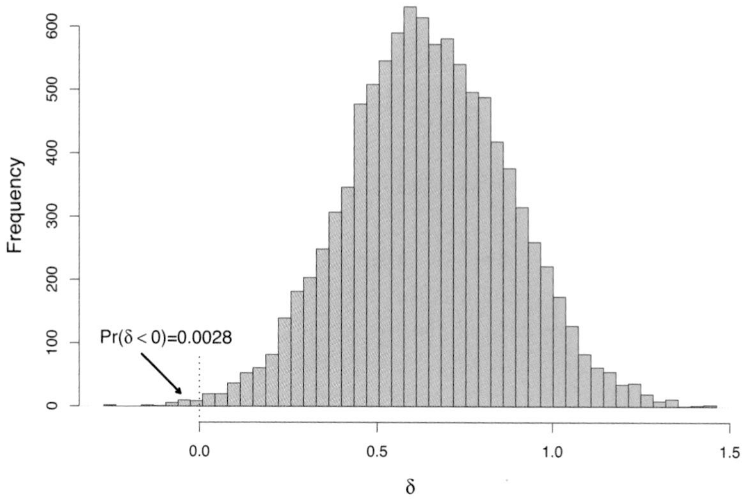

Figure 13.6 B =10,000 Gibbs samples for "Bayes t-statistic" (13.73) comparing $y = 1$ with $y = 0$ values for vasoconstriction data.

B =10,000 values $\delta^{(b)}$ were less than 0, giving a "Bayesian t-test" estimate

$$P\{\delta < 0 | \bar{x}_1, \bar{x}_0, T_1, T_0\} = 0.0028. \tag{13.74}$$

(The usual t-test yielded one-sided p-value 0.0047 against the null hypothesis $\mu_0 = \mu_1$.) An appealing feature of Gibbs sampling is that having obtained $\boldsymbol{\theta}^{(1)}, \boldsymbol{\theta}^{(2)}, \ldots, \boldsymbol{\theta}^{(B)}$ (13.59) the posterior distribution of any parameter $\gamma = t(\boldsymbol{\theta})$ is obtained directly from the B values $\gamma^{(b)} = t(\boldsymbol{\theta}^{(b)})$.

Gibbs sampling requires the ability to sample from the full conditional distributions (13.61). A more general Markov chain Monte Carlo method, commonly referred to as MCMC, makes clearer the basic idea. Suppose the space of possible $\boldsymbol{\theta}$ values is finite, say $\{\boldsymbol{\theta}(1), \boldsymbol{\theta}(2), \ldots, \boldsymbol{\theta}(M)\}$, and we wish to simulate samples from a posterior distribution putting probability $p(i)$ on $\boldsymbol{\theta}(i)$,

$$\boldsymbol{p} = (p(1), p(2), \ldots, p(M)). \tag{13.75}$$

The MCMC algorithm begins with the choice of a "candidate" probability distribution $q(i, j)$ for moving from $\boldsymbol{\theta}(i)$ to $\boldsymbol{\theta}(j)$; in theory $q(i, j)$ can be almost anything, for instance $q(i, j) = 1/(M - 1)$ for $j \neq i$. The simulated samples $\boldsymbol{\theta}^{(b)}$ are obtained by a random walk: if $\boldsymbol{\theta}^{(b)}$ equals $\boldsymbol{\theta}(i)$, then $\boldsymbol{\theta}^{(b+1)}$ equals $\boldsymbol{\theta}(j)$ with probability[7]

$$Q(i,j) = q(i,j) \cdot \min\left\{\frac{p(j)q(j,i)}{p(i)q(i,j)}, 1\right\} \tag{13.76}$$

for $j \neq i$, while with probability

$$Q(i,i) = 1 - \sum_{j \neq i} Q(i,j) \tag{13.77}$$

$\boldsymbol{\theta}^{(b+1)} = \boldsymbol{\theta}^{(b)} = \boldsymbol{\theta}(i)$. Markov chain theory then says that, under quite general conditions, the empirical distribution of the random walk values $\boldsymbol{\theta}^{(b)}$ will approach the desired distribution $\boldsymbol{p}$ as b gets large.

A heuristic argument for why this happens begins by supposing that $\boldsymbol{\theta}^{(1)}$ was in fact generated by sampling from the target distribution $\boldsymbol{p}$, $\Pr\{\boldsymbol{\theta}^{(1)} = i\} = p(i)$, and then $\boldsymbol{\theta}^{(2)}$ was obtained according to transition probabilities (13.76)–(13.77). A little algebra shows that (13.76) implies

$$p(i)Q(i,j) = p(j)Q(j,i), \tag{13.78}$$

the so-called *balance equations*. This results in

$$\begin{aligned}\Pr\left\{\boldsymbol{\theta}^{(2)} = i\right\} &= p(i)Q(i,i) + \sum_{j \neq i} p(j)Q(j,i) \\ &= p(i)\sum_{i=1}^{M} Q(i,j) = p(i).\end{aligned} \tag{13.79}$$

In other words, if $\boldsymbol{\theta}^{(1)}$ has distribution $\boldsymbol{p}$ then so will $\boldsymbol{\theta}^{(2)}$, and likewise $\boldsymbol{\theta}^{(3)}, \boldsymbol{\theta}^{(4)}, \ldots$; $\boldsymbol{p}$ is the *equilibrium distribution* of the Markov chain random walk defined by transition probabilities Q. Under reasonable con-
†9 ditions,† $\boldsymbol{\theta}^{(b)}$ must asymptotically attain distribution $\boldsymbol{p}$ no matter how $\boldsymbol{\theta}^{(1)}$ is initially selected.

13.5 Example: Modeling Population Admixture

MCMC has had a big impact in statistical genetics, where Bayesian modeling is popular and useful for representing the complex evolutionary processes. Here we illustrate its use in demography and modeling *admixture*—estimating the contributions from ancestral populations in an individual

[7] In Bayes applications, $p(i) = g(\boldsymbol{\theta}(i)|\boldsymbol{x}) = g(\boldsymbol{\theta}(i))f_{\boldsymbol{\theta}(i)}(\boldsymbol{x})/f(\boldsymbol{x})$ (13.56). However, $f(\boldsymbol{x})$ is not needed since it cancels out of (13.76), a considerable advantage in complicated situations when $f(\boldsymbol{x})$ is often unavailable, and a prime reason for the popularity of MCMC.

genome. For example, we might consider human ancestry, and for each individual wish to estimate the proportion of their genome coming from **European**, **African**, and **Asian** origins. The procedure we describe here is unsupervised—a type of soft clustering—but we will see it can be very informative with regard to such questions. We have a sample of n individuals, and we assume each arose from possible admixture among J parent populations, each with their own characteristic vector of allele frequencies. For us $J = 3$, and let $Q_i \in \mathcal{S}_3$ denote a probability vector for individual i representing the proportions of their heritage coming from populations $j \in \{1, 2, 3\}$ (see Section 5.4). We have genomic measurements for each individual, in our case SNPs (single-nucleotide polymorphisms) at each of M well-spaced loci, and hence can assume they are in linkage equilibrium. At each SNP we have a measurement that identifies the two alleles (one per chromosome), where each can be either the wild-type A or the mutation a. That is, we have the genotype G_{im} at SNP m for individual i: a three-level factor with levels {**AA**, **Aa**, **aa**} which we code as $0, 1, 2$. Table 13.6 shows some examples.

Table 13.6 *A subset of the genotype data on 197 individuals, each with genotype measurements at 100 SNPs. In this case the* **ethnicity** *is known for each individual, one of* **Japanese**, **African**, **European**, *or* **African American**. *For example, individual* **NA12239** *has* **genotype Aa** *for SNP1,* **NA19247** *has* **AA**, *and* **NA20126** *has* **aa**.

Subject	SNP$_1$	SNP$_2$	SNP$_3$	···	SNP$_{97}$	SNP$_{98}$	SNP$_{99}$	SNP$_{100}$
NA10852	1	1	0	···	1	1	0	0
NA12239	1	1	0	···	1	1	0	0
NA19072	0	0	0	···	0	0	0	0
NA19247	0	0	2	···	0	0	0	2
NA20126	2	0	0	···	2	0	0	0
NA18868	0	0	1	···	0	0	0	1
NA19257	0	0	0	···	0	0	0	0
NA19079	0	1	0	···	0	1	0	0
NA19067	0	0	0	···	0	0	0	0
NA19904	0	0	1	···	0	0	0	1

Let P_j be the (unknown) M-vector of minor allele frequencies (proportions actually) in population j. We have available a sample of n individuals, and for each sample we have their genomic information measured at each of the M loci. Some of the individuals might appear to have pure ancestral origins, but many do not. Our goal is to estimate Q_i, $i = 1, \ldots, n$, and P_j, $j \in \{1, 2, 3\}$.

For this purpose it is useful to pose a generative model. We first create a pair of variables $X_{im} = (X_{im}^{(1)}, X_{im}^{(2)})$ corresponding to each G_{im}, to which we allocate the two alleles (in arbitrary order). For example, if $G_{im} = 1$ (corresponding to Aa), then we might set $X_{im}^{(1)} = 0$ and $X_{im}^{(2)} = 1$ (or vice versa). If $G_{im} = 0$ they are both 0, and if $G_{im} = 2$, they are both 1. Let $Z_{im} \in \{1, 2, 3\}^2$ represent the ancestral origin for individual i of each of these allele copies X_{im} at locus m, again a two-vector with elements $Z_{im} = (Z_{im}^{(1)}, Z_{im}^{(2)})$. Then our generative model goes as follows.

1 $Z_{im}^{(c)} \sim \text{Mult}(1, Q_i)$, independently at each m, for each copy $c = 1, 2$. That is, we select the ancestral origin of each chromosome at locus m according to the individual's mixture proportions Q_i.

2 $X_{im}^{(c)} \sim \text{Bi}(1, \; P_{jm})$ if $Z_{im}^{(c)} = j$, for each copy $c = 1, 2$. What this means is that, for each of the two ancestral picks at locus m (one for each arm of the chromosome), we draw a binomial with the appropriate allele frequency.

To complete the Bayesian specification, we need to supply priors for the Q_i and also for P_{jm}. Although one can get fancy here, we resort to the recommended flat priors, which are

- $Q_i \sim D(\lambda, \lambda, \lambda)$, a flat three-component Dirichlet, independently for
†10 each subject i† and
- $P_{jm} \sim D(\gamma, \gamma)$ independently for each population j, and each locus m (the beta distribution; see †10 in the end notes).

We use the least-informative values $\lambda = \gamma = 1$. In practice, these could get updated as well, but for the purposes of this demonstration we leave them fixed at these values.

Let $\boldsymbol{X}$ be the $n \times M \times 2$ array of observed alleles for all n samples. We wish to estimate the posterior distribution $\Pr(P, Q|\boldsymbol{X})$, referring collectively to all the elements of P and Q.

For this purpose we use Gibbs sampling, which amounts to the following sequence.

0 Initialize $Z^{(0)}, \; P^{(0)}, Q^{(0)}$.

1 Sample $Z^{(b)}$ from the conditional distribution $\Pr(Z|\boldsymbol{X}, P^{(b-1)}, Q^{(b-1)})$.

2 Sample $P^{(b)}, Q^{(b)}$ from the conditional distribution $\Pr(P, Q|\boldsymbol{X}, Z^{(b)})$.

Gibbs is effective when one can sample efficiently from these conditional distributions, which is the case here.

In step 2, we can sample P and Q separately. It can be seen that for each (j, m) we should sample P_{jm} from

$$P_{jm}|\boldsymbol{X}, Z \sim D(\lambda + n_{jm}^{(0)}, \ \lambda + n_{jm}^{(1)}), \tag{13.80}$$

where $Z = Z^{(b)}$ and

$$\begin{aligned} n_{jm}^{(0)} &= \#\{(i, c) \ : \ X_{im}^{(c)} = 0 \text{ and } Z_{im}^{(c)} = j\}, \\ n_{jm}^{(1)} &= \#\{(i, c) \ : \ X_{im}^{(c)} = 1 \text{ and } Z_{im}^{(c)} = j\}. \end{aligned} \tag{13.81}$$

This follows from the conjugacy of the two-component Dirichlet (beta) with the binomial distribution, Table 13.1.

Updating Q_i involves simulating from

$$Q_i|\boldsymbol{X}, Z \sim D(\gamma + m_{i1}, \gamma + m_{i2}, \gamma + m_{i3}), \tag{13.82}$$

where m_{ij} is the number of allele copies in individual i that originated (according to $Z = Z^{(b)}$) in population j:

$$m_{ij} = \#\{(c, m) \ : \ Z_{im}^{(c)} = j\}. \tag{13.83}$$

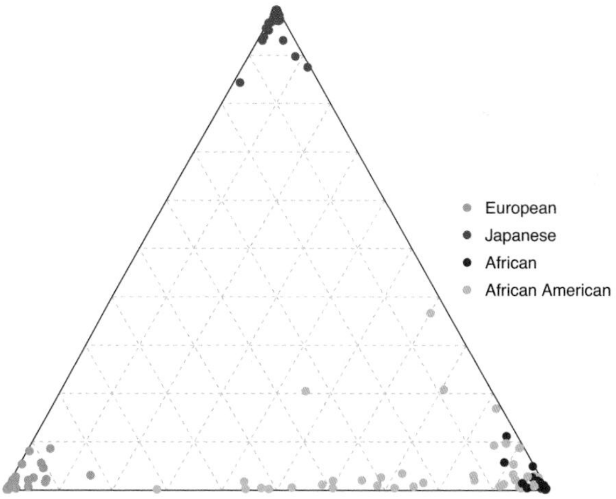

Figure 13.7 Barycentric coordinate plot for the estimated posterior means of the Q_i based on MCMC sampling.

Step 1 can be performed by simulating $Z_{im}^{(c)}$ independently, for each i, m, and c from

$$\Pr(Z_{im}^{(c)} = j \mid X, P, Q) = \frac{Q_{ij} \Pr(X_{im}^{(c)} \mid P, Z_{im}^{(c)} = j)}{\sum_{\ell=1}^{3} Q_{i\ell} \Pr(X_{im}^{(c)} \mid P, Z_{im}^{(c)} = \ell)}. \tag{13.84}$$

The probabilities on the right refer back to our generative distribution described earlier.

Figure 13.7 shows a triangle plot that summarizes the result of running the MCMC algorithm on our 197 subjects. We used a burn in of 1000 complete iterations, and then a further 2000 to estimate the the distribution of the parameters of interest, in this case the Q_i. Each dot in the figure represents a three-component probability vector, and is the posterior mean of the sampled Q_i for each subject. The points are colored according to the known ethnicity. Although this algorithm is unsupervised, we see that the ethnic groups cluster nicely in the corners of the simplex, and allow us to identify these clusters. The **African American** group is spread between the **African** and **European** clusters (with a little movement toward the **Japanese**).

Markov chain methods are versatile tools that have proved their value in Bayesian applications. There are some drawbacks.

- The algorithms are not universal in the sense of maximum likelihood, requiring some individual ingenuity with each application.
- As a result, applications, especially of Gibbs sampling, have favored a small set of convenient priors, mainly Jeffreys and conjugates, that simplify the calculations. This can cast doubt on the relevance of the resulting Bayes inferences.
- Successive realizations $\boldsymbol{\theta}^{(b)}$ are highly correlated with each other, making the convergence of estimates such as $\bar{\boldsymbol{\theta}} = \sum \boldsymbol{\theta}^{(b)}/B$ slow.
- The correlation makes it difficult to assign a standard error to $\bar{\boldsymbol{\theta}}$. Actual applications ignore an initial B_0 of the $\boldsymbol{\theta}^{(b)}$ values (as a "burn-in" period) and go on to large enough B such that estimates like $\bar{\boldsymbol{\theta}}$ appear to settle down. However, neither the choice of B_0 nor that of B may be clear.

Objective Bayes offers a paradigm of our book's theme, the effect of electronic computation on statistical inference: ingenious new algorithms facilitated Bayesian applications over a wide class of applied problems and, in doing so, influenced the dominant philosophy of the whole area.

13.6 Notes and Details

The books by Savage (1954) and de Finetti (1972), summarizing his earlier work, served as foundational texts for the subjective Bayesian school of inference. Highly influential, they championed a framework for Bayesian applications based on coherent behavior and the careful elucidation of personal probabilities. A current leading text on Bayesian methods, Carlin and Louis (2000), does not reference either Savage or de Finetti. Now Jeffreys (1961), again following earlier works, claims foundational status. The change of direction has not gone without protest from the subjectivists—see Adrian Smith's discussion of O'Hagan (1995)—but is nonetheless almost a complete rout.

Metropolis *et al.* (1953), as part of nuclear weapons research, developed the first MCMC algorithm. A vigorous line of work on Markov chain methods for solving difficult probability problems has continued to flourish under such names as particle filtering and sequential Monte Carlo; see Gerber and Chopin (2015) and its enthusiastic discussion.

Modeling population admixture (Pritchard *et al.*, 2000) is one of several applications of hierarchical Bayesian models and MCMC in genetics. Other applications include haplotype estimation and motif finding, as well as estimation of phylogenetic trees. The examples in this section were developed with the kind help of Hua Tang and David Golan, both from the Stanford Genetics department. Hua suggested the example and provided helpful guidance; David provided the data, and ran the MCMC algorithm using the STRUCTURE program in the Pritchard lab.

†1 [p. 236] *Uninformative priors.* A large catalog of possible uninformative priors has been proposed, thoroughly surveyed by Kass and Wasserman (1996). One approach is to use the likelihood from a small part of the data, say just one or two data points out of n, as the prior, as with the "intrinsic priors" of Berger and Pericchi (1996), or O'Hagan's (1995) "fractional Bayes factors." Another approach is to minimize some mathematical measure of prior information, as with Bernardo's (1979) "reference priors" or Jaynes' (1968) "maximum entropy" criterion. Kass and Wasserman list a dozen more possibilities.

†2 [p. 236] *Coverage matching priors.* Welch and Peers (1963) showed that, for a multiparameter family $f_\mu(x)$ and real-valued parameter of interest $\theta = t(\mu)$, there exist priors $g(\mu)$ such that the Bayes credible interval of coverage α has frequentist coverage $\alpha + O(1/n)$, with n the sample size. In other words, the credible intervals are "second-order accurate" confidence intervals. Tibshirani (1989), building on Stein's (1985) work, produced the

nice formulation (13.9). Stein's paper developed the *least-favorable family*, the one-parameter subfamily of $f_\mu(x)$ that does not inappropriately increase the amount of Fisher information for estimating θ. Cox and Reid's (1987) *orthogonal parameters* form (13.8) is formally equivalent to the least favorable family construction.

Least favorable family versions of reference priors and intrinsic priors have been proposed to avoid the difficulty with general-purpose uninformative priors seen in Figure 11.7. They do so, but at the price of requiring a different prior for each choice of $\theta = t(\mu)$—which begins to sound more frequentistic than Bayesian.

†3 [p. 238] *Conjugate families theorem.* Theorem 13.1, (13.16)–(13.18), is rigorously derived in Diaconis and Ylvisaker (1979). Families other than (13.14) have conjugate-like properties, but not the neat posterior expectation result (13.18).

†4 [p. 238] *Poisson formula* (13.20). This follows immediately from (13.14), using $\alpha = \log(\mu)$, $\psi(\alpha) = \mu$, and $V(\mu) = \mu$ for the Poisson.

†5 [p. 239] *Inverse gamma and chi-square distributions.* A G_ν variate (13.22) has density $\mu^{\nu-1}e^{-\mu} / \Gamma(\nu)$. An *inverse gamma* variate $1/G_\nu$ has density $\mu^{-(\nu+1)}\, e^{-1/\mu}/\Gamma(\nu)$, so

$$g_{n_0,x_0}(\mu) = c\mu^{-(n_0x_0+2)}e^{-n_0x_0\nu/\mu} \tag{13.85}$$

is the gamma conjugate density in Table 13.1. The gamma results can be restated in terms of chi-squared variates:

$$x_i \sim \mu\frac{\chi^2_m}{m} = \mu\frac{G_{m/2}}{m/2} \tag{13.86}$$

has conjugate prior

$$g_{n_0,x_0}(\mu) \sim n_0x_0m/\chi^2_{n_0m+2}, \tag{13.87}$$

an *inverse chi-squared* distribution.

†6 [p. 240] *Vasoconstriction data.* Efron and Gous (2001) use this data to illustrate a theory connecting Bayes factors with Fisherian hypothesis testing. It is part of a larger data set appearing in Finney (1947), also discussed in Kass and Raftery (1995).

†7 [p. 245] *Jeffreys' and Fisher's scales of evidence.* Jeffreys' scale as it appears in Table 13.3 is taken from the slightly amended form in Kass and Raftery (1995). Efron and Gous (2001) compare it with Fisher's scale for the contradictory results of Table 13.5. Fisher and Jeffreys worked in different scientific contexts—small-sample agricultural experiments versus

hard-science geostatistics—which might explain Jeffreys' more stringent conception of what constitutes significant evidence.

†8 [p. 246] *The Bayesian information criterion.* The BIC was proposed by Schwarz (1978). Kass and Wasserman (1996) provide an extended discussion of the BIC and model selection. "Proofs" of (13.37) ultimately depend on sample size coherency (13.49), as in Efron and Gous (2001). Quotation marks are used here to indicate the basically qualitative nature of BIC: if we think of the data points as being collected in pairs then n becomes $n/2$ in (13.38), etc., so it doesn't pay to put too fine a point on the criterion.

†9 [p. 256] *MCMC convergence.* Suppose we begin the MCMC random walk (13.76)–(13.77) by choosing $\boldsymbol{\theta}^{(1)}$ according to some arbitrary starting distribution $\boldsymbol{p}^{(1)}$. Let $\boldsymbol{p}^{(b)}$ be the distribution of $\boldsymbol{\theta}^{(b)}$, obtained after b steps of the random walk. Markov chain theory says that, under certain broad conditions on $Q(i, j)$, $\boldsymbol{p}^{(b)}$ will converge to the target distribution $\boldsymbol{p}$ (13.75). Moreover, the convergence is geometric in the L_1 norm $\sum |p_k^{(b)} - p_k|$, successive discrepancies eventually decreasing by a multiplicative factor. A proof appears in Tanner and Wong (1987). Unfortunately, the factor won't be known in most applications, and the actual convergence may be quite slow.

†10 [p. 258] *Dirichlet distribution.* The Dirichlet is a multivariate generalization of the beta distribution (Section 5.1), typically used to represent prior distributions for the multinomial distribution. For $x = (x_1, x_2, \ldots, x_k)'$, with $x_j \in (0, 1)$, $\sum_j x_j = 1$, the $D(\nu)$ density is defined as

$$f_\nu(x) = \frac{1}{B(\nu)} \prod_{j=1}^{k} x_j^{\nu_j - 1}, \tag{13.88}$$

where $B(\nu) = \prod_j \Gamma(\nu_j) / \Gamma(\sum_j \nu_j)$.

14

Statistical Inference and Methodology in the Postwar Era

The fundamentals of statistical inference—frequentist, Bayesian, Fisherian—were set in place by the end of the first half of the twentieth century, as discussed in Part I of this book. The postwar era witnessed a massive expansion of statistical methodology, responding to the data-driven demands of modern scientific technology. We are now at the end of Part II, "Early Computer-Age Methods," having surveyed the march of new statistical algorithms and their inferential justification from the 1950s through the 1990s.

This was a time of opportunity for the discipline of statistics, when the speed of computation increased by a factor of a thousand, and then another thousand. As we said before, a land bridge had opened to a new continent, but not everyone was eager to cross. We saw a mixed picture: the computer played a minor or negligible role in the development of some influential topics such as empirical Bayes, but was fundamental to others such as the bootstrap.

Fifteen major topics were examined in Chapters 6 through 13. What follows is a short scorecard of their inferential affinities, Bayesian, frequentist, or Fisherian, as well as an assessment of the computer's role in their development. None of this is very precise, but the overall picture, illustrated in Figure 14.1, is evocative.

Empirical Bayes

Robbins' original development of formula (6.5) was frequentistic, but most statistical researchers were frequentists in the postwar era so that could be expected. The obvious Bayesian component of empirical Bayes arguments is balanced by their frequentist emphasis on (nearly) unbiased estimation of Bayesian estimators, as well as the restriction to using only current data for inference. Electronic computation played hardly any role in the theory's development (as indicated by blue coloring in the figure). Of course mod-

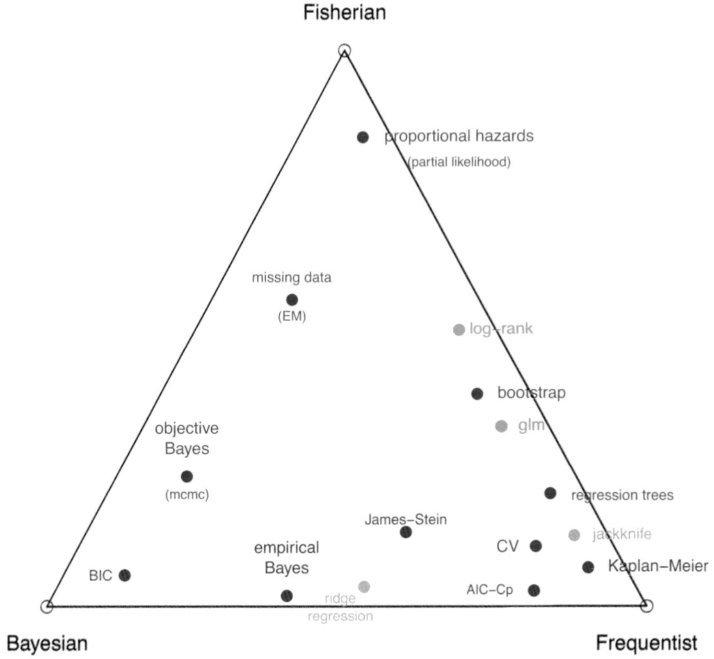

Figure 14.1 Bayesian, frequentist, and Fisherian influences, as described in the text, on 15 major topics, 1950s through 1990s. Colors indicate the importance of electronic computation in their development: red, crucial; violet, very important; green, important; light blue, less important; blue, negligible.

ern empirical Bayes applications are heavily computational, but that is the case for most methods now.

James–Stein and Ridge Regression

The frequentist roots of James–Stein estimation are more definitive, especially given the force of the James–Stein theorem (7.16). Nevertheless, the empirical Bayes interpretation (7.13) lends James–Stein some Bayesian credibility. Electronic computation played no role in its development. This was less true for ridge regression, colored light blue in the figure, where the matrix calculation (7.36) would have been daunting in the pre-electronic age. The Bayesian justification (7.37)–(7.39) of ridge regression carries more weight than for James–Stein, given the absence of a strong frequentist theorem.

Generalized Linear Models

GLM development began with a pronounced Fisherian emphasis on likelihood[1] modeling, but settled down to more or less standard frequentist regression theory. A key operational feature, low-dimensional sufficient statistics, limited its computational demands, but GLM theory could not have developed before the age of electronic computers (as indicated by green coloring).

Regression Trees

Model building by means of regression trees is a computationally intensive enterprise, indicated by its red color in Figure 14.1. Its justification has been mainly in terms of asymptotic frequentist properties.

Survival Analysis

The Kaplan–Meier estimate, log-rank test, and proportional hazards model move from the frequentist pole of the diagram toward the Fisherian pole as the conditioning arguments in Sections 9.2 through 9.4 become more elaborate. The role of computation in their development increases in the same order. Kaplan–Meier estimates can be done by hand (and were), while it is impossible to contemplate proportional hazards analysis without the computer. Partial likelihood, the enabling argument for the theory, is a quintessential Fisherian device.

Missing Data and the EM Algorithm

The imputation of missing data has a Bayesian flavor of *indirect evidence*, but the "fake data" principle (9.44)–(9.46) has Fisherian roots. Fast computation was important to the method's development, particularly so for the EM algorithm.

Jackknife and Bootstrap

The purpose of the jackknife was to calculate frequentist standard errors and biases. Electronic computation was of only minor importance in its

[1] More explicitly, *quasilikelihoods*, an extension to a wider class of exponential family models.

development. By contrast, the bootstrap is the archetype for computer-intensive statistical inference. It combines frequentism with Fisherian devices: plug-in estimation of accuracy estimates, as in (10.18)–(10.19), and correctness arguments for bootstrap confidence intervals, (11.79)–(11.83).

Cross-Validation

The renaissance of interest in cross-validation required fast computation, especially for assessing modern computer-intensive prediction algorithms. As pointed out in the text following Figure (12.3), cross-validation is a *strongly* frequentist procedure.

BIC, AIC, and C_p

These three algorithms were designed to *avoid* computation, BIC for Bayesian model selection, Section (13.3), AIC and C_p for unbiased estimation of frequentist prediction error, (12.76) and (12.50).

Objective Bayes and MCMC

In addition to their Bayesian provenance, objective Bayes methods have some connection with fiducial ideas and the bootstrap, as discussed in Section 11.5. (An argument can be made that they are at least as frequentist as they are Bayesian—see the notes below—though that has not been acted upon in coloring the figure.) Gibbs sampling and MCMC, the enabling algorithms, epitomize modern computer-intensive inference.

Notes

Figure 14.1 is an updated version of Figure 8 in Efron (1998), "R. A. Fisher in the 21st Century." There the difficulty of properly placing *objective Bayes* is confessed, with Erich Lehmann arguing for a more frequentist (decision-theoretic) location: "In fact, the concept of uninformative prior is philosophically close to Wald's least favorable distribution, and the two often coincide."

Figure 14.1 shows a healthy mixture of philosophical and computational tactics at work, with all three edges (but not the center) of the triangle in play. All new points will be red (computer-intensive) as we move into the twenty-first century in Part III. Our triangle will have to struggle to accommodate some major developments based on machine learning, a philosophically atheistic approach to statistical inference.

Part III

Twenty-First-Century Topics

15

Large-Scale Hypothesis Testing and False-Discovery Rates

By the final decade of the twentieth century, electronic computation fully dominated statistical practice. Almost all applications, classical or otherwise, were now performed on a suite of computer platforms: SAS, SPSS, Minitab, Matlab, S (later R), and others.

The trend accelerates when we enter the twenty-first century, as statistical methodology struggles, most often successfully, to keep up with the vastly expanding pace of scientific data production. This has been a two-way game of pursuit, with statistical algorithms chasing ever larger data sets, while inferential analysis labors to rationalize the algorithms.

Part III of our book concerns topics in twenty-first-century[1] statistics. The word "topics" is intended to signal selections made from a wide catalog of possibilities. Part II was able to review a large portion (though certainly not all) of the important developments during the postwar period. Now, deprived of the advantage of hindsight, our survey will be more illustrative than definitive.

For many statisticians, *microarrays* provided an introduction to large-scale data analysis. These were revolutionary biomedical devices that enabled the assessment of individual activity for thousands of genes at once—and, in doing so, raised the need to carry out thousands of simultaneous hypothesis tests, done with the prospect of finding only a few interesting genes among a haystack of null cases. This chapter concerns large-scale hypothesis testing and the *false-discovery rate*, the breakthrough in statistical inference it elicited.

[1] Actually what historians might call "the long twenty-first century" since we will begin in 1995.

15.1 Large-Scale Testing

The `prostate` cancer data, Figure 3.4, came from a microarray study of $n = 102$ men, 52 prostate cancer patients and 50 normal controls. Each man's gene expression levels were measured on a panel of $N = 6033$ genes, yielding a 6033×102 matrix of measurements x_{ij},

$$x_{ij} = \text{activity of } i\text{th gene for } j\text{th man.} \tag{15.1}$$

For each gene, a two-sample t statistic (2.17) t_i was computed comparing gene i's expression levels for the 52 patients with those for the 50 controls. Under the null hypothesis H_{0i} that the patients' and the controls' responses come from the same normal distribution of gene i expression levels, t_i will follow a standard Student t distribution with 100 degrees of freedom, t_{100}. The transformation

$$z_i = \Phi^{-1}\left(F_{100}(t_i)\right), \tag{15.2}$$

where F_{100} is the cdf of a t_{100} distribution and Φ^{-1} the inverse function of a standard normal cdf, makes z_i standard normal under the null hypothesis:

$$H_{0i} : z_i \sim \mathcal{N}(0, 1). \tag{15.3}$$

Of course the investigators were hoping to spot some *non-null* genes, ones for which the patients and controls respond differently. It can be
†1 shown that a reasonable model for both null and non-null genes is[2] †

$$z_i \sim \mathcal{N}(\mu_i, 1), \tag{15.4}$$

μ_i being the *effect size* for gene i. Null genes have $\mu_i = 0$, while the investigators hoped to find genes with large positive or negative μ_i effects.

Figure 15.1 shows the histogram of the 6033 z_i values. The red curve is the scaled $\mathcal{N}(0, 1)$ density that would apply if in fact *all* of the genes were null, that is if all of the μ_i equaled zero.[3] We can see that the curve is a little too high near the center and too low in the tails. Good! Even though most of the genes appear null, the discrepancies from the curve suggest that there are some non-null cases, the kind the investigators hoped to find.

Large-scale testing refers exactly to this situation: having observed a large number N of test statistics, how should we decide which if any of the null hypotheses to reject? Classical testing theory involved only a single case, $N = 1$. A theory of multiple testing arose in the 1960s, "multiple"

[2] This is model (3.28), with z_i now replacing the notation x_i.

[3] It is $ce^{-z^2/2}/\sqrt{2\pi}$ with c chosen to make the area under the curve equal the area of the histogram.

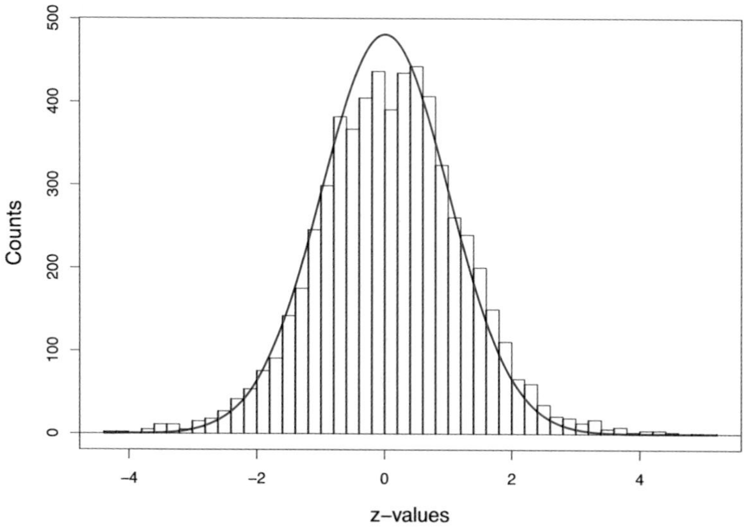

Figure 15.1 Histogram of $N = 6033$ z-values, one for each gene in the prostate cancer study. If all genes were null (15.3) the histogram would track the red curve. For which genes can we reject the null hypothesis?

meaning N between 2 and perhaps 20. The microarray era produced data sets with N in the hundreds, thousands, and now even millions. This sounds like piling difficulty upon difficulty, but in fact there are some inferential advantages to the large-N framework, as we will see.

The most troubling fact about large-scale testing is how easy it is to be fooled. Running 100 separate hypothesis tests at significance level 0.05 will produce about five "significant" results even if each case is actually null. The classical *Bonferroni bound* avoids this fallacy by strengthening the threshold of evidence required to declare an individual case significant (i.e., non-null). For an overall significance level α, perhaps $\alpha = 0.05$, with N simultaneous tests, the Bonferroni bound rejects the ith null hypothesis H_{0i} only if it attains individual significance level α/N. For $\alpha = 0.05$, $N = 6033$, and $H_{0i} : z_i \sim \mathcal{N}(0, 1)$, the one-sided Bonferroni threshold for significance is $-\Phi^{-1}(0.05/N) = 4.31$ (compared with 1.645 for $N = 1$). Only four of the prostate study genes surpass this threshold.

Classic hypothesis testing is usually phrased in terms of *significance levels* and *p-values*. If test statistic z has cdf $F_0(z)$ under the null hypothesis then[4]

$$p = 1 - F_0(z) \tag{15.5}$$

is the right-sided p-value, larger z giving smaller p-value. "Significance level" refers to a prechosen threshold value, e.g., $\alpha = 0.05$. The null hypothesis is "rejected at level α" if we observe $p \le \alpha$. Table 13.4 on page 245 (where "coverage level" means one minus the significance level) shows Fisher's scale for interpreting p-values.

A level-α test for a single null hypothesis H_0 satisfies, by definition,

$$\alpha = \Pr\{\text{reject true } H_0\}. \tag{15.6}$$

For a collection of N null hypotheses H_{0i}, the *family-wise error rate* is the probability of making even one false rejection,

$$\text{FWER} = \Pr\{\text{reject any true } H_{0i}\}. \tag{15.7}$$

Bonferroni's procedure controls FWER at level α: let I_0 be the indices of the *true* H_{0i}, having say N_0 members. Then

$$\begin{aligned} \text{FWER} &= \Pr\left\{\bigcup_{I_0}\left(p_i \le \frac{\alpha}{N}\right)\right\} \le \sum_{I_0}\Pr\left\{p_i \le \frac{\alpha}{N}\right\} \\ &= N_0\frac{\alpha}{N} \le \alpha, \end{aligned} \tag{15.8}$$

the top line following from Boole's inequality (which doesn't require even independence among the p_i).

The Bonferroni bound is quite conservative: for $N = 6033$ and $\alpha = 0.05$ we reject only those cases having $p_i \le 8.3 \cdot 10^{-6}$. One can do only a
†2 little better under the FWER constraint. "Holm's procedure,"† which offers modest improvement over Bonferroni, goes as follows.

- Order the observed p-values from smallest to largest,

$$p_{(1)} \le p_{(2)} \le p_{(3)} \le \ldots \le p_{(i)} \le \ldots \le p_{(N)}, \tag{15.9}$$

  with $H_{0(i)}$ denoting the corresponding null hypotheses.
- Let i_0 be the smallest index i such that

$$p_{(i)} > \alpha/(N - i + 1). \tag{15.10}$$

- *Reject* all null hypotheses $H_{0(i)}$ for $i < i_0$ and *accept* all with $i \ge i_0$.

[4] The left-sided p-value is $p = F_0(z)$. We will avoid two-sided p-values in this discussion.

It can be shown that Holm's procedure controls FWER at level α, while being slightly more generous than Bonferroni in declaring rejections.

15.2 False-Discovery Rates

The FWER criterion aims to control the probability of making even *one* false rejection among N simultaneous hypothesis tests. Originally developed for small-scale testing, say $N \leq 20$, FWER usually proved too conservative for scientists working with N in the thousands. A quite different and more liberal criterion, false-discovery rate (FDR) control, has become standard.

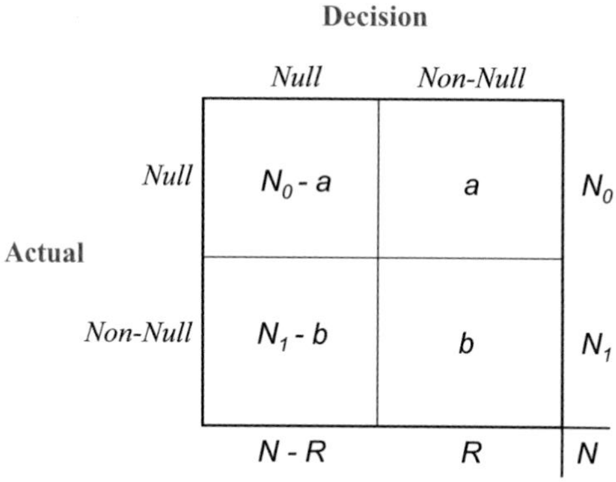

Figure 15.2 A decision rule $\mathcal{D}$ has rejected R out of N null hypotheses; a of these decisions were incorrect, i.e., they were "false discoveries," while b of them were "true discoveries." The false-discovery proportion Fdp equals a/R.

Figure 15.2 diagrams the outcome of a hypothetical decision rule $\mathcal{D}$ applied to the data for N simultaneous hypothesis-testing problems, N_0 null and $N_1 = N - N_0$ non-null. An omniscient oracle has reported the rule's results: R null hypotheses have been rejected; a of these were cases of *false discovery*, i.e., valid null hypotheses, for a "false-discovery proportion" (Fdp) of

$$\text{Fdp}(\mathcal{D}) = a/R. \tag{15.11}$$

(We define Fdp $= 0$ if $R = 0$.) Fdp is unobservable—without the oracle we cannot see a—but under certain assumptions we can control its expectation.

Define

$$\text{FDR}(\mathcal{D}) = E\left\{\text{Fdp}(\mathcal{D})\right\}. \tag{15.12}$$

A decision rule $\mathcal{D}$ controls FDR at level q, with q a prechosen value between 0 and 1, if

$$\text{FDR}(\mathcal{D}) \leq q. \tag{15.13}$$

It might seem difficult to find such a rule, but in fact a quite simple but ingenious recipe does the job. Ordering the observed p-values from smallest to largest as in (15.9), define i_{max} to be the largest index for which

$$p_{(i)} \leq \frac{i}{N}q, \tag{15.14}$$

and let $\mathcal{D}_q$ be the rule[5] that rejects $H_{0(i)}$ for $i \leq i_{\text{max}}$, accepting otherwise.
†3 A proof of the following theorem is referenced in the chapter endnotes.†

Theorem (Benjamini–Hochberg FDR Control) *If the p-values corresponding to valid null hypotheses are independent of each other, then*

$$\text{FDR}(\mathcal{D}_q) = \pi_0 q \leq q, \qquad \textit{where } \pi_0 = N_0/N. \tag{15.15}$$

In other words $\mathcal{D}_q$ controls FDR at level $\pi_0 q$. The null proportion π_0 is unknown (though estimable), so the usual claim is that $\mathcal{D}_q$ controls FDR at level q. Not much is sacrificed: large-scale testing problems are most often fishing expeditions in which most of the cases are null, putting π_0 near 1, identification of a few non-null cases being the goal. The choice $q = 0.1$ is typical practice.

The popularity of FDR control hinges on the fact that it is more generous than FWER in declaring significance.[6] Holm's procedure (15.10) rejects null hypothesis $H_{0(i)}$ if

$$p_{(i)} \leq \text{Threshold(Holm's)} = \frac{\alpha}{N - i + 1}, \tag{15.16}$$

while $\mathcal{D}_q$ (15.13) has threshold

$$p_{(i)} \leq \text{Threshold}(\mathcal{D}_q) = \frac{q}{N}i. \tag{15.17}$$

[5] Sometimes denoted "BH$_q$" after its inventors Benjamini and Hochberg; see the chapter endnotes.

[6] The classic term "significant" for a non-null identification doesn't seem quite right for FDR control, especially given the Bayesian connections of Section 15.3, and we will sometimes use "interesting" instead.

In the usual range of interest, large N and small i, the ratio

$$\frac{\text{Threshold}(\mathcal{D}_q)}{\text{Threshold(Holm's)}} = \frac{q}{\alpha}\left(1 - \frac{i-1}{N}\right) i \tag{15.18}$$

increases almost linearly with i.

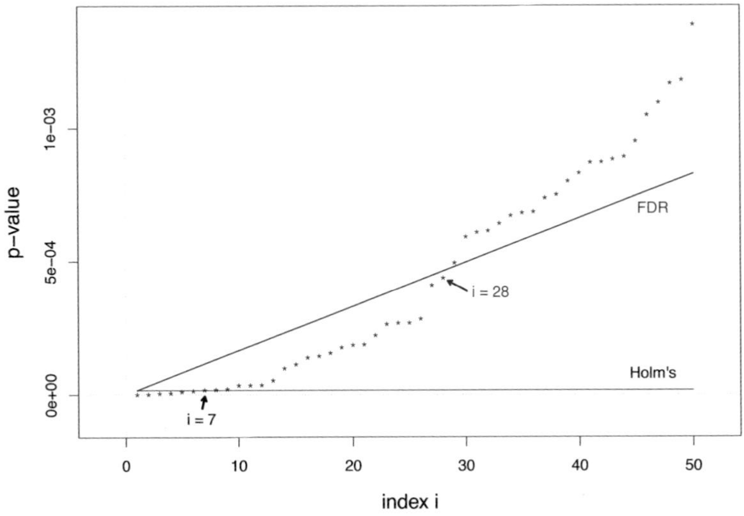

Figure 15.3 Ordered p-values $p_{(i)} = 1 - \Phi(z_{(i)})$ plotted versus i for the 50 largest z-values from the **prostate** data in Figure 15.1. The FDR control boundary (algorithm $\mathcal{D}_q$, $q = 0.1$) rejects $H_{0(i)}$ for the 28 smallest values $p_{(i)}$, while Holm's FWER procedure ($\alpha = 0.1$) rejects for only the 7 smallest values. (The upward slope of Holm's boundary (15.16) is too small to see here.)

Figure 15.3 illustrates the comparison for the right tail of the prostate data of Figure 15.1, with $p_i = 1 - \Phi(z_i)$ (15.3), (15.5), and $\alpha = q = 0.1$. The FDR procedure rejects $H_{0(i)}$ for the 28 largest z-values ($z_{(i)} \geq 3.33$), while FWER control rejects only the 7 most extreme z-values ($z_{(i)} \geq 4.14$).

Hypothesis testing has been a traditional stronghold of frequentist decision theory, with "Type 1" error control being strictly enforced, very often at the 0.05 level. It is surprising that a new control criterion, FDR, has taken hold in large-scale testing situations. A critic, noting FDR's relaxed rejection standards in Figure 15.3, might raise some pointed questions.

1 Is controlling a *rate* (i.e., FDR) as meaningful as controlling a *probability* (of Type 1 error)?
2 How should q be chosen?
3 The control theorem depends on independence among the p-values. Isn't this unlikely in situations such as the prostate study?
4 The FDR significance for gene i_0, say one with $z_{i_0} = 3$, depends on the results of all the other genes: the more "other" z_i values exceed 3, the more interesting gene i_0 becomes (since that increases i_0's index i in the ordered list (15.9), making it more likely that p_{i_0} lies below the $\mathcal{D}_q$ threshold (15.14)). Does this make inferential sense?

A Bayes/empirical Bayes restatement of the $\mathcal{D}_q$ algorithm helps answer these questions, as discussed next.

15.3 Empirical Bayes Large-Scale Testing

In practice, single-case hypothesis testing has been a frequentist preserve. Its methods demand little from the scientist—only the choice of a test statistic and the calculation of its null distribution—while usually delivering a clear verdict. By contrast, Bayesian model selection, whatever its inferential virtues, raises the kinds of difficult modeling questions discussed in Section 13.3.

It then comes as a pleasant surprise that things are different for large-scale testing: Bayesian methods, at least in their empirical Bayes manifestation, no longer demand heroic modeling efforts, and can help untangle the interpretation of simultaneous test results. This is particularly true for the FDR control algorithm $\mathcal{D}_q$ of the previous section.

A simple Bayesian framework for simultaneous testing is provided by the *two-groups model*: each of the N cases (the genes for the prostate study) is either null with prior probability π_0 or non-null with probability $\pi_1 = 1 - \pi_0$; the resulting observation z then has density either $f_0(z)$ or $f_1(z)$,

$$\begin{aligned} \pi_0 &= \Pr\{\text{null}\} & f_0(z) &\text{ density if null,} \\ \pi_1 &= \Pr\{\text{non-null}\} & f_1(z) &\text{ density if non-null.} \end{aligned} \tag{15.19}$$

For the prostate study, π_0 is nearly 1, and $f_0(z)$ is the standard normal density $\phi(z) = \exp(-z^2/2)/\sqrt{2\pi}$ (15.3), while the non-null density remains to be estimated.

Let $F_0(z)$ and $F_1(z)$ be the cdf values corresponding to $f_0(z)$ and $f_1(z)$, with "survival curves"

$$S_0(z) = 1 - F_0(z) \quad \text{and} \quad S_1(z) = 1 - F_1(z), \tag{15.20}$$

$S_0(z_0)$ being the probability that a null z-value exceeds z_0, and similarly for $S_1(z)$. Finally, define $S(z)$ to be the mixture survival curve

$$S(z) = \pi_0 S_0(z) + \pi_1 S_1(z). \tag{15.21}$$

The *mixture density*

$$f(z) = \pi_0 f_0(z) + \pi_1 f_1(z) \tag{15.22}$$

determines $S(z)$,

$$S(z_0) = \int_{z_0}^{\infty} f(z)\, dz. \tag{15.23}$$

Suppose now that observation z_i for case i is seen to exceed some threshold value z_0, perhaps $z_0 = 3$. Bayes' rule gives

$$\begin{aligned} \text{Fdr}(z_0) &\equiv \Pr\{\text{case } i \text{ is null} | z_i \geq z_0\} \\ &= \pi_0 S_0(z_0)/S(z_0), \end{aligned} \tag{15.24}$$

the correspondence with (3.5) on page 23 being $\pi_0 = g(\mu)$, $S_0(z_0) = f_\mu(x)$, and $S(z_0) = f(x)$. Fdr is the "Bayes false-discovery rate," as contrasted with the frequentist quantity FDR (15.12).

In typical applications, $S_0(z_0)$ is assumed known[7] (equaling $1 - \Phi(z_0)$ in the prostate study), and π_0 is assumed to be near 1. The denominator $S(z_0)$ in (15.24) is unknown, but—and this is the crucial point—it has an obvious estimate in large-scale testing situations, namely

$$\hat{S}(z_0) = N(z_0)/N, \qquad \text{where } N(z_0) = \#\{z_i \geq z_0\}. \tag{15.25}$$

(By the definition of the two-group model, each z_i has marginal density $f(z)$, making $\hat{S}(z_0)$ the usual empirical estimate of $S(z_0)$ (15.23).) Plugging into (15.24) yields an empirical Bayes estimate of the Bayes false-discovery rate

$$\widehat{\text{Fdr}}(z_0) = \pi_0 S_0(z_0) \big/ \hat{S}(z_0). \tag{15.26}$$

The connection with FDR control is almost immediate. First of all, from definitions (15.5) and (15.20) we have $p_i = S_0(z_i)$; also for the ith from the largest z-value we have $\hat{S}(z_{(i)}) = i/N$ (15.25). Putting these together, condition (15.14), $p_{(i)} \leq (i/N)q$, becomes

$$S_0(z_{(i)}) \leq \hat{S}(z_{(i)}) \cdot q, \tag{15.27}$$

[7] But see Section 15.5.

or $S_0(z_{(i)})/\hat{S}(z_{(i)}) \leq q$, which can be written as

$$\widehat{\text{Fdr}}(z_{(i)}) \leq \pi_0 q \tag{15.28}$$

(15.26). In other words, the $\mathcal{D}_q$ algorithm, which rejects those null hypotheses having[8] $p_{(i)} \leq (i/N)q$, is in fact rejecting those cases for which the empirical Bayes posterior probability of nullness is too small, as defined by (15.28). The Bayesian nature of FDR control offers a clear advantage to the investigating scientist, who gets a numerical assessment of the probability that he or she will be wasting time following up any one of the selected cases.

We can now respond to the four questions at the end of the previous section:

1 FDR control *does* relate to a probability—the Bayes posterior probability of nullness.
2 The choice of q for $\mathcal{D}_q$ amounts to setting the maximum tolerable amount of Bayes risk of nullness[9] (usually after taking $\pi_0 = 1$ in (15.28)).
3 Most often the z_i, and hence the p_i, will be correlated with each other. Even under correlation, however, $\hat{S}(z_0)$ in (15.25) is still unbiased for $S(z_0)$, making $\widehat{\text{Fdr}}(z_0)$ (15.26) nearly unbiased for $\text{Fdr}(z_0)$ (15.24). There *is* a price to be paid for correlation, which increases the *variance* of $S_0(z_0)$ and $\widehat{\text{Fdr}}(z_0)$.
4 In the Bayes two-groups model (15.19), all of the non-null z_i are i.i.d. observations from the non-null density $f_1(z)$, with survival curve $S_1(z)$. The number of null cases z_i exceeding some threshold z_0 has *fixed* expectation $N\pi_0 S_0(z_0)$. Therefore an increase in the number of observed values z_i exceeding z_0 must come from a heavier right tail for $f_1(z)$, implying a greater posterior probability of non-nullness $\text{Fdr}(z_0)$ (15.24). This point is made more clearly in the *local false-discovery* framework of the next section. It emphasizes the "learning from the experience of others" aspect of empirical Bayes inference, Section 7.4. The question of "Which others?" is returned to in Section 15.6.

Figure 15.4 illustrates the two-group model (15.19). The N cases are

[8] The algorithm, as stated just before the FDR control theorem (15.15), is actually a little more liberal in allowing rejections.

[9] For a case of particular interest, the calculation can be reversed: if the case has ordered index i then, according to (15.14), the value $q = Np_i/i$ puts it exactly on the boundary of rejection, making this its q-value. The 50th largest z-value for the prostate data has $z_i = 2.99$, $p_i = 0.00139$, and q-value 0.168, that being both the frequentist boundary for rejection and the empirical Bayes probability of nullness.

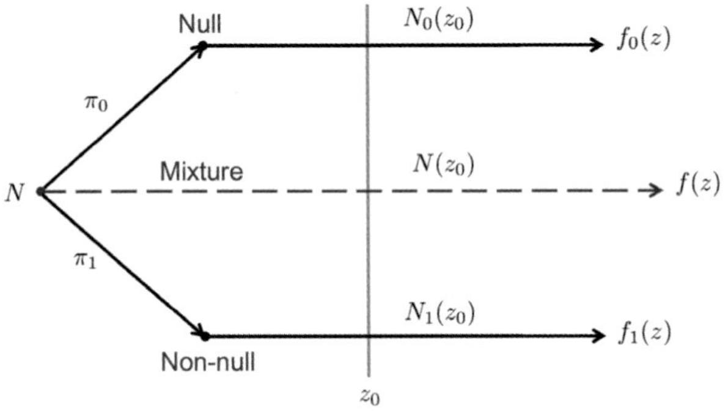

Figure 15.4 A diagram of the two-groups model (15.19). Here the statistician observes values z_i from a mixture density $f(z) = \pi_0 f_0(z) + \pi_1 f_1(z)$ and decides to reject or accept the null hypothesis H_{0i} depending on whether z_i exceeds or is less than the threshold value z_0.

randomly dispatched to the two arms in proportions π_0 and π_1, at which point they produce z-values according to either $f_0(z)$ or $f_1(z)$. Suppose we are using a simple decision rule $\mathcal{D}$ that rejects the ith null hypothesis if z_i exceeds some threshold z_0, and accepts otherwise,

$$\mathcal{D}: \begin{cases} \text{Reject } H_{0i} & \text{if } z_i > z_0 \\ \text{Accept } H_{0i} & \text{if } z_i \leq z_0. \end{cases} \tag{15.29}$$

The oracle of Figure 15.2 knows that $N_0(z_0) = a$ of the null case z-values exceeded z_0, and similarly $N_1(z_0) = b$ of the non-null cases, leading to

$$N(z_0) = N_0(z_0) + N_1(z_0) = R \tag{15.30}$$

total rejections. The false-discovery proportion (15.11) is

$$\text{Fdp} = \frac{N_0(z_0)}{N(z_0)} \tag{15.31}$$

but this is unobservable since we see only $N(z_0)$.

The clever inferential strategy of false-discovery rate theory substitutes the *expectation* of $N_0(z_0)$,

$$E\{N_0(z_0)\} = N\pi_0 S_0(z_0), \tag{15.32}$$

for $N_0(z_0)$ in (15.31), giving

$$\widehat{\text{Fdp}} = \frac{N\pi_0 S_0(z_0)}{N(z_0)} = \frac{\pi_0 S_0(z_0)}{\hat{S}(z_0)} = \widehat{\text{Fdr}}(z_0), \tag{15.33}$$

using (15.25) and (15.26). Starting from the two-groups model, $\widehat{\text{Fdr}}(z_0)$ is an obvious empirical (i.e., frequentist) estimate of the Bayesian probability $\text{Fdr}(z_0)$, as well as of Fdp.

If placed in the Bayes–Fisher–frequentist triangle of Figure 14.1, false-discovery rates would begin life near the frequentist corner but then migrate at least part of the way toward the Bayes corner. There are remarkable parallels with the James–Stein estimator of Chapter 7. Both theories began with a striking frequentist theorem, which was then inferentially rationalized in empirical Bayes terms. Both rely on the use of indirect evidence—learning from the experience of others. The difference is that James–Stein estimation always aroused controversy, while FDR control has been quickly welcomed into the pantheon of widely used methods. This could reflect a change in twenty-first-century attitudes or, perhaps, only that the $\mathcal{D}_q$ rule better conceals its Bayesian aspects.

15.4 Local False-Discovery Rates

Tail-area statistics (p-values) were synonymous with classic one-at-a-time hypothesis testing, and the $\mathcal{D}_q$ algorithm carried over p-value interpretation to large-scale testing theory. But tail-area calculations are neither necessary nor desirable from a Bayesian viewpoint, where, having observed test statistic z_i equal to some value z_0, we should be more interested in the probability of nullness given $z_i = z_0$ than given $z_i \geq z_0$.

To this end we define the *local false-discovery rate*

$$\text{fdr}(z_0) = \Pr\{\text{case } i \text{ is null} | z_i = z_0\} \tag{15.34}$$

as opposed to the tail-area false-discovery rate $\text{Fdr}(z_0)$ (15.24). The main point of what follows is that reasonably accurate empirical Bayes estimates of fdr are available in large-scale testing problems.

As a first try, suppose that $\mathcal{Z}_0$, a proposed region for rejecting null hypotheses, is a small interval centered at z_0,

$$\mathcal{Z}_0 = \left[z_0 - \frac{d}{2}, z_0 + \frac{d}{2}\right], \tag{15.35}$$

with d perhaps 0.1. We can redraw Figure 15.4, now with $N_0(\mathcal{Z}_0)$, $N_1(\mathcal{Z}_0)$, and $N(\mathcal{Z}_0)$ the null, non-null, and total number of z-values in $\mathcal{Z}_0$. The local false-discovery proportion,

$$\text{fdp}(z_0) = N_0(\mathcal{Z}_0)/N(\mathcal{Z}_0) \tag{15.36}$$

is unobservable, but we can replace $N_0(\mathcal{Z}_0)$ with $N\pi_0 f_0(z_0)d$, its approximate expectation as in (15.31)–(15.33), yielding the estimate[10]

$$\widehat{\text{fdr}}(z_0) = N\pi_0 f_0(z_0)d/N(\mathcal{Z}_0). \tag{15.37}$$

Estimate (15.37) would be needlessly noisy in practice; z-value distributions tend to be smooth, allowing the use of regression estimates for $\text{fdr}(z_0)$. Bayes' theorem gives

$$\text{fdr}(z) = \pi_0 f_0(z)/f(z) \tag{15.38}$$

in the two-groups model (15.19) (with μ in (3.5) now the indicator of null or non-null states, and x now z). Drawing a smooth curve $\hat{f}(z)$ through the histogram of the z-values yields the more efficient estimate

$$\widehat{\text{fdr}}(z_0) = \pi_0 f_0(z_0)/\hat{f}(z_0); \tag{15.39}$$

the null proportion π_0 can be estimated—see Section 15.5—or set equal to 1.

Figure 15.5 shows $\widehat{\text{fdr}}(z)$ for the prostate study data of Figure 15.1, where $\hat{f}(z)$ in (15.39) has been estimated as described below. The curve hovers near 1 for the 93% of the cases having $|z_i| \leq 2$, sensibly suggesting that there is no involvement with prostate cancer for most genes. It declines quickly for $|z_i| \geq 3$, reaching the conventionally "interesting" threshold

$$\widehat{\text{fdr}}(z) \leq 0.2 \tag{15.40}$$

for $z_i \geq 3.34$ and $z_i \leq -3.40$. This was attained for 27 genes in the right tail and 25 in the left, these being reasonable candidates to flag for follow-up investigation.

The curve $\hat{f}(z)$ used in (15.39) was obtained from a fourth-degree log polynomial Poisson regression fit to the histogram in Figure 15.1, as in Figure 10.5 (10.52)–(10.56). Log polynomials of degree 2 through 6 were fit by maximum likelihood, giving total residual deviances (8.35) shown in Table 15.1. An enormous improvement in fit is seen in going from degree 3 to 4, but nothing significant after that, with decreases less than the null value 2 suggested by (12.75).

[10] Equation (15.37) makes argument (4) of the previous section clearer: having more "other" z-values fall into $\mathcal{Z}_0$ increases $N(\mathcal{Z}_0)$, decreasing $\widehat{\text{fdr}}(z_0)$ and making it more likely that $z_i = z_0$ represents a non-null case.

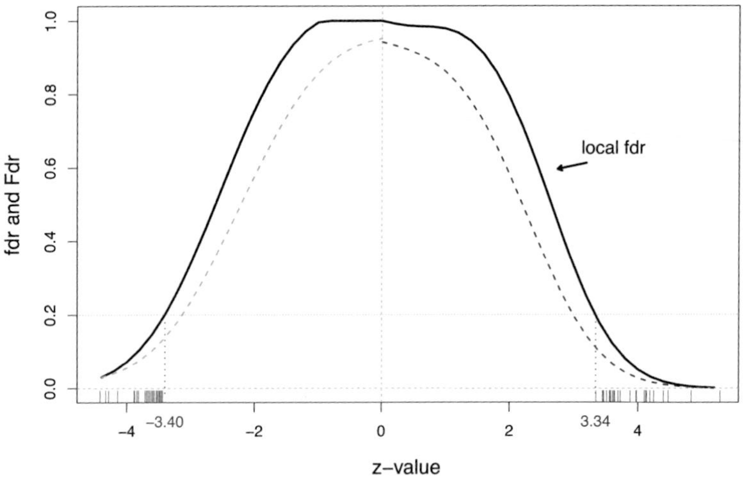

Figure 15.5 Local false-discovery rate estimate $\widehat{\text{fdr}}(z)$ (15.39) for prostate study of Figure 15.1; 27 genes on the right and 25 on the left, indicated by dashes, have $\widehat{\text{fdr}}(z_i) \leq 0.2$; light dashed curves are the left and right tail-area estimates $\widehat{\text{Fdr}}(z)$ (15.26).

Table 15.1 *Total residual deviances from log polynomial Poisson regressions of the prostate data, for polynomial degrees 2 through 6; degree 4 is preferred.*

Degree	2	3	4	5	6
Deviance	138.6	137.1	65.3	64.3	63.8

The points in Figure 15.6 represent the log bin counts from the histogram in Figure 15.1 (excluding zero counts), with the solid curve showing the 4th-degree MLE polynomial fit. Also shown is the standard normal log density

$$\log f_0(z) = -\frac{1}{2}z^2 + \text{constant}. \tag{15.41}$$

It fits reasonably well for $|z| < 2$, emphasizing the null status of the gene majority.

The cutoff $\widehat{\text{fdr}}(z) \leq 0.2$ for declaring a case interesting is not completely arbitrary. Definitions (15.38) and (15.22), and a little algebra, show that it

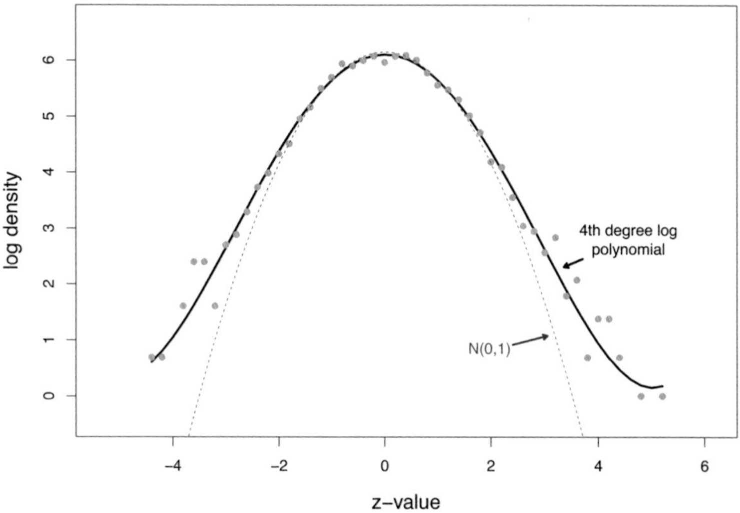

Figure 15.6 Points are log bin counts for Figure 15.1's histogram. The solid black curve is a fourth-degree log-polynomial fit used to calculate $\widehat{\text{fdr}}(z)$ in Figure 15.5. The dashed red curve, the log null density (15.41), provides a reasonable fit for $|z| \leq 2$.

is equivalent to

$$\frac{f_1(z)}{f_0(z)} \geq 4\frac{\pi_0}{\pi_1}. \tag{15.42}$$

If we assume $\pi_0 \geq 0.90$, as is reasonable in most large-scale testing situations, this makes the Bayes factor $f_1(z)/f_0(z)$ quite large,

$$\frac{f_1(z)}{f_0(z)} \geq 36, \tag{15.43}$$

"strong evidence" against the null hypothesis in Jeffreys' scale, Table 13.3.

There is a simple relation between the local and tail-area false-discovery rates:† †4

$$\text{Fdr}(z_0) = E\left\{\text{fdr}(z)|z \geq z_0\right\}; \tag{15.44}$$

so $\text{Fdr}(z_0)$ is the average value of $\text{fdr}(z)$ for z greater than z_0. In interesting situations, $\text{fdr}(z)$ will be a decreasing function for large values of z, as on the right side of Figure 15.5, making $\text{Fdr}(z_0) < \text{fdr}(z_0)$. This accounts

for the conventional significance cutoff $\widehat{\text{Fdr}}(z) \leq 0.1$ being smaller than
†5 $\widehat{\text{fdr}}(z) \leq 0.2$ (15.40).†

The Bayesian interpretation of local false-discovery rates carries with it the advantages of Bayesian coherency. We don't have to change definitions as with left-sided and right-sided tail-area $\widehat{\text{Fdr}}$ estimates, since $\widehat{\text{fdr}}(z)$ applies without change to both tails.[11] Also, we don't need a separate theory for "true-discovery rates," since

$$\text{tdr}(z_0) \equiv 1 - \text{fdr}(z_0) = \pi_1 f_1(z_0)/f(z_0) \tag{15.45}$$

is the conditional probability that case i is *non-null* given $z_i = z_0$.

15.5 Choice of the Null Distribution

The null distribution, $f_0(z)$ in the two-groups model (15.19), plays a crucial role in large-scale testing, just as it does in the classic single-case theory. Something different however happens in large-scale problems: with thousands of z-values to examine at once, it can become clear that the conventional theoretical null is inappropriate for the situation at hand. Put more positively, large-scale applications may allow us to empirically determine a more realistic null distribution.

The `police` data of Figure 15.7 illustrates what can happen. Possible racial bias in pedestrian stops was assessed for $N = 2749$ New York City police officers in 2006. Each officer was assigned a score z_i, large positive scores suggesting racial bias. The z_i values were summary scores from a complicated logistic regression model intended to compensate for differences in the time of day, location, and context of the stops. Logistic regression theory suggested the theoretical null distribution

$$H_{0i} : z_i \sim \mathcal{N}(0, 1) \tag{15.46}$$

for the absence of racial bias.

The trouble is that the center of the z-value histogram in Figure 15.7, which should track the $\mathcal{N}(0, 1)$ curve applying to the presumably large fraction of null-case officers, is much too wide. (Unlike the situation for the prostate data in Figure 15.1.) An MLE fitting algorithm discussed below produced the *empirical null*

$$H_{0i} : z_i \sim \mathcal{N}(0.10, 1.40^2) \tag{15.47}$$

[11] Going further, z in the two-groups model could be multidimensional. Then tail-area false-discovery rates would be unavailable, but (15.38) would still legitimately define $\text{fdr}(z)$.

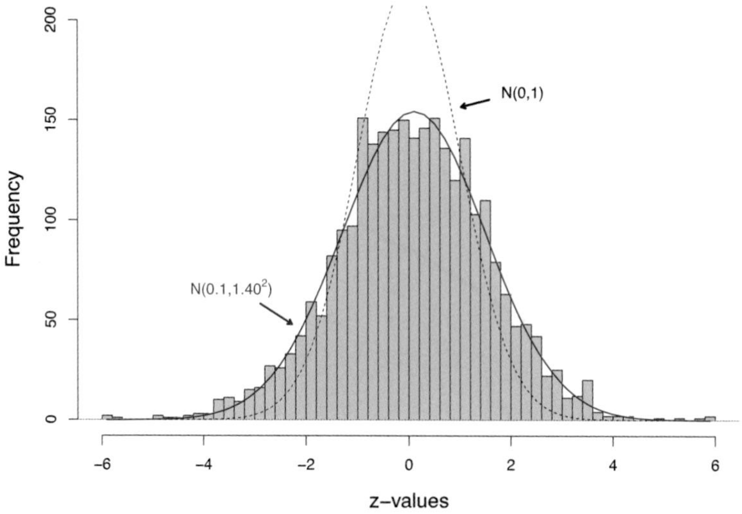

Figure 15.7 Police data; histogram of z scores for $N = 2749$ New York City police officers, with large z_i suggesting racial bias. The center of the histogram is too wide compared with the theoretical null distribution $z_i \sim \mathcal{N}(0, 1)$. An MLE fit to central data gave $\mathcal{N}(0.10, 1.40^2)$ as empirical null.

as appropriate here. This is reinforced by a QQ plot of the z_i values shown in Figure 15.8, where we see most of the cases falling nicely along a $\mathcal{N}(0.09, 1.42^2)$ line, with just a few outliers at both extremes.

There is a lot at stake here. Based on the empirical null (15.47) only four officers reached the "probably racially biased" cutoff $\widehat{\text{fdr}}(z_i) \leq 0.2$, the four circled points at the far right of Figure 15.8; the fifth point had $\widehat{\text{fdr}} = 0.38$ while all the others exceeded 0.80. The theoretical $\mathcal{N}(0, 1)$ null was much more severe, assigning $\widehat{\text{fdr}} \leq 0.2$ to the 125 officers having $z_i \geq 2.50$. One can imagine the difference in newspaper headlines.

From a classical point of view it seems heretical to question the theoretical null distribution, especially since there is no substitute available in single-case testing. Once alerted by data sets like the police study, however, it is easy to list reasons for doubt:

- *Asymptotics* Taylor series approximations go into theoretical null calculations such as (15.46), which can lead to inaccuracies, particularly in the crucial tails of the null distribution.
- *Correlations* False-discovery rate methods are correct *on the average*,

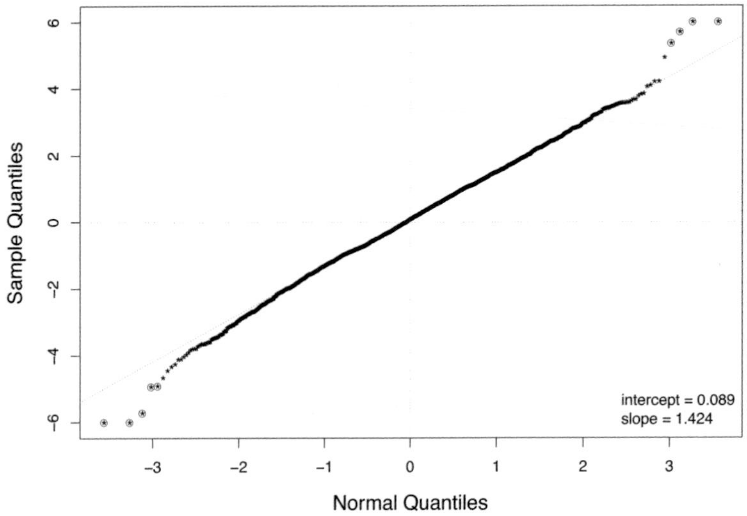

Figure 15.8 QQ plot of police data z scores; most scores closely follow the $\mathcal{N}(0.09, 1.42^2)$ line with a few outliers at either end. The circled points are cases having local false-discovery estimate $\widehat{\text{fdr}}(z_i) \leq 0.2$, based on the empirical null. Using the theoretical $\mathcal{N}(0, 1)$ null gives 216 cases with $\widehat{\text{fdr}}(z_i) \leq 0.2$, 91 on the left and 125 on the right.

even with correlations among the N z-values. However, severe correlation destabilizes the z-value histogram, which can become randomly wider or narrower than theoretically predicted, undermining theoretical null results
†6 for the data set at hand.†

- *Unobserved covariates* The police study was *observational*: individual encounters were not assigned at random to the various officers but simply observed as they happened. Observed covariates such as the time of day and the neighborhood were included in the logistic regression model, but one can never rule out the possibility of influential unobserved covariates.
- *Effect size considerations* The hypothesis-testing setup, where a large fraction of the cases are truly null, may not be appropriate. An *effect size* model, with $\mu_i \sim g(\cdot)$ and $z_i \sim \mathcal{N}(\mu_i, 1)$, might apply, with the prior $g(\mu)$ *not* having an atom at $\mu = 0$. The nonatomic choice $g(\mu) \sim \mathcal{N}(0.10, 0.63^2)$ provides a good fit to the QQ plot in Figure 15.8.

Empirical Null Estimation

Our point of view here is that the theoretical null (15.46), $z_i \sim \mathcal{N}(0, 1)$, is not completely wrong but needs adjustment for the data set at hand. To this end we assume the two-groups model (15.19), with $f_0(z)$ normal but not necessarily $\mathcal{N}(0, 1)$, say

$$f_0(z) \sim \mathcal{N}(\delta_0, \sigma_0^2). \tag{15.48}$$

In order to compute the local false-discovery rate $\text{fdr}(z) = \pi_0 f_0(z)/f(z)$ we want to estimate the three numerator parameters $(\delta_0, \sigma_0, \pi_0)$, the mean and standard deviation of the null density and the proportion of null cases. (The denominator $f(z)$ is estimated as in Section 15.4.)

Our key assumptions (besides (15.48)) are that π_0 is large, say $\pi_0 \geq 0.90$, and that most of the z_i near 0 are null cases. The algorithm `locfdr` †begins by selecting a set $\mathcal{A}_0$ near $z = 0$ in which it is assumed that *all* the †7
z_i in $\mathcal{A}_0$ are null; in terms of the two-groups model, the assumption can be stated as

$$f_1(z) = 0 \text{ for } z \in \mathcal{A}_0. \tag{15.49}$$

Modest violations of (15.49), which are to be expected, produce small biases in the empirical null estimates. Maximum likelihood based on the number and values of the z_i observed in $\mathcal{A}_0$ yield the empirical null estimates† $(\hat{\delta}_0, \hat{\sigma}_0, \hat{\pi}_0)$. †8

Applied to the police data, `locfdr` chose $\mathcal{A}_0 = [-1.8, 2.0]$ and produced estimates

$$\left(\hat{\delta}_0, \hat{\sigma}_0, \hat{\pi}_0\right) = (0.10, 1.40, 0.989). \tag{15.50}$$

Two small simulation studies described in Table 15.2 give some idea of the variabilities and biases inherent in the `locfdr` estimation process.

The third method, somewhere between the theoretical and empirical null estimates but closer to the former, relies on permutations. The vector $\boldsymbol{z}$ of 6033 z-values for the `prostate` data of Figure 15.1 was obtained from a study of 102 men, 52 cancer patients and 50 controls. Randomly permuting the men's data, that is randomly choosing 50 of the 102 to be "controls" and the remaining 52 to be "patients," and then carrying through steps (15.1)–(15.2) gives a vector $\boldsymbol{z}^*$ in which any actual cancer/control differences have been suppressed. A histogram of the z_i^* values (perhaps combining several permutations) provides the "permutation null." Here we are extending Fisher's original permutation idea, Section 4.4, to large-scale testing.

Ten permutations of the prostate study data produced an almost perfect

Table 15.2 *Means and standard deviations of* $(\hat{\delta}_0, \hat{\sigma}_0, \hat{\pi}_0)$ *for two simulation studies of empirical null estimation using* `locfdr`. $N = 5000$ *cases each trial with* $(\delta_0, \sigma_0, \pi_0)$ *as shown; 250 trials; two-groups model* (15.19) *with non-null density* $f_1(z)$ *equal to* $\mathcal{N}(3, 1)$ *(left side) or* $\mathcal{N}(4.2, 1)$ *(right side).*

	δ_0	σ_0	π_0	δ_0	σ_0	π_0
true	**0**	**1.0**	**.95**	**.10**	**1.40**	**.95**
mean	.015	1.017	.962	.114	1.418	.958
st dev	.019	.017	.005	.025	.029	.006

$\mathcal{N}(0, 1)$ permutation null. (This is as expected from the classic theory of permutation t-tests.) Permutation methods reliably overcome objection 1 to the theoretical null distribution, over-reliance on asymptotic approximations, but cannot cure objections 2, 3, and 4.† †9

Whatever the cause of disparity, the operational difference between the theoretical and empirical null distribution is clear: with the latter, the significance of an outlying case is judged relative to the dispersion of the majority, not by a theoretical yardstick as with the former. This was persuasive for the police data, but the story isn't one-sided. Estimating the null distribution adds substantially to the variability of $\widehat{\text{fdr}}$ or $\widehat{\text{Fdr}}$. For situations such as the prostate data, when the theoretical null looks nearly correct,[12] it is reasonable to stick with it.

The very large data sets of twenty-first-century applications encourage self-contained methodology that proceeds from just the data at hand using a minimum of theoretical constructs. False-discovery rate empirical Bayes analysis of large-scale testing problems, with data-based estimation of $\hat{\pi}_0$, $\hat{f}_0$, and $\hat{f}$, comes close to the ideal in this sense.

15.6 Relevance

False-discovery rates return us to the purview of *indirect evidence*, Sections 6.4 and 7.4. Our interest in any one gene in the prostate cancer study depends on its own z score of course, but also on the other genes' scores—"learning from the experience of others," in the language used before.

The crucial question we have been avoiding is "Which others?" Our tacit answer has been "All the cases that arrive in the same data set," all the genes

[12] The `locfdr` algorithm gave $(\hat{\delta}_0, \hat{\sigma}_0, \hat{\pi}_0) = (0.00, 1.06, 0.984)$ for the prostate data.

in the prostate study, all the officers in the police study. Why this can be a dangerous tactic is shown in our final example.

A **DTI** (diffusion tensor imaging) study compared six dyslexic children with six normal controls. Each **DTI** scan recorded fluid flows at $N = 15{,}443$ "voxels," i.e., at 15,443 three-dimensional brain coordinates. A score z_i comparing dyslexics with normal controls was calculated for each voxel i, calibrated such that the theoretical null distribution of "no difference" was

$$H_{0i} : z_i \sim \mathcal{N}(0, 1) \tag{15.51}$$

as at (15.3).

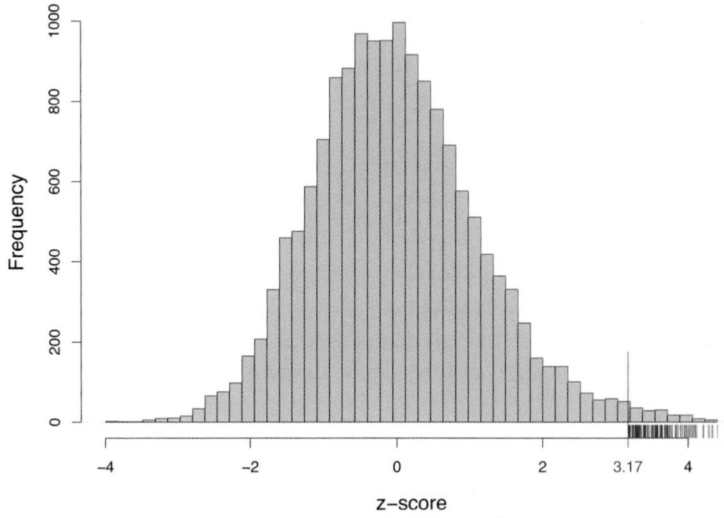

Figure 15.9 Histogram of z scores for the **DTI** study, comparing dyslexic versus normal control children at 15,443 brain locations. A FDR analysis based on the empirical null distribution gave 149 voxels with $\widehat{\text{fdr}}(z_i) \leq 0.20$, those having $z_i \geq 3.17$ (indicated by red dashes).

Figure 15.9 shows the histogram of all 15,443 z_i values, normal-looking near the center and with a heavy right tail; **locfdr** gave empirical null parameters

$$\left(\hat{\delta}_0, \hat{\sigma}_0, \hat{\pi}_0\right) = (-0.12, 1.06, 0.984), \tag{15.52}$$

the 149 voxels with $z_i \geq 3.17$ having $\widehat{\text{fdr}}$ values ≤ 0.20. Using the theoretical null (15.51) yielded only modestly different results, now the 177 voxels with $z_i \geq 3.07$ having $\widehat{\text{fdr}}_i \leq 0.20$.

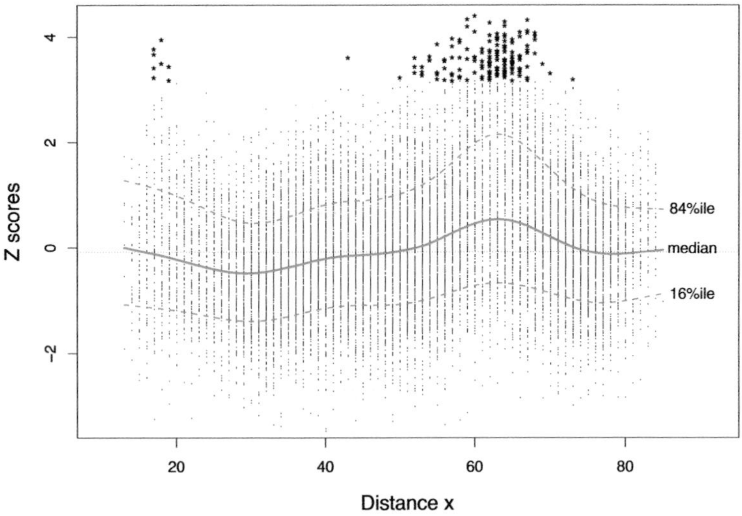

Figure 15.10 A plot of 15,443 z_i scores from a **DTI** study (vertical axis) and voxel distances x_i from the back of the brain (horizontal axis). The starred points are the 149 voxels with $\widehat{\text{fdr}}(z_i) \leq 0.20$, which occur mostly for x_i in the interval $[50, 70]$.

In Figure 15.10 the voxel scores z_i, graphed vertically, are plotted versus x_i, the voxel's distance from the back of the brain. Waves of differing response are apparent. Larger values occur in the interval $50 \leq x \leq 70$, where the entire z-value distribution—low, medium, and high—is pushed up. Most of the 149 voxels having $\widehat{\text{fdr}}_i \leq 0.20$ occur at the top of this wave.

Figure 15.10 raises the problem of fair comparison. Perhaps the 4,653 voxels with x_i between 50 and 70 should be compared only with each other, and not with all 15,443 cases. Doing so gave

$$\left(\hat{\delta}_0, \hat{\sigma}_0, \hat{\pi}_0\right) = (0.23, 1.18, 0.970), \tag{15.53}$$

only 66 voxels having $\widehat{\text{fdr}}_i \leq 0.20$, those with $z_i \geq 3.57$.

All of this is a question of *relevance*: which other voxels i are relevant to the assessment of significance for voxel i_0? One might argue that this is a question for the scientist who gathers the data and not for the statistical analyst, but that is unlikely to be a fruitful avenue, at least not without a lot of back-and-forth collaboration. Standard Bayesian analysis solves the problem by dictate: the assertion of a prior is also an assertion of its relevance. Empirical Bayes situations expose the dangers lurking in such assertions.

Relevance was touched upon in Section 7.4, where the limited translation rule (7.47) was designed to protect extreme cases from being shrunk too far toward the bulk of ordinary ones. One could imagine having a "relevance function" $\rho(x_i, z_i)$ that, given the covariate information x_i and response z_i for case$_i$, somehow adjusts an ensemble false-discovery rate estimate to correctly apply to the case of interest—but such a theory barely exists.† †10

Summary

Large-scale testing, particularly in its false-discovery rate implementation, is not at all the same thing as the classic Fisher–Neyman–Pearson theory:

- Frequentist single-case hypothesis testing depends on the theoretical long-run behavior of samples from the theoretical null distribution. With data available from say $N = 5000$ simultaneous tests, the statistician has his or her own "long run" in hand, diminishing the importance of theoretical modeling. In particular, the data may cast doubt on the theoretical null, providing a more appropriate empirical null distribution in its place.
- Classic testing theory is purely frequentist, whereas false-discovery rates combine frequentist and Bayesian thinking.
- In classic testing, the attained significance level for case i depends only on its own score z_i, while $\widehat{\text{fdr}}(z_i)$ or $\widehat{\text{Fdr}}(z_i)$ also depends on the observed z-values for other cases.
- Applications of single-test theory usually hope for *rejection* of the null hypothesis, a familiar prescription being 0.80 power at size 0.05. The opposite is true for large-scale testing, where the usual goal is to *accept* most of the null hypotheses, leaving just a few interesting cases for further study.
- Sharp null hypotheses such as $\mu = 0$ are less important in large-scale applications, where the statistician is happy to accept a hefty proportion of uninterestingly small, but nonzero, effect sizes μ_i.
- False-discovery rate hypothesis testing involves a substantial amount of estimation, blurring the line beteen the two main branches of statistical inference.

15.7 Notes and Details

The story of false-discovery rates illustrates how developments in scientific technology (microarrays in this case) can influence the progress of statistical inference. A substantial theory of simultaneous inference was developed between 1955 and 1995, mainly aimed at the frequentist control of family-wise error rates in situations involving a small number of hypothesis tests, maybe up to 20. Good references are Miller (1981) and Westfall and Young (1993).

Benjamini and Hochberg's seminal 1995 paper introduced false-discovery rates at just the right time to catch the wave of large-scale data sets, now involving thousands of simultaneous tests, generated by microarray applications. Most of the material in this chapter is taken from Efron (2010), where the empirical Bayes nature of Fdr theory is emphasized. The police data is discussed and analyzed at length in Ridgeway and MacDonald (2009).

†1 [p. 272] *Model* (15.4). Section 7.4 of Efron (2010) discusses the following result for the non-null distribution of z-values: a transformation such as (15.2) that produces a z-value (i.e., a standard normal random variable $z \sim \mathcal{N}(0,1)$) under the null hypothesis gives, to a good approximation, $z \sim \mathcal{N}(\mu, \sigma_\mu^2)$ under reasonable alternatives. For the specific situation in (15.2), Student's t with 100 degrees of freedom, $\sigma_\mu^2 \doteq 1$ as in (15.4).

†2 [p. 274] *Holm's procedure.* Methods of FWER control, including Holm's procedure, are surveyed in Chapter 3 of Efron (2010). They display a large amount of mathematical ingenuity, and provided the background against which FDR theory developed.

†3 [p. 276] *FDR control theorem.* Benjamini and Hochberg's striking control theorem (15.15) was rederived by Storey *et al.* (2004) using martingale theory. The basic idea of false discoveries, as displayed in Figure 15.2, goes back to Soric (1989).

†4 [p. 285] *Formula* (15.44). Integrating $\text{fdr}(z) = \pi_0 f_0(z)/f(z)$ gives

$$E\{\text{fdr}(z)|z \geq z_0\} = \int_{z_0}^{\infty} \pi_0 f_0(z)\,dz \Bigg/ \int_{z_0}^{\infty} f(z)\,dz \qquad (15.54)$$
$$= \pi_0 S_0(z_0)/S(z_0) = \text{Fdr}(z_0).$$

†5 [p. 286] *Thresholds for Fdr and fdr.* Suppose the survival curves $S_0(z)$ and $S_1(z)$ (15.20) satisfy the "Lehmann alternative" relationship

$$\log S_1(z) = \gamma \log S_0(z) \qquad (15.55)$$

for large values of z, where γ is a positive constant less than 1. (This is a reasonable condition for the non-null density $f_1(z)$ to produce larger positive values of z than does the null density $f_0(z)$.) Differentiating (15.55) gives

$$\frac{\pi_0}{\pi_1}\frac{f_0(z)}{f_1(z)} = \frac{1}{\gamma}\frac{\pi_0}{\pi_1}\frac{S_0(z)}{S_1(z)}, \tag{15.56}$$

after some rearrangement. But $\text{fdr}(z) = \pi_0 f_0(z)/(\pi_0 f_0(z) + \pi_1 f_1(z))$ is algebraically equivalent to

$$\frac{\text{fdr}(z)}{1-\text{fdr}(z)} = \frac{\pi_0}{\pi_1}\frac{f_0(z)}{f_1(z)}, \tag{15.57}$$

and similarly for $\text{Fdr}(z)/(1-\text{Fdr}(z))$, yielding

$$\frac{\text{fdr}(z)}{1-\text{fdr}(z)} = \frac{1}{\gamma}\frac{\text{Fdr}(z)}{1-\text{Fdr}(z)}. \tag{15.58}$$

For large z, both $\text{fdr}(z)$ and $\text{Fdr}(z)$ go to zero, giving the asymptotic relationship

$$\text{fdr}(z) \doteq \text{Fdr}(z)/\gamma. \tag{15.59}$$

If $\gamma = 1/2$ for instance, $\text{fdr}(z)$ will be about twice $\text{Fdr}(z)$ where z is large. This motivates the suggested relative thresholds $\widehat{\text{fdr}}(z_i) \le 0.20$ compared with $\widehat{\text{Fdr}}(z_i) \le 0.10$.

†6 [p. 288] *Correlation effects.* The Poisson regression method used to estimate $\hat{f}(z)$ in Figure 15.5 proceeds as if the components of the N-vector of z_i values $\boldsymbol{z}$ are independent. Approximation (10.54), that the kth bin count $y_k \,\dot{\sim}\, \text{Poi}(\mu_k)$, requires independence. If not, it can be shown that $\text{var}(y_k)$ increases above the Poisson value μ_k as

$$\text{var}(y_k) \doteq \mu_k + \alpha^2 c_k. \tag{15.60}$$

Here c_k is a fixed constant depending on $f(z)$, while α^2 is the root mean square correlation between all pairs z_i and z_j,

$$\alpha^2 = \left[\sum_{i=1}^{N}\sum_{j\neq i}\text{cov}(z_i, z_j)^2\right] \Bigg/ N(N-1). \tag{15.61}$$

Estimates like $\widehat{\text{fdr}}(z)$ in Figure 15.5 remain nearly unbiased under correlation, but their sampling variability increases as a function of α. Chapters 7 and 8 of Efron (2010) discuss correlation effects in detail.

Often, α can be estimated. Let $\boldsymbol{X}$ be the 6033×50 matrix of gene expression levels measured for the control subject in the prostate study. Rows

i and j provide an unbiased estimate of $\text{cor}(z_i, z_j)^2$. Modern computation is sufficiently fast to evaluate all $N(N-1)/2$ pairs (though that isn't necessary, sampling is faster) from which estimate $\hat{\alpha}$ is obtained. It equaled 0.016 ± 0.001 for the control subjects, and 0.015 ± 0.001 for the 6033×52 matrix of the cancer patients. Correlation is not much of a worry for the prostate study, but other microarray studies show much larger $\hat{\alpha}$ values. Sections 6.4 and 8.3 of Efron (2010) discuss how correlations can undercut inferences based on the theoretical null even when it is correct for all the null cases.

†7 [p. 289] *The program* `locfdr`. Available from CRAN, this is an R program that provides fdr and Fdr estimates, using both the theoretical and empirical null distributions.

†8 [p. 289] *ML estimation of the empirical null.* Let $\mathcal{A}_0$ be the "zero set" (15.49), $\boldsymbol{z}_0$ the set of z_i observed to be in $\mathcal{A}_0$, $\mathcal{I}_0$ their indices, and N_0 the number of z_i in $\mathcal{A}_0$. Also define

$$\phi_{\delta_0,\sigma_0}(z) = e^{-\frac{1}{2}\left(\frac{z-\delta_0}{\sigma_0}\right)^2} \Big/ \sqrt{2\pi\sigma_0^2},$$
$$P(\delta_0, \sigma_0) = \int_{\mathcal{A}_0} \phi_{\delta_0,\sigma_0}(z)\, dz \quad \text{and} \quad \theta = \pi_0 P(\delta_0, \sigma_0). \tag{15.62}$$

(So $\theta = \Pr\{z_i \in \mathcal{A}_0\}$ according to (15.48)–(15.49).) Then $\boldsymbol{z}_0$ has density and likelihood

$$f_{\delta_0,\sigma_0,\pi_0}(\boldsymbol{z}_0) = \left[\binom{N}{N_0}\theta^{N_0}(1-\theta)^{N-N_0}\right]\left[\prod_{\mathcal{I}_0} \frac{\phi_{\delta_0,\sigma_0}(z_i)}{P_{\delta_0,\sigma_0}}\right], \tag{15.63}$$

the first factor being the binomial probability of seeing N_0 of the z_i in $\mathcal{A}_0$, and the second the conditional probability of those z_i falling within $\mathcal{A}_0$. The second factor is numerically maximized to give $(\hat{\delta}_0, \hat{\sigma}_0)$, while $\hat{\theta} = N_0/N$ is obtained from the first, and then $\hat{\pi}_0 = \hat{\theta}/P(\hat{\delta}_0, \hat{\sigma}_0)$. This is a partial likelihood argument, as in Section 9.4; `locfdr` centers $\mathcal{A}_0$ at the median of the N z_i values, with width about twice the interquartile range estimate of σ_0.

†9 [p. 290] *The permutation null.* An impressive amount of theoretical effort concerned the "permutation t-test": in a single-test two-sample situation, permuting the data and computing the t statistic gives, after a great many repetitions, a histogram dependably close to that of the standard t distribution; see Hoeffding (1952). This was Fisher's justification for using the standard t-test on nonnormal data.

The argument cuts both ways. Permutation methods tend to recreate the theoretical null, even in situations like that of Figure 15.7 where it isn't appropriate. The difficulties are discussed in Section 6.5 of Efron (2010).

†10 [p. 293] *Relevance theory.* Suppose that in the **DTI** example shown in Figure 15.10 we want to consider only voxels with $x = 60$ as relevant to an observed z_i with $x_i = 60$. Now there may not be enough relevant cases to adequately estimate $\text{fdr}(z_i)$ or $\text{Fdr}(z_i)$. Section 10.1 of Efron (2010) shows how the complete-data estimates $\widehat{\text{fdr}}(z_i)$ or $\widehat{\text{Fdr}}(z_i)$ can be efficiently modified to conform to this situation.

16

Sparse Modeling and the Lasso

The amount of data we are faced with keeps growing. From around the late 1990s we started to see *wide* data sets, where the number of variables far exceeds the number of observations. This was largely due to our increasing ability to measure a large amount of information automatically. In genomics, for example, we can use a high-throughput experiment to automatically measure the expression of tens of thousands of genes in a sample in a short amount of time. Similarly, sequencing equipment allows us to genotype millions of SNPs (single-nucleotide polymorphisms) cheaply and quickly. In document retrieval and modeling, we represent a document by the presence or count of each word in the dictionary. This easily leads to a feature vector with 20,000 components, one for each distinct vocabulary word, although most would be zero for a small document. If we move to bi-grams or higher, the feature space gets really large.

In even more modest situations, we can be faced with hundreds of variables. If these variables are to be predictors in a regression or logistic regression model, we probably do not want to use them all. It is likely that a subset will do the job well, and including all the redundant variables will degrade our fit. Hence we are often interested in identifying a good subset of variables. Note also that in these wide-data situations, even linear models are over-parametrized, so some form of reduction or regularization is essential.

In this chapter we will discuss some of the popular methods for model selection, starting with the time-tested and worthy forward-stepwise approach. We then look at the lasso, a popular modern method that does selection and shrinkage via convex optimization. The LARs algorithm ties these two approaches together, and leads to methods that can deliver paths of solutions.

Finally, we discuss some connections with other modern big- and wide-data approaches, and mention some extensions.

16.1 Forward Stepwise Regression

Stepwise procedures have been around for a very long time. They were originally devised in times when data sets were quite modest in size, in particular in terms of the number of variables. Originally thought of as the poor cousins of "best-subset" selection, they had the advantage of being much cheaper to compute (and in fact *possible* to compute for large p). We will review best-subset regression first.

Suppose we have a set of n observations on a response y_i and a vector of p predictors $x_i' = (x_{i1}, x_{i2}, \ldots, x_{ip})$, and we plan to fit a linear regression model. The response could be quantitative, so we can think of fitting a linear model by least squares. It could also be binary, leading to a linear logistic regression model fit by maximum likelihood. Although we will focus on these two cases, the same ideas transfer exactly to other generalized linear models, the Cox model, and so on. The idea is to build a model using a subset of the variables; in fact the smallest subset that adequately explains the variation in the response is what we are after, both for inference and for prediction purposes. Suppose our loss function for fitting the linear model is L (e.g. sum of squares, negative log-likelihood). The method of best-subset regression is simple to describe, and is given in Algorithm 16.1. Step 3 is easy to state, but requires a lot of computation. For

---

Algorithm 16.1 BEST-SUBSET REGRESSION.

---

1 Start with $m = 0$ and the null model $\hat{\eta}_0(x) = \hat{\beta}_0$, estimated by the mean of the y_i.
2 At step $m = 1$, pick the single variable j that fits the response best, in terms of the loss L evaluated on the training data, in a univariate regression $\hat{\eta}_1(x) = \hat{\beta}_0 + x_j'\hat{\beta}_j$. Set $\mathcal{A}_1 = \{j\}$.
3 For each subset size $m \in \{2, 3, \ldots, M\}$ (with $M \leq \min(n-1, p)$) identify the best subset $\mathcal{A}_m$ of size m when fitting a linear model $\hat{\eta}_m(x) = \hat{\beta}_0 + x_{\mathcal{A}_m}'\hat{\beta}_{\mathcal{A}_m}$ with m of the p variables, in terms of the loss L.
4 Use some external data or other means to select the "best" amongst these M models.

---

p much larger than about 40 it becomes prohibitively expensive to perform exactly—a so-called "N-P complete" problem because of its combinatorial complexity (there are 2^p subsets). Note that the subsets need not be nested:

the best subset of size $m = 3$, say, need not include both or any of the variables in the best subset of size $m = 2$.

In step 4 there are a number of methods for selecting m. Originally the C_p criterion of Chapter 12 was proposed for this purpose. Here we will favor K-fold cross-validation, since it is applicable to all the methods discussed in this chapter.

It is interesting to digress for a moment on how cross-validation works here. We are using it to select the subset size m on the basis of prediction performance (on future data). With $K = 10$, we divide the n training observations randomly into 10 equal size groups. Leaving out say group $k = 1$, we perform steps 1–3 on the 9/10ths, and for each of the chosen models, we summarize the prediction performance on the group-1 data. We do this $K = 10$ times, each time with group k left out. We then average the 10 performance measures for each m, and select the value of m corresponding to the best performance. Notice that for each m, the 10 models $\hat{\eta}_m(x)$ might involve different subsets of variables! This is not a concern, since we are trying to find a good value of m for the method. Having identified $\hat{m}$, we rerun steps 1–3 on the entire training set, and deliver the chosen model $\hat{\eta}_{\hat{m}}(x)$.

As hinted above, there are problems with best-subset regression. A primary issue is that it works exactly only for relatively small p. For example, we cannot run it on the **spam** data with 57 variables (at least not in 2015 on a Macbook Pro!). We may also think that even if we could do the computations, with such a large search space the variance of the procedure might be too high.

As a result, more manageable stepwise procedures were invented. Forward stepwise regression, Algorithm 16.2, is a simple modification of best-subset, with the modification occurring in step 3. Forward stepwise regression produces a *nested* sequence of models $\emptyset \ldots \subset \mathcal{A}_{m-1} \subset \mathcal{A}_m \subset \mathcal{A}_{m+1} \ldots$. It starts with the null model, here an intercept, and adds variables one at a time. Even with large p, identifying the best variable to add at each step is manageable, and can be distributed if clusters of machines are available. Most importantly, it is feasible for large p. Figure 16.1 shows the coefficient profiles for forward-stepwise linear regression on the **spam** training data. Here there are 57 input variables (relative prevalence of particular words in the document), and an "official" (train, test) split of (3065, 1536) observations. The response is coded as +1 if the email was spam, else -1. The figure caption gives the details. We saw the **spam** data earlier, in Table 8.3, Figure 8.7 and Figure 12.2.

Fitting the entire forward-stepwise linear regression path as in the figure

Algorithm 16.2 FORWARD STEPWISE REGRESSION.

1. Start with $m = 0$ and the null model $\hat{\eta}_0(x) = \hat{\beta}_0$, estimated by the mean of the y_i.
2. At step $m = 1$, pick the single variable j that fits the response best, in terms of the loss L evaluated on the training data, in a univariate regression $\hat{\eta}_1(x) = \hat{\beta}_0 + x_j'\hat{\beta}_j$. Set $\mathcal{A}_1 = \{j\}$.
3. For each subset size $m \in \{2, 3, \ldots, M\}$ (with $M \leq \min(n-1, p)$) identify the variable k that when augmented with $\mathcal{A}_{m-1}$ to form $\mathcal{A}_m$, leads to the model $\hat{\eta}_m(x) = \hat{\beta}_0 + x_{\mathcal{A}_m}'\hat{\beta}_{\mathcal{A}_m}$ that performs best in terms of the loss L.
4. Use some external data or other means to select the "best" amongst these M models.

(when $n > p$) has essentially the same cost as a single least squares fit on all the variables. This is because the sequence of models can be updated each time a variable is added.† However, this is a consequence of the linear model and squared-error loss. †1

Suppose instead we run a forward stepwise logistic regression. Here updating does not work, and the entire fit has to be recomputed by maximum likelihood each time a variable is added. Identifying which variable to add in step 3 in principle requires fitting an $(m+1)$-variable model $p - m$ times, and seeing which one reduces the deviance the most. In practice, we can use score tests which are much cheaper to evaluate.† These amount †2 to using the quadratic approximation to the log-likelihood from the final iteratively reweighted least-squares (IRLS) iteration for fitting the model with m terms. The score test for a variable not in the model is equivalent to testing for the inclusion of this variable in the weighted least-squares fit. Hence identifying the next variable is almost back to the previous cases, requiring $p - m$ simple regression updates.† Figure 16.2 shows the test †3 misclassification error for forward-stepwise linear regression and logistic regression on the `spam` data, as a function of the number of steps. They both level off at around 25 steps, and have a similar shape. However, the logistic regression gives more accurate classifications.[1]

Although forward-stepwise methods are possible for large p, they get tedious for very large p (in the thousands), especially if the data could sup-

[1] For this example we can halve the gap between the curves by optimizing the prediction threshold for linear regression.

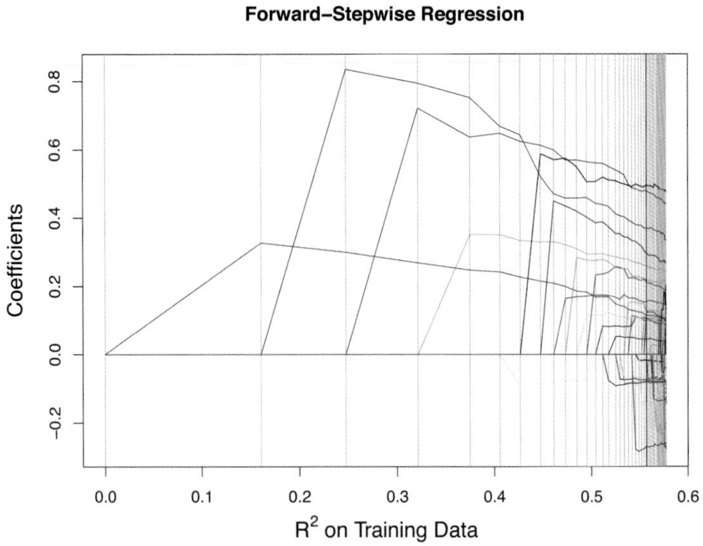

Figure 16.1 Forward stepwise linear regression on the **spam** data. Each curve corresponds to a particular variable, and shows the progression of its coefficient as the model grows. These are plotted against the training R^2, and the vertical gray bars correspond to each step. Starting at the left at step 1, the first selected variable explains $R^2 = 0.16$; adding the second increases R^2 to 0.25, etc. What we see is that early steps have a big impact on the R^2, while later steps hardly have any at all. The vertical black line corresponds to step 25 (see Figure 16.2), and we see that after that the step-wise improvements in R^2 are negligible.

port a model with many variables. However, if the ideal active set is fairly small, even with many thousands of variables forward-stepwise selection is a viable option.

Forward-stepwise selection delivers a sequence of models, as seen in the previous figures. One would generally want to select a single model, and as discussed earlier, we often use cross-validation for this purpose. Figure 16.3 illustrates using stepwise linear regression on the **spam** data. Here the sequence of models are fit using squared-error loss on the binary response variable. However, cross-validation scores each model for *misclassification error*, the ultimate goal of this modeling exercise. This highlights one of the advantages of cross-validation in this context. A con-

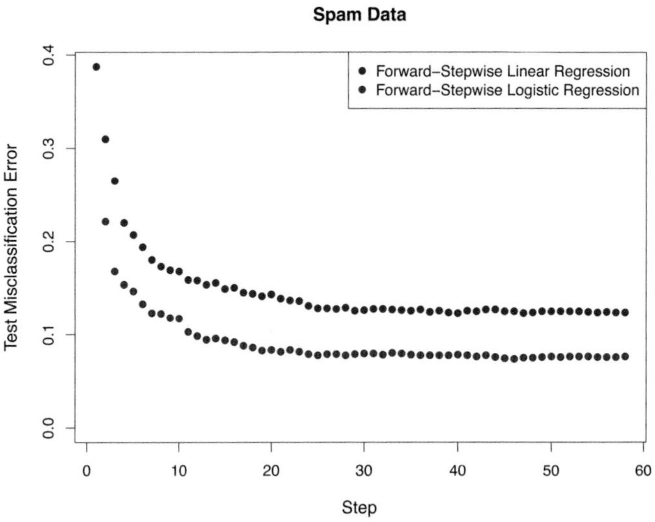

Figure 16.2 Forward-stepwise regression on the `spam` data. Shown is the misclassification error on the test data, as a function of the number of steps. The brown dots correspond to linear regression, with the response coded as -1 and +1; a prediction greater than zero is classified as +1, one less than zero as -1. The blue dots correspond to logistic regression, which performs better. We see that both curves essentially reach their minima after 25 steps.

venient (differentiable and smooth) loss function is used to fit the sequence of models. However, we can use any performance measure to evaluate the sequence of models; here misclassification error is used. In terms of the parameters of the linear model, misclassification error would be a difficult and discontinuous loss function to use for parameter estimation. All we need to use it for here is pick the best model size. There appears to be little benefit in going beyond 25–30 terms.

16.2 The Lasso

The stepwise model-selection methods of the previous section are useful if we anticipate a model using a relatively small number of variables, even if the pool of available variables is very large. If we expect a moderate number of variables to play a role, these methods become cumbersome.

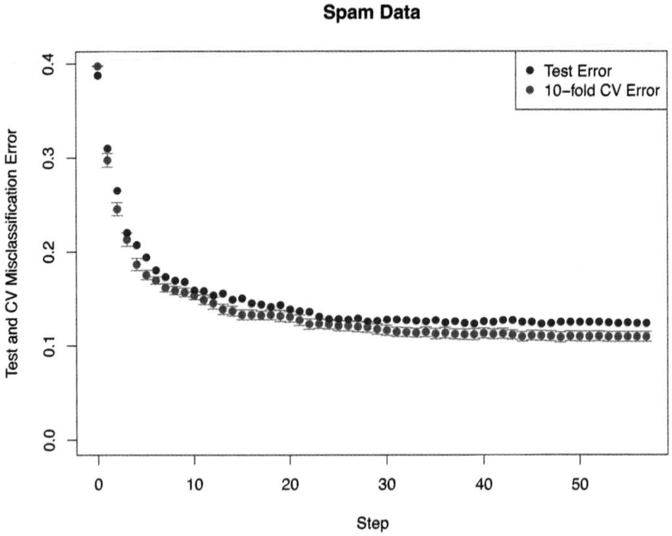

Figure 16.3 Ten-fold cross-validated misclassification errors (green) for forward-stepwise regression on the `spam` data, as a function of the step number. Since each error is an average of 10 numbers, we can compute a (crude) standard error; included in the plot are pointwise standard-error bands. The brown curve is the misclassification error on the test data.

Another black mark against forward-stepwise methods is that the sequence of models is derived in a *greedy* fashion, without any claimed optimality. The methods we describe here are derived from a more principled procedure; indeed they solve a *convex optimization*, as defined below.

We will first present the lasso for squared-error loss, and then the more general case later. Consider the constrained linear regression problem

$$\underset{\beta_0 \in \mathbb{R},\, \beta \in \mathbb{R}^p}{\text{minimize}} \frac{1}{n} \sum_{i=1}^{n} (y_i - \beta_0 - x_i'\beta)^2 \text{ subject to } \|\beta\|_1 \leq t, \tag{16.1}$$

where $\|\beta\|_1 = \sum_{j=1}^{p} |\beta_j|$, the ℓ_1 norm of the coefficient vector. Since both the loss and the constraint are convex in β, this is a convex optimization problem, and it is known as the *lasso*. The constraint $\|\beta\|_1 \leq t$ restricts the coefficients of the model by pulling them toward zero; this has the effect of reducing their variance, and prevents overfitting. Ridge regression is an earlier great uncle of the lasso, and solves a similar problem to (16.1), ex-

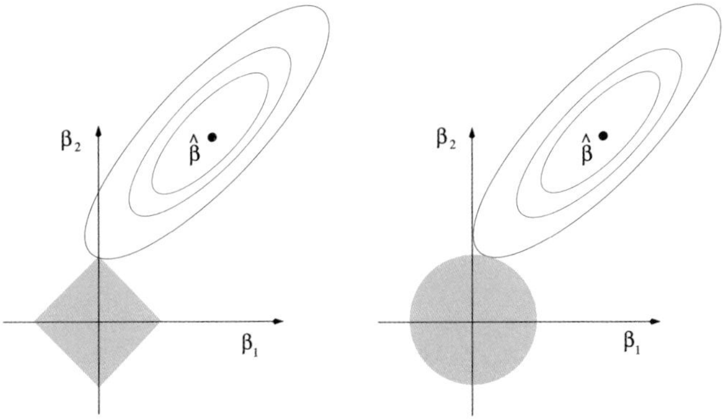

Figure 16.4 An example with $\beta \in \mathbb{R}^2$ to illustrate the difference between ridge regression and the lasso. In both plots, the red contours correspond to the squared-error loss function, with the unrestricted least-squares estimate $\hat{\beta}$ in the center. The blue regions show the constraints, with the lasso on the left and ridge on the right. The solution to the constrained problem corresponds to the value of β where the expanding loss contours first touch the constraint region. Due to the shape of the lasso constraint, this will often be at a corner (or an edge more generally), as here, which means in this case that the minimizing β has $\beta_1 = 0$. For the ridge constraint, this is unlikely to happen.

cept the constraint is $\|\beta\|_2 \leq t$; ridge regression bounds the quadratic ℓ_2 norm of the coefficient vector. It also has the effect of pulling the coefficients toward zero, in an apparently very similar way. Ridge regression is discussed in Section 7.3.[2] Both the lasso and ridge regression are shrinkage methods, in the spirit of the James–Stein estimator of Chapter 7.

A big difference, however, is that for the lasso, the solution typically has many of the β_j equal to zero, while for ridge they are all nonzero. Hence the lasso does variable selection and shrinkage, while ridge only shrinks. Figure 16.4 illustrates this for $\beta \in \mathbb{R}^2$. In higher dimensions, the ℓ_1 norm has sharp edges and corners, which correspond to coefficient estimates zero in β.

Since the constraint in the lasso treats all the coefficients equally, it usually makes sense for all the elements of x to be in the same units. If not, we

[2] Here we use the "bound" form of ridge regression, while in Section 7.3 we use the "Lagrange" form. They are equivalent, in that for every "Lagrange" solution, there is a corresponding bound solution.

typically standardize the predictors beforehand so that each has variance one.

Two natural *boundary* values for t in (16.1) are $t = 0$ and $t = \infty$. The former corresponds to the constant model (the fit is the mean of the y_i,)[3] and the latter corresponds to the unrestricted least-squares fit. In fact, if $n > p$, and $\hat{\beta}$ is the least-squares estimate, then we can replace ∞ by
†4 $\|\hat{\beta}\|_1$, and any value of $t \geq \|\hat{\beta}\|_1$ is a non-binding constraint.† Figure 16.5

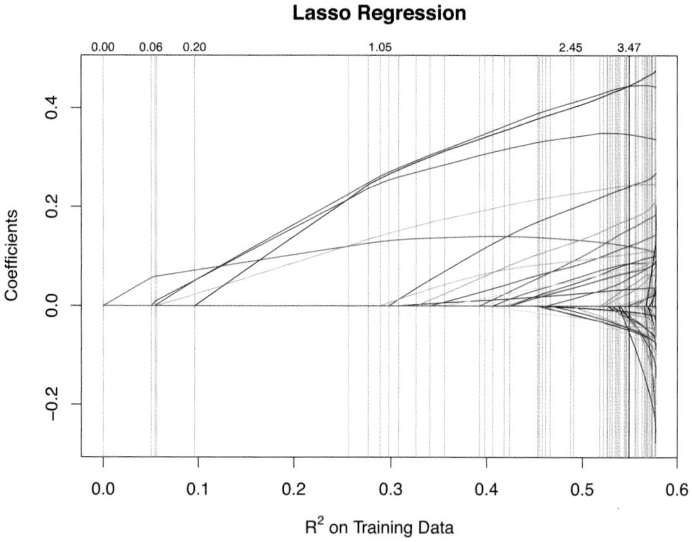

Figure 16.5 The lasso linear regression regularization path on the `spam` data. Each curve corresponds to a particular variable, and shows the progression of its coefficient as the regularization bound t grows. These curves are plotted against the training R^2 rather than t, to make the curves comparable with the forward-stepwise curves in Figure 16.1. Some values of t are indicated at the top. The vertical gray bars indicate changes in the active set of nonzero coefficients, typically an inclusion. Here we see clearly the role of the ℓ_1 penalty; as t is relaxed, coefficients become nonzero, but in a smoother fashion than in forward stepwise.

shows the regularization path[4] for the lasso linear regression problem on

[3] We typically do not restrict the intercept in the model.
[4] Also known as the *homotopy* path.

the **spam** data; that is, the solution path for all values of t. This can be computed exactly, as we will see in Section 16.4, because the coefficient profiles are piecewise linear in t. It is natural to compare this coefficient profile with the analagous one in Figure 16.1 for forward-stepwise regression. Because of the control of $\|\hat{\beta}(t)\|_1$, we don't see the same range as in forward stepwise, and observe somewhat smoother behavior. Figure 16.6 contrasts

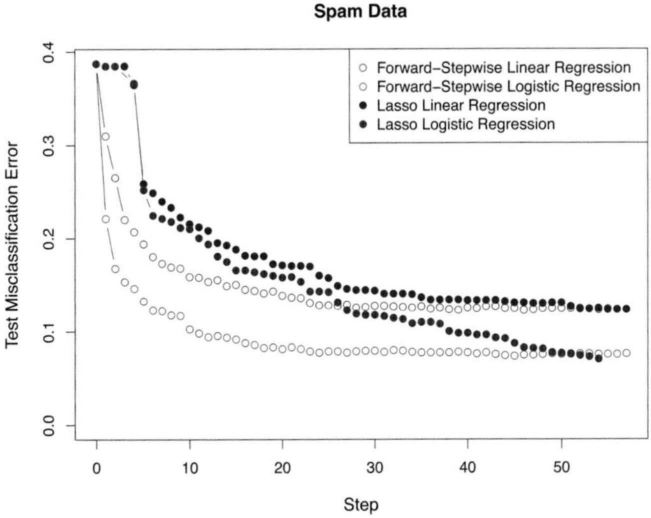

Figure 16.6 Lasso versus forward-stepwise regression on the **spam** data. Shown is the misclassification error on the test data, as a function of the number of variables in the model. Linear regression is coded brown, logistic regression blue; hollow dots forward stepwise, solid dots lasso. In this case it appears stepwise and lasso achieve the same performance, but lasso takes longer to get there, because of the shrinkage.

the prediction performance on the **spam** data for lasso regularized models (linear regression and logistic regression) versus forward-stepwise models. The results are rather similar at the end of the path; here forward stepwise can achieve classification performance similar to that of lasso regularized logistic regression with about half the terms. Lasso logistic regression (and indeed any likelihood-based linear model) is fit by penalized maximum likelihood:

$$\underset{\beta_0\in\mathbb{R},\ \beta\in\mathbb{R}^p}{\text{minimize}}\ \frac{1}{n}\sum_{i=1}^{n} L(y_i, \beta_0 + \beta' x_i) \text{ subject to } \|\beta\|_1 \leq t. \tag{16.2}$$

Here L is the negative of the log-likelihood function for the response distribution.

16.3 Fitting Lasso Models

The lasso objectives (16.1) or (16.2) are differentiable and convex in β and β_0, and the constraint is convex in β. Hence solving these problems is a convex optimization problem, for which standard packages are available. It turns out these problems have special structure that can be exploited to yield efficient algorithms for fitting the entire path of solutions as in Figures 16.1 and 16.5. We will start with problem (16.1), which we rewrite in the more convenient *Lagrange* form:

$$\underset{\beta\in\mathbb{R}^p}{\text{minimize}}\ \frac{1}{2n}\|\boldsymbol{y} - \boldsymbol{X}\beta\|^2 + \lambda\|\beta\|_1. \tag{16.3}$$

Here we have centered $\boldsymbol{y}$ and the columns of $\boldsymbol{X}$ beforehand, and hence the intercept has been omitted. The Lagrange and constraint versions are equivalent, in the sense that any solution $\hat{\beta}(\lambda)$ to (16.3) with $\lambda \geq 0$ corresponds to a solution to (16.1) with $t = \|\hat{\beta}(\lambda)\|_1$. Here large values of λ will encourage solutions with small ℓ_1 norm coefficient vectors, and vice-versa; $\lambda = 0$ corresponds to the ordinary least squares fit.

The solution to (16.3) satisfies the subgradient condition

$$-\frac{1}{n}\langle \boldsymbol{x}_j, \boldsymbol{y} - \boldsymbol{X}\hat{\beta}\rangle + \lambda s_j = 0, \quad j = 1,\ldots,p, \tag{16.4}$$

where $s_j \in \text{sign}(\hat{\beta}_j),\ j = 1,\ldots,p$. This notation means $s_j = \text{sign}(\hat{\beta}_j)$ if $\hat{\beta}_j \neq 0$, and $s_j \in [-1,1]$ if $\hat{\beta}_j = 0$.) We use the inner-product notation $\langle a, b\rangle = a'b$ in (16.4), which leads to more evocative expressions. These subgradient conditions are the modern way of characterizing solutions to problems of this kind, and are equivalent to the Karush–Kuhn–Tucker optimality conditions. From these conditions we can immediately learn some properties of a lasso solution.

- $\frac{1}{n}|\langle \boldsymbol{x}_j, \boldsymbol{y} - \boldsymbol{X}\hat{\beta}\rangle| = \lambda$ for all members of the *active set*; i.e., each of the variables in the model (with nonzero coefficient) has the same covariance with the residuals (in absolute value).

- $\frac{1}{n}|\langle \boldsymbol{x}_k, \boldsymbol{y} - \boldsymbol{X}\hat{\beta}\rangle| \leq \lambda$ for all variables not in the active set (i.e. with coefficients zero).

These conditions are interesting and have a big impact on computation. Suppose we have the solution $\hat{\beta}(\lambda_1)$ at λ_1, and we decrease λ by a small amount to $\lambda_2 < \lambda_1$. The coefficients and hence the residuals change, in such a way that the covariances all remain tied at the smaller value λ_2. If in the process the active set has not changed, and nor have the signs of their coefficients, then we get an important consequence: $\hat{\beta}(\lambda)$ is *linear* for $\lambda \in [\lambda_2, \lambda_1]$. To see this, suppose $\mathcal{A}$ indexes the active set, which is the same at λ_1 and λ_2, and let $s_{\mathcal{A}}$ be the constant sign vector. Then we have

$$\begin{aligned} \boldsymbol{X}'_{\mathcal{A}}(\boldsymbol{y} - \boldsymbol{X}\hat{\beta}(\lambda_1)) &= s_{\mathcal{A}}\lambda_1, \\ \boldsymbol{X}'_{\mathcal{A}}(\boldsymbol{y} - \boldsymbol{X}\hat{\beta}(\lambda_2)) &= s_{\mathcal{A}}\lambda_2. \end{aligned}$$

By subtracting and solving we get

$$\hat{\beta}_{\mathcal{A}}(\lambda_2) - \hat{\beta}_{\mathcal{A}}(\lambda_1) = (\lambda_1 - \lambda_2)(\boldsymbol{X}'_{\mathcal{A}}\boldsymbol{X}_{\mathcal{A}})^{-1}s_{\mathcal{A}}, \tag{16.5}$$

and the remaining coefficients (with indices not in $\mathcal{A}$) are all zero. This shows that the full coefficient vector $\hat{\beta}(\lambda)$ is linear for $\lambda \in [\lambda_2, \lambda_1]$. In fact, the coefficient profiles for the lasso are continuous and piecewise linear over the entire range of λ, with *knots* occurring whenever the active set changes, or the signs of the coefficients change.

Another consequence is that we can easily determine λ_{max}, the smallest value for λ such that the solution $\hat{\beta}(\lambda_{max}) = 0$. From (16.4) this can be seen to be $\lambda_{max} = \max_j \frac{1}{n}|\langle \boldsymbol{x}_j, \boldsymbol{y}\rangle|$.

These two facts plus a few more details enable us to compute the exact solution path for the squared-error-loss lasso; that is the topic of the next section.

16.4 Least-Angle Regression

We have just seen that the lasso coefficient profile $\hat{\beta}(\lambda)$ is piecewise linear in λ, and that the elements of the active set are tied in their absolute covariance with the residuals. With $\boldsymbol{r}(\lambda) = \boldsymbol{y} - \boldsymbol{X}\hat{\beta}(\lambda)$, the covariance between $\boldsymbol{x}_j$ and the evolving residual is $c_j(\lambda) = |\langle x_j, \boldsymbol{r}(\lambda)\rangle|$. Hence these also change in a piecewise linear fashion, with $c_j(\lambda) = \lambda$ for $j \in \mathcal{A}$, and $c_j(\lambda) \leq \lambda$ for $j \notin \mathcal{A}$. This inspires the *Least-Angle Regression* algorithm, given in Algorithm 16.3, which exploits this linearity to fit the entire lasso regularization path.

Algorithm 16.3 LEAST-ANGLE REGRESSION.

1 Standardize the predictors to have mean zero and unit ℓ_2 norm. Start with the residual $\boldsymbol{r}_0 = \boldsymbol{y} - \bar{\boldsymbol{y}}$, $\beta^0 = (\beta_1, \beta_2, \ldots, \beta_p) = 0$.

2 Find the predictor $\boldsymbol{x}_j$ most correlated with $\boldsymbol{r}_0$; i.e., with largest value for $|\langle \boldsymbol{x}_j, \boldsymbol{r}_0 \rangle|$. Call this value λ_0, define the active set $\mathcal{A} = \{j\}$, and $\boldsymbol{X}_{\mathcal{A}}$, the matrix consisting of this single variable.

3 For $k = 1, 2, \ldots, K = \min(n-1, p)$ do:

 (a) Define the least-squares direction $\delta = \frac{1}{\lambda_{k-1}}(\boldsymbol{X}_{\mathcal{A}}'\boldsymbol{X}_{\mathcal{A}})^{-1}\boldsymbol{X}_{\mathcal{A}}'\boldsymbol{r}_{k-1}$, and define the p-vector Δ such that $\Delta_{\mathcal{A}} = \delta$, and the remaining elements are zero.

 (b) Move the coefficients β from β^{k-1} in the direction Δ toward their least-squares solution on $\boldsymbol{X}_{\mathcal{A}}$: $\beta(\lambda) = \beta^{k-1} + (\lambda_{k-1} - \lambda)\Delta$ for $0 < \lambda \le \lambda_{k-1}$, keeping track of the evolving residuals $\boldsymbol{r}(\lambda) = \boldsymbol{y} - \boldsymbol{X}\beta(\lambda) = \boldsymbol{r}_{k-1} - (\lambda_{k-1} - \lambda)\boldsymbol{X}_{\mathcal{A}}\Delta$.

 (c) Keeping track of $|\langle \boldsymbol{x}_\ell, \boldsymbol{r}(\lambda) \rangle|$ for $\ell \notin \mathcal{A}$, identify the largest value of λ at which a variable "catches up" with the active set; if the variable has index ℓ, that means $|\langle \boldsymbol{x}_\ell, \boldsymbol{r}(\lambda) \rangle| = \lambda$. This defines the next "knot" λ_k.

 (d) Set $\mathcal{A} = \mathcal{A} \cup \ell$, $\beta^k = \beta(\lambda_k) = \beta^{k-1} + (\lambda_{k-1} - \lambda_k)\Delta$, and $\boldsymbol{r}_k = \boldsymbol{y} - \boldsymbol{X}\beta^k$.

4 Return the sequence $\{\lambda_k, \beta^k\}_0^K$.

In step 3(a) $\delta = (\boldsymbol{X}_{\mathcal{A}}'\boldsymbol{X}_{\mathcal{A}})^{-1}s_{\mathcal{A}}$ as in (16.5). We can think of the LAR algorithm as a democratic version of forward-stepwise regression. In forward-stepwise regression, we identify the variable that will improve the fit the most, and then move all the coefficients toward the new least-squares fit. As described in endnote †4, this is sometimes done by computing the inner products of each (unadjusted) variable with the residual, and picking the largest in absolute value. In step 3 of Algorithm 16.3, we move the coefficients for the variables in the active set $\mathcal{A}$ *toward* their least-squares fit (keeping their inner products tied), but stop when a variable not in $\mathcal{A}$ catches up in inner product. At that point, it is invited into the club, and the process continues.

Step 3(c) can be performed efficiently because of the linearity of the evolving inner products; for each variable not in $\mathcal{A}$, we can determine exactly when (in λ time) it would catch up, and hence which catches up first and when. Since the path is piecewise linear, and we know the slopes, this

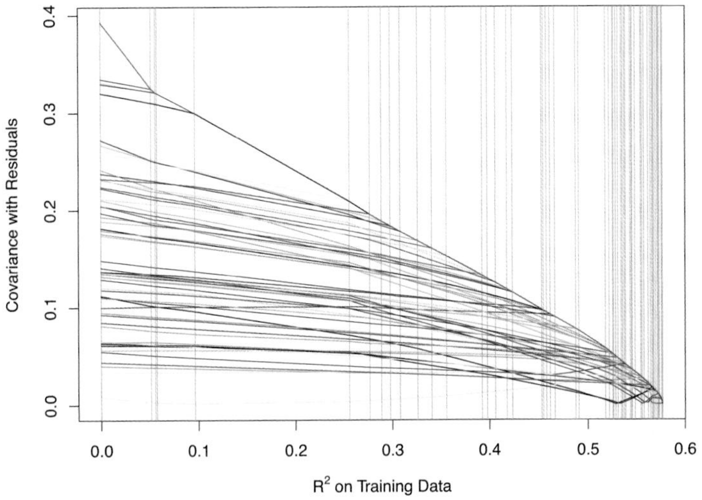

Figure 16.7 Covariance evolution on the `spam` data. As variables tie for maximal covariance, they become part of the active set. These occasions are indicated by the vertical gray bars, again plotted against the training R^2 as in Figure 16.5.

means we know the path *exactly* without further computation between λ_{k-1} and the newly found λ_k.

The name "least-angle regression" derives from the fact that in step 3(b) the fitted vector evolves in the direction $\boldsymbol{X}\Delta = \boldsymbol{X}_{\mathcal{A}}\delta$, and its inner product with each active vector is given by $\boldsymbol{X}'_{\mathcal{A}}\boldsymbol{X}_{\mathcal{A}}\delta = s_{\mathcal{A}}$. Since all the columns of $\boldsymbol{X}$ have unit norm, this means the angles between each active vector and the evolving fitted vector are equal and hence minimal.

The main computational burden in Algorithm 16.3 is in step 3(a), computing the new direction, each time the active set is updated. However, this is easily performed using standard updating of a QR decomposition, and hence the computations for the entire path are of the same order as that of a single least-squares fit using all the variables.

The vertical gray lines in Figure 16.5 show when the active set changes. We see the slopes change at each of these transitions. Compare with the corresponding Figure 16.1 for forward-stepwise regression.

Figure 16.7 shows the the decreasing covariance during the steps of the

LAR algorithm. As each variable joins the active set, the covariances become tied. At the end of the path, the covariances are all zero, because this is the unregularized ordinary least-squares solution.

It turns out that the LAR algorithm is not quite the lasso path; variables can *drop out* of the active set as the path evolves. This happens when a coefficient curve passes through zero. The subgradient equations (16.4) imply that the sign of each active coefficient matches the sign of the gradient. However, a simple addition to step 3(c) in Algorithm 16.3 takes care of the issue:

3(c)+ *lasso modification*: If a nonzero coefficient crosses zero before the next variable enters, drop it from $\mathcal{A}$ and recompute the joint least-squares direction Δ using the reduced set.

Figure 16.5 was computed using the `lars` package in `R`, with the `lasso` option set to accommodate step 3(c)+; in this instance there was no need for dropping. Dropping tends to occur when some of the variables are highly correlated.

Lasso and Degrees of Freedom

We see in Figure 16.6 (left panel) that forward-stepwise regression is more aggressive than the lasso, in that it brings down the training MSE faster. We can use the covariance formula for df from Chapter 12 to quantify the amount of fitting at each step.

In the right panel we show the results of a simulation for estimating the df of forward-stepwise regression and the lasso for the `spam` data. Recall the covariance formula

$$df = \frac{1}{\sigma^2} \sum_{i=1}^{n} \operatorname{cov}(y_i, \hat{y}_i). \tag{16.6}$$

These covariances are of course with respect to the sampling distribution of the y_i, which we do not have access to since these are real data. So instead we simulate from fitted values from the full least-squares fit, by adding Gaussian errors with the appropriate (estimated) standard deviation. (This is the parametric bootstrap calculation (12.64).)

It turns out that each step of the LAR algorithm spends one df, as is evidenced by the brown curve in the right plot of Figure 16.8. Forward stepwise spends more df in the earlier stages, and can be erratic.

Under some technical conditions on the X matrix (that guarantee that

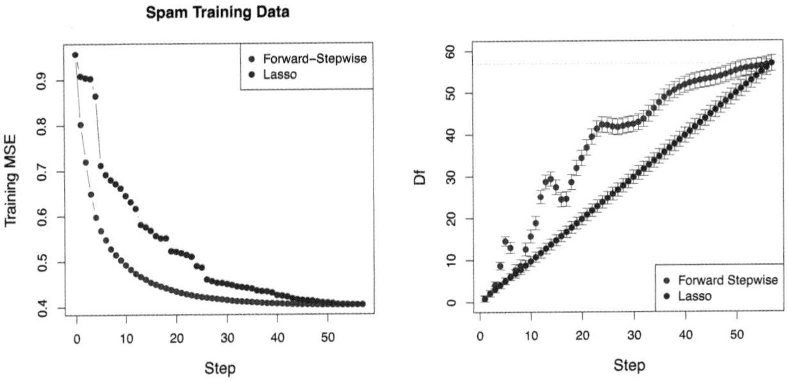

Figure 16.8 Left: Training mean-squared error (MSE) on the `spam` data, for forward-stepwise regression and the lasso, as a function of the size of the active set. Forward stepwise is more aggressive than the lasso, in that it (over-)fits the training data more quickly. Right: Simulation showing the degrees of freedom or df of forward-stepwise regression versus lasso. The lasso uses one df per step, while forward stepwise is greedier and uses more, especially in the early steps. Since these df were computed using 5000 random simulated data sets, we include standard-error bands on the estimates.

LAR delivers the lasso path), one can show that the df is *exactly* one per step. More generally, for the lasso, if we define $\widehat{df}(\lambda) = |\mathcal{A}(\lambda)|$ (the size of the active set at λ), we have that $E[\widehat{df}(\lambda)] = df(\lambda)$. In other words, the size of the active set is an unbiased estimate of df.

Ordinary least squares with a predetermined sequence of variables spends one df per variable. Intuitively forward stepwise spends more, because it pays a price (in some extra df) for searching.† Although the lasso does search for the next variable, it does not fit the new model all the way, but just until the next variable enters. At this point, one new df has been spent. †5

16.5 Fitting Generalized Lasso Models

So far we have focused on the lasso for squared-error loss, and exploited the piecewise-linearity of its coefficient profile to efficiently compute the entire path. Unfortunately this is not the case for most other loss functions, so obtaining the coefficient path is potentially more costly. As a case in point, we will use logistic regression as an example; in this case in (16.2) L represents the negative binomial log-likelihood. Writing the loss explicitly and using the Lagrange form for the penalty, we wish to solve

$$\underset{\beta_0\in\mathbb{R},\,\beta\in\mathbb{R}^p}{\text{minimize}} -\left[\frac{1}{n}\sum_{i=1}^{n} y_i \log\mu_i + (1-y_i)\log(1-\mu_i)\right] + \lambda\|\beta\|_1. \quad (16.7)$$

Here we assume the $y_i \in \{0,1\}$ and μ_i are the fitted probabilities

$$\mu_i = \frac{e^{\beta_0+x_i'\beta}}{1+e^{\beta_0+x_i'\beta}}. \quad (16.8)$$

Similar to (16.4), the solution satisfies the subgradient condition

$$\frac{1}{n}\langle \boldsymbol{x}_j, \boldsymbol{y}-\boldsymbol{\mu}\rangle - \lambda s_j = 0, \quad j=1,\ldots,p, \quad (16.9)$$

where $s_j \in \text{sign}(\beta_j)$, $j = 1,\ldots,p$, and $\boldsymbol{\mu}' = (\mu_1,\ldots,\mu_n)$.[5] However, the nonlinearity of μ_i in β_j results in piecewise nonlinear coefficient profiles. Instead we settle for a solution path on a sufficiently fine grid of values for λ. It is once again easy to see that the largest value of λ we need consider is

$$\lambda_{max} = \max_j |\langle \boldsymbol{x}_j, \boldsymbol{y} - \bar{y}\mathbf{1}\rangle|, \quad (16.10)$$

since this is the smallest value of λ for which $\hat{\beta} = 0$, and $\hat{\beta}_0 = \text{logit}(\bar{y})$. A reasonable sequence is 100 values $\lambda_1 > \lambda_2 > \ldots > \lambda_{100}$ equally spaced on the log-scale from λ_{max} down to $\epsilon\lambda_{max}$, where ϵ is some small fraction such as 0.001.

An approach that has proven to be surprisingly efficient is *path-wise coordinate descent*.

- For each value λ_k, solve the lasso problem for one β_j only, holding all the others fixed. Cycle around until the estimates stabilize.
- By starting at λ_1, where all the parameters are zero, we use *warm starts* in computing the solutions at the decreasing sequence of λ values. The warm starts provide excellent initializations for the sequence of solutions $\hat{\beta}(\lambda_k)$.
- The active set grows slowly as λ decreases. Computational hedges that guess the active set prove to be particularly efficient. If the guess is good (and correct), one iterates coordinate descent using only those variables,

[5] The equation for the intercept is $\frac{1}{n}\sum_{i=1}^{n} y_i = \frac{1}{n}\sum_{i=1}^{n}\mu_i$.

until convergence. One more sweep through all the variables confirms the hunch.

The **R** package **glmnet** employs a *proximal-Newton* strategy at each value λ_k.

1 Compute a weighted least squares (quadratic) approximation to the log-likelihood L at the current estimate for the solution vector $\hat{\beta}(\lambda_k)$; This produces a *working response* and observation weights, as in a regular GLM.
2 Solve the weighted least-squares lasso at λ_k by coordinate descent, using warm starts and active-set iterations.

We now give some details, which illustrate why these particular strategies are effective. Consider the weighted least-squares problem

$$\underset{\beta_j}{\text{minimize}} \frac{1}{2n} \sum_{i=1}^{n} w_i (z_i - \beta_0 - x_i'\beta)^2 + \lambda \|\beta\|_1, \tag{16.11}$$

with all but β_j fixed at their current values. Writing $r_i = z_i - \beta_0 - \sum_{\ell \neq j} x_{i\ell}\beta_\ell$, we can recast (16.11) as

$$\underset{\beta_j}{\text{minimize}} \frac{1}{2n} \sum_{i=1}^{n} w_i (r_i - x_{ij}\beta_j)^2 + \lambda |\beta_j|, \tag{16.12}$$

a one-dimensional problem. The subgradient equation is

$$\frac{1}{n} \sum_{i=1}^{n} w_i x_{ij} (r_i - x_{ij}\beta_j) - \lambda \cdot \text{sign}(\beta_j) = 0. \tag{16.13}$$

The simplest form of the solution occurs if each variable is standardized to have weighted mean zero and variance one, and the weights sum to one; in that case we have a two-step solution.

1 Compute the weighted simple least-squares coefficient

$$\tilde{\beta}_j = \langle \boldsymbol{x}_j, \boldsymbol{r} \rangle_w = \sum_{i=1}^{n} w_i x_{ij} r_i. \tag{16.14}$$

2 *Soft-threshold* $\tilde{\beta}_j$ to produce $\hat{\beta}_j$:

$$\hat{\beta}_j = \begin{cases} 0 & \text{if } |\tilde{\beta}| < \lambda; \\ \text{sign}(\tilde{\beta}_j)(|\tilde{\beta}_j| - \lambda) & \text{otherwise.} \end{cases} \tag{16.15}$$

Without the standardization, the solution is almost as simple but less intuitive.

Hence each coordinate-descent update essentially requires an inner product, followed by the soft thresholding operation. This is especially convenient for x_{ij} that are stored in *sparse-matrix* format, since then the inner products need only visit the nonzero values. If the coefficient is zero before the step, and remains zero, one just moves on, otherwise the model is updated.

Moving from the solution at λ_k (for which $|\langle \boldsymbol{x}_j, \boldsymbol{r}\rangle_w| = \lambda_k$ for all the nonzero coefficients $\hat{\beta}_j$), down to the smaller λ_{k+1}, one might expect all variables for which $|\langle \boldsymbol{x}_j, \boldsymbol{r}\rangle_w| \geq \lambda_{k+1}$ would be natural candidates for the new active set. The *strong rules* lower the bar somewhat, and include any variables for which $|\langle \boldsymbol{x}_j, \boldsymbol{r}\rangle_w| \geq \lambda_{k+1} - (\lambda_k - \lambda_{k+1})$; this tends to rarely make mistakes, and still leads to considerable computational savings.

Apart from variations in the loss function, other penalties are of interest as well. In particular, the *elastic net* penalty bridges the gap between the lasso and ridge regression. That penalty is defined as

$$P_\alpha(\beta) = \frac{1}{2}(1-\alpha)\|\beta\|_2^2 + \alpha\|\beta\|_1, \tag{16.16}$$

where the factor $1/2$ in the first term is for mathematical convenience. When the predictors are excessively correlated, the lasso performs somewhat poorly, since it has difficulty in choosing among the correlated cousins. Like ridge regression, the elastic net shrinks the coefficients of correlated variables toward each other, and tends to select correlated variables in groups. In this case the co-ordinate descent update is almost as simple as in (16.15)

$$\hat{\beta}_j = \begin{cases} 0 & \text{if } |\tilde{\beta}| < \alpha\lambda; \\ \frac{\text{sign}(\tilde{\beta}_j)(|\tilde{\beta}_j|-\alpha\lambda)}{1+(1-\alpha)\lambda} & \text{otherwise,} \end{cases} \tag{16.17}$$

again assuming the observations have weighted variance equal to one. When $\alpha = 0$, the update corresponds to a coordinate update for ridge regression.

Figure 16.9 compares lasso with forward-stepwise logistic regression on the `spam` data, here using all binarized variables and their pairwise interactions. This amounts to 3061 variables in all, once degenerate variables have been excised. Forward stepwise takes a long time to run, since it enters one variable at a time, and after each one has been selected, a new GLM must be fit. The lasso path, as fit by `glmnet`, includes many new variables at each step (λ_k), and is extremely fast (6 s for the entire path). For very large

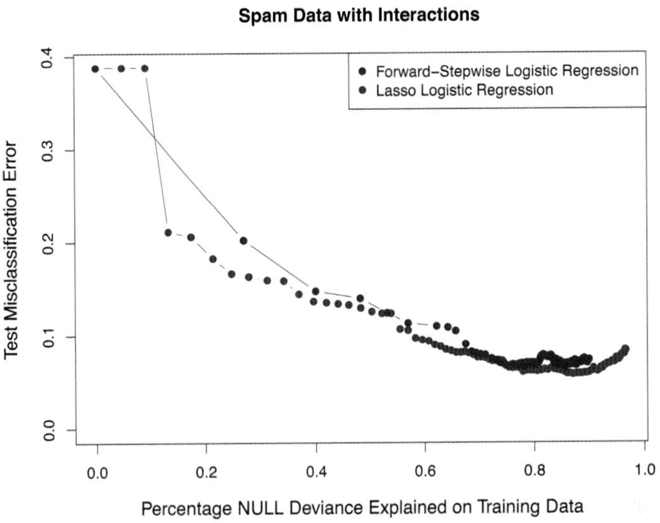

Figure 16.9 Test misclassification error for lasso versus forward-stepwise logistic regression on the `spam` data, where we consider pairwise interactions as well as main effects (3061 predictors in all). Here the minimum error for lasso is 0.057 versus 0.064 for stepwise logistic regression, and 0.071 for the main-effects-only lasso logistic regression model. The stepwise models went up to 134 variables before encountering convergence issues, while the lasso had a largest active set of size 682.

and wide modern data sets (millions of examples and millions of variables), the lasso path algorithm is feasible and attractive.

16.6 Post-Selection Inference for the Lasso

This chapter is mostly about building interpretable models for prediction, with little attention paid to inference; indeed, inference is generally difficult for adaptively selected models.

Suppose we have fit a lasso regression model with a particular value for λ, which ends up selecting a subset $\mathcal{A}$ of size $|\mathcal{A}| = k$ of the p available variables. The question arises as to whether we can assign p-values to these selected variables, and produce confidence intervals for their coefficients. A recent burst of research activity has made progress on these important problems. We give a very brief survey here, with references ap-

†6 pearing in the notes. † We discuss post-selection inference more generally in Chapter 20.

One question that arises is whether we are interested in making inferences about the population regression parameters using the full set of p predictors, or whether interest is restricted to the population regression parameters using only the subset $\mathcal{A}$.

For the first case, it has been proposed that one can view the coefficients of the selected model as an efficient but biased estimate of the full population coefficient vector. The idea is to then *debias* this estimate, allowing inference for the full vector of coefficients. Of course, sharper inference will be available for the stronger variables that were selected in the first place.

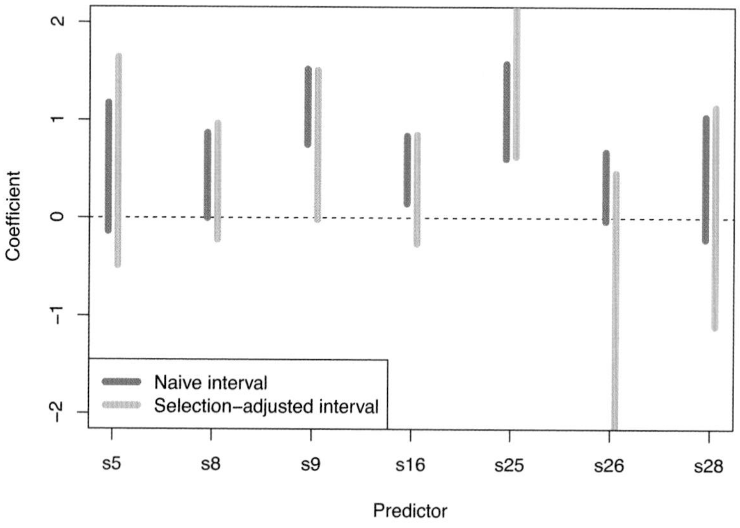

Figure 16.10 HIV data. Linear regression of drug resistance in HIV-positive patients on seven sites, indicators of mutations at particular genomic locations. These seven sites were selected from a total of 30 candidates, using the lasso. The naive 95% confidence intervals (dark) use standard linear-regression inference, ignoring the selection event. The light intervals are 95% confidence intervals, using linear regression, but conditioned on the selection event.

For the second case, the idea is to *condition* on the selection event(s) and hence the set $\mathcal{A}$ itself, and then perform conditional inference on the unrestricted (i.e. not lasso-shrunk) regression coefficients of the response on only the variables in $\mathcal{A}$. For the case of a lasso with squared-error loss, it turns out that the set of response vectors $\boldsymbol{y} \in \mathbb{R}^N$ that would lead to a particular subset $\mathcal{A}$ of variables in the active set form a convex polytope in $\mathbb{R}^N$ (if we condition on the signs of the coefficients as well; ignoring the signs leads to a finite union of such polytopes). This, along with delicate Gaussian conditioning arguments, leads to truncated Gaussian and t-distrubtions for parameters of interest.

Figure 16.10 shows the results of using the lasso to select variables in an HIV study. The outcome Y is a measure of the resistence to an HIV-1 treatment (nucleoside reverse transcriptase inhibitor), and the 30 predictors are indicators of whether mutations had occurred at particular genomic sites. Lasso regression with 10-fold cross-validation selected a value of $\lambda = 0.003$ and the seven sites indicated in the figure had nonzero coefficients. The dark bars in the figure indicate standard 95% confidence intervals for the coefficients of the selected variables, using linear regression, and ignoring the fact that the lasso was used to select the variables. Three variables are significant, and two more nearly so. The lighter bars are confidence intervals in a similar regression, but conditioned on the selection event. †We see that they are generally wider, and only variable $s25$ remains †7
significant.

16.7 Connections and Extensions

There are interesting connections between lasso models and other popular approaches to the prediction problem. We will briefly cover two of these here, namely support-vector machines and boosting.

Lasso Logistic Regression and the SVM

We show in Section 19.3 that ridged logistic regression has a lot in common with the linear support-vector machine. For separable data the limit as $\lambda \downarrow 0$ in ridged logistic regression coincides with the SVM. In addition their loss functions are somewhat similar. The same holds true for ℓ_1 regularized logistic regression versus the ℓ_1 SVM—their end-path limits are the same. In fact, due to the similarity of the loss functions, their solutions are not too different elsewhere along the path. However, the end-path behavior is a little more complex. They both converge to the ℓ_∞ maximizing margin separator—that is, the margin is measured with respect to the ℓ_∞ distance of points to the decision boundary, or maximum absolute coordinate.† †8

Lasso and Boosting

In Chapter 17 we discuss boosting, a general method for building a complex prediction model using simple building components. In its simplest form (regression) boosting amounts to the following simple iteration:

1 Inititialize $b = 0$ and $F^0(x) := 0$.
2 For $b = 1, 2, \ldots, B$:

(a) compute the residuals $r_i = y_i - F(x_i),\ i = 1, \ldots, n$;
(b) fit a small regression tree to the observations $(x_i, r_i)_1^n$, which we can think of as estimating a function $g^b(x)$; and
(c) update $F^b(x) = F^{b-1}(x) + \epsilon \cdot g^b(x)$.

The "smallness" of the tree limits the interaction order of the model (e.g. a tree with only two splits involves at most two variables). The number of terms B and the shrinkage parameter ϵ are both tuning parameters that control the rate of learning (and hence overfitting), and need to be set, for example by cross-validation.

In words this algorithm performs a search in the space of trees for the one most correlated with the residual, and then moves the fitted function F^b a small amount in that direction—a process known as *forward-stagewise fitting*. One can paraphrase this simple algorithm in the context of linear regression, where in step 2(b) the space of small trees is replaced by linear functions.

1 Inititialize $\beta^0 = 0$, and standardize all the variables $\boldsymbol{x}_j,\ j = 1, \ldots, p$.
2 For $b = 1, 2, \ldots, B$:

(a) compute the residuals $\boldsymbol{r} = \boldsymbol{y} - \boldsymbol{X}\beta^b$;
(b) find the predictor $\boldsymbol{x}_j$ most correlated with the residual vector $\boldsymbol{r}$; and
(c) update β^b to β^{b+1}, where $\beta_j^{b+1} = \beta_j^b + \epsilon \cdot s_j$ (s_j being the sign of the correlation), leaving all the other components alone.

For small ϵ the solution paths for this least-squares boosting and the lasso are very similar. It is natural to consider the limiting case or *infinitesimal forward stagewise* fitting, which we will abbreviate *i*FS. One can imagine a scenario where a number of variables are vying to win the competition in step 2(b), and once they are tied their coefficients move in concert as they each get incremented. This was in fact the inspiration for the LAR algorithm 16.3, where $\mathcal{A}$ represents the set of tied variables, and δ is the relative number of turns they each have in getting their coefficients updated. It turns out that *i*FS is often but not always exactly the lasso; it can
†9 instead be characterized as a type of monotone lasso.†

Not only do these connections inspire new insights and algorithms for the lasso, they also offer insights into boosting. We can think of boosting as fitting a monotone lasso path in the high-dimensional space of variables defined by all possible trees of a certain size.

Extensions of the Lasso

The idea of using ℓ_1 regularization to induce sparsity has taken hold, and variations of these ideas have spread like wildfire in applied statistical modeling. Along with advances in convex optimization, hardly any branch of applied statistics has been left untouched. We dont go into detail here, but refer the reader to the references in the endnotes. Instead we will end this section with a (non-exhaustive) list of such applications, which may entice the reader to venture into this domain.

- The group lasso penalty $\sum_{k=1}^{K} \|\theta_k\|_2$ applies to vectors θ_k of parameters, and selects whole groups at a time. Armed with these penalties, one can derive lasso-like schemes for including multilevel factors in linear models, as well as hierarchical schemes for including low-order interactions.
- The graphical lasso applies ℓ_1 penalties in the problem of edge selection in dependence graphs.
- Sparse principal components employ ℓ_1 penalties to produce components with many loadings zero. The same ideas are applied to discriminant analysis and canonical correlation analysis.
- The nuclear norm of a matrix is the sum of its singular values—a lasso penalty on matrices. Nuclear-norm regularization is popular in matrix completion for estimating missing entries in a matrix.

16.8 Notes and Details

Classical regression theory aimed for an unbiased estimate of each predictor variable's effect. Modern *wide* data sets, often with enormous numbers of predictors p, make that an untenable goal. The methods described here, by necessity, use shrinkage methods, biased estimation, and sparsity.

The lasso was introduced by Tibshirani (1996), and has spawned a great deal of research. The recent monograph by Hastie *et al.* (2015) gives a compact summary of some of the areas where the lasso and sparsity have been applied. The regression version of boosting was given in Hastie *et al.* (2009, Chapter 16), and inspired the *least-angle regression* algorithm (Efron *et al.*, 2004)—a new and more democratic version of forward-stepwise regression, as well as a fast algorithm for fitting the lasso. These authors showed under some conditions that each step of the LAR agorithm corresponds to one df; Zou *et al.* (2007) show that, with a fixed λ, the size of the active set is unbiased for the df for the lasso. Hastie *et al.* (2009) also view boosting as fitting a lasso regularization path in the high-dimensional space of trees.

Friedman *et al.* (2010) developed the pathwise coordinate-descent algorithm for generalized lasso problems, and provide the `glmnet` package for `R` (Friedman *et al.*, 2009). Strong rules for lasso screening are due to Tibshirani *et al.* (2012). Hastie *et al.* (2015, Chapter 3) show the similiarity between the ℓ_1 SVM and lasso logistic regression.

We now give some particular technical details on topics covered in the chapter.

†1 [p. 301] *Forward-stepwise computations.* Building up the forward-stepwise model can be seen as a guided Gram–Schmidt orthogonalization (QR decomposition). After step r, all $p-r$ variables not in the model are orthogonal to the r in the model, and the latter are in QR form. Then the next variable to enter is the one most correlated with the residuals. This is the one that will reduce the residual sum-of-squares the most, and one requires $p-r$ n-vector inner products to identify it. The regression is then updated trivially to accommodate the chosen one, which is then regressed out of the $p-r-1$ remaining variables.

†2 [p. 301] *Iteratively reweighted least squares (IRLS).* Generalized linear models (Chapter 8) are fit by maximum-likelihood, and since the log-likelihood is differentiable and concave, typically a Newton algorithm is used. The Newton algorithm can be recast as an iteratively reweighted linear regression algorithm (McCullagh and Nelder, 1989). At each iteration one computes a *working response* variable z_i, and a weight per observation w_i (both of which depend on the current parameter vector $\hat{\beta}$). Then the Newton update for $\hat{\beta}$ is obtained by a weighted least-squares fit of the z_i on the x_i with weights w_i (Hastie *et al.*, 2009, Section 4.4.1).

†3 [p. 301] *Forward-stepwise logistic regression computations.* Although the current model is in the form of a weighted least-squares fit, the $p-r$ variables not in the model cannot be kept orthogonal to those in the model (the weights keep changing!). However, since our current model will have performed a weighted QR decomposition (say), this orthogonalization can be obtained without too much cost. We will need $p-r$ multiplications of an $r \times n$ matrix with an n vector—$O((p-r) \cdot r \cdot n$ computations. An even simpler alternative for the selection is to use the size of the gradient of the

log-likelihood, which simply requires an inner product $|\langle \boldsymbol{y} - \hat{\boldsymbol{\mu}}_r, \boldsymbol{x}_j \rangle|$ for each omitted variable $\boldsymbol{x}_j$ (assuming all the variables are standardized to unit variance).

†4 [p. 306] *Best ℓ_1 interpolant.* If $p > n$, then another boundary solution becomes interesting for the lasso. For t sufficiently large, we will be able to achieve a perfect fit to the data, and hence a zero residual. There will be many such solutions, so it becomes interesting to find the perfect-fit solution with smallest value of t: the minimum-ℓ_1-norm perfect-fit solution. This requires solving a separate convex-optimization problem.

†5 [p. 313] *More on df.* When the search is easy in that a variable stands out as far superior, LAR takes a big step, and forward stepwise spends close to a unit df. On the other hand, when there is close competition, the LAR steps are small, and a unit df is spent for little progress, while forward stepwise can spend a fair bit more than a unit df (the price paid for searching). In fact, the df_j curve for forward stepwise can exceed p for $j < p$ (Jansen *et al.*, 2015).

†6 [p. 318] *Post-selection inference.* There has been a lot of activity around post-selection inference for lasso and related methods, all of it since 2012. To a large extent this was inspired by the work of Berk *et al.* (2013), but more tailored to the particular selection process employed by the lasso. For the debiasing approach we look to the work of Zhang and Zhang (2014), van de Geer *et al.* (2014) and Javanmard and Montanari (2014). The conditional inference approach began with Lockhart *et al.* (2014), and then was developed further in a series of papers (Lee *et al.*, 2016; Taylor *et al.*, 2015; Fithian *et al.*, 2014), with many more in the pipeline.

†7 [p. 319] *Selective inference software.* The example in Figure 16.10 was produced using the `R` package `selectiveInference` (Tibshirani *et al.*, 2016). Thanks to Rob Tibshirani for providing this example.

†8 [p. 319] *End-path behavior of ridge and lasso logistic regression for separable data.* The details here are somewhat technical, and rely on dual norms. Details are given in Hastie *et al.* (2015, Section 3.6.1).

†9 [p. 320] *LAR and boosting.* Least-squares boosting moves the "winning" coefficient in the direction of the correlation of its variable with the residual. The direction δ computed in step 3(a) of the LAR algorithm may have some components whose signs do not agree with their correlations, especially if the variables are very correlated. This can be fixed by a particular nonnegative least-squares fit to yield an exact path algorithm for iFS; details can be found in Efron *et al.* (2004).

17

Random Forests and Boosting

In the modern world we are often faced with enormous data sets, both in terms of the number of observations n and in terms of the number of variables p. This is of course good news—we have always said the more data we have, the better predictive models we can build. Well, we are there now—we have tons of data, and must figure out how to use it.

Although we can scale up our software to fit the collection of linear and generalized linear models to these behemoths, they are often too modest and can fall way short in terms of predictive power. A need arose for some general purpose tools that could scale well to these bigger problems, and exploit the large amount of data by fitting a much richer class of functions, almost automatically. Random forests and boosting are two relatively recent innovations that fit the bill, and have become very popular as "out-the-box" learning algorithms that enjoy good predictive performance. Random forests are somewhat more automatic than boosting, but can also suffer a small performance hit as a consequence.

These two methods have something in common: they both represent the fitted model by a sum of regression trees. We discuss trees in some detail in Chapter 8. A single regression tree is typically a rather *weak* prediction model; it is rather amazing that an ensemble of trees leads to the state of the art in black-box predictors!

We can broadly describe both these methods very simply.

Random forest Grow many deep regression trees to randomized versions of the training data, and average them. Here "randomized" is a wide-ranging term, and includes bootstrap sampling and/or subsampling of the observations, as well as subsampling of the variables.

Boosting Repeatedly grow shallow trees to the residuals, and hence build up an additive model consisting of a sum of trees.

The basic mechanism in random forests is variance reduction by averaging. Each deep tree has a high variance, and the averaging brings the variance down. In boosting the basic mechanism is bias reduction, although different flavors include some variance reduction as well. Both methods inherit all the good attributes of trees, most notable of which is variable selection.

17.1 Random Forests

Suppose we have the usual setup for a regression problem, with a training set consisting of an $n \times p$ data matrix $\boldsymbol{X}$ and an n-vector of responses $\boldsymbol{y}$. A tree (Section 8.4) fits a piecewise constant surface $\hat{r}(x)$ over the domain $\mathcal{X}$ by recursive partitioning. The model is built in a greedy fashion, each time creating two daughter nodes from a terminal node by defining a binary split using one of the available variables. The model can hence be represented by a binary tree. Part of the art in using regression trees is to know how deep to grow the tree, or alternatively how much to prune it back. Typically that is achieved using left-out data or cross-validation. Figure 17.1 shows a tree fit to the `spam` training data. The splitting variables and split points are indicated. Each node is labeled as `spam` or `ham` (not spam; see footnote 7 on page 115). The numbers beneath each node show misclassified/total. The overall misclassification error on the test data is 9.3%, which compares poorly with the performance of the lasso (Figure 16.9: 7.1% for linear lasso, 5.7% for lasso with interactions). The surface $\hat{r}(x)$ here is clearly complex, and by its nature represents a rather high-order interaction (the deepest branch is eight levels, and involves splits on eight different variables). Despite the promise to deliver interpretable models, this bushy tree is not easy to interpret. Nevertheless, trees have some desirable properties. The following lists some of the good and bad properties of trees.

▲ Trees automatically select variables; only variables used in defining splits are *in* the model.
▲ Tree-growing algorithms scale well to large n; growing a tree is a divide-and-conquer operation.
▲ Trees handle mixed features (quantitative/qualitative) seamlessly, and can deal with missing data.
▲ Small trees are easy to interpret.
▼ Large trees are not easy to interpret.
▼ Trees do not generally have good prediction performance.

Trees are inherently high-variance function estimators, and the bushier they are, the higher the variance. The early splits dictate the architecture of

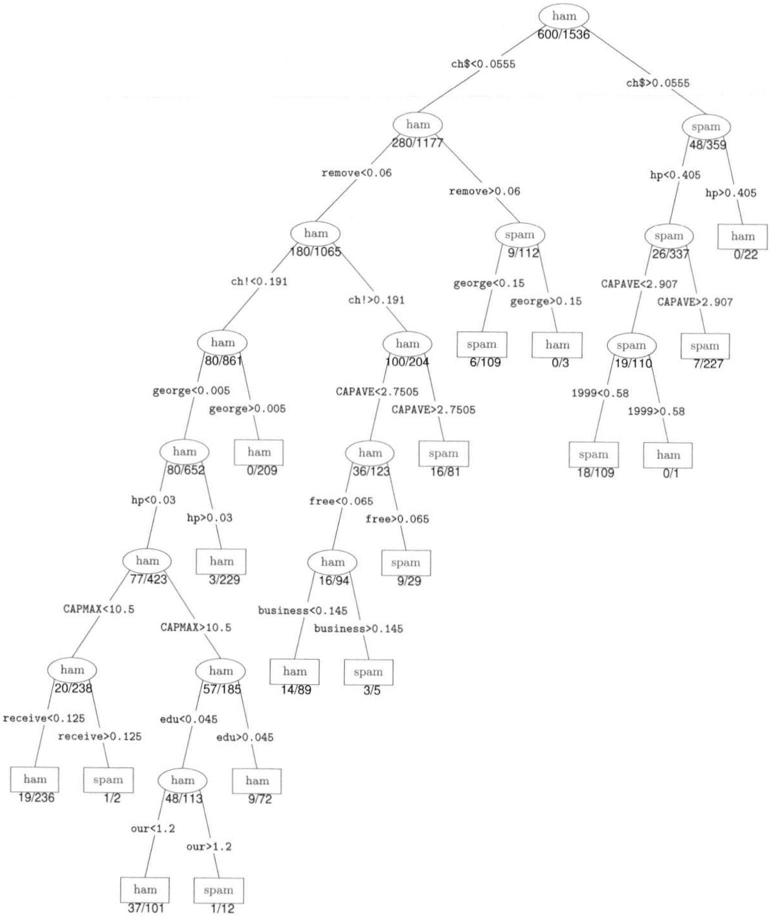

Figure 17.1 Regression tree fit to the binary **spam** data, a bigger version of Figure 8.7. The initial trained tree was far bushier than the one displayed; it was then *optimally* pruned using 10-fold cross-validation.

the tree. On the other hand, deep bushy trees localize the training data (using the variables that matter) to a relatively small region around the target point. This suggests low bias. The idea of random forests (and its predecessor bagging) is to grow many very bushy trees, and get rid of the variance by averaging. In order to benefit from averaging, the individual trees should not be too correlated. This is achieved by injecting some randomness into the tree-growing process. Random forests achieve this in two ways.

1 Bootstrap: each tree is grown to a bootstrap resampled training data set, which makes them different and somewhat decorrelates them.
2 Split-variable randomization: each time a split is to be performed, the search for the split variable is limited to a random subset of m of the p variables. Typical values of m are $\sqrt{p}$ or $p/3$.

When $m = p$, the randomization amounts to using only step 1, and was an earlier ancestor of random forests called *bagging*. In most examples the second level of randomization pays dividends.

Algorithm 17.1 RANDOM FOREST.

---

1 Given training data set $\boldsymbol{d} = (\boldsymbol{X}, \boldsymbol{y})$. Fix $m \leq p$ and the number of trees B.
2 For $b = 1, 2, \ldots, B$, do the following.
   (a) Create a bootstrap version of the training data $\boldsymbol{d}_b^*$, by randomly sampling the n rows with replacement n times. The sample can be represented by the bootstrap frequency vector $\boldsymbol{w}_b^*$.
   (b) Grow a maximal-depth tree $\hat{r}_b(x)$ using the data in $\boldsymbol{d}_b^*$, sampling m of the p features at random prior to making each split.
   (c) Save the tree, as well as the bootstrap sampling frequencies for each of the training observations.
3 Compute the random-forest fit at any prediction point x_0 as the average

$$\hat{r}_{\text{rf}}(x_0) = \frac{1}{B} \sum_{b=1}^{B} \hat{r}_b(x_0).$$

4 Compute the OOB$_i$ error for each response observation y_i in the training data, by using the fit $\hat{r}_{\text{rf}}^{(i)}$, obtained by averaging only those $\hat{r}_b(x_i)$ for which observation i was *not* in the bootstrap sample. The overall OOB error is the average of these OOB$_i$.

Algorithm 17.1 gives some of the details; some more are given in the technical notes.† †1

Random forests are easy to use, since there is not much tuning needed. The package `randomForest` in **R** sets as a default $m = \sqrt{p}$ for classification trees, and $m = p/3$ for regression trees, but one can use other values. With $m = 1$ the split variable is completely random, so all variables get a chance. This will decorrelate the trees the most, but can create bias, somewhat similar to that in ridge regression. Figure 17.2 shows the

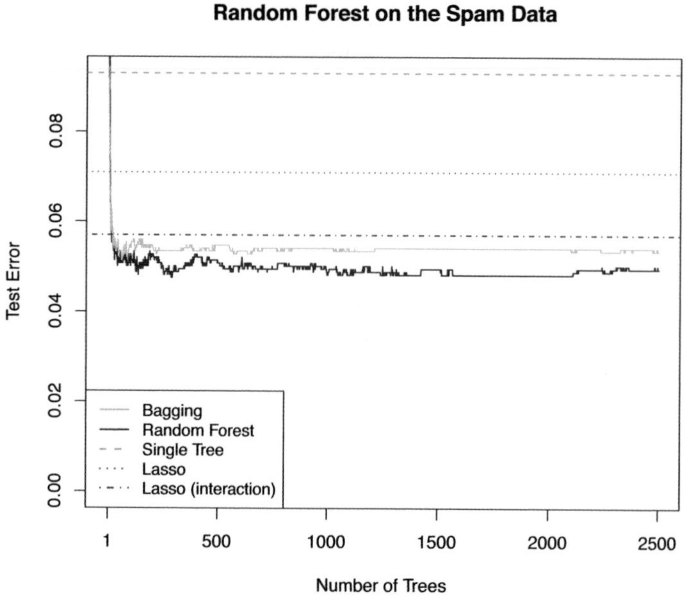

Figure 17.2 Test misclassification error of random forests on the **spam** data, as a function of the number of trees. The red curve selects $m = 7$ of the $p = 57$ features at random as candidates for the split variable, each time a split is made. The blue curve uses $m = 57$, and hence amounts to bagging. Both bagging and random forests outperform the lasso methods, and a single tree.

misclassification performance of a random forest on the **spam** test data, as a function of the number of trees averaged. We see that in this case, after a relatively small number of trees (500), the error levels off. The number B of trees averaged is not a real tuning parameter; as with the bootstrap (Chapters 10 and 11), we need a sufficient number for the estimate to stabilize, but cannot overfit by having too many.

Random forests have been described as adaptive nearest-neighbor estimators—adaptive in that they select predictors. A k-nearest-neighbor estimate finds the k training observations closest in feature space to the target point x_0, and averages their responses. Each tree in the random forest drills down by recursive partitioning to pure terminal nodes, often consisiting of a single observation. Hence, when evaluating the prediction from each tree, $\hat{r}_b(x_0) = y_\ell$ for some ℓ, and for many of the trees this could be the same

ℓ. From the whole collection of B trees, the number of distinct ℓs can be fairly small. Since the partitioning that reaches the terminal nodes involves only a subset of the predictors, the neighborhoods so defined are adaptive.

Out-of-Bag Error Estimates

Random forests deliver cross-validated error estimates at virtually no extra cost. The idea is similar to the bootstrap error estimates discussed in Chapter 10. The computation is described in step 4 of Algorithm 17.1. In making the prediction for observation pair (x_i, y_i), we average all the random-forest trees $\hat{r}_b(x_i)$ for which that pair is not in the corresponding bootstrap sample:

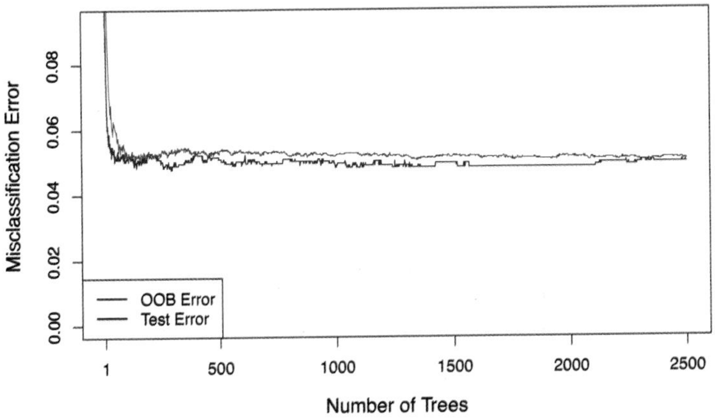

Figure 17.3 Out-of-bag misclassification error estimate for the `spam` data (blue) versus the test error (red), as a function of the number of trees.

$$\hat{r}_{\text{rf}}^{(i)}(x_i) = \frac{1}{B_i} \sum_{b\,:\,w_{bi}^*=0} \hat{r}_b(x_i), \tag{17.1}$$

where B_i is the number of times observation i was not in the bootstrap sample (with expected value $e^{-1}B \approx 0.37B$). We then compute the OOB error estimate

$$\text{err}_{\text{OOB}} = \frac{1}{n}\sum_{i=1}^{n} L[y_i, \hat{r}_{\text{rf}}^{(i)}(x_i)], \tag{17.2}$$

where L is the loss function of interest, such as misclassification or squared-error loss. If B is sufficiently large (about three times the number needed for the random forest to stabilize), we can see that the OOB error estimate is equivalent to leave-one-out cross-validation error.

Standard Errors

We can use very similar ideas to estimate the variance of a random-forest prediction, using the *jackknife* variance estimator (see (10.6) in Chapter 10). If $\hat{\theta}$ is a statistic estimated using all n training observations, then the jackknife estimate of the variance of $\hat{\theta}$ is given by

$$\widehat{\mathrm{V}}_{\mathrm{jack}}(\hat{\theta}) = \frac{n-1}{n}\sum_{i=1}^{n}\left(\hat{\theta}_{(i)} - \hat{\theta}_{(\cdot)}\right)^2, \qquad (17.3)$$

where $\hat{\theta}_{(i)}$ is the estimate using all but observation i, and $\hat{\theta}_{(\cdot)} = \frac{1}{n}\sum_i \hat{\theta}_{(i)}$.

The natural jackknife variance estimate for a random-forest prediction at x_0 is obtained by simply plugging into this formula:

$$\widehat{\mathrm{V}}_{\mathrm{jack}}(\hat{r}_{\mathrm{rf}}(x_0)) = \frac{n-1}{n}\sum_{i=1}^{n}\left(\hat{r}_{\mathrm{rf}}^{(i)}(x_0) - \hat{r}_{\mathrm{rf}}(x_0)\right)^2. \qquad (17.4)$$

This formula is derived under the $B = \infty$ setting, in which case $\hat{r}_{\mathrm{rf}}(x_0)$ is an expectation under bootstrap sampling, and hence is free of Monte Carlo variability. This also makes the distinction clear: we are estimating the sampling variability of a random-forest prediction $\hat{r}_{\mathrm{rf}}(x_0)$, as distinct from any Monte Carlo variation. In practice B is finite, and expression (17.4) will have Monte Carlo bias and variance. All of the $\hat{r}_{\mathrm{rf}}^{(i)}(x_0)$ are based on B bootstrap samples, and they are hence noisy versions of their expectations. Since the n quantities summed in (17.4) are squared, by Jensen's inequality we will have positive bias (and it turns out that this bias dominates the Monte Carlo variance). Hence one would want to use a much larger value of B when estimating variances, than was used in the original random-forest fit. Alternatively, one can use the same B bootstrap samples as were used to fit the random forest, along with a *bias-corrected* version of the
†2 jackknife variance estimate:†

$$\widehat{\mathrm{V}}^{\mathrm{u}}_{\mathrm{jack}}(\hat{r}_{\mathrm{rf}}(x_0)) = \widehat{\mathrm{V}}_{\mathrm{jack}}(\hat{r}_{\mathrm{rf}}(x_0)) - (e-1)\frac{n}{B}\hat{v}(x_0), \qquad (17.5)$$

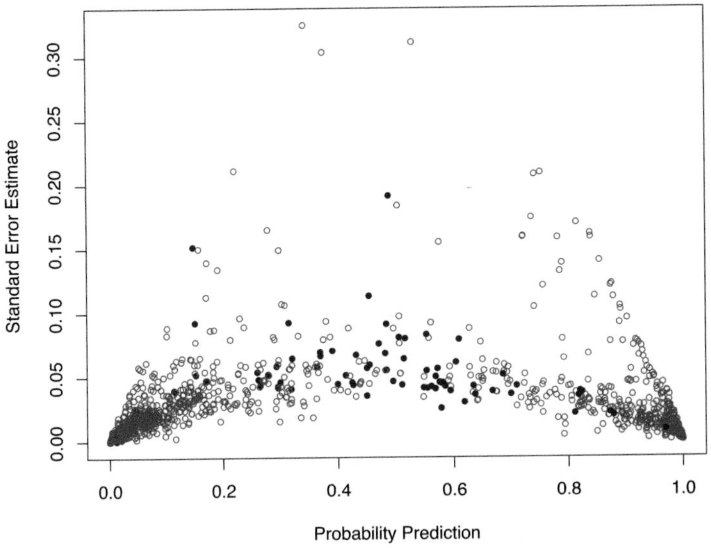

Figure 17.4 Jackknife Standard Error estimates (with bias correction) for the probability estimates in the `spam` test data. The points labeled red were misclassifications, and tend to concentrate near the decision boundary (0.5).

where $e = 2.718\ldots$, and

$$\hat{v}(x_0) = \frac{1}{B}\sum_{b=1}^{B}\left(\hat{r}_b(x_0) - \hat{r}_{\text{rf}}(x_0)\right)^2, \tag{17.6}$$

the bootstrap estimate of the variance of a single random-forest tree. All these quantities are easily computed from the output of a random forest, so they are immediately available. Figure 17.4 shows the predicted probabilities and their jackknife estimated standard errors for the `spam` test data. The estimates near the decision boundary tend to have higher standard errors.

Variable-Importance Plots

A random forest is something of a black box, giving good predictions but usually not much insight into the underlying surface it has fit. Each random-forest tree $\hat{r}_b$ will have used a subset of the predictors as splitting variables, and each tree is likely to use overlapping but not necessarily

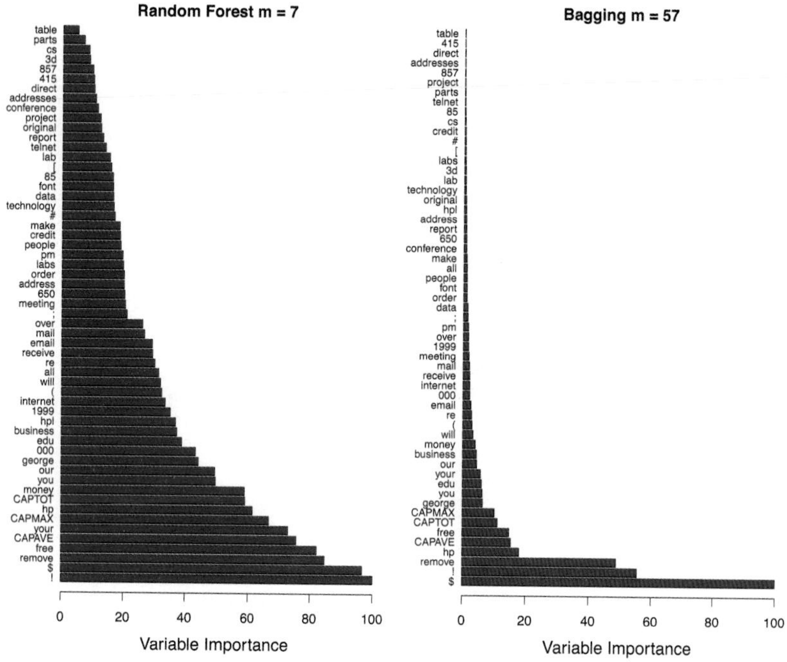

Figure 17.5 Variable-importance plots for random forests fit to the spam data. On the left we have the $m = 7$ random forest; due to the split-variable randomization, it spreads the importance among the variables. On the right is the $m = 57$ random forest or bagging, which focuses on a smaller subset of the variables.

identical subsets. One might conclude that any variable never used in any of the trees is unlikely to be important, but we would like a method of assessing the relative importance of variables that are included in the ensemble. Variable-importance plots fit this bill. Whenever a variable is used in a tree, the algorithm logs the decrease in the split-criterion due to this split. These are accumulated over all the trees, for each variable, and summarized as relative importance measures. Figure 17.5 demonstrates this on the `spam` data. We see that the $m = 7$ random forest, by virtue of the split-variable randomization, spreads the importance out much more than bagging, which always gets to pick the best variable for splitting. In this sense small m has some similarity to ridge regression, which also tends to share the coefficients evenly among correlated variables.

17.2 Boosting with Squared-Error Loss

Boosting was originally proposed as a means for improving the performance of "weak learners" in binary classification problems. This was achieved through resampling training points—giving more weight to those which had been misclassified—to produce a new classifier that would boost the performance in previously problematic areas of feature space. This process is repeated, generating a stream of classifiers, which are ultimately combined through voting[1] to produce the final classifier. The prototypical weak learner was a decision tree.

Boosting has evolved since this earliest invention, and different flavors are popular in statistics, computer science, and other areas of pattern recognition and prediction. We focus on the version popular in statistics—gradient boosting—and return to this early version later in the chapter. Algorithm 17.2

Algorithm 17.2 GRADIENT BOOSTING WITH SQUARED-ERROR LOSS.

---

1 Given a training sample $\boldsymbol{d} = (\boldsymbol{X}, \boldsymbol{y})$. Fix the number of steps B, the shrinkage factor ϵ and the tree depth d. Set the initial fit $\widehat{G}_0 \equiv 0$, and the residual vector $\boldsymbol{r} = \boldsymbol{y}$.

2 For $b = 1, 2, \ldots, B$ repeat:

  (a) Fit a regression tree $\tilde{g}_b$ to the data $(\boldsymbol{X}, \boldsymbol{r})$, grown best-first to depth d: this means the total number of splits are d, and each successive split is made to that terminal node that yields the biggest reduction in residual sum of squares.

  (b) Update the fitted model with a shrunken version of $\tilde{g}_b$: $\widehat{G}_b = \widehat{G}_{b-1} + \hat{g}_b$, with $\hat{g}_b = \epsilon \cdot \tilde{g}_b$.

  (c) Update the residuals accordingly: $r_i = r_i - \hat{g}_b(x_i)$, $i = 1, \ldots, n$.

3 Return the sequence of fitted functions $\widehat{G}_b$, $b = 1, \ldots, B$.

---

gives the most basic version of gradient boosting, for squared-error loss. This amounts to building a model by repeatedly fitting a regression tree to the residuals. Importantly, the tree is typically quite small, involving a small number d of splits—it is indeed a *weak learner*. After each tree has been grown to the residuals, it is shrunk down by a factor ϵ before it is added to the current model; this is a means of slowing the learning process. Despite the obvious similarities with a random forest, boosting is different in a fundamental way. The trees in a random forest are identically

[1] Each classifier $\hat{c}_b(x_0)$ predicts a class label, and the class with the most "votes" wins.

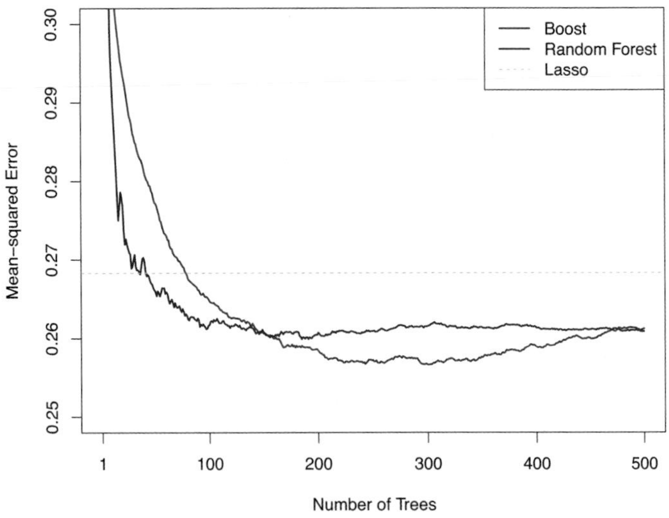

Figure 17.6 Test performance of a boosted regression-tree model fit to the **ALS** training data, with $n = 1197$ and $p = 369$. Shown is the mean-squared error on the 625 designated test observations, as a function of the number of trees. Here the depth $d = 4$ and $\epsilon = 0.02$. Boosting achieves a lower test MSE than a random forest. We see that as the number of trees B gets large, the test error for boosting starts to increase—a consequence of overfitting. The random forest does not overfit. The dotted blue horizontal line shows the best performance of a linear model, fit by the lasso. The differences are less dramatic than they appear, since the vertical scale does not extend to zero.

distributed—the same (random) treatment is repeatedly applied to the same data. With boosting, on the other hand, each tree is trying to amend errors made by the ensemble of previously grown trees. The number of terms B is important as well, because unlike random forests, a boosted regression model can overfit if B is too large. Hence there are three tuning parameters, B, d and ϵ, and each can change the performance of a boosted model, sometimes considerably.

†3 Figure 17.6 shows the test performance of boosting on the **ALS** data.† These data represent measurements on patients with *amyotrophic lateral sclerosis* (Lou Gehrig's disease). The goal is to predict the rate of progression of an ALS functional rating score (**FRS**). There are 1197 training

measurements on 369 predictors and the response, with a corresponding test set of size 625 observations.

As is often the case, boosting slightly outperforms a random forest here, but at a price. Careful tuning of boosting requires considerable extra work, with time-costly rounds of cross-validation, whereas random forests are almost automatic. In the following sections we explore in more detail some of the tuning parameters. The `R` package `gbm` implements gradient boosting, with some added bells and whistles. By default it grows each new tree on a 50% random sub-sample of the training data. Apart from speeding up the computations, this has a similar effect to bagging, and results in some variance reduction in the ensemble.

We can also compute a variable-importance plot, as we did for random forests; this is displayed in Figure 17.7 for the `ALS` data. Only 267 of the 369 variables were ever used, with one variable `Onset.Delta` standing out ahead of the others. This measures the amount of time that has elapsed since the patient was first diagnosed with ALS, and hence a larger value will indicate a slower progression rate.

Tree Depth and Interaction Order

Tree depth d is an important parameter for gradient boosted models, and the right choice will depend on the data at hand. Here depth $d = 4$ appears to be a good choice on the test data. Without test data, we could use cross-validation to make the selection. Apart from a general complexity measure, tree depth also controls the interaction order of the model.[2] The easiest case is with $d = 1$, where each tree consists of a single split (a stump). Suppose we have a fitted boosted model $\widehat{G}_B(x)$, using B trees. Denote by $\mathcal{B}_j \subseteq \mathcal{B} = \{1, 2, \ldots, B\}$ the indices of the trees that made the single split using variable j, for $j = 1, \ldots, p$. These $\mathcal{B}_j$ are disjoint (some $\mathcal{B}_\ell$ can be

[2] A $(k-1)$th-order interaction is also known as a k-way interaction. Hence an order-one interaction model has two-way interactions, and an order-zero model is additive.

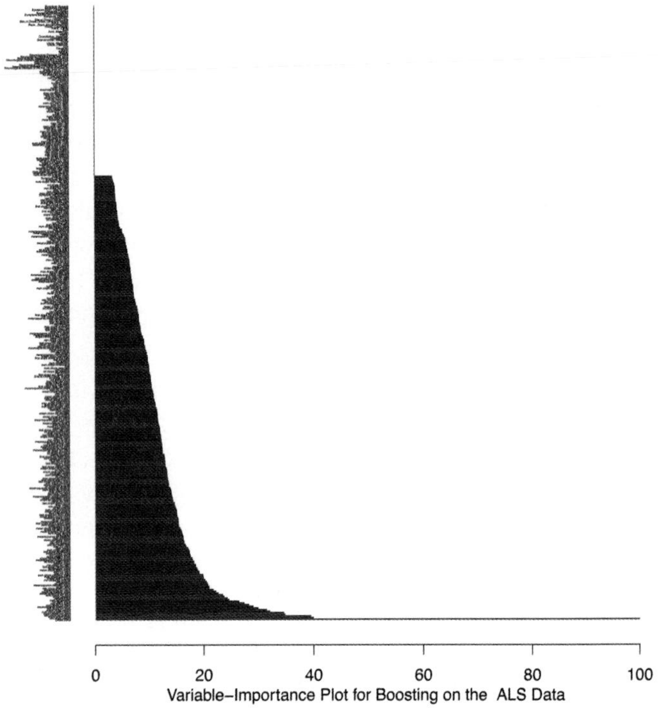

Figure 17.7 Variable importance plot for the **ALS** data. Here 267 of the 369 variables were used in the ensemble. There are too many variables for the labels to be visible, so this plot serves as a visual guide. Variable **Onset.Delta** has relative importance 100 (the lowest red bar), more than double the next two at around 40 (**last.slope.weight** and **alsfrs.score.slope**). However, the importances drop off slowly, suggesting that the model requires a significant fraction of the variables.

empty), and $\bigcup_{j=1}^{p} \mathcal{B}_j = \mathcal{B}$. Then we can write

$$\begin{aligned} \widehat{G}_B(x) &= \sum_{b=1}^{B} \hat{g}_b(x) \\ &= \sum_{j=1}^{p} \sum_{b \in \mathcal{B}_j} \hat{g}_b(x) \\ &= \sum_{j=1}^{p} \hat{f}_j(x_j). \end{aligned} \tag{17.7}$$

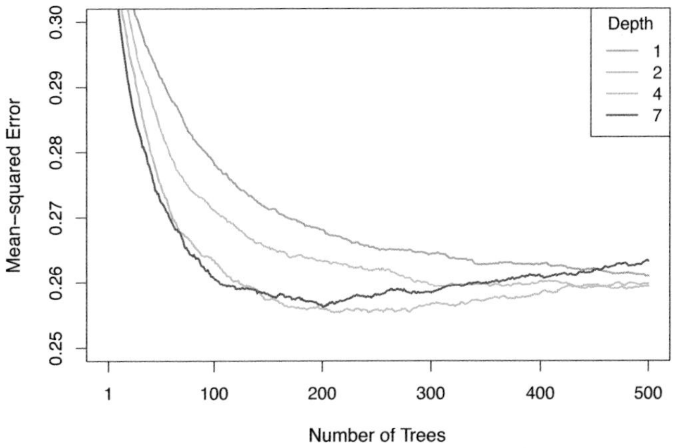

Figure 17.8 `ALS` test error for boosted models with different depth parameters d, and all using the same shrinkage parameter $\epsilon = 0.02$. It appears that $d = 1$ is inferior to the rest, with $d = 4$ about the best. With $d = 7$, overfitting begins around 200 trees, with $d = 4$ around 300, while neither of the other two show evidence of overfitting by 500 trees.

Hence boosted stumps fits an additive model, but in a fully adaptive way. It selects variables, and also selects how much action to devote to each variable. We return to additive models in Section 17.5. Figure 17.9 shows the three functions with highest relative importance. The first function confirms that a longer time since diagnosis (more negative `Onset.Delta`) predicts a slower decline. `last.slope.weight` is the difference in body weight at the last two visits—again positive is good. Likewise for `alsfrs.score.slope`, which measures the local slope of the `FRS` score after the first two visits.

In a similar way, boosting with $d = 2$ fits a two-way interaction model; each tree involves at most two variables. In general, boosting with $d = k$ leads to a $(k - 1)$th-order interaction model. Interaction order is perhaps a more natural way to think of model complexity.

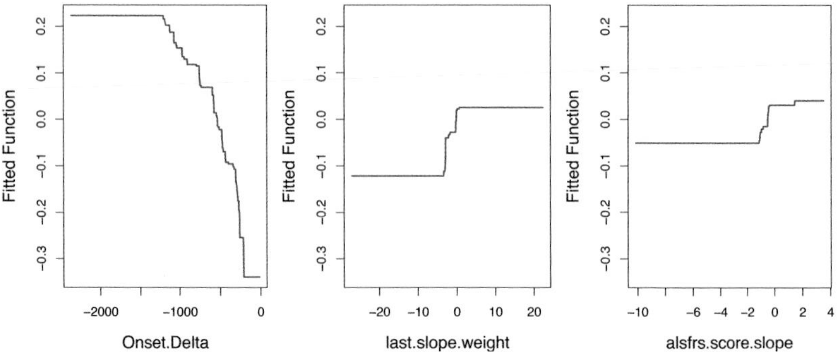

Figure 17.9 Three of the fitted functions (17.7) for the ALS data, in a boosted stumps model ($d = 1$), each centered to average zero over the training data. In terms of the outcome, bigger is better (slower decline in FRS). The first function confirms that a longer time since diagnosis (more negative value of `Onset.Delta`) predicts a slower decline. The variable `last.slope.weight` is the difference in body weight at the last two visits—again positive is good. Likewise for `alsfrs.score.slope`, which measures the local slope of the FRS score after the first two visits.

Shrinkage

The shrinkage parameter ϵ controls the rate at which boosting fits—and hence overfits—the data. Figure 17.10 demonstrates the effect of shrinkage on the ALS data. The under-shrunk ensemble (red) quickly overfits the data, leading to poor validation error. The blue ensemble uses a shrinkage parameter 20 times smaller, and reaches a lower validation error. The downside of a very small shrinkage parameter is that it can take many trees to adequately fit the data. On the other hand, the shrunken fits are smoother, take much longer to overfit, and hence are less sensitive to the stopping point B.

17.3 Gradient Boosting

We now turn our attention to boosting models using other than square-error loss. We focus on the family of generalized models generated by the exponential family of response distributions (see Chapter 8). The most popular and relevant in this class is logistic regression, where we are interested in modeling $\mu(x) = \Pr(Y = 1|X = x)$ for a Bernoulli response variable.

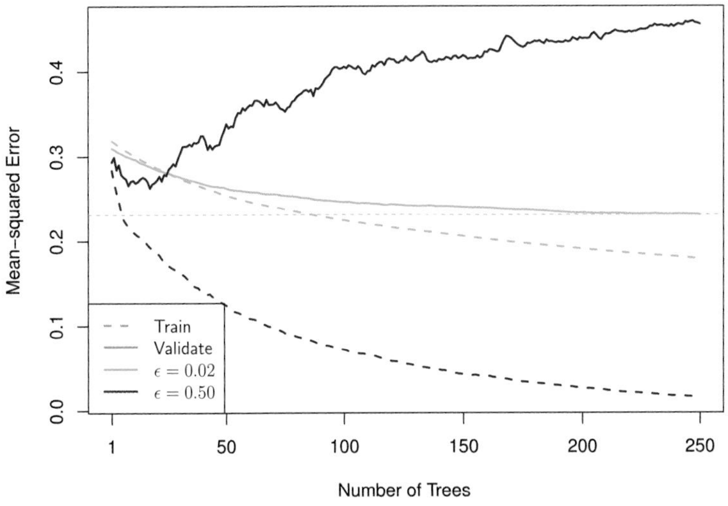

Figure 17.10 Boosted $d = 3$ models with different shrinkage parameters, fit to a subset of the ALS data. The solid curves are validation errors, the dashed curves training errors, with red for $\epsilon = 0.5$ and blue for $\epsilon = 0.02$. With $\epsilon = 0.5$, the training error drops rapidly with the number of trees, but the validation error starts to increase rapidly after an initial decrease. With $\epsilon = 0.02$ (25 times smaller), the training error drops more slowly. The validation error also drops more slowly, but reaches a lower minimum (the horizontal dotted line) than the $\epsilon = 0.5$ case. In this case, the slower learning has paid off.

The idea is to fit a model of the form

$$\lambda(x) = G_B(x) = \sum_{b=1}^{B} g_b(x; \gamma_b), \tag{17.8}$$

where $\lambda(x)$ is the natural parameter in the conditional distribution of $Y|X = x$, and the $g_b(x;\gamma_b)$ are simple functions such as shallow trees. Here we have indexed each function by a parameter vector γ_b; for trees these would capture the identity of the split variables, their split values, and the constants in the terminal nodes. In the case of the Bernoulli response, we have

$$\lambda(x) = \log\left(\frac{\Pr(Y = 1|X = x)}{\Pr(Y = 0|X = x)}\right), \tag{17.9}$$

the logit link function that relates the mean to the natural parameter. In general, if $\mu(x) = \mathrm{E}(Y|X = x)$ is the conditional mean, we have $\eta[\mu(x)] = \lambda(x)$, where η is the monotone *link function*.

Algorithm 17.3 outlines a general strategy for building a model by forward stagewise fitting. L is the loss function, such as the negative log-likelihood for Bernoulli responses, or squared-error for Gaussian responses. Although we are thinking of trees for the simple functions $g(x;\gamma)$, the ideas generalize. This algorithm is easier to state than to implement. For

---

Algorithm 17.3 GENERALIZED BOOSTING BY FORWARD-STAGEWISE FITTING

---

1 Define the class of functions $g(x;\gamma)$. Start with $\hat{G}_0(x) = 0$, and set B and the shrinkage parameter $\epsilon > 0$.

2 For $b = 1, \ldots, B$ repeat the following steps.

(a) Solve

$$\hat{\gamma}_b = \arg\min_{\gamma} \sum_{i=1}^{n} L\left(y_i, \widehat{G}_{b-1}(x_i) + g(x_i;\gamma)\right)$$

(b) Update $\widehat{G}_b(x) = \widehat{G}_{b-1}(x) + \hat{g}_b(x)$, with $\hat{g}_b(x) = \epsilon \cdot g(x;\hat{\gamma}_b)$.

3 Return the sequence $\widehat{G}_b(x),\ b = 1, \ldots, B$.

---

squared-error loss, at each step we need to solve

$$\underset{\gamma}{\text{minimize}} \sum_{i=1}^{n} (r_i - g(x_i;\gamma))^2, \tag{17.10}$$

with $r_i = y_i - \widehat{G}_{b-1}(x_i),\ i = 1, \ldots, n$. If $g(\cdot;\gamma)$ represents a depth-d tree, (17.10) is still difficult to solve. But here we can resort to the usual greedy heuristic, and grow a depth-d tree to the residuals by the usual top-down splitting, as in step 2(a) of Algorithm 17.2. Hence in this case, we have exactly the squared-error boosting Algorithm 17.2. For more general loss functions, we rely on one more heuristic for solving step 2(a), inspired by gradient descent. Algorithm 17.4 gives the details. The idea is to perform functional gradient descent on the loss function, in the n-dimensional space of the fitted vector. However, we want to be able to evaluate our new function everywhere, not just at the n original values x_i. Hence once the (negative) gradient vector has been computed, it is approximated by a depth-d tree (which *can* be evaluated everywhere). Taking a step of length ϵ down

the gradient amounts to adding ϵ times the tree to the current function.† †4
Gradient boosting is quite general, and can be used with any differentiable

Algorithm 17.4 GRADIENT BOOSTING

1 Start with $\hat{G}_0(x) = 0$, and set B and the shrinkage parameter $\epsilon > 0$.
2 For $b = 1, \ldots, B$ repeat the following steps.
   (a) Compute the pointwise negative gradient of the loss function at the current fit:
$$r_i = -\left.\frac{\partial L(y_i, \lambda_i)}{\partial \lambda_i}\right|_{\lambda_i = \hat{G}_{b-1}(x_i)}, \quad 1 = 1, \ldots, n.$$
   (b) Approximate the negative gradient by a depth-d tree by solving
$$\underset{\gamma}{\text{minimize}} \sum_{i=1}^{n} (r_i - g(x_i; \gamma))^2 .$$
   (c) Update $\hat{G}_b(x) = \hat{G}_{b-1}(x) + \hat{g}_b(x)$, with $\hat{g}_b(x) = \epsilon \cdot g(x; \hat{\gamma}_b)$.
3 Return the sequence $\hat{G}_b(x),\ b = 1, \ldots, B$.

loss function. The `R` package `gbm` implements Algorithm 17.4 for a variety of loss functions, including squared-error, binomial (Bernoulli), Laplace (ℓ_1 loss), multinomial, and others. Included as well is the partial likelihood for the Cox proportional hazards model (Chapter 9). Figure 17.11 compares the misclassification error of boosting on the `spam` data, with that of random forests and bagging. Since boosting has more tuning parameters, a careful comparison must take these into account. Using the McNemar test we would conclude that boosting and random forest are not significantly different from each other, but both outperform bagging.

17.4 Adaboost: the Original Boosting Algorithm

The original proposal for boosting looked quite different from what we have presented so far. Adaboost was developed for the two-class classification problem, where the response is coded as -1/1. The idea was to fit a sequence of classifiers to modified versions of the training data, where the modifications give more weight to misclassified points. The final classification is by weighted majority vote. The details are rather specific, and are given in Algorithm 17.5. Here we distinguish a classifier $C(x) \in \{-1, 1\}$, which returns a class label, rather than a probability. Algorithm 17.5 gives

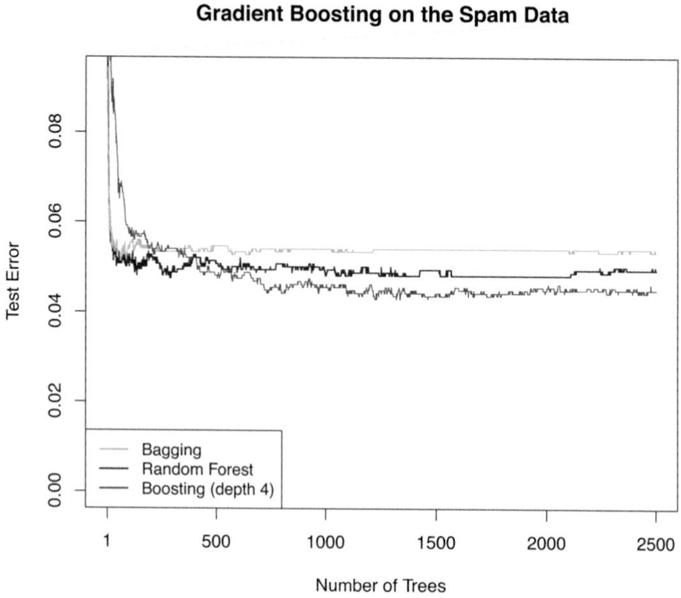

Figure 17.11 Test misclassification for gradient boosting on the `spam` data, compared with a random forest and bagging. Although boosting appears to be better, it requires crossvaldiation or some other means to estimate its tuning parameters, while the random forest is essentially automatic.

the *Adaboost.M1* algorithm. Although the classifier in step 2(a) can be arbitrary, it was intended for *weak learners* such as shallow trees. Steps 2(c)–(d) look mysterious. Its easy to check that, with the reweighted points, the classifier $\hat{c}_b$ just learned would have weighted error 0.5, that of a coin flip. We also notice that, although the individual classifiers $\hat{c}_b(x)$ produce values ± 1, the ensemble $\widehat{G}_b(x)$ takes values in $\mathbb{R}$.

It turns out that the Adaboost Algorithm 17.5 fits a logistic regression model via a version of the general boosting Algorithm 17.3, using an exponential loss function. The functions $\widehat{G}_b(x)$ output in step 3 of Algorithm 17.5 are estimates of (half) the logit function $\lambda(x)$.

To show this, we first motivate the exponential loss, a somewhat unusual choice, and show how it is linked to logistic regression. For a -1/1 response y and function $f(x)$, the exponential loss is defined as $L_E(y, f(x)) = \exp[-yf(x)]$. A simple calculation shows that the solution to the (condi-

Algorithm 17.5 Adaboost

1 Initialize the observation weights $w_i = 1/n,\ i = 1, \ldots, n$.
2 For $b = 1, \ldots, B$ repeat the following steps.
  (a) Fit a classifier $\hat{c}_b(x)$ to the training data, using observation weights w_i.
  (b) Compute the weighted misclassification error for $\hat{c}_b$:
$$\text{err}_b = \frac{\sum_{i=1}^n w_i I[y_i \neq \hat{c}_b(x_i)]}{\sum_{i=1}^n w_i}.$$
  (c) Compute $\alpha_b = \log[(1 - \text{err}_b)/\text{err}_b]$.
  (d) Update the weights $w_i \leftarrow w_i \cdot \exp\left(\alpha_b \cdot I[y_i \neq c_b(x_i)]\right),\ i = 1, \ldots, n$.
3 Output the sequence of functions $\widehat{G}_b(x) = \sum_{\ell=1}^b \alpha_m \hat{c}_\ell(x)$ and corresponding classifiers $\widehat{C}_b(x) = \text{sign}\left[\widehat{G}_b(x)\right],\ b = 1, \ldots, B$.

tional) population minimization problem

$$\underset{f(x)}{\text{minimize}}\, \text{E}[e^{-yf(x)} \mid x] \tag{17.11}$$

is given by

$$f(x) = \frac{1}{2} \log\left(\frac{\Pr(y = +1|x)}{\Pr(y = -1|x)}\right). \tag{17.12}$$

Inverting, we get

$$\Pr(y = +1|x) = \frac{e^{f(x)}}{e^{-f(x)} + e^{f(x)}} \text{ and } \Pr(y = -1|x) = \frac{e^{-f(x)}}{e^{-f(x)} + e^{f(x)}}, \tag{17.13}$$

a perfectly reasonable (and symmetric) model for a probability. The quantity $yf(x)$ is known as the margin (see also Chapter 19); if the margin is positive, the classification using $C_f(x) = \text{sign}(f(x))$ is correct for y, else it is incorrect if the margin is negative. The magnitude of $yf(x)$ is proportional to the (signed) distance of x from the classification boundary (exactly for linear models, approximately otherwise). For -1/1 data, we can also write the (negative) binomial log-likelihood in terms of the margin.

Using (17.13) we have

$$\begin{aligned} L_B(y, f(x)) &= -\{I(y = -1)\log \Pr(y = -1|x) \\ &\qquad + I(y = +1)\log \Pr(y = +1|x)\} \\ &= \log\left(1 + e^{-2yf(x)}\right). \end{aligned} \tag{17.14}$$

$\mathrm{E}\left[\log\left(1 + e^{-2yf(x)}\right) \mid x\right]$ also has population minimizer $f(x)$ equal to half the logit (17.12).[3] Figure 17.12 compares the exponential loss function with this binomial loss. They both asymptote to zero in the right tail—the area of correct classification. In the left tail, the binomial loss asymptotes to a linear function, much less severe than the exponential loss.

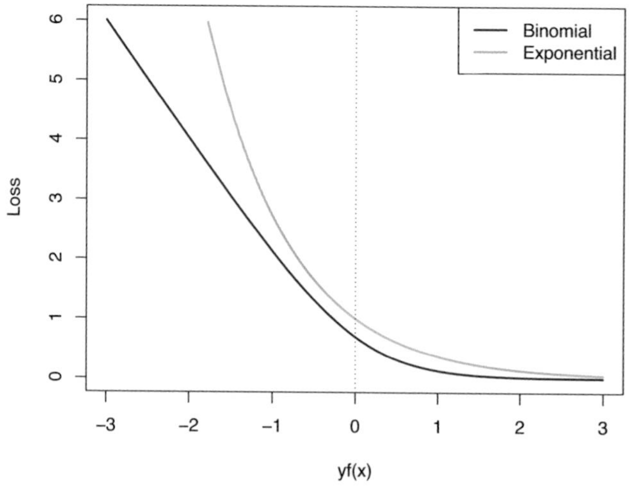

Figure 17.12 Exponential loss used in Adaboost, versus the binomial loss used in the usual logistic regression. Both estimate the logit function. The exponential left tail, which punishes misclassifications, is much more severe than the asymptotically linear tail of the binomial.

The exponential loss simplifies step 2(a) in the gradient boosting Algo-

[3] The half comes from the symmetric representation we use.

rithm 17.3.

$$\begin{aligned}\sum_{i=1}^{n} L_E\left(y_i, \widehat{G}_{b-1}(x_i) + g(x_i;\gamma)\right) &= \sum_{i=1}^{n} \exp[-y_i(\widehat{G}_{b-1}(x_i) + g(x_i;\gamma)] \\ &= \sum_{i=1}^{n} w_i \exp[-y_i g(x_i;\gamma)] \qquad (17.15) \\ &= \sum_{i=1}^{n} w_i L_E\left(y_i, g(x_i;\gamma)\right),\end{aligned}$$

with $w_i = \exp[-y_i \widehat{G}_{b-1}(x_i)]$. This is just a weighted exponential loss with the past history encapsulated in the observation weight w_i (see step 2(a) in Algorithm 17.5). We give some more details in the chapter endnotes on how this reduces to the Adaboost algorithm.† †5

The Adaboost algorithm achieves an error rate on the `spam` data comparable to binomial gradient boosting.

17.5 Connections and Extensions

Boosting is a general nonparametric function-fitting algorithm, and shares attributes with a variety of existing methods. Here we relate boosting to two different approaches: generalized additive models and the lasso of Chapter 16.

Generalized Additive Models

Boosting fits additive, low-order interaction models by a forward stagewise strategy. Generalized additive models (GAMs) are a predecessor, a semi-parametric approach toward nonlinear function fitting. A GAM has the form

$$\lambda(x) = \sum_{j=1}^{p} f_j(x_j), \qquad (17.16)$$

where again $\lambda(x) = \eta[\mu(x)]$ is the natural parameter in an exponential family. The attraction of a GAM is that the components are interpretable and can be visualized, and they can move us a big step up from a linear model.

There are many ways to specify and fit additive models. For the f_j, we could use parametric functions (e.g. polynomials), fixed-knot regression splines, or even linear functions for some terms. Less parametric options are smoothing splines and local regression (see Section 19.8). In the case of squared-error loss (the Gaussian case), there is a natural set of *backfitting* equations for fitting a GAM:

$$\hat{f}_j \leftarrow \mathcal{S}_j(\boldsymbol{y} - \sum_{\ell \neq j} \hat{\boldsymbol{f}}_\ell), \; j = 1, \ldots, p. \tag{17.17}$$

Here $\hat{\boldsymbol{f}}_\ell = [\hat{f}_\ell(x_{1\ell}), \ldots, (\hat{f}_\ell(x_{n\ell})]'$ is the n-vector of fitted values for the current estimate of function f_ℓ. Hence the term in parentheses is a *partial residual*, removing all the current function fits from $\boldsymbol{y}$ except the one about to be updated. $\mathcal{S}_j$ is a smoothing operator derived from variable $\boldsymbol{x}_j$ that gets applied to this residual and delivers the next estimate for function f_ℓ. Backfitting starts with all the functions zero, and then cycles through these equations for $j = 1, 2, \ldots, p, 1, 2, \ldots$ in a block-coordinate fashion, until all the functions stabilize.

The first pass through all the variables is similar to the regression boosting Algorithm 17.2, where each new function takes the residuals from the past fits, and models them using a tree (for $\mathcal{S}_j$). The difference is that boosting never goes back and fixes up past functions, but fits in a forward-stagewise fashion, leaving all past functions alone. Of course, with its adaptive fitting mechanism, boosting can select the same variables as used before, and thereby update that component of the fit. Boosting with stumps (single-split trees, see the discussion on tree depth on 335 in Section 17.2) can hence be seen as an adaptive way for fitting an additive model, that simultaneously performs variable selection and allows for different amounts of smoothing for different variables.

Boosting and the Lasso

In Section 16.7 we drew attention to the close connection between the forward-stagewise fitting of boosting (with shrinkage) and the lasso, via infinitesimal forward-stagewise regression. Here we take this a step further, by using the lasso as a post-processor for boosting (or random forests).

Boosting with shrinkage does a good job in building a prediction model, but at the end of the day can involve a lot of trees. Because of the shrinkage, many of these trees could be similar to each other. The idea here is to use the lasso to select a subset of these trees, reweight them, and hence produce a prediction model with far fewer trees and, one hopes, comparable accuracy. Suppose boosting has produced a sequence of fitted trees $\hat{g}_b(x),\; b = 1, \ldots, B$. We then solve the lasso problem

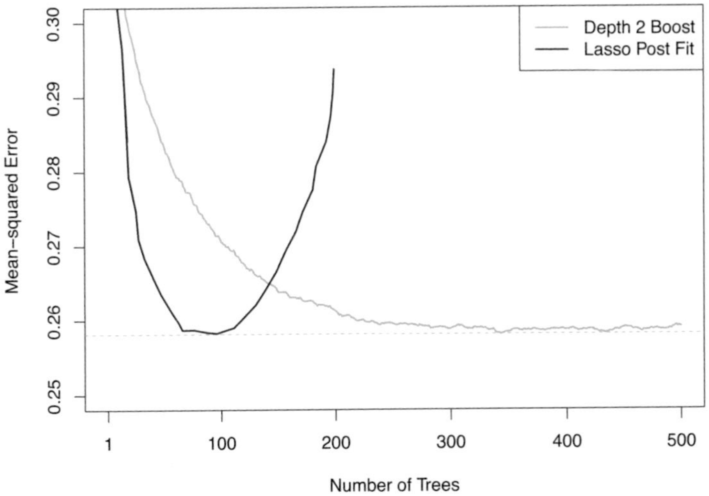

Figure 17.13 Post-processing of the trees produced by boosting on the **ALS** data. Shown is the test prediction error as a function of the number of trees selected by the (nonnegative) lasso. We see that the lasso can do as good a job with one-third the number of trees, although selecting the correct number is critical.

$$\underset{\{\beta_b\}_1^B}{\text{minimize}} \sum_{i=1}^{n} L\left[y_i, \sum_{b=1}^{B} \hat{g}_b(x_i)\beta_b\right] + \lambda \sum_{b=1}^{B} |\beta_b| \tag{17.18}$$

for different values of λ. This model selects some of the trees, and assigns differential weights to them. A reasonable variant is to insist that the weights are nonnegative. Figure 17.13 illustrates this approach on the **ALS** data. Here we could use one-third of the trees. Often the savings are much more dramatic.

17.6 Notes and Details

Random forests and boosting live at the cutting edge of modern prediction methodology. They fit models of breathtaking complexity compared with classical linear regression, or even with standard GLM modeling as practiced in the late twentieth century (Chapter 8). They are routinely used as prediction engines in a wide variety of industrial and scientific applications. For the more cautious, they provide a terrific benchmark for how well a traditional parametrized model is performing: if the random forests does much better, you probably have some work to do, by including some important interactions and the like.

The regression and classification trees discussed in Chapter 8 (Breiman *et al.*, 1984) took traditional models to a new level, with their ability to adapt to the data, select variables, and so on. But their prediction performance is somewhat lacking, and so they stood the risk of falling by the wayside. With their new use as building blocks in random forests and boosting, they have reasserted themselves as critical elements in the modern toolbox.

Random forests and bagging were introduced by Breiman (2001), and boosting by Schapire (1990) and Freund and Schapire (1996). There has been much discussion on why boosting works (Breiman, 1998; Friedman *et al.*, 2000; Schapire and Freund, 2012); the statistical interpretation given here can also be found in Hastie *et al.* (2009), and led to the gradient boosting algorithm (Friedman, 2001). Adaboost was first described in Freund and Schapire (1997). Hastie *et al.* (2009, Chapter 15) is devoted to random forests. For the examples in this chapter we used the `randomForest` package in `R` (Liaw and Wiener, 2002), and for boosting the `gbm` (Ridgeway, 2005) package. The lasso post-processing idea is due to Friedman and Popescu (2005), which we implemented using `glmnet` (Friedman *et al.*, 2009). Generalized additive models are described in Hastie and Tibshirani (1990).

We now give some particular technical details on topics covered in the chapter.

†1 [p. 327] *Averaging trees.* A maximal-depth tree splits every node until it is *pure*, meaning all the responses are the same. For very large n this might be unreasonable; in practice, one can put a lower bound on the minimum count in a terminal node. We are deliberately vague about the response type in Algorithm 17.1. If it is quantitative, we would fit a regression tree. If it is binary or multilevel qualitative, we would fit a classification tree. In this case at the averaging stage, there are at least two strategies. The original random-forest paper (Breiman, 2001) proposed that each tree should make a classification, and then the ensemble uses a plurality vote. An alternative reasonable strategy is to average the class probabilities produced by the trees; these procedures are identical if the trees are grown to maximal depth.

†2 [p. 330] *Jackknife variance estimate.* The jackknife estimate of variance for a random forest, and the bias-corrected version, is described in Wager *et al.* (2014). The jackknife formula (17.3) is applied to the $B = \infty$ ver-

sion of the random forest, but of course is estimated by plugging in finite B versions of the quantities involved. Replacing $\hat{r}_{\mathrm{rf}}^{(\cdot)}(x_0)$ by its expectation $\hat{r}_{\mathrm{rf}}(x_0)$ is not the problem; its that each of the $\hat{r}_{\mathrm{rf}}^{(i)}(x_0)$ vary about their bootstrap expectations, compounded by the square in expression (17.4). Calculating the bias requires some technical derivations, which can be found in that reference.

They also describe the infinitesimal jackknife estimate of variance, given by

$$\widehat{\mathrm{V}}_{\mathrm{IJ}}(\hat{r}_{\mathrm{rf}}(x_0)) = \sum_{i=1}^{n} \widehat{\mathrm{cov}}_i^2, \tag{17.19}$$

with

$$\widehat{\mathrm{cov}}_i = \widehat{\mathrm{cov}}(w^*, \hat{r}_*(x_0)) = \frac{1}{B}\sum_{b=1}^{B}(w_{bi}^* - 1)(\hat{r}_b(x_0) - \hat{r}_{\mathrm{rf}}(x_0)), \tag{17.20}$$

as discussed in Chapter 20. It too has a bias-corrected version, given by

$$\widehat{\mathrm{V}}_{\mathrm{IJ}}^{u}(\hat{r}_{\mathrm{rf}}(x_0)) = \widehat{\mathrm{V}}_{\mathrm{IJ}}(\hat{r}_{\mathrm{rf}}(x_0)) - \frac{n}{B}\hat{v}(x_0), \tag{17.21}$$

similar to (17.5).

†3 [p. 334] *The* **ALS** *data.* These data were kindly provided by Lester Mackey and Lilly Fang, who won the DREAM challenge prediction prize in 2012 (Kuffner *et al.*, 2015). It includes some additional variables created by them. Their winning entry used Bayesian trees, not too different from random forests.

†4 [p. 341] *Gradient-boosting details.* In Friedman's gradient-boosting algorithm (Hastie *et al.*, 2009, Chapter 10, for example), a further refinement is implemented. The tree in step 2(b) of Algorithm 17.4 is used to define the structure (split variables and splits), but the values in the terminal nodes are left to be updated. We can think of partitioning the parameters $\gamma = (\gamma_s, \gamma_t)$, and then represent the tree as $g(x;\gamma) = T(x;\gamma_s)'\gamma_t$. Here $T(x;\gamma_s)$ is a vector of $d+1$ binary basis functions that indicate the terminal node reached by input x, and γ_t are the $d+1$ values of the terminal nodes of the tree. We learn $\hat{\gamma}_s$ by approximating the gradient in step 2(b) by a tree, and then (re-)learn the terminal-node parameters $\hat{\gamma}_t$ by solving the optimization problem

$$\underset{\gamma_t}{\text{minimize}} \sum_{i=1}^{n} L\left(y_i, \hat{G}_{b-1}(x_i) + T(x_i;\hat{\gamma}_s)'\gamma_t\right). \tag{17.22}$$

Solving (17.22) amounts to fitting a simple GLM with an *offset*.

†5 [p. 345] *Adaboost and gradient boosting.* Hastie *et al.* (2009, Chapter 10) derive Adaboost as an instance of Algorithm 17.3. One detail is that the trees $g(x;\gamma)$ are replaced by a simplified scaled classifier $\alpha \cdot c(x;\gamma')$. Hence, from (17.15), in step 2(a) of Algorithm 17.3 we need to solve

$$\underset{\alpha,\gamma'}{\text{minimize}} \sum_{i=1}^{n} w_i \exp[-y_i \alpha c(x_i;\gamma')]. \tag{17.23}$$

The derivation goes on to show that

- minimizing (17.23) for any value of $\alpha > 0$ can be achieved by fitting a classification tree $c(x;\hat{\gamma}')$ to minimize the weighted misclassification error

$$\sum_{i=1}^{n} w_i I[y_i \neq c(x_i,\gamma')];$$

- given $c(x;\hat{\gamma}')$, α is estimated as in step 2(c) of Algorithm 17.5 (and is non-negative);
- the weight-update scheme in step 2(d) of Algorithm 17.5 corresponds exactly to the weights as computed in (17.15).

18

Neural Networks and Deep Learning

Something happened in the mid 1980s that shook up the applied statistics community. Neural networks (NNs) were introduced, and they marked a shift of predictive modeling towards computer science and machine learning. A neural network is a highly parametrized model, inspired by the architecture of the human brain, that was widely promoted as a *universal approximator*—a machine that with enough data could learn any smooth predictive relationship.

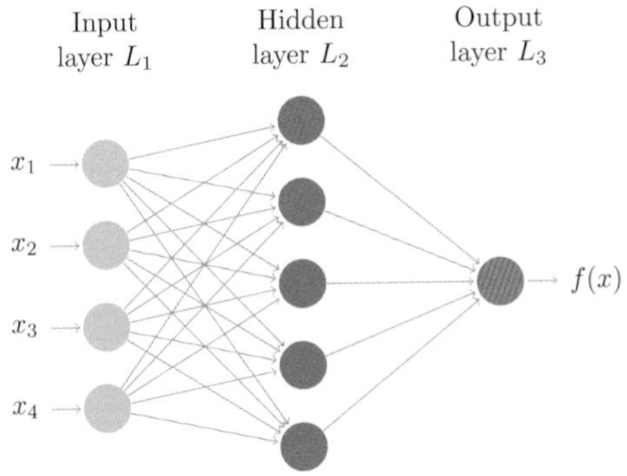

Figure 18.1 Neural network diagram with a single hidden layer. The hidden layer derives transformations of the inputs—nonlinear transformations of linear combinations—which are then used to model the output.

Figure 18.1 shows a simple example of a *feed-forward* neural network diagram. There are four predictors or inputs x_j, five hidden units $a_\ell = g(w_{\ell 0}^{(1)} + \sum_{j=1}^{4} w_{\ell j}^{(1)} x_j)$, and a single output unit $o = h(w_0^{(2)} + \sum_{\ell=1}^{5} w_\ell^{(2)} a_\ell)$. The language associated with NNs is colorful: memory units or *neurons* automatically learn new features from the data through a process called *supervised learning*. Each neuron a_l is connected to the input layer via a vector of parameters or *weights* $\{w_{\ell j}^{(1)}\}_1^p$ (the (1) refers to the first layer and ℓj refers to the jth variable and ℓth unit). The intercept terms $w_{\ell 0}^{(1)}$ are called a *bias*, and the function g is a nonlinearity, such as the sigmoid function $g(t) = 1/(1 + e^{-t})$. The idea was that each neuron will learn a simple binary on/off function; the sigmoid function is a smooth and differentiable compromise. The final or output layer also has weights, and an output function h. For quantitative regression h is typically the identity function, and for a binary response it is once again the sigmoid. Note that without the nonlinearity in the hidden layer, the neural network would reduce to a generalized linear model (Chapter 8). Typically neural networks are fit by maximum likelihood, usually with a variety of forms of regularization.

The knee-jerk response from statisticians was "What's the big deal? A neural network is just a nonlinear model, not too different from many other generalizations of linear models."

While this may be true, neural networks brought a new energy to the field. They could be scaled up and generalized in a variety of ways: many hidden units in a layer, multiple hidden layers, weight sharing, a variety of colorful forms of regularization, and innovative learning algorithms for massive data sets. And most importantly, they were able to solve problems on a scale far exceeding what the statistics community was used to. This was part computing scale and expertise, part liberated thinking and creativity on the part of this computer science community. New journals were
†1 devoted to the field,[†] and several popular annual conferences (initially at ski resorts) attracted their denizens, and drew in members of the statistics community.

After enjoying considerable popularity for a number of years, neural networks were somewhat sidelined by new inventions in the mid 1990s, such as boosting (Chapter 17) and SVMs (Chapter 19). Neural networks were *passé*. But then they re-emerged with a vengeance after 2010—the reincarnation now being called *deep learning*. This renewed enthusiasm is a result of massive improvements in computer resources, some innovations, and the ideal niche learning tasks such as image and video classification, and speech and text processing.

18.1 Neural Networks and the Handwritten Digit Problem

Neural networks really cut their baby teeth on an optical character recognition (OCR) task: automatic reading of handwritten digits, as in a zipcode. Figure 18.2 shows some examples, taken from the **MNIST** corpus.[†] The †2 idea is to build a classifier $C(x) \in \{0, 1, \ldots, 9\}$ based on the input image $x \in \mathbb{R}^{28 \times 28}$, a 28×28 grid of image intensities. In fact, as is often the case, it is more useful to learn the probability function $\Pr(y = j \,|\, x)$, $j = 0, 1, 2, \ldots, 9$; this is indeed the target for our neural network. Figure 18.3

Figure 18.2 Examples of handwritten digits from the **MNIST** corpus. Each digit is represented by a 28×28 grayscale image, derived from normalized binary images of different shapes and sizes. The value stored for each pixel in an image is a nonnegative eight-bit representation of the amount of gray present at that location. The 784 pixels for each image are the predictors, and the 0–9 class labels the response. There are 60,000 training images in the full data set, and 10,000 in the test set.

shows a neural network with three hidden layers, a successful configuration for this digit classification problem. In this case the output layer has 10 nodes, one for each of the possible class labels. We use this example to walk the reader through some of the aspects of the configuration of a network, and fitting it to training data. Since all of the layers are functions of their previous layers, and finally functions of the input vector x, the network represents a somewhat complex function $f(x; \mathcal{W})$, where $\mathcal{W}$ represents the entire collection of weights. Armed with a suitable loss function, we could simply barge right in and throw it at our favorite optimizer. In the early days this was not computationally feasible, especially when special

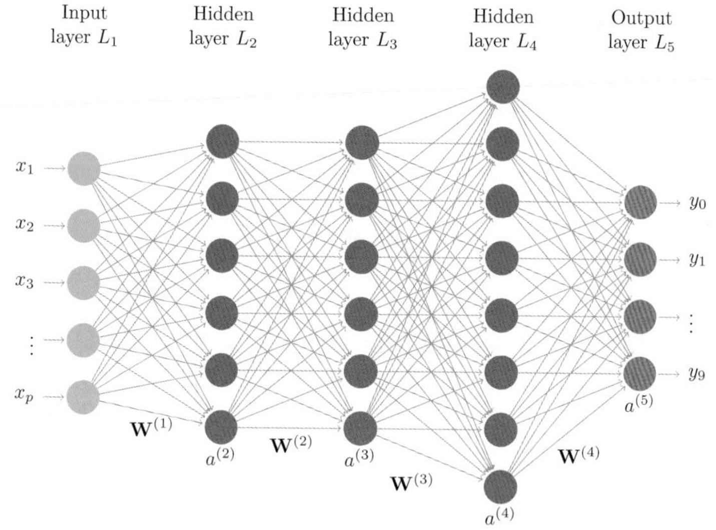

Figure 18.3 Neural network diagram with three hidden layers and multiple outputs, suitable for the **MNIST** handwritten-digit problem. The input layer has $p = 784$ units. Such a network with hidden layer sizes $(1024, 1024, 2048)$, and particular choices of tuning parameters, achieves the state-of-the art error rate of 0.93% on the "official" test data set. This network has close to four million weights, and hence needs to be heavily regularized.

structure is imposed on the weight vectors. Today there are fairly automatic systems for setting up and fitting neural networks, and this view is not too far from reality. They mostly use some form of gradient descent, and rely on an organization of parameters that leads to a manageable calculation of the gradient.

The network in Figure 18.3 is complex, so it is essential to establish a convenient notation for referencing the different sets of parameters. We continue with the notation established for the single-layer network, but with some additional annotations to distinguish aspects of different layers. From the first to the second layer we have

$$z_\ell^{(2)} = w_{\ell 0}^{(1)} + \sum_{j=1}^{p} w_{\ell j}^{(1)} x_j, \tag{18.1}$$

$$a_\ell^{(2)} = g^{(2)}(z_\ell^{(2)}). \tag{18.2}$$

We have separated the linear transformations $z_\ell^{(2)}$ of the x_j from the nonlinear transformation of these, and we allow for layer-specific nonlinear transformations $g^{(k)}$. More generally we have the transition from layer $k-1$ to layer k:

$$z_\ell^{(k)} = w_{\ell 0}^{(k-1)} + \sum_{j=1}^{p_{k-1}} w_{\ell j}^{(k-1)} a_j^{(k-1)}, \tag{18.3}$$

$$a_\ell^{(k)} = g^{(k)}(z_\ell^{(k)}). \tag{18.4}$$

In fact (18.3)–(18.4) can serve for the input layer (18.1)–(18.2) if we adopt the notation that $a_\ell^{(1)} \equiv x_\ell$ and $p_1 = p$, the number of input variables. Hence each of the arrows in Figure 18.3 is associated with a weight parameter.

It is simpler to adopt a vector notation

$$z^{(k)} = \boldsymbol{W}^{(k-1)} a^{(k-1)} \tag{18.5}$$

$$a^{(k)} = g^{(k)}(z^{(k)}), \tag{18.6}$$

where $\boldsymbol{W}^{(k-1)}$ represents the matrix of weights that go from layer L_{k-1} to layer L_k, $a^{(k)}$ is the entire vector of activations at layer L_k, and our notation assumes that $g^{(k)}$ operates elementwise on its vector argument. We have also absorbed the bias parameters $w_{\ell 0}^{(k-1)}$ into the matrix $\boldsymbol{W}^{(k-1)}$, which assumes that we have augmented each of the activation vectors $a^{(k)}$ with a constant element 1.

Sometimes the nonlinearities $g^{(k)}$ at the inner layers are the same function, such as the function σ defined earlier. In Section 18.5 we present a network for natural color image classification, where a number of different activation functions are used.

Depending on the response, the final transformation $g^{(K)}$ is usually special. For M-class classification, such as here with $M = 10$, one typically uses the *softmax* function

$$g^{(K)}(z_m^{(K)}; z^{(K)}) = \frac{e^{z_m^{(K)}}}{\sum_{\ell=1}^{M} e^{z_\ell^{(K)}}}, \tag{18.7}$$

which computes a number (probability) between zero and one, and all M of them sum to one.[1]

[1] This is a symmetric version of the inverse link function used for multiclass logistic regression.

18.2 Fitting a Neural Network

As we have seen, a neural network model is a complex, hierarchical function $f(x; \mathcal{W})$ of the the feature vector x, and the collection of weights $\mathcal{W}$. For typical choices for the $g^{(k)}$, this function will be differentiable. Given a training set $\{x_i, y_i\}_1^n$ and a loss function $L[y, f(x)]$, along familiar lines we might seek to solve

$$\underset{\mathcal{W}}{\text{minimize}} \left\{ \frac{1}{n} \sum_{i=1}^{n} L[y_i, f(x_i; \mathcal{W})] + \lambda J(\mathcal{W}) \right\}, \tag{18.8}$$

where $J(\mathcal{W})$ is a nonnegative regularization term on the elements of $\mathcal{W}$, and $\lambda \geq 0$ is a tuning parameter. (In practice there may be multiple regularization terms, each with their own λ.) For example an early popular penalty is the quadratic

$$J(\mathcal{W}) = \frac{1}{2} \sum_{k=1}^{K-1} \sum_{j=1}^{p_k} \sum_{\ell=1}^{p_{k+1}} \left\{ w_{\ell j}^{(k)} \right\}^2, \tag{18.9}$$

as in ridge regression (7.41). Also known as the *weight-decay* penalty, it pulls the weights toward zero (typically the biases are not penalized). Lasso penalties (Chapter 16) are also popular, as are mixtures of these (an elastic net).

For binary classification we could take L to be binomial deviance (8.14), in which case the neural network amounts to a penalized logistic regression, Section 8.1, albeit a highly parametrized and penalized one. Loss functions are usually convex in f, but not in the elements of $\mathcal{W}$, so solving (18.8) is difficult, and at best we seek good local optima. Most methods are based on some form of gradient descent, with many associated bells and whistles. We briefly discuss some elements of the current practice in finding good solutions to (18.8).

Computing the Gradient: Backpropagation

The elements of $\mathcal{W}$ occur in layers, since $f(x; \mathcal{W})$ is defined as a series of compositions, starting from the input layer. Computing the gradient is also done most naturally in layers (the chain rule for differentiation; see for example (18.10) in Algorithm 18.1 below), and our notation makes this easier to describe in a recursive fashion. We will consider computing the derivative of $L[y, f(x; \mathcal{W}]$ with respect to any of the elements of $\mathcal{W}$, for a generic input–output pair x, y; since the loss part of the objective is a sum, the overall gradient will be the sum of these individual gradient elements over the training pairs (x_i, y_i).

The intuition is as follows. Given a training generic pair (x, y), we first make a forward pass through the network, which creates activations at each of the nodes $a_\ell^{(k)}$ in each of the layers, including the final output layer. We would then like to compute an error term $\delta_\ell^{(k)}$ that measures the responsibility of each node for the error in predicting the true output y. For the output activations $a_\ell^{(K)}$ these errors are easy: either residuals or generalized residuals, depending on the loss function. For activations at inner layers, $\delta_\ell^{(k)}$ will be a weighted sum of the errors terms of nodes that use $a_\ell^{(k)}$ as inputs. The *backpropagation* Algorithm 18.1 gives the details for computing the gradient for a single input–output pair x, y. We leave it to the reader to verify that this indeed implements the chain rule for differentiation.

Algorithm 18.1 BACKPROPAGATION

1 Given a pair x, y, perform a "feedforward pass," computing the activations $a_\ell^{(k)}$ at each of the layers $L_2, L_3, \ldots, L_K$; i.e. compute $f(x; \mathcal{W})$ at x using the current $\mathcal{W}$, saving each of the intermediary quantities along the way.

2 For each output unit ℓ in layer L_K, compute

$$\begin{aligned}\delta_\ell^{(K)} &= \frac{\partial L[y, f(x, \mathcal{W})]}{\partial z_\ell^{(K)}} \\ &= \frac{\partial L[y, f(x; \mathcal{W})]}{\partial a_\ell^{(K)}} \dot{g}^{(K)}(z_\ell^{(K)}),\end{aligned} \tag{18.10}$$

where $\dot{g}$ denotes the derivative of $g(z)$ wrt z. For example for $L(y, f) = \frac{1}{2}\|y - f\|_2^2$, (18.10) becomes $-(y_\ell - f_\ell) \cdot \dot{g}^{(K)}(z_\ell^{(K)})$.

3 For layers $k = K-1,\ K-2, \ldots, 2$, and for each node ℓ in layer k, set

$$\delta_\ell^{(k)} = \left(\sum_{j=1}^{p_{k+1}} w_{j\ell}^{(k)} \delta_j^{(k+1)} \right) \dot{g}^{(k)}(z_\ell^{(k)}). \tag{18.11}$$

4 The partial derivatives are given by

$$\frac{\partial L[y, f(x; \mathcal{W})]}{\partial w_{\ell j}^{(k)}} = a_j^{(k)} \delta_\ell^{(k+1)}. \tag{18.12}$$

One again matrix–vector notation simplifies these expressions a bit:

(18.10) becomes (for squared-error loss)

$$\delta^{(K)} = -(y - a^{(K)}) \circ \dot{g}^{(K)}(z^{(K)}), \tag{18.13}$$

where $\circ$ denotes the Hadamard (elementwise) product;
(18.11) becomes

$$\delta^{(k)} = \left(\boldsymbol{W}^{(k)\prime}\delta^{(k+1)}\right) \circ \dot{g}^{(k)}(z^{(k)}); \tag{18.14}$$

(18.12) becomes

$$\frac{\partial L[y, f(x; \mathcal{W})]}{\partial \boldsymbol{W}^{(k)}} = \delta^{(k+1)} a^{(k)\prime}. \tag{18.15}$$

Backpropagation was considered a breakthrough in the early days of neural networks, since it made fitting a complex model computationally manageable.

Gradient Descent

Algorithm 18.1 computes the gradient of the loss function at a single generic pair (x, y); with n training pairs the gradient of the first part of (18.8) is given by

$$\Delta \boldsymbol{W}^{(k)} = \frac{1}{n} \sum_{i=1}^{n} \frac{\partial L[y_i, f(x_i; \mathcal{W}]}{\partial \boldsymbol{W}^{(k)}}. \tag{18.16}$$

With the quadratic form (18.9) for the penalty, a gradient-descent update is

$$\boldsymbol{W}^{(k)} \leftarrow \boldsymbol{W}^{(k)} - \alpha\left(\Delta \boldsymbol{W}^{(k)} + \lambda \boldsymbol{W}^{(k)}\right), \quad k = 1, \ldots, K-1, \tag{18.17}$$

where $\alpha \in (0, 1]$ is the *learning rate*.

Gradient descent requires starting values for all the weights $\mathcal{W}$. Zero is not an option, because each layer is symmetric in the weights flowing to the different neurons, hence we rely on starting values to break the symmetries. Typically one would use random starting weights, close to zero; random uniform or Gaussian weights are common.

There are a multitude of "tricks of the trade" in fitting or "learning" a neural network, and many of them are connected with gradient descent. Here we list some of these, without going into great detail.

Stochastic Gradient Descent

Rather than process all the observations before making a gradient step, it can be more efficient to process smaller batches at a time—even batches

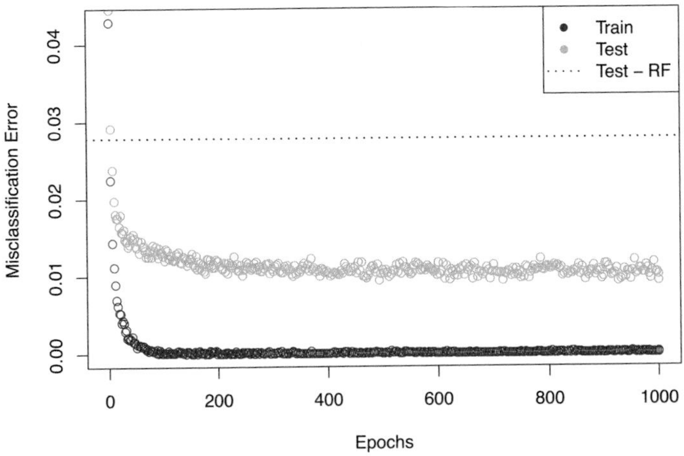

Figure 18.4 Training and test misclassification error as a function of the number of epochs of training, for the **MNIST** digit classification problem. The architecture for the network is shown in Figure 18.3. The network was fit using accelerated gradient descent with adaptive rate control, a rectified linear activation function, and dropout regularization (Section 18.5). The horizontal broken line shows the error rate of a random forest (Section 17.1). A logistic regression model (Section 8.1) achieves only 0.072 (off the scale).

of size one! These batches can be sampled at random, or systematically processed. For large data sets distributed on multiple computer cores, this can be essential for reasons of efficiency. An *epoch* of training means that all n training samples have been used in gradient steps, irrespective of how they have been grouped (and hence how many gradient steps have been made).

Accelerated Gradient Methods

The idea here is to allow previous iterations to build up momentum and influence the current iterations. The iterations have the form

$$\mathcal{V}_{t+1} = \mu\mathcal{V}_t - \alpha(\Delta\mathcal{W}_t + \lambda\mathcal{W}_t), \tag{18.18}$$

$$\mathcal{W}_{t+1} = \mathcal{W}_t + \mathcal{V}_{t+1}, \tag{18.19}$$

using $\mathcal{W}_t$ to represent the entire collection of weights at iteration t. $\mathcal{V}_t$ is a *velocity vector* that accumulates gradient information from previous iterations, and is controlled by an additional momentum parameter μ. When correctly tuned, accelerated gradient descent can achieve much faster convergence rates; however, tuning tends to be a difficult process, and is typically done adaptively.

Rate Annealing

A variety of creative methods have been proposed to adapt the learning rate to avoid jumping across good local minima. These tend to be a mixture of principled approaches combined with ad-hoc adaptations that tend to work
†3 well in practice.† Figure 18.4 shows the performance of our neural net on the **MNIST** digit data. This achieves state-of-the art misclassification error rates on these data (just under 0.093% errors), and outperforms random forests (2.8%) and a generalized linear model (7.2%). Figure 18.5 shows the 93 misclassified digits.

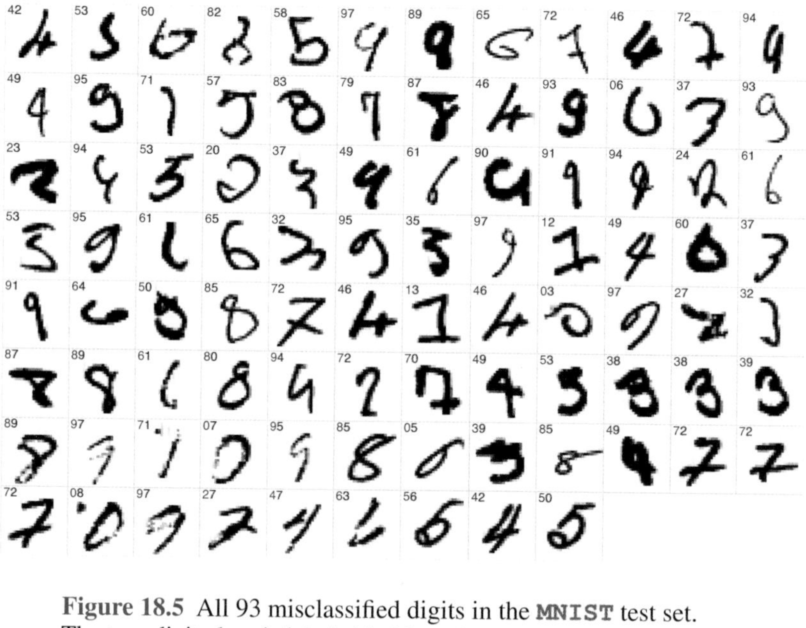

Figure 18.5 All 93 misclassified digits in the **MNIST** test set. The true digit class is labeled in blue, the predicted in red.

Other Tuning Parameters

Apart from the many details associated with gradient descent, there are several other important structural and operational aspects of neural networks that have to be specified.

Number of Hidden Layers, and Their Sizes

With a single hidden layer, the number of hidden units determines the number of parameters. In principle, one could treat this number as a tuning parameter, which could be adjusted to avoid overfitting. The current collective wisdom suggests it is better to have an abundant number of hidden units, and control the model complexity instead by weight regularization. Having deeper networks (more hidden layers) increases the complexity as well. The correct number tends to be task specific; having two hidden layers with the digit recognition problem leads to competitive performance.

Choice of Nonlinearities

There are a number of activation functions $g^{(k)}$ in current use. Apart from the sigmoid function, which transforms its input to a values in (0, 1), other popular choices are

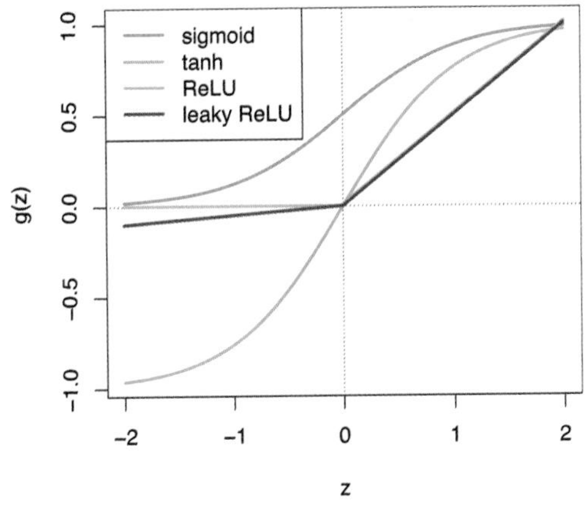

Figure 18.6 Activation functions. ReLU is a rectified linear (unit).

$$\text{tanh:} \quad g(z) = \frac{e^z - e^{-z}}{e^z + e^{-z}},$$

which delivers values in $(-1, 1)$.

$$\text{rectified linear:} \quad g(z) = z_+,$$

or the positive-part function. This has the advantage of making the gradient computations cheaper to compute.

$$\text{leaky rectified linear:} \quad g_\alpha(z) = z_+ - \alpha z_-,$$

for α nonnegative and close to zero. The rectified linear tends to have flat spots, because of the many zero activations; this is an attempt to avoid these and the accompanying zero gradients.

Choice of Regularization

Typically this is a mixture of ℓ_2 and ℓ_1 regularization, each of which requires a tuning parameter. As in lasso and regression applications, the bias terms (intercepts) are usually not regularized. The weight regularization is typically light, and serves several roles. The ℓ_2 reduces problems with collinearity, the ℓ_1 can ignore irrelevant features, and both slow the rate of overfitting, especially with deep (over-parametrized) networks.

Early Stopping

Neural nets are typically over-parametrized, and hence are prone to overfitting. Originally early stopping was set up as the primary tuning parameter, and the stopping time was determined using a held-out set of validation data. In modern networks the regularization is tuned adaptively to avoid overfitting, and hence it is less of a problem. For example, in Figure 18.4 we see that the test misclassification error has flattened out, and does not rise again with increasing number of epochs.

18.3 Autoencoders

An autoencoder is a special neural network for computing a type of nonlinear principal-component decomposition.

The linear principal component decomposition is a popular and effective linear method for reducing a large set of correlated variables to a typically smaller number of linear combinations that capture most of the variance in the original set. Hence, given a collection of n vectors $x_i \in \mathbb{R}^p$ (assumed to have mean zero), we produce a derived set of uncorrelated features $z_i \in \mathbb{R}^q$

($q \leq p$, and typically smaller) via $z_i = \boldsymbol{V}'x_i$. The columns of $\boldsymbol{V}$ are orthonormal, and are derived such that the first component of z_i has maximal variance, the second has the next largest variance and is uncorrelated with the first, and so on. It is easy to show that the columns of $\boldsymbol{V}$ are the leading q eigenvectors of the sample covariance matrix $\boldsymbol{S} = \frac{1}{n}\boldsymbol{X}'\boldsymbol{X}$.

Principal components can also be derived in terms of a best-approximating linear subspace, and it is this version that leads to the nonlinear generalization presented here. Consider the optimization problem

$$\underset{\boldsymbol{A}\in\mathbb{R}^{p\times q},\ \{\gamma_i\}_1^n\in\mathbb{R}^{q\times n}}{\text{minimize}} \sum_{i=1}^{n} \|x_i - \boldsymbol{A}\gamma_i\|_2^2, \tag{18.20}$$

for $q < p$. The subspace is defined by the column space of $\boldsymbol{A}$, and for each point x_i we wish to locate its best approximation in the subspace (in terms of Euclidean distance). Without loss of generality, we can assume $\boldsymbol{A}$ has orthonormal columns, in which case $\hat{\gamma}_i = \boldsymbol{A}'x_i$ for each i (n separate linear regressions). Plugging in, (18.20) reduces to

$$\underset{\boldsymbol{A}\in\mathbb{R}^{p\times q},\ \boldsymbol{A}'\boldsymbol{A}=\boldsymbol{I}_q}{\text{minimize}} \sum_{i=1}^{n} \|x_i - \boldsymbol{A}\boldsymbol{A}'x_i\|_2^2. \tag{18.21}$$

A solution is given by $\hat{\boldsymbol{A}} = \boldsymbol{V}$, the matrix above of the first q principal-component direction vectors computed from the x_i. By analogy, a single-layer autoencoder solves a nonlinear version of this problem:

$$\underset{\boldsymbol{W}\in\mathbb{R}^{q\times p}}{\text{minimize}} \sum_{i=1}^{n} \|x_i - \boldsymbol{W}'g(\boldsymbol{W}x_i)\|_2^2, \tag{18.22}$$

for some nonlinear activation function g; see Figure 18.7 (left panel). If g is the identity function, these solutions coincide (with $\boldsymbol{W} = \boldsymbol{V}'$).

Figure 18.7 (right panel) represents the learned row of $\boldsymbol{W}$ as images, when the autoencoder is fit to the **MNIST** digit database. Since autoencoders do not require a response (the class labels in this case), this decomposition is unsupervised. It is often expensive to label images, for example, while unlabeled images are abundant. Autoencoders provide a means for extracting potentially useful features from such data, which can then be used with labeled data to train a classifier. In fact, they are often used as *warm starts* for the weights when fitting a supervised neural network.

Once again there are a number of bells and whistles that make autoencoders more effective.

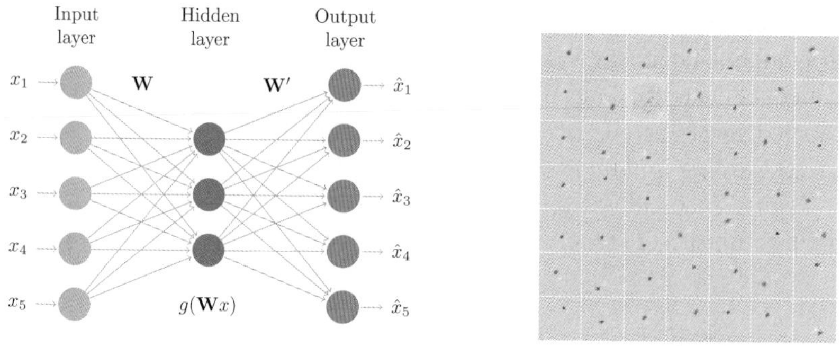

Figure 18.7 Left: Network representation of an autoencoder used for unsupervised learning of nonlinear principal components. The middle layer of hidden units creates a bottleneck, and learns nonlinear representations of the inputs. The output layer is the transpose of the input layer, so the network tries to reproduce the input data using this restrictive representation. Right: Images representing the estimated rows of $\boldsymbol{W}$ using the `MNIST` database; the images can be seen as filters that detect local gradients in the image pixels. In each image, most of the weights are zero, and the nonzero weights are localized in the two-dimensional image space.

- ℓ_1 regularization applied to the rows of $\boldsymbol{W}$ lead to sparse weight vectors, and hence local features, as was the case in our example.
- Denoising is a process where noise is added to the input layer (but not the output), resulting in features that do not focus on isolated values, such as pixels, but instead have some *volume*. We discuss denoising further in Section 18.5.
- With regularization, the bottleneck is not necessary, as in the figure or in principal components. In fact we can learn many more than p components.
- Autoencoders can also have multiple layers, which are typically learned sequentially. The activations learned in the first layer are treated as the input (and output) features, and a model like (18.22) is fit to them.

18.4 Deep Learning

Neural networks were reincarnated around 2010 with "deep learning" as a flashier name, largely a result of much faster and larger computing systems, plus a few new ideas. They have been shown to be particularly successful in the difficult task of classifying natural images, using what is known as a convolutional architecture. Initially autoencoders were considered a crucial aspect of deep learning, since unlabeled images are abundant. However, as labeled corpi become more available, the word on the street is that supervised learning is sufficient.

Figure 18.8 shows examples of natural images, each with a class label such as `beaver`, `sunflower`, `trout` etc. There are 100 class labels in

Figure 18.8 Examples of natural images. The `CIFAR-100` database consists of 100 color image classes, with 600 examples in each class (500 train, 100 test). Each image is $32 \times 32 \times 3$ (red, green, blue). Here we display a randomly chosen image from each class. The classes are organized by hierarchical structure, with 20 coarse levels and five subclasses within each. So, for example, the first five images in the first column are `aquatic mammals`, namely `beaver`, `dolphin`, `otter`, `seal` and `whale`.

all, and 500 training images and 100 test images per class. The goal is to build a classifier to assign a label to an image. We present the essential details of a deep-learning network for this task—one that achieves a respectable classification performance of 35% errors on the designated test set.[2] Figure 18.9 shows a typical deep-learning architecture, with many

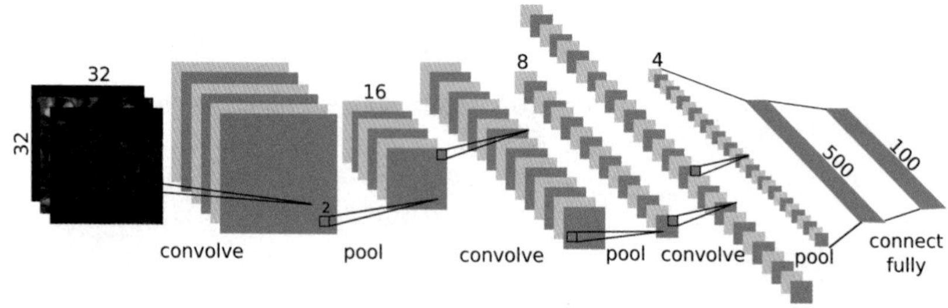

Figure 18.9 Architecture of a deep-learning network for the `CIFAR-100` image classification task. The input layer and hidden layers are all represented as images, except for the last hidden layer, which is "flattened" (vectorized). The input layer consists of the $p_1 = 3$ color (red, green, and blue) versions of an input image (unlike earlier, here we use the p_k to refer to the number of images rather than the totality of pixels). Each of these color panes is 32×32 pixels in dimension. The first hidden layer computes a convolution using a bank of p_2 distinct $q \times q \times p_1$ learned filters, producing an array of images of dimension $p_2 \times 32 \times 32$. The next *pool* layer reduces each non-overlapping block of $\ell \times \ell$ numbers in each pane of the first hidden layer to a single number using a "max" operation. Both q and ℓ are typically small; each was 2 for us. These convolve and pool layers are repeated here three times, with changing dimensions (in our actual implementation, there are 13 layers in total). Finally the 500 derived features are flattened, and a fully connected layer maps them to the 100 classes via a "softmax" activation.

hidden layers. These consist of two special types of layers: "convolve" and "pool." We describe each in turn.

Convolve Layer

Figure 18.10 illustrates a convolution layer, and some details are given in

[2] Classification becomes increasingly difficult as the number of classes grows. With equal representation in each class, the NULL or random error rate for K classes is $(K-1)/K$; 50% for two classes, 99% for 100.

the caption. If an image x is represented by a $k \times k$ matrix, and a filter f

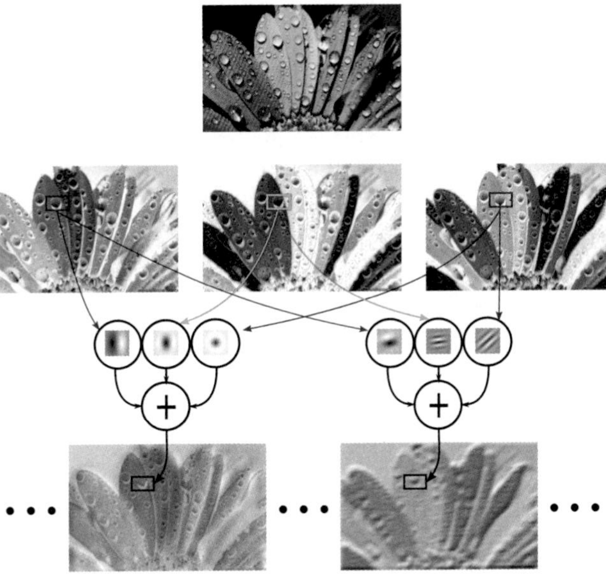

Figure 18.10 Convolution layer for the input images. The input image is split into its three color components. A single filter is a $q \times q \times p_1$ array (here one $q \times q$ for each of the $p_1 = 3$ color panes), and is used to compute an inner product with a correspondingly sized subimage in each pane, and summed across the p_1 panes. We used $q = 2$, and small values are typical. This is repeated over all (overlapping) $q \times q$ subimages (with boundary padding), and hence produces an image of the same dimension as one of the input panes. This is the convolution operation. There are p_2 different versions of this filter, and hence p_2 new panes are produced. Each of the p_2 filters has $p_1 q^2$ weights, which are learned via backpropagation.

is a $q \times q$ matrix with $q \ll k$, the convolved image is another $k \times k$ matrix $\tilde{x}$ with elements $\tilde{x}_{i,j} = \sum_{\ell=1}^{q} \sum_{\ell'=1}^{q} x_{i+\ell,j+\ell'} f_{\ell,\ell'}$ (with edge padding to achieve a full-sized $k \times k$ output image). In our application we used 2×2, but other sizes such as 3×3 are popular. It is most natural to represent the structure in terms of these images as in Figure 18.9, but they could all be vectorized into a massive network diagram as in Figures 18.1 and 18.3. However, the weights would have special sparse structure, with most being zero, and the nonzero values repeated ("weight sharing").

Pool Layer

The pool layer corresponds to a kind of nonlinear activation. It reduces each nonoverlapping block of $r \times r$ pixels ($r = 2$ for us) to a single number by computing their maximum. Why maximum? The convolution filters are themselves small image patches, and are looking to identify similar patches in the target image (in which case the inner product will be high). The max operation introduces an element of local translation invariance. The pool operation reduces the size of each image by a factor r in each dimension. To compensate, the number of tiles in the next convolution layer is typically increased accordingly. Also, as these tiles get smaller, the effective weights resulting from the convolution operator become denser. Eventually the tiles are the same size as the convolution filter, and the layer becomes fully connected.

18.5 Learning a Deep Network

Despite the additional structure imposed by the convolution layers, deep networks are learned by gradient descent. The gradients are computed by backpropagation as before, but with special care taken to accommodate the tied weights in the convolution filters. However, a number of additional tricks have been introduced that appear to improve the performance of modern deep learning networks. These are mostly aimed at regularization; indeed, our 100-class image network has around 50 million parameters, so regularization is essential to avoid overfitting. We briefly discuss some of these.

Dropout

This is a form of regularization that is performed when learning a network, typically at different rates at the different layers. It applies to all networks, not just convolutional; in fact, it appears to work better when applied at the deeper, denser layers. Consider computing the activation $z_\ell^{(k)}$ in layer k as in (18.3) for a single observation during the feed-forward stage. The idea is to randomly set each of the p_{k-1} nodes $a_j^{(k-1)}$ to zero with probability ϕ, and inflate the remaining ones by a factor $1/(1-\phi)$. Hence, for this observation, those nodes that survive have to *stand in* for those omitted. This can be shown to be a form of ridge regularization, and when done correctly
†4 improves performance.[†] The fraction ϕ omitted is a tuning parameter, and for convolutional networks it appears to be better to use different values at

different layers. In particular, as the layers become denser, ϕ is increased: from 0 in the input layer to 0.5 in the final, fully connected layer.

Input Distortion

This is another form of regularization that is particularly suitable for tasks like image classification. The idea is to augment the training set with many distorted copies of an input image (but of course the same label). These distortions can be location shifts and other small affine transformations, but also color and shading shifts that might appear in natural images. We show

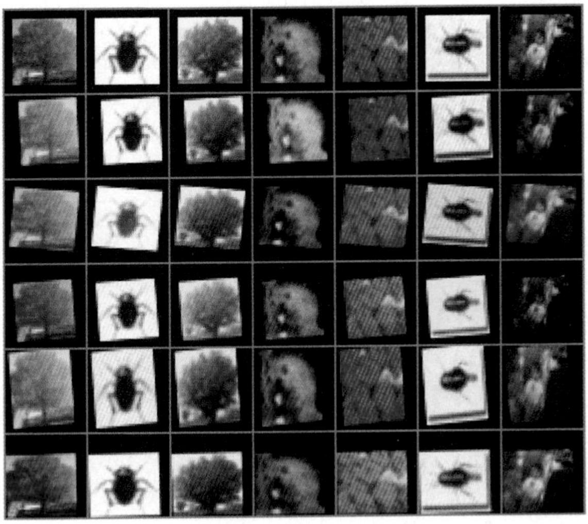

Figure 18.11 Each column represents distorted versions of an input image, including affine and color distortions. The input images are padded on the boundary to increase the size, and hence allow space for some of the distortions.

some distorted versions of input images in Figure 18.11. The distortions are such that a human would have no trouble identifying any of the distorted images if they could identify the original.[†] This both enriches the training †5
data with *hints*, and also prevents overfitting to the original image. One could also apply distortions to a test image, and then "poll" the results to produce a final classification.

Configuration

Designing the correct architecture for a deep-learning network, along with the various choices at each layer, appears to require experience and trial and error. We summarize the third and final architecture which we built for classifying the **CIFAR-100** data set in Algorithm 18.2.† In addition to these size parameters for each layer, we must select the activation functions and additional regularization. In this case we used the leaky rectified linear functions $g_\alpha(z)$ (Section 18.2), with α increasing from 0.05 in layer 5 up to 0.5 in layer 13. In addition a type of ℓ_2 regularization was imposed on the weights, restricting all incoming weight vectors to a node to have ℓ_2 norm bounded by one. Figure 18.12 shows both the progress of the optimization objective (red) and the test misclassification error (blue) as the gradient-descent algorithm proceeds. The accelerated gradient method maintains a memory, which we can see was restarted twice to get out of local minima. Our network achieved a test error rate of 35% on the 10,000 test images (100 images per class). The best reported error rate we have seen is 25%, so apparently we have some way to go!

†6

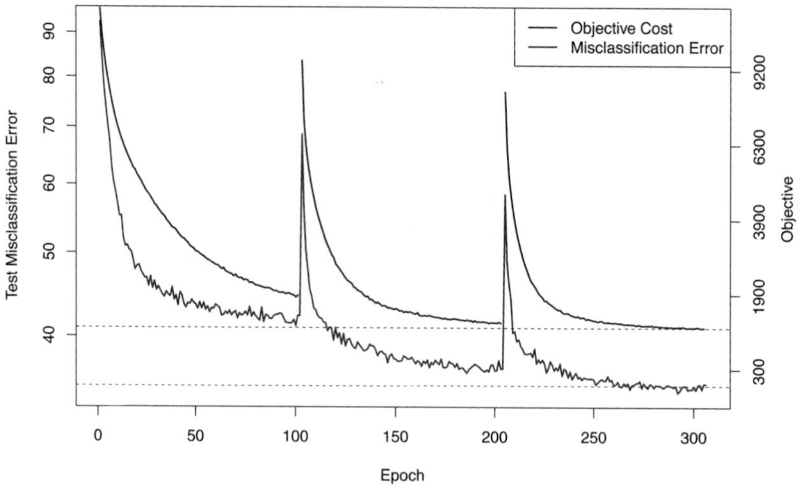

Figure 18.12 Progress of the algorithm as a function of the number of epochs. The accelerated gradient algorithm is "restarted" every 100 epochs, meaning the long-term memory is forgotten, and a new trail is begun, starting at the current solution. The red curve shows the objective (negative penalized log-likelihood on the training data). The blue curve shows test-set misclassification error. The vertical axis is on the log scale, so zero cannot be included.

Algorithm 18.2 CONFIGURATION PARAMETERS FOR DEEP-LEARNING NETWORK USED ON THE **CIFAR-100** DATA.

Layer 1: 100 convolution maps each with $2 \times 2 \times 3$ kernel (the 3 for three colors). The input image is padded from 32×32 to 40×40 to accommodate input distortions.

Layers 2 and 3: 100 convolution maps each $2 \times 2 \times 100$. Compositions of convolutions are roughly equivalent to convolutions with a bigger bandwidth, and the smaller ones have fewer parameters.

Layer 4: Max pool 2×2 layer, pooling nonoverlapping 2×2 blocks of pixels, and hence reducing the images to size 20×20.

Layer 5: 300 convolution maps each $2 \times 2 \times 100$, with dropout learning with rate $\phi_5 = 0.05$.

Layer 6: Repeat of Layer 5.

Layer 7: Max pool 2×2 layer (down to 10×10 images).

Layer 8: 600 convolution maps each $2 \times 2 \times 300$, with dropout rate $\phi_8 = 0.10$.

Layer 9: 800 convolution maps each $2 \times 2 \times 600$, with dropout rate $\phi_9 = 0.10$.

Layer 10: Max pool 2×2 layer (down to 5×5 images).

Layer 11: 1600 convolution maps, each $1 \times 1 \times 800$. This is a pixelwise weighted sum across the 800 images from the previous layer.

Layer 12: 2000 fully connected units, with dropout rate $\phi_{12} = 0.25$.

Layer 13: Final 100 output units, with softmax activation, and dropout rate $\phi_{13} = 0.5$.

18.6 Notes and Details

The reader will notice that probability models have disappeared from the development here. Neural nets are elaborate regression methods aimed solely at prediction—not estimation or explanation in the language of Section 8.4. In place of parametric optimality criteria, the machine learning community has focused on a set of specific prediction data sets, like the digits **MNIST** corpus and **CIFAR-100**, as benchmarks for measuring performance.

There is a vast literature on neural networks, with hundreds of books and thousands of papers. With the resurgence of deep learning, this literature is again growing. Two early statistical references on neural networks are Ripley (1996) and Bishop (1995), as well as Hastie *et al.* (2009) devote one chapter to the topic. Part of our description of backpropagation in Section 18.2 was guided by Andrew Ng's online Stanford lecture notes (Ng, 2015). Bengio *et al.* (2013) provide a useful review of autoencoders. LeCun *et al.* (2015) give a brief overview of deep learning, written by three pioneers of this field: Yann LeCun, Yoshua Bengio and Geoffrey Hinton; we also benefited from reading Ngiam *et al.* (2010). Dropout learning (Srivastava *et al.*, 2014) is a relatively new idea, and its connections with ridge regression were most usefully described in Wager *et al.* (2013). The most popular version of accelerated gradient descent is due to Nesterov (2013). Learning with hints is due to Abu-Mostafa (1995). The material in Sections 18.4 and 18.5 benefited greatly from discussions with Rakesh Achanta (Achanta and Hastie, 2015), who produced some of the color images and diagrams, and designed and fit the deep-learning network to the `CIFAR-100` data.

†1 [p. 352] The Neural Information Processing Systems (NIPS) conferences started in late Fall 1987 in Denver, Colorado, and post-conference workshops were held at the nearby ski resort at Vail. These are still very popular today, although the venue has changed over the years. The NIPS proceedings are refereed, and NIPS papers count as publications in most fields, especially Computer Science and Engineering. Although neural networks were initially the main topic of the conferences, a modern NIPS conference covers all the latest ideas in machine learning.

†2 [p. 353] `MNIST` is a curated database of images of handwritten digits (LeCun and Cortes, 2010). There are 60,000 training images, and 10,000 test images, each a 28×28 grayscale image. These data have been used as a testbed for many different learning algorithms, so the reported best error rates might be optimistic.

†3 [p. 360] *Tuning parameters.* Typical neural network implementations have dozens of tuning parameters, and many of these are associated with the fine tuning of the descent algorithm. We used the `h2o.deepLearning` function in the `R` package `h2o` to fit our model for the `MNIST` data set. It has around 20 such parameters, although most default to factory-tuned constants that have been found to work well on many examples. Arno Candel was very helpful in assisting us with the software.

†4 [p. 368] *Dropout and ridge regression.* Dropout was originally proposed in Srivastava *et al.* (2014), and reinterpreted in Wager *et al.* (2013). Dropout was inspired by the random selection of variables at each tree split in a random forest (Section 17.1). Consider a simple version of dropout for the linear regression problem with squared-error loss. We have an $n \times p$ regression matrix $\mathbf{X}$, and a response n-vector $\mathbf{y}$. For simplicity we assume all variables have mean zero, so we can ignore intercepts. Consider the

following random least-squares criterion:

$$L_I(\beta) = \frac{1}{2}\sum_{i=1}^{n}\left(y_i - \sum_{j=1}^{p} x_{ij} I_{ij} \beta_j\right)^2.$$

Here the I_{ij} are i.i.d variables $\forall i, j$ with

$$I_{ij} = \begin{cases} 0 & \text{with probability } \phi, \\ 1/(1-\phi) & \text{with probability } 1-\phi, \end{cases}$$

(this particular form is used so that $E[I_{ij}] = 1$). Using simple probability it can be shown that the expected score equations can be written

$$\mathrm{E}\left[\frac{\partial L_I(\beta)}{\partial \beta}\right] = -\boldsymbol{X}'\boldsymbol{y} + \boldsymbol{X}'\boldsymbol{X}\beta + \frac{\phi}{1-\phi}\boldsymbol{D}\beta = 0, \tag{18.23}$$

with $\boldsymbol{D} = \mathrm{diag}\{\|\boldsymbol{x}_1\|^2, \|\boldsymbol{x}_2\|^2, \ldots, \|\boldsymbol{x}_p\|^2\}$. Hence the solution is given by

$$\hat{\beta} = \left(\boldsymbol{X}'\boldsymbol{X} + \frac{\phi}{1-\phi}\boldsymbol{D}\right)^{-1}\boldsymbol{X}'\boldsymbol{y}, \tag{18.24}$$

a generalized ridge regression. If the variables are standardized, the term $\boldsymbol{D}$ becomes a scalar, and the solution is identical to ridge regression. With a nonlinear activation function, the interpretation changes slightly; see Wager *et al.* (2013) for details.

†5 [p. 369] *Distortion and ridge regression.* We again show in a simple example that input distortion is similar to ridge regression. Assume the same setup as in the previous example, except a different randomized version of the criterion:

$$L_N(\beta) = \frac{1}{2}\sum_{i=1}^{n}\left(y_i - \sum_{j=1}^{p}(x_{ij} + n_{ij})\beta_j\right)^2.$$

Here we have added random noise to the prediction variables, and we assume this noise is i.i.d $(0, \lambda)$. Once again the expected score equations can be written

$$\mathrm{E}\left[\frac{\partial L_N(\beta)}{\partial \beta}\right] = -\boldsymbol{X}'\boldsymbol{y} + \boldsymbol{X}'\boldsymbol{X}\beta + \lambda\beta = 0, \tag{18.25}$$

because of the independence of all the n_{ij} and $\mathrm{E}(n_{ij}^2) = \lambda$. Once again this leads to a ridge regression. So replacing each observation pair x_i, y_i by the collection $\{x_i^{*b}, y_i\}_{b=1}^{B}$, where each x_i^{*b} is a noisy version of x_i, is approximately equivalent to a ridge regression on the original data.

†6 [p. 370] *Software for deep learning.* Our deep learning convolutional network for the `CIFAR-100` data was constructed and run by Rakesh Achanta in `Theano`, a Python-based system (Bastien *et al.*, 2012; Bergstra *et al.*, 2010). `Theano` has a user-friendly language for specifying the host of parameters for a deep-learning network, and uses symbolic differentiation for computing the gradients needed in stochastic gradient descent. In 2015 Google announced an open-source version of their `TensorFlow` software for fitting deep networks.

19

Support-Vector Machines and Kernel Methods

While linear logistic regression has been the mainstay in biostatistics and epidemiology, it has had a mixed reception in the machine-learning community. There the goal is often classification accuracy, rather than statistical inference. Logistic regression builds a classifier in two steps: fit a conditional probability model for $\Pr(Y = 1|X = x)$, and then classify as a one if $\widehat{\Pr}(Y = 1|X = x) \geq 0.5$. SVMs bypass the first step, and build a classifier directly.

Another rather awkward issue with logistic regression is that it fails if the training data are linearly separable! What this means is that, in the feature space, one can separate the two classes by a linear boundary. In cases such as this, maximum likelihood fails and some parameters march off to infinity. While this might have seemed an unlikely scenario to the early users of logistic regression, it becomes almost a certainty with modern *wide* genomics data. When $p \gg n$ (more features than observations), we can typically always find a separating hyperplane. Finding an *optimal separating hyperplane* was in fact the launching point for SVMs. As we will see, they have more than this to offer, and in fact live comfortably alongside logistic regression.

SVMs pursued an age-old approach in statistics, of enriching the feature space through nonlinear transformations and basis expansions; a classical example being augmenting a linear regression with interaction terms. A linear model in the enlarged space leads to a nonlinear model in the ambient space. This is typically achieved via the "kernel trick," which allows the computations to be performed in the n-dimensional space for an arbitrary number of predictors p. As the field matured, it became clear that in fact this kernel trick amounted to estimation in a reproducing-kernel Hilbert space.

Finally, we contrast the kernel approach in SVMs with the nonparametric regression techniques known as kernel smoothing.

19.1 Optimal Separating Hyperplane

Figure 19.1 shows a small sample of points in $\mathbb{R}^2$, each belonging to one of two classes (blue or orange). Numerically we would score these classes as +1 for say blue, and -1 for orange.[1] We define a two-class linear classifier via a function $f(x) = \beta_0 + x'\beta$, with the convention that we classify a point x_0 as +1 if $f(x_0) > 0$, and as -1 if $f(x_0) < 0$ (on the fence we flip a coin). Hence the classifier itself is $C(x) = \text{sign}[f(x)]$. The decision

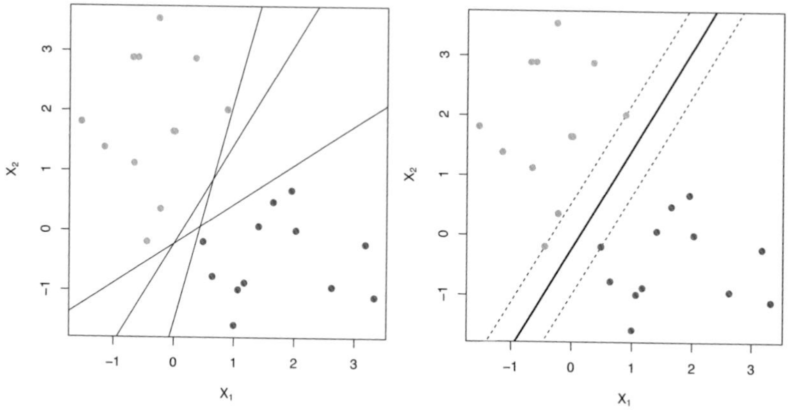

Figure 19.1 Left panel: data in two classes in $\mathbb{R}^2$. Three potential decision boundaries are shown; each separate the data perfectly. Right panel: the optimal separating hyperplane (a line in $\mathbb{R}^2$) creates the biggest margin between the two classes.

boundary is the set $\{x \mid f(x) = 0\}$. We see three different classifiers in the left panel of Figure 19.1, and they all classifier the points perfectly. The optimal separating hyperplane is the linear classifier that creates the largest *margin* between the two classes, and is shown in the right panel (it is also known as an optimal-margin classifier). The underlying hope is that, by making a big margin on the training data, it will also classify future observations well.

†1 Some elementary geometry† shows that the (signed) Euclidean distance from a point x_0 to the linear decision boundary defined by f is given by

$$\frac{1}{\|\beta\|_2} f(x_0). \tag{19.1}$$

With this in mind, for a separating hyperplane the quantity $\frac{1}{\|\beta\|_2} y_i f(x_i)$ is

[1] In this chapter, the ± 1 scoring leads to convenient notation.

the distance of x_i from the decision boundary.[2] This leads to an optimization problem for creating the optimal margin classifier:

$$\underset{\beta_0,\,\beta}{\text{maximize}}\ M \qquad (19.2)$$
$$\text{subject to } \frac{1}{\|\beta\|_2} y_i(\beta_0 + x'\beta) \geq M,\ i = 1, \ldots, n.$$

A rescaling argument reduces this to the simpler form

$$\underset{\beta_0,\,\beta}{\text{minimize}}\ \|\beta\|_2 \qquad (19.3)$$
$$\text{subject to } y_i(\beta_0 + x'\beta) \geq 1,\ i = 1, \ldots, n.$$

This is a quadratic program, which can be solved by standard techniques in convex optimization.† One noteworthy property of the solution is that †2

$$\hat{\beta} = \sum_{i \in \mathcal{S}} \hat{\alpha}_i x_i, \qquad (19.4)$$

where $\mathcal{S}$ is the *support set*. We can see in Figure 19.1 that the margin touches three points (vectors); in this case there are $|\mathcal{S}| = 3$ support vectors, and clearly the orientation of $\hat{\beta}$ is determined by them. However, we still have to solve the optimization problem to identify the three points in $\mathcal{S}$, and their coefficients $\alpha_i,\ i \in \mathcal{S}$. Figure 19.2 shows an optimal-margin classifier fit to *wide* data, that is data where $p \gg n$. These are gene-expression measurements on $p = 3571$ genes measured on blood samples from $n = 72$ leukemia patients (first seen in Chapter 1). They were classified into two classes, 47 acute lymphoblastic leukemia (**ALL**) and 25 myeloid leukemia (**AML**). In cases like this, we are typically guaranteed a separating hyperplane[3]. In this case 42 of the 72 points are support points. One might be justified in thinking that this solution is overfit to this small amount of data. Indeed, when broken into a training and test set, we see that the test data encroaches well into the margin region, but in this case none are misclassified. Such classifiers are very popular in the wide-data world of genomics, largely because they seem to work very well. They offer a simple alternative to logistic regression, in a situation where the latter fails. However, sometimes the solution is overfit, and a modification is called for. This same modification takes care of nonseparable situations as well.

[2] Since all the points are correctly classified, the sign of $f(x_i)$ agrees with y_i, hence this quantity is always positive.

[3] If $n \leq p + 1$ we can always find a separating hyperplane, unless there are exact feature ties across the class barrier!

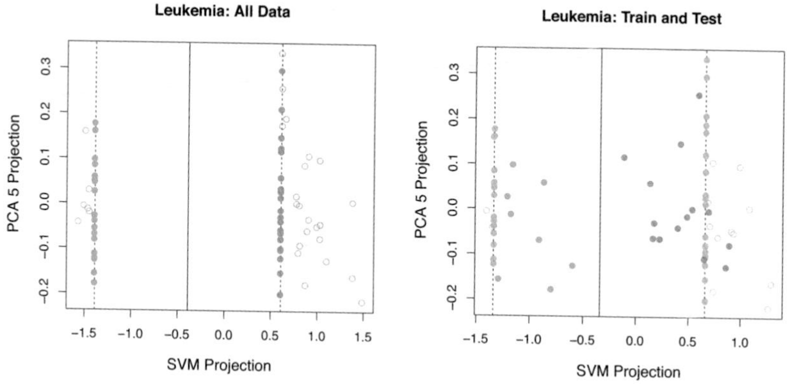

Figure 19.2 Left panel: optimal margin classifier fit to **leukemia** data. There are 72 observations from two classes—47 **ALL** and 25 **AML**—and 3571 gene-expression variables. Of the 72 observations, 42 are support vectors, sitting on the margin. The points are plotted against their fitted classifier function $\hat{f}(x)$, labeled SVM projection, and the fifth principal component of the data (chosen for display purposes, since it has low correlation with the former). Right panel: here the optimal margin classifier was fit to a random subset of 50 of the 72 observations, and then used to classify the remaining 22 (shown in color). Although these points fall on the wrong sides of their respective margins, they are all correctly classified.

19.2 Soft-Margin Classifier

Figure 19.3 shows data in $\mathbb{R}^2$ that are not separable. The generalization to a *soft* margin allows points to violate their margin. Each of the violators has a line segment connecting it to its margin, showing the extent of the violation. The soft-margin classifier solves

$$\begin{aligned} &\underset{\beta_0,\,\beta}{\text{minimize}}\ \|\beta\|_2 \\ &\text{subject to } y_i(\beta_0 + x_i'\beta) \geq 1 - \epsilon_i, \\ &\epsilon_i \geq 0,\ i = 1, \ldots, n, \text{ and } \sum_{i=1}^{n} \epsilon_i \leq B. \end{aligned} \tag{19.5}$$

Here B is the budget for the total amount of overlap. Once again, the solution has the form (19.4), except now the support set $\mathcal{S}$ includes any vectors on the margin as well as those that violate the margin. The bigger B, the

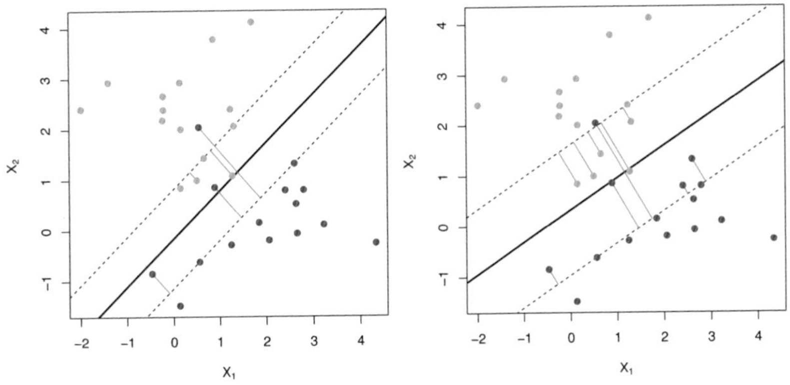

Figure 19.3 For data that are not separable, such as here, the soft-margin classifier allows margin violations. The budget B for the total measure of violation becomes a tuning parameter. The bigger the budget, the wider the soft margin and the more support points there are involved in the fit.

bigger the support set, and hence the more points that have a say in the solution. Hence bigger B means more stability and lower variance. In fact, even for separable data, allowing margin violations via B lets us regularize the solution by tuning B.

19.3 SVM Criterion as Loss Plus Penalty

It turns out that one can reformulate (19.5) and (19.3) in more traditional terms as the minimization of a loss plus a penalty:

$$\underset{\beta_0,\,\beta}{\text{minimize}} \sum_{i=1}^{n} [1 - y_i(\beta_0 + x_i'\beta)]_+ + \lambda \|\beta\|_2^2. \tag{19.6}$$

Here the *hinge* loss $L_H(y, f(x)) = [1 - yf(x)]_+$ operates on the margin quantity $yf(x)$, and is piecewise linear as in Figure 19.4.[†] The same margin [†3] quantity came up in boosting in Section 17.4. The quantity $[1 - y_i(\beta_0 + x_i'\beta)]_+$ is the cost for x_i being on the wrong side of its margin (the cost is zero if it's on the correct side). The correspondence between (19.6) and (19.5) is exact; large λ corresponds to large B, and this formulation makes explicit the form of regularization. For separable data, the optimal separating hyperplane solution (19.3) corresponds to the limiting minimum-norm solution as $\lambda \downarrow 0$. One can show that the population minimizer of the

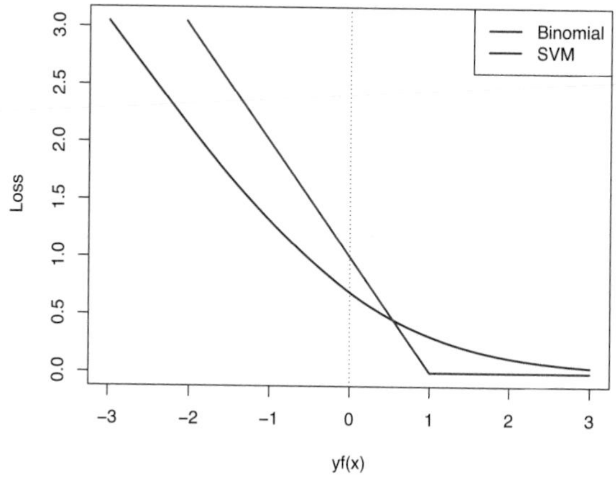

Figure 19.4 The hinge loss penalizes observation margins $yf(x)$ less than $+1$ linearly, and is indifferent to margins greater than $+1$. The negative binomial log-likelihood (deviance) has the same asymptotes, but operates in a smoother fashion near the *elbow* at $yf(x) = 1$.

hinge loss is in fact the Bayes classifier.[4] This shows that the SVM is in
†4 fact directly estimating the classifier $C(x) \in \{-1, +1\}$.†

The red curve in Figure 19.4 is (half) the binomial deviance for logistic regression (i.e. $f(x) = \beta_0 + x'\beta$ is now modeling logit $\Pr(Y = +1|X = x)$). With $Y = \pm 1$, the deviance can also be written in terms of the margin, and the ridged logistic regression corresponding to (19.6) has the form

$$\underset{\beta_0,\,\beta}{\text{minimize}} \sum_{i=1}^{n} \log[1 + e^{-y_i(\beta_0 + x_i'\beta)}] + \lambda\|\beta\|_2^2. \tag{19.7}$$

Logistic regression is discussed in Section 8.1, as well as Sections 16.5 and 17.4. This form of the binomial deviance is derived in (17.13) on page 343. These loss functions have some features in common, as can be seen in the figure. The binomial loss asymptotes to zero for large positive margins, and to a linear loss for large negative margins, matching the hinge loss in this regard. The main difference is that the hinge has a sharp elbow at +1, while the binomial bends smoothly. A consequence of this is that the binomial solution involves all the data, via weights $p_i(1 - p_i)$ that fade smoothly with distance from the decision boundary, as apposed to the binary nature

[4] The Bayes classifier $C(x)$ for a two-class problem using equal costs for misclassification errors assigns x to the class for which $\Pr(y|x)$ is largest.

of support points. Also, as seen in Section 17.4 as well, the population minimizer of the binomial deviance is the logit of the class probability

$$\lambda(x) = \log\left(\frac{\Pr(y=+1|x)}{\Pr(y=-1|x)}\right), \tag{19.8}$$

while that of the hinge loss is its sign $C(x) = \text{sign}[\lambda(x)]$. Interestingly, as $\lambda \downarrow 0$ the solution direction $\hat{\beta}$ to the ridged logistic regression problem (19.7) converges to that of the SVM.† †5

These forms immediately suggest other generalizations of the linear SVM. In particular, we can replace the ridge penalty $\|\beta\|_2^2$ by the sparsity-inducing lasso penalty $\|\beta\|_1$, which will set some coefficients to zero and hence perform feature selection. Publicly available software (e.g. package `liblineaR` in `R`) is available for fitting such lasso-regularized support-vector classifiers.

19.4 Computations and the Kernel Trick

The form of the solution $\hat{\beta} = \sum_{i \in \mathcal{S}} \hat{\alpha}_i x_i$ for the optimal- and soft-margin classifier has some important consequences. For starters, we can write the fitted function evaluated at a point x as

$$\begin{aligned} \hat{f}(x) &= \hat{\beta}_0 + x'\hat{\beta} \\ &= \hat{\beta}_0 + \sum_{i \in \mathcal{S}} \hat{\alpha}_i \langle x, x_i \rangle, \end{aligned} \tag{19.9}$$

where we have deliberately replaced the transpose notation with the more suggestive inner product. Furthermore, we show in (19.23) in Section 19.9 that the Lagrange dual involves the data only through the n^2 pairwise inner products $\langle x_i, x_j \rangle$ (the elements of the $n \times n$ *gram* matrix $\boldsymbol{XX}'$). This means that the computations for computing the SVM solution scale linearly with p, although potentially cubic[5] in n. With very large p (in the tens of thousands and even millions as we will see), this can be convenient.

It turns out that all ridge-regularized linear models with wide data can be reparametrized in this way. Take ridge regression, for example:

$$\underset{\beta}{\text{minimize}} \; \|\boldsymbol{y} - \boldsymbol{X}\beta\|_2^2 + \lambda\|\beta\|_2^2. \tag{19.10}$$

This has solution $\hat{\beta} = (\boldsymbol{X}'\boldsymbol{X} + \lambda \boldsymbol{I}_p)^{-1}\boldsymbol{X}'\boldsymbol{y}$, and with p large requires inversion of a $p \times p$ matrix. However, it can be shown that $\hat{\beta} = \boldsymbol{X}'\hat{\alpha} =$

[5] In practice $O(n^2|\mathcal{S}|)$, and, with modern approximate solutions, much faster than that.

$\sum_{i=1}^{n} \hat{\alpha}_i x_i$, with $\hat{\alpha} = (\boldsymbol{X}\boldsymbol{X}' + \lambda \boldsymbol{I}_n)^{-1}\boldsymbol{y}$, which means the solution can be obtained in $O(n^2 p)$ rather than $O(np^2)$ computations. Again the gram
†6 matrix has played a role, and $\hat{\beta}$ has the same form as for the SVM.†

We now imagine expanding the p-dimensional feature vector x into a potentially much larger set $h(x) = [h_1(x), h_2(x), \ldots, h_m(x)]$; for an example to latch onto, think polynomial basis of total degree d. As long as we have an efficient way to compute the inner products $\langle h(x), h(x_j)\rangle$ for any x, we can compute the SVM solution in this enlarged space just as easily as in the original. It turns out that convenient *kernel* functions exist that do just that. For example $K_d(x, z) = (1 + \langle x, z\rangle)^d$ creates a basis expansion
†7 h_d of polynomials of total degree d, and $K_d(x, z) = \langle h_d(x), h_d(z)\rangle$.†

The polynomial kernels are mainly useful as existence proofs; in practice other more useful kernels are used. Probably the most popular is the radial kernel

$$K(x, z) = e^{-\gamma \|x-z\|_2^2}. \tag{19.11}$$

This is a positive definite function, and can be thought of as computing an inner product in some feature space. Here the feature space is in principle infinite-dimensional, but of course effectively finite.[6] Now one can think of the representation (19.9) in a different light;

$$\hat{f}(x) = \hat{\alpha}_0 + \sum_{i \in \mathcal{S}} \hat{\alpha}_i K(x, x_i), \tag{19.12}$$

an expansion of radial basis functions, each centered on one of the training examples. Figure 19.5 illustrates such an expansion in $\mathbb{R}^1$. Using such nonlinear kernels expands the scope of SVMs considerably, allowing one to fit classifiers with nonlinear decision boundaries.

One may ask what objective is being optimized when we move to this kernel representation. This is covered in the next section, but as a sneak preview we present the criterion

$$\underset{\alpha_0, \alpha}{\text{minimize}} \sum_{j=1}^{n} \left[1 - y_j \left(\alpha_0 + \sum_{i=1}^{n} \alpha_i K(x_j, x_i)\right)\right]_+ + \lambda \alpha' \boldsymbol{K} \alpha, \tag{19.13}$$

where the $n \times n$ matrix $\boldsymbol{K}$ has entries $K(x_j, x_i)$.

As an illustrative example in $\mathbb{R}^2$ (so we can visualize the nonlinear boundaries), we generated the data in Figure 19.6. We show two SVM

[6] A bivariate function $K(x, z)$ ($\mathbb{R}^p \times \mathbb{R}^p \mapsto \mathbb{R}^1$) is positive-definite if, for every q, every $q \times q$ matrix $\boldsymbol{K} = \{K(x_i, x_j)\}$ formed using distinct entries $x_1, x_2, \ldots, x_q$ is positive definite. The feature space is defined in terms of the eigen-functions of the kernel.

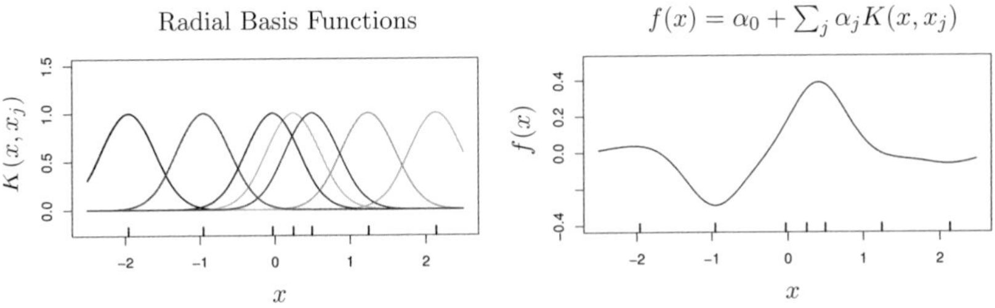

Figure 19.5 Radial basis functions in $\mathbb{R}^1$. The left panel shows a collection of radial basis functions, each centered on one of the seven observations. The right panel shows a function obtained from a particular linear expansion of these basis functions.

solutions, both using a radial kernel. In the left panel, some margin errors are committed, but the solution looks reasonable. However, with the flexibility of the enlarged feature space, by decreasing the budget B we can typically overfit the training data, as is the case in the right panel. A separate little blue island was created to accommodate the one blue point in a sea of brown.

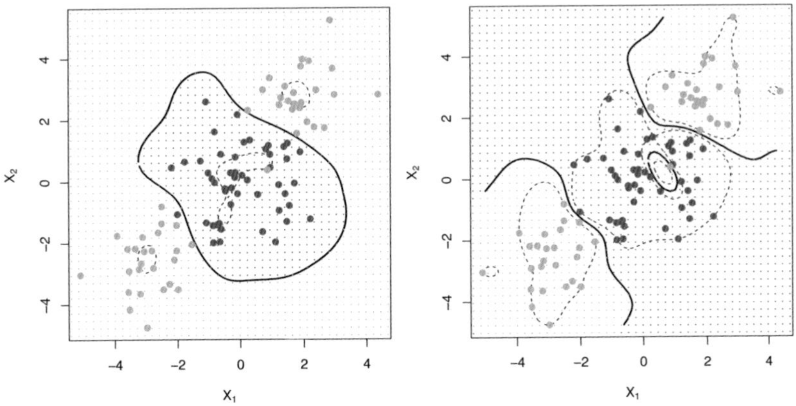

Figure 19.6 Simulated data in two classes in $\mathbb{R}^2$, with SVM classifiers computed using the radial kernel (19.11). The left panel uses a larger value of B than the right. The solid lines are the decision boundaries in the original space (linear boundaries in the expanded feature space). The dashed lines are the projected margins in both cases.

19.5 Function Fitting Using Kernels

The analysis in the previous section is heuristic—replacing inner products by kernels that compute inner products in some (implicit) feature space. Indeed, this is how kernels were first introduced in the SVM world. There is however a rich literature behind such approaches, which goes by the name *function fitting in reproducing-kernel Hilbert spaces (RKHSs)*. We give a very brief overview here. One starts with a bivariate positive-definite kernel $K : \mathbb{R}^p \times \mathbb{R}^p \to \mathbb{R}^1$, and we consider a space $\mathcal{H}_K$ of functions $f : \mathbb{R}^p \to \mathbb{R}^1$ generated by the kernel: $f \in \text{span}\{K(\cdot, z),\ z \in \mathbb{R}^p\}$[7] The
†8 kernel also induces a norm on the space $\|f\|_{\mathcal{H}_K}$,† which can be thought of as a roughness measure.

We can now state a very general optimization problem for fitting a function to data, when restricted to this class;

$$\underset{f \in \mathcal{H}_K}{\text{minimize}} \left\{ \sum_{i=1}^{n} L(y_i, \alpha_0 + f(x_i)) + \lambda \|f\|^2_{\mathcal{H}_K} \right\}, \tag{19.14}$$

a search over a possibly infinite-dimensional function space. Here L is an arbitrary loss function. The "magic" of these spaces in the context of this problem is that one can show that the solution is finite-dimensional:

$$\hat{f}(x) = \sum_{i=1}^{n} \hat{\alpha}_i K(x, x_i), \tag{19.15}$$

a linear basis expansion with basis functions $k_i(x) = K(x, x_i)$ anchored at each of the observed "vectors" x_i in the training data. Moreover, using the "reproducing" property of the kernel in this space, one can show that the penalty reduces to

$$\|\hat{f}\|^2_{\mathcal{H}_K} = \sum_{i=1}^{n} \sum_{j=1}^{n} \hat{\alpha}_i \hat{\alpha}_j K(x_i, x_j) = \hat{\alpha}' \boldsymbol{K} \hat{\alpha}. \tag{19.16}$$

Here $\boldsymbol{K}$ is the $n \times n$ *gram* matrix of evaluations of the kernel, equivalent to the $\boldsymbol{XX}'$ matrix for the linear case.

Hence the abstract problem (19.14) reduces to the generalized ridge problem

$$\underset{\alpha \in \mathbb{R}^n}{\text{minimize}} \left\{ \sum_{i=1}^{n} L\left(y_i, \alpha_0 + \sum_{j=1}^{n} \alpha_i K(x_i, x_j) \right) + \lambda \alpha' \boldsymbol{K} \alpha \right\}. \tag{19.17}$$

[7] Here $k_z = K(\cdot, z)$ is considered a function of the first argument, and the second argument is a parameter.

Indeed, if L is the hinge loss as in (19.6), this is the equivalent "loss plus penalty" criterion being fit by the kernel SVM. Alternatively, if L is the binomial deviance loss as in (19.7), this would fit a kernel version of logistic regression. Hence most fitting methods can be generalized to accommodate kernels.

This formalization opens the door to a wide variety of applications, depending on the kernel function used. Alternatively, as long as we can compute suitable similarities between objects, we can build sophisticated classifiers and other models for making predictions about other attributes of the objects.[8] In the next section we consider a particular example.

19.6 Example: String Kernels for Protein Classification

One of the important problems in computational biology is to classify proteins into functional and structural classes based on their sequence similarities. Protein molecules can be thought of as strings of amino acids, and differ in terms of length and composition. In the example we consider, the lengths vary between 75 and 160 amino-acid molecules, each of which can be one of 20 different types, labeled using the letters of the alphabet.

Here follow two protein examples x_1 and x_2, of length 110 and 153 respectively:

IPTSALVKETLALLSTHRTLLIANETLRIPVPVHKNHQLCTEEIFQGIGTLESQTVQGGTV
ERLFKNLSLIKKYIDGQKKKCGEERRRVNQFLDY**LQE**FLGVMNTEWI

PHRRDLCSRSIWLARKIRSDLTALTESYVKHQGLWSELTEAER**LQE**NLQAYRTFHVLLA
RLLEDQQVHFTPTEGDFHQAIHTLLLQVAAFAYQIEELMILLEYKIPRNEADGMLFEKK
LWGLKVL**QE**LSQWTVRSIHDLRFISSHQTGIP

We treat the proteins x as documents consisting of letters, with a dictionary of size 20. Our feature vector $h^m(x)$ will consist of the counts for all m-grams in the protein—that is, distinct sequences of consecutive letters of length m. As an illustration, we use $m = 3$, which results in 20^3 =8,000 possible sub-sequences; hence $h^3(x)$ will be a vector of length 8,000, with each element the number of times that particular sub-sequence occurs in the protein x. In our example, the sub-sequence **LQE** occurs once in the first, and twice in the second protein, so $h^3_{\text{LQE}}(x_1) = 1$ and $h^3_{\text{LQE}}(x_2) = 2$.

The number of possible sequences of length m is 20^m, which can be very

[8] As long as the similarities behave like inner products; i.e. they form positive semi-definite matrices.

large for moderate m. Also the vast majority of the sub-sequences do not match the strings in our training set, which means $h^m(x)$ will be sparse. It turns out that we can compute the $n \times n$ inner product matrix or *string kernel* $\boldsymbol{K}_m(x_1, x_2) = \langle h^m(x_1), h^m(x_2) \rangle$ efficiently using tree structures, without
†9 actually computing the individual vectors.[†] Armed with the kernel, we

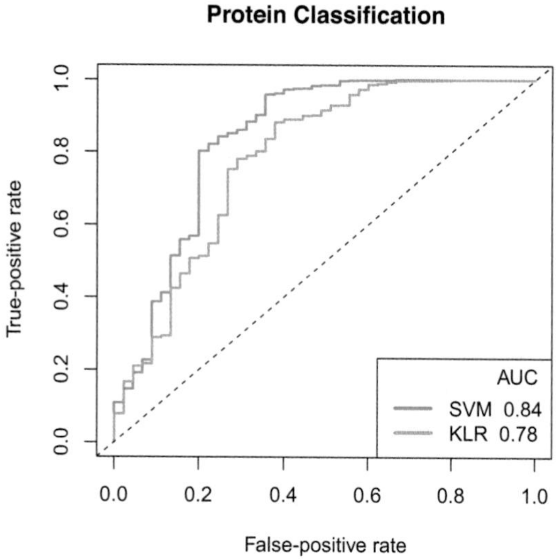

Figure 19.7 ROC curves for two classifiers fit to the protein data. The ROC curves were computed using 10-fold cross-validation, and trace the tradeoff between false-positive and true-positive error rates as the classifier threshold is varied. The area under the curve (AUC) summarises the overall performance of each classifier. Here the SVM is slightly superior to kernel logistic regression.

can now use it to fit a regularized SVM or logistic regression model, as outlined in the previous section. The data consist of 1708 proteins in two classes—negative (1663) and positive (45). We fit both the kernel SVM and kernel logistic regression models. For both methods, cross-validation suggested a very small value for λ. Figure 19.7 shows the ROC tradeoff curve for each, using 10-fold cross-validation. Here the SVM outperforms logistic regression.

19.7 SVMs: Concluding Remarks

SVMs have been wildly successful, and are one of the "must have" tools in any machine-learning toolbox. They have been extended to cover many different scenarios, other than two-class classification, with some awkwardness in cases. The extension to nonlinear function-fitting via kernels (inspiring the "machine" in the name) generated a mini industry. Kernels are parametrized, learned from data, with special problem-specific structure, and so on.

On the other hand, we know that fitting high-dimensional nonlinear functions is intrinsically difficult (the "curse of dimensionality"), and SVMs are not immune. The quadratic penalty implicit in kernel methodology means all features are included in the model, and hence sparsity is generally not an option. Why then this unbridled enthusiasm? Classifiers are far less sensitive to bias–variance tradeoffs, and SVMs are mostly popular for their classification performance. The ability to define a kernel for measuring similarities between abstract objects, and then train a classifier, is a novelty added by these approaches that was missed in the past.

19.8 Kernel Smoothing and Local Regression

The phrase "kernel methodology" might mean something a little different to statisticians trained in the 1970–90 period. Kernel smoothing represents a broad range of tools for performing non- and semi-parametric regression. Figure 19.8 shows a Gaussian kernel smooth fit to some artificial data $\{x_i, y_i\}_1^n$. It computes at each point x_0 a weighted average of the y-values of neighboring points, with weights given by the height of the kernel. In its simplest form, this estimate can be written as

$$\hat{f}(x_0) = \sum_{i=1}^{n} y_i K_\gamma(x_0, x_i), \tag{19.18}$$

where $K_\gamma(x_0, x_i)$ represents the radial kernel with width parameter γ.[9] Notice the similarity to (19.15); here the $\hat{\alpha}_i = y_i$, and the complexity of the model is controlled by γ. Despite this similarity, and the use of the same kernel, these methodologies are rather different.

The focus here is on local estimation, and the kernel does the localizing. Expression (19.18) is almost a weighted average—almost because

[9] Here $K_\gamma(x, \mu)$ is the normalized Gaussian density with mean μ and variance $1/\gamma$.

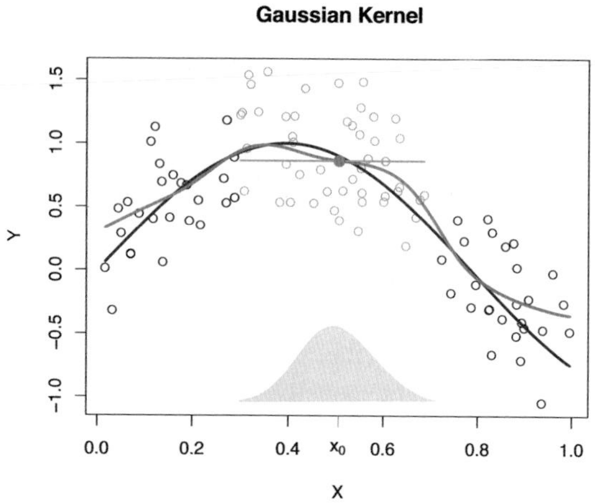

Figure 19.8 A Gaussian kernel smooth of simulated data. The points come from the blue curve with added random errors. The kernel smoother fits a weighted mean of the observations, with the weighting kernel centered at the target point, x_0 in this case. The points shaded orange contribute to the fit at x_0. As x_0 moves across the domain, the smoother traces out the green curve. The width of the kernel is a tuning parameter. We have depicted the Gaussian weighting kernel in this figure for illustration; in fact its vertical coordinates are all positive and integrate to one.

$\sum_{i=1}^n K_\gamma(x_0, x_i) \approx 1$. In fact, the Nadaraya–Watson estimator is more explicit:

$$\hat{f}_{NW}(x_0) = \frac{\sum_{i=1}^n y_i K_\gamma(x_0, x_i)}{\sum_{i=1}^n K_\gamma(x_0, x_i)}. \tag{19.19}$$

Although Figure 19.8 is one-dimensional, the same formulation applies to x in higher dimensions.

Weighting kernels other than the Gaussian are typically favored; in particular, near-neighbor kernels with compact support. For example, the tricube kernel used by the `lowess` smoother in **R** is defined as follows:

1 Define $d_i = \|x_0 - x_i\|_2$, $i = 1, \ldots, n$, and let $d_{(m)}$ be the mth smallest (the distance of the mth nearest neighbor to x_0). Let $u_i = d_i / d_{(m)}$, $i = 1, \ldots, n$.

2 The tricube kernel is given by

$$K_s(x_0, x_i) = \begin{cases} \left(1 - u_i^3\right)^3 & \text{if } u_i \le 1; \\ 0 & \text{otherwise,} \end{cases} \qquad (19.20)$$

where $s = m/n$, the *span* of the kernel. Near-neighbor kernels such as this adapt naturally to the local density of the x_i; wider in low-density regions, narrower in high-density regions. A tricube kernel is illustrated in Figure 19.9.

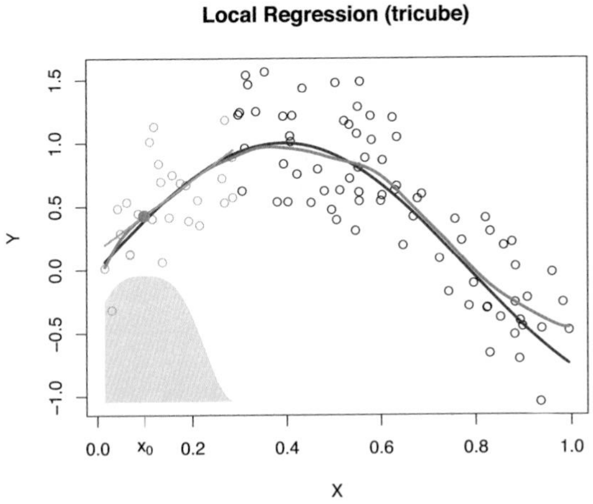

Figure 19.9 Local regression fit to the simulated data. At each point x_0, we fit a locally weighted linear least-squares model, and use the fitted value to estimate $\hat{f}_{LR}(x_0)$. Here we use the tricube kernel (19.20), with a span of 25%. The orange points are in the weighting neighborhood, and we see the orange linear fit computed by kernel weighted least squares. The green dot is the fitted value at x_0 from this local linear fit.

Weighted means suffer from boundary bias—we can see in Figure 19.8 that the estimate appears biased upwards at both boundaries. The reason is that, for example on the left, the estimate for the function on the boundary averages points always to the right, and since the function is locally increasing, there is an upward bias. *Local linear regression* is a natural generalization that fixes such problems. At each point x_0 we solve the following weighted least-squares problem

$$(\hat{\beta}_0(x_0), \hat{\beta}(x_0)) = \arg\min_{\beta_0,\beta} \sum_{i=1}^{n} K_s(x_0, x_i)(y_i - \beta_0 - x_i\beta)^2. \qquad (19.21)$$

Then $\hat{f}_{LR}(x_0) = \hat{\beta}_0(x_0) + x_0\hat{\beta}(x_0)$. One can show that, to first order,
†10 $\hat{f}_{LR}(x_0)$ removes the boundary bias exactly.[†]

Figure 19.9 illustrates the procedure on our simulated data, using the tricube kernel with a span of 25% of the data. In practice, the width of the kernel (the span here) has to be selected by some means; typically we use cross-validation.

Local regression works in any dimension; that is, we can fit two- or higher-dimensional surfaces using exactly the same technique. Here the ability to remove boundary bias really pays off, since the boundaries can be complex. These are referred to as *memory-based methods*, since there is no fitted model. We have to save all the training data, and recompute the local fit every time we make a prediction.

Like kernel SVMs and their relatives, kernel smoothing and local regression break down in high dimensions. Here the near neighborhoods become so wide that they are no longer local.

19.9 Notes and Details

In the late 1980s and early 1990s, machine-learning research was largely driven by prediction problems, and the neural-network community at AT&T Bell laboratories was amongst the leaders. The problem of the day was the US Post-Office handwritten zip-code OCR challenge—a 10-class image classification problem. Vladimir Vapnik was part of this team, and along with colleagues invented a more direct approach to classification, the support-vector machine. This started with the seminal paper by Boser *et al.* (1992), which introduced the optimal margin classifier (optimal separating hyperplane); see also Vapnik (1996). The ideas took off quite rapidly, attracting a large cohort of researchers, and evolved into the more general class of "kernel" methods—that is, models framed in reproducing-kernel Hilbert spaces. A good general reference is Schölkopf and Smola (2001).

†1 [p. 376] *Geometry of separating hyperplanes.* Let $f(x) = \beta' x + \beta_0$ define a linear decision boundary $\{x \mid f(x) = 0\}$ in $\mathbb{R}^p$ (an affine set of codimension one). The unit vector normal to the boundary is $\beta/\|\beta\|_2$, where $\|\cdot\|_2$ denotes the ℓ_2 or Euclidean norm. How should one compute the distance from a point x to this boundary? If x_0 is any point on the boundary (i.e.

$f(x_0) = 0$), we can project $x - x_0$ onto the normal, giving us

$$\frac{\beta'(x - x_0)}{\|\beta\|_2} = \frac{1}{\|\beta\|_2} f(x),$$

as claimed in (19.1). Note that this is the signed distance, since $f(x)$ will be positive or negative depending on what side of the boundary it lies on.

†2 [p. 377] *The "support" in SVM.* The Lagrange primal problem corresponding to (19.3) can be written as

$$\underset{\beta_0,\,\beta}{\text{minimize}} \left\{ \frac{1}{2}\beta'\beta + \sum_{i=1}^{n} \gamma_i [1 - y_i(\beta_0 + x_i'\beta)] \right\}, \tag{19.22}$$

where $\gamma_i \geq 0$ are the Lagrange multipliers. On differentiating we find that $\beta = \sum_{i=1}^{n} \gamma_i y_i x_i$ and $\sum_{i=1}^{n} y_i \gamma_i = 0$. With $\alpha_i = y_i \gamma_i$, we get (19.4), and note that the positivity constraint on γ_i will lead to some of the α_i being zero. Plugging into (19.22) we obtain the Lagrange dual problem

$$\begin{aligned} \underset{\{\gamma_i\}_1^n}{\text{maximize}} & \left\{ \sum_{i=1}^{n} \gamma_i - \frac{1}{2} \sum_{i=1}^{n} \sum_{j=1}^{n} \gamma_i \gamma_j y_i y_j x_i' x_j \right\} \\ & \text{subject to } \gamma_i \geq 0, \quad \sum_{i=1}^{n} y_i \gamma_i = 0. \end{aligned} \tag{19.23}$$

†3 [p. 379] *The SVM loss function.* The constraint in (19.5) can be succinctly captured via the expression

$$\sum_{i=1}^{n} [1 - y_i(\beta_0 + x_i'\beta)]_+ \leq B. \tag{19.24}$$

We only require a (positive) ϵ_i if our margin is less than 1, and we get charged for the sum of these ϵ_i. We now use a Lagrange multiplier to enforce the constraint, leading to

$$\underset{\beta_0,\,\beta}{\text{minimize}} \|\beta\|_2^2 + \gamma \sum_{i=1}^{n} [1 - y_i(\beta_0 + x_i'\beta)]_+. \tag{19.25}$$

Multiplying by $\lambda = 1/\gamma$ gives us (19.6).

†4 [p. 380] *The SVM estimates a classifier.* The following derivation is due to Wahba *et al.* (2000). Consider

$$\underset{f(x)}{\text{minimize}}\; E_{Y|X=x} \{[1 - Yf(x)]_+\}. \tag{19.26}$$

Dropping the dependence on x, the objective can be written as $P_+[1 - f]_+ + P_-[1+f]_+$, where $P_+ = \Pr(Y = +1|X = x)$, and $P_- = \Pr(Y = -1|X = x) = 1 - P_+$. From this we see that

$$f = \begin{cases} +1 & \text{if } P_+ > \frac{1}{2} \\ -1 & \text{if } P_- < \frac{1}{2}. \end{cases} \tag{19.27}$$

†5 [p. 381] *SVM and ridged logistic regression.* Rosset *et al.* (2004) show that the limiting solution as $\lambda \downarrow 0$ to (19.7) for separable data coincides with that of the SVM, in the sense that $\hat{\beta}/\|\hat{\beta}\|_2$ converges to the same quantity for the SVM. However, because of the required normalization for logistic regression, the SVM solution is preferable. On the other hand, for overlapped situations, the logistic-regression solution has some advantages, since its target is the logit of the class probabilities.

†6 [p. 382] *The kernel trick.* The trick here is to observe that from the score equations we have $-\boldsymbol{X}'(\boldsymbol{y} - \boldsymbol{X}\beta) + \lambda\beta = 0$, which means we can write $\hat{\beta} = \boldsymbol{X}'\alpha$ for some α. We now plug this into the score equations, and some simple manipulation gives the result. A similar result holds for ridged logistic regression, and in fact any linear model with a ridge penalty on the coefficients (Hastie and Tibshirani, 2004).

†7 [p. 382] *Polynomial kernels.* Consider $K_2(x,z) = (1 + \langle x, z\rangle)^2$, for x (and z) in $\mathbb{R}^2$. Expanding we get

$$K_2(x,z) = 1 + 2x_1z_1 + 2x_2z_2 + 2x_1x_2z_1z_2 + x_1^2z_1^2 + x_2^2z_2^2.$$

This corresponds to $\langle h_2(x), h_2(z)\rangle$ with

$$h_2(x) = (1, \sqrt{2}x_1, \sqrt{2}x_2, \sqrt{2}x_1x_2, x_1^2, x_2^2).$$

The same is true for $p > 2$ and for degree $d > 2$.

†8 [p. 384] *Reproducing kernel Hilbert spaces.* Suppose K has eigen expansion $K(x,z) = \sum_{i=1}^{\infty} \gamma_i\phi_i(x)\phi_i(z)$, with $\gamma_i \geq 0$ and $\sum_{i=1}^{\infty} \gamma_i < \infty$. Then we say $f \in \mathcal{H}_K$ if $f(x) = \sum_{i=1}^{\infty} c_i\phi_i(x)$, with

$$\|f\|^2_{\mathcal{H}_K} \equiv \sum_{i=1}^{\infty} \frac{c_i^2}{\gamma_i} < \infty. \tag{19.28}$$

Often $\|f\|_{\mathcal{H}_K}$ behaves like a roughness penalty, in that it penalizes unlikely members in the span of $K(\cdot, z)$ (assuming that these correspond to "rough" functions). If f has some high loadings c_j on functions ϕ_j with small eigenvalues γ_j (i.e. not prominent members of the span), the norm becomes large. Smoothing splines and their generalizations correspond to function fitting in a RKHS (Wahba, 1990).

†9 [p. 386] This methodology and the data we use in our example come from Leslie *et al.* (2003).

†10 [p. 390] *Local regression and bias reduction.* By expanding the unknown true $f(x)$ in a first-order Taylor expansion about the target point x_0, one can show that $E\hat{f}_{LR}(x_0) \approx f(x_0)$ (Hastie and Loader, 1993).

20

Inference After Model Selection

The classical theory of model selection focused on "F tests" performed within Gaussian regression models. Inference after model selection (for instance, assessing the accuracy of a fitted regression curve) was typically done ignoring the model selection process. This was a matter of necessity: the combination of discrete model selection and continuous regression analysis was too awkward for simple mathematical description. Electronic computation has opened the door to a more honest analysis of estimation accuracy, one that takes account of the variability induced by data-based model selection.

Figure 20.1 displays the `cholesterol` data, an example we will use for illustration in what follows: cholestyramine, a proposed cholesterol-lowering drug, was administered to $n = 164$ men for an average of seven years each. The response variable d_i was the ith man's decrease in cholesterol level over the course of the experiment. Also measured was c_i, his compliance or the proportion of the intended dose actually taken, ranging from 1 for perfect compliers to zero for the four men who took none at all. Here the 164 c_i values have been transformed to approximately follow a standard normal distribution,

$$c_i \dot\sim \mathcal{N}(0, 1). \tag{20.1}$$

We wish to predict cholesterol decrease from compliance. Polynomial regression models, with d_i a Jth-order polynomial in c_i, were considered, for degrees $J = 1, 2, 3, 4, 5$, or 6. The C_p criterion (12.51) was applied and selected a cubic model, $J = 3$, as best. The curve in Figure 20.1 is the OLS (ordinary least squares) cubic regression curve fit to the cholesterol data set

$$\{(c_i, d_i),\ i = 1, 2, \ldots, 164\}. \tag{20.2}$$

We are interested in answering the following question: how accurate is the

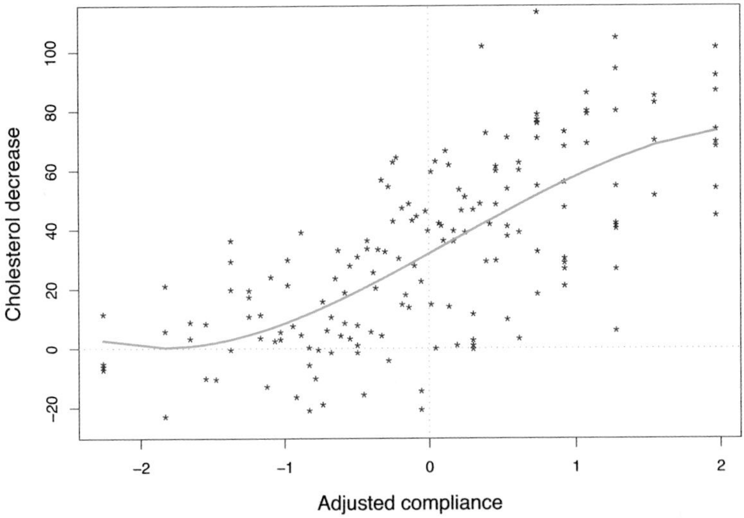

Figure 20.1 `Cholesterol` data: cholesterol decrease plotted versus adjusted compliance for 164 men taking `cholestyramine`. The green curve is OLS cubic regression, with "cubic" selected by the C_p criterion. How accurate is the fitted curve?

fitted curve, taking account of C_p selection as well as OLS estimation? (See Section 20.2 for an answer.)

Currently, there is no overarching theory for inference after model selection. This chapter, more modestly, presents a short series of vignettes that illustrate promising analyses of individual situations. See also Section 16.6 for a brief report on progress in post-selection inference for the lasso.

20.1 Simultaneous Confidence Intervals

In the early 1950s, just before the beginnings of the computer revolution, substantial progress was made on the problem of setting simultaneous confidence intervals. "Simultaneous" here means that there exists a catalog of parameters of possible interest,

$$\boldsymbol{C} = \{\theta_1, \theta_2, \ldots, \theta_J\}, \tag{20.3}$$

and we wish to set a confidence interval for each of them with some fixed probability, typically 0.95, that *all* of the intervals will contain their respective parameters.

As a first example, we return to the **diabetes** data of Section 7.3: $n = 442$ diabetes patients each have had $p = 10$ medical variables measured at baseline, with the goal of predicting **prog**, disease progression one year later. Let $\boldsymbol{X}$ be the 442×10 matrix with ith row x_i' the 10 measurements for patient i; $\boldsymbol{X}$ has been standardized so that each of its columns has mean 0 and sum of squares 1. Also let $\boldsymbol{y}$ be the 442-vector of centered **prog** measurements (that is, subtracting off the mean of the **prog** values).

Ordinary least squares applied to the normal linear model,

$$\boldsymbol{y} \sim \mathcal{N}_n(\boldsymbol{X}\beta, \sigma^2 \boldsymbol{I}), \tag{20.4}$$

yields MLE

$$\hat{\beta} = (\boldsymbol{X}'\boldsymbol{X})^{-1}\boldsymbol{X}'\boldsymbol{y}, \tag{20.5}$$

satisfying

$$\hat{\beta} \sim \mathcal{N}_p(\beta, \sigma^2 \boldsymbol{V}), \qquad \boldsymbol{V} = (\boldsymbol{X}'\boldsymbol{X})^{-1}, \tag{20.6}$$

as at (7.34).

The 95% Student-t confidence interval (11.49) for β_j, the jth component of β, is

$$\hat{\beta}_j \pm \hat{\sigma} V_{jj}^{1/2} t_q^{.975}, \tag{20.7}$$

where $\hat{\sigma} = 54.2$ is the usual unbiased estimate of σ,

$$\hat{\sigma}^2 = \|\boldsymbol{y} - \boldsymbol{X}\hat{\beta}\|^2/q, \qquad q = n - p = 432, \tag{20.8}$$

and $t_q^{.975} = 1.97$ is the 0.975 quantile of a Student-t distribution with q degrees of freedom.

The catalog $\boldsymbol{C}$ in (20.3) is now $\{\beta_1, \beta_2, \ldots, \beta_{10}\}$. The individual intervals (20.7), shown in Table 20.1, each have 95% coverage, but they are not simultaneous: there is a greater than 5% chance that at least one of the β_j values lies outside its claimed interval.

Valid 95% simultaneous intervals for the 10 parameters appear on the right side of Table 20.1. These are the *Scheffé intervals*

$$\hat{\beta}_j \pm \hat{\sigma} V_{jj}^{1/2} k_{p,q}^{(\alpha)}, \tag{20.9}$$

discussed next. The crucial constant $k_{p,q}^{(\alpha)}$ equals 4.30 for $p = 10$, $q = 432$, and $\alpha = 0.95$. That makes the Scheffé intervals wider than the t intervals (20.7) by a factor of 2.19. One expects to pay an extra price for simultaneous coverage, but a factor greater than two induces sticker shock.

Scheffé's method depends on the pivotal quantity

$$Q = \left(\hat{\beta} - \beta\right)' V^{-1} \left(\hat{\beta} - \beta\right) \Big/ \hat{\sigma}^2, \tag{20.10}$$

Table 20.1 *Maximum likelihood estimates $\hat{\beta}$ for 10 diabetes predictor variables* (20.6)*; separate 95% Student-t confidence limits, also simultaneous 95% Scheffé intervals. The Scheffé intervals are wider by a factor of 2.19.*

		Student-t		Scheffé	
	$\hat{\beta}$	Lower	Upper	Lower	Upper
age	−0.5	−6.1	5.1	−12.7	11.8
sex	−11.4	−17.1	−5.7	−24.0	1.1
bmi	24.8	18.5	31.0	11.1	38.4
map	15.4	9.3	21.6	2.1	28.8
tc	−37.7	−76.7	1.2	−123.0	47.6
ldl	22.7	−9.0	54.4	−46.7	92.1
hdl	4.8	−15.1	24.7	−38.7	48.3
tch	8.4	−6.7	23.5	−24.6	41.5
ltg	35.8	19.7	51.9	0.6	71.0
glu	3.2	−3.0	9.4	−10.3	16.7

which under model (20.4) has a scaled "F distribution,"[1]

$$Q \sim pF_{p,q}. \tag{20.11}$$

If $k_{p,q}^{(\alpha)2}$ is the αth quantile of a $pF_{p,q}$ distribution then $\Pr\{Q \leq k_{p,q}^{(\alpha)2}\} = \alpha$ yields

$$\Pr\left\{\frac{\left(\beta - \hat{\beta}\right)' V^{-1} \left(\beta - \hat{\beta}\right)}{\hat{\sigma}^2} \leq k_{p,q}^{(\alpha)2}\right\} = \alpha \tag{20.12}$$

for any choice of β and σ in model (20.4). Having observed $\hat{\beta}$ and $\hat{\sigma}$, (20.12) defines an elliptical confidence region $\mathcal{E}$ for the parameter vector β.

Suppose we are interested in a particular linear combination of the coordinates of β, say

$$\beta_c = c'\beta, \tag{20.13}$$

[1] $F_{p,q}$ is distributed as $(\chi_p^2/p)/(\chi_q^2/q)$, the two chi-squared variates being independent. Calculating the percentiles of $F_{p,q}$ was a major project of the pre-war period.

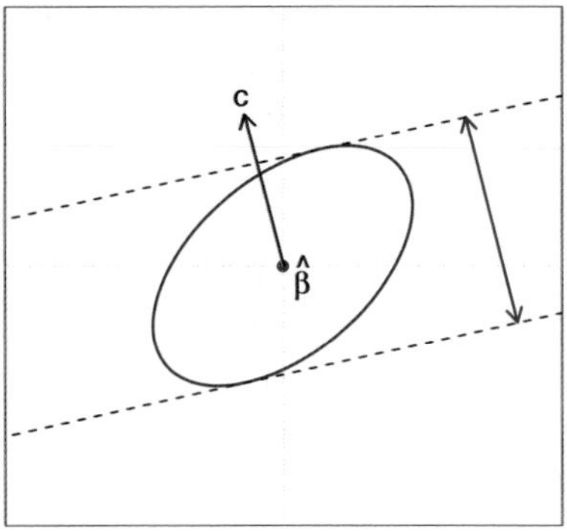

Figure 20.2 Ellipsoid of possible vectors β defined by (20.12) determines confidence intervals for $\beta_c = c'\beta$ according to the "bounding hyperplane" construction illustrated. The red line shows the confidence interval for β_c if c is a unit vector, $c'Vc = 1$.

where c is a fixed p-dimensional vector. If β exists in $\mathcal{E}$ then we must have

$$\beta_c \in \left[\min_{\beta \in \mathcal{E}}(c'\beta),\ \max_{\beta \in \mathcal{E}}(c'\beta)\right], \tag{20.14}$$

†1 which turns out† to be the interval centered at $\hat{\beta}_c = c'\hat{\beta}$,

$$\beta_c \in \hat{\beta}_c \pm \hat{\sigma}(c'Vc)^{1/2} k_{p,q}^{(\alpha)}. \tag{20.15}$$

(This agrees with (20.9) where c is the jth coordinate vector $(0, \ldots, 0, 1, 0, \ldots, 0)'$.) The construction is illustrated in Figure 20.2.

Theorem (Scheffé) *If $\hat{\beta} \sim \mathcal{N}_p(\beta, \sigma^2 V)$ independently of $\hat{\sigma}^2 \sim \sigma^2\chi_q^2/q$, then with probability α the confidence statement* (20.15) *for $\beta_c = c'\beta$ will be simultaneously true for all choices of the vector c.*

Here we can think of "model selection" as the choice of the linear combination of interest $\theta_c = c'\beta$. Scheffé's theorem allows "data snooping": the statistician can examine the data and *then* choose which θ_c (or many θ_c's) to estimate, without invalidating the resulting confidence intervals.

An important application has the $\hat{\beta}_j$'s as independent estimates of efficacy for competing treatments—perhaps different experimental drugs for the same target disease:

$$\hat{\beta}_j \overset{\text{ind}}{\sim} \mathcal{N}(\beta_j, \sigma^2/n_j), \qquad \text{for } j = 1, 2, \ldots, J, \tag{20.16}$$

the n_j being known sample sizes. In this case the catalog $\boldsymbol{C}$ might comprise all pairwise differences $\beta_i - \beta_j$, as the statistician tries to determine which treatments are better or worse than the others.

The fact that Scheffé's limits apply to *all* possible linear combinations $c'\beta$ is a blessing and a curse, the curse being their very large width, as seen in Table 20.1. Narrower simultaneous limits† are possible if we restrict the catalog $\boldsymbol{C}$, for instance to just the pairwise differences $\beta_i - \beta_j$. †2

A serious objection, along Fisherian lines, is that the Scheffé confidence limits are *accurate* without being *correct*. That is, the intervals have the claimed overall frequentist coverage probability, but may be misleading when applied to individual cases. Suppose for instance that $\sigma^2/n_j = 1$ for $j = 1, 2, \ldots, J$ in (20.16) and that we observe $\hat{\beta}_1 = 10$, with $|\hat{\beta}_j| < 2$ for all the others. Even if we looked at the data before singling out $\hat{\beta}_1$ for attention, the usual Student-t interval (20.7) seems more appropriate than its much longer Scheffé version (20.9). This point is made more convincingly in our next vignette.

••———••———••———••

A familiar but pernicious abuse of model selection concerns multiple hypothesis testing. Suppose we observe N independent normal variates z_i, each with its own *effect size* μ_i,

$$z_i \overset{\text{ind}}{\sim} \mathcal{N}(\mu_i, 1) \qquad \text{for } i = 1, 2, \ldots, N, \tag{20.17}$$

and, as in Section 15.1, we wish to test the null hypotheses

$$H_{0i} : \mu_i = 0. \tag{20.18}$$

Being alert to the pitfalls of simultaneous testing, we employ a false-discovery rate control algorithm (15.14), which rejects R of the N null hypotheses, say for cases $i_1, i_2, \ldots, i_R$. (R equaled 28 in the example of Figure 15.3.)

So far so good. The "familiar abuse" comes in then setting the usual confidence intervals

$$\mu_i \in \hat{\mu}_i \pm 1.96 \tag{20.19}$$

(95% coverage) for the R selected cases. This ignores the model selection process: the data-based selection of the R cases must be taken into account in making legitimate inferences, even if R is only 1 so multiplicity is not a concern.

This problem is addressed by the theory of *false-coverage control*. Suppose algorithm $\mathcal{A}$ sets confidence intervals for R of the N cases, of which

r are actually false coverages, i.e., ones not containing the true effect size μ_i. The false-coverage rate (FCR) of $\mathcal{A}$ is the expected proportion of non-coverages

$$\text{FCR}(\mathcal{A}) = E\{r/R\}, \tag{20.20}$$

the expectation being with respect to model (20.17). The goal, as with the FDR theory of Section 15.2, is to construct algorithm $\mathcal{A}$ to control FCR below some fixed value q.

The BY_q algorithm[2] controls FCR below level q in three easy steps, beginning with model (20.17).

1 Let p_i be the p-value corresponding to z_i,

$$p_i = \Phi(z_i) \tag{20.21}$$

for left-sided significance testing, and order the $p_{(i)}$ values in ascending order,

$$p_{(1)} \le p_{(2)} \le p_{(3)} \le \dots \le p_{(N)}. \tag{20.22}$$

2 Calculate $R = \max\{i : p_{(i)} \le i \cdot q/N\}$, and (as in the BH_q algorithm (15.14)–(15.15)) declare the R corresponding null hypotheses false.

3 For each of the R cases, construct the confidence interval

$$\mu_i \in z_i \pm z^{(\alpha_R)}, \qquad \text{where } \alpha_R = 1 - Rq/N \tag{20.23}$$

($z^{(\alpha)} = \Phi^{-1}(\alpha)$).

Theorem 20.1 *Under model* (20.17), *BY_q has FCR $\le q$; moreover, none of the intervals* (20.23) *contain* $\mu_i = 0$.

A simulated example of BY_q was run according to these specifications:

$$\begin{aligned} N = 10{,}000, \quad q = 0.05, \quad & z_i \sim \mathcal{N}(\mu_i, 1) \\ \mu_i = 0 \qquad & \text{for } i = 1, 2, \dots, 9000, \\ \mu_i \sim \mathcal{N}(-3, 1) \qquad & \text{for } i = 9001, \dots, 10{,}000. \end{aligned} \tag{20.24}$$

In this situation we have 9000 null cases and 1000 non-null cases (all but 2 of which had $\mu_i < 0$).

Because this is a simulation, we can plot the pairs (z_i, μ_i) to assess the BY_q algorithm's performance. This is done in Figure 20.3 for the 1000 non-null cases (the green points). BY_q declared $R = 565$ cases non-null, those having $z_i \le -2.77$ (the circled points); 14 of the 565 declarations

[2] Short for "Benjamini–Yekutieli;" see the chapter endnotes.

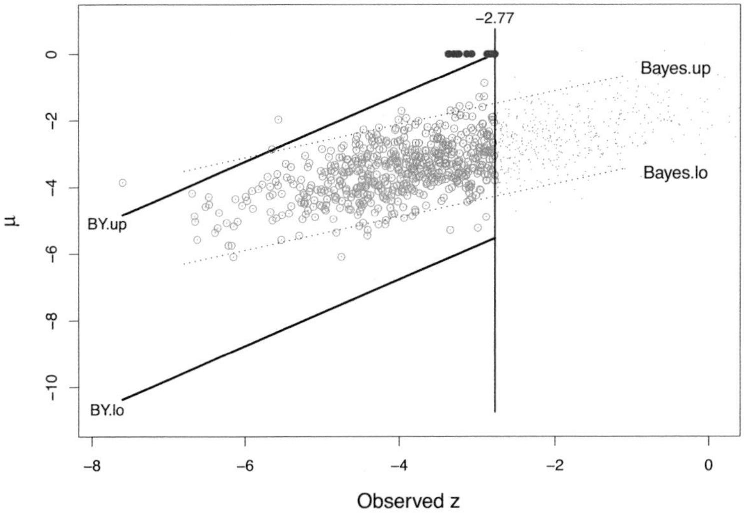

Figure 20.3 Simulation experiment (20.24) with N =10,000 cases, of which 1000 are non-null; the green points (z_i, μ_i) are these non-null cases. The FDR control algorithm BH_q $(q = 0.05)$ declared the 565 circled cases having $z_i \leq -2.77$ to be non-null, of which the 14 red points were actually null. The heavy black lines show BY_q 95% confidence intervals for the 565 cases, only 17 of which failed to contain μ_i. Actual Bayes posterior 95% intervals for non-null cases (20.26), dotted lines, have half the width and slope of BY_q limits.

were actually null cases (the red circled points), giving false-discovery proportion $14/565 = 0.025$. The heavy black lines trace the BY_q confidence limits (20.23) as a function of $z \leq -2.77$.

The first thing to notice is that FCR control has indeed been achieved: only 17 of the declared cases lie outside their limits (the 14 nulls and 3 non-nulls), for a false-coverage rate of $17/565 = 0.030$, safely less than $q = 0.05$. The second thing, however, is that the BY_q limits provide a misleading idea of the location of μ_i given z_i: they are much too wide and slope too low, especially for more negative z_i values.

In this situation we can describe precisely the posterior distribution of μ_i given z_i for the non-null cases,

$$\mu_i | z_i \sim \mathcal{N}\left(\frac{z_i - 3}{2}, \frac{1}{2}\right), \tag{20.25}$$

this following from $\mu_i \sim \mathcal{N}(-3, 1)$, $z_i|\mu_i \sim \mathcal{N}(\mu_i, 1)$, and Bayes' rule (5.20)–(5.21). The Bayes credible 95% limits

$$\mu_i \in \frac{z_i - 3}{2} \pm \frac{1}{\sqrt{2}} 1.96 \tag{20.26}$$

are indicated by the dotted lines in Figure 20.3. They are half as wide as the BY_q limits, and have slope $1/2$ rather than 1.

In practice, of course, we would only see the z_i, not the μ_i, making (20.26) unavailable to us. We return to this example in Chapter 21, where empirical Bayes methods will be seen to provide a good approximation to the Bayes limits. (See Figure 31.5.)

As with Scheffé's method, the BY_q intervals can be accused of being accurate but not correct. "Correct" here has a Bayesian/Fisherian flavor that is hard to pin down, except perhaps in large-scale applications, where empirical Bayes analyses can suggest appropriate inferences.

20.2 Accuracy After Model Selection

The cubic regression curve for the `cholesterol` data seen in Figure 20.1 was selected according to the C_p criterion of Section 12.3. Polynomial regression models, predicting cholesterol decrease d_i in terms of powers ("degrees") of adjusted compliance c_i, were fit by ordinary least squares for degrees $0, 1, 2, \ldots, 6$. Table 20.2 shows C_p estimates (12.51) being minimized at degree 3.

Table 20.2 *C_p table for cholesterol data of Figure 20.1, comparing OLS polynomial models of degrees 0 through 6. The cubic model, degree $= 3$, is the minimizer (80,000 subtracted from the C_p values for easier comparison; assumes $\sigma = 22.0$).*

Degree	C_p
0	71887
1	1132
2	1412
3	667
4	1591
5	1811
6	2758

We wish to assess the accuracy of the fitted curve, taking account of both the C_p model selection method and the OLS fitting process. The bootstrap is a natural candidate for the job. Here we will employ the nonparametric bootstrap of Section 10.2 (rather than the parametric bootstrap of Section 10.4, though this would be no more difficult to carry out).

The **cholesterol** data set (20.2) comprises $n = 164$ pairs $x_i = (c_i, d_i)$; a nonparametric bootstrap sample $\boldsymbol{x}$ (10.13) consists of 164 pairs chosen at random and *with* replacement from the original 164. Let $t(\boldsymbol{x}^*)$ be the curve obtained by applying the C_p/OLS algorithm to the original data set $\boldsymbol{x}^*$ and likewise $t(\boldsymbol{x}^*)$ for the algorithm applied to $\boldsymbol{x}^*$; and for a given point c on the compliance scale let

$$\hat{\theta}_c^* = t(c, \boldsymbol{x}^*) \tag{20.27}$$

be the value of $t(\boldsymbol{x}^*)$ evaluated at compliance $= c$.

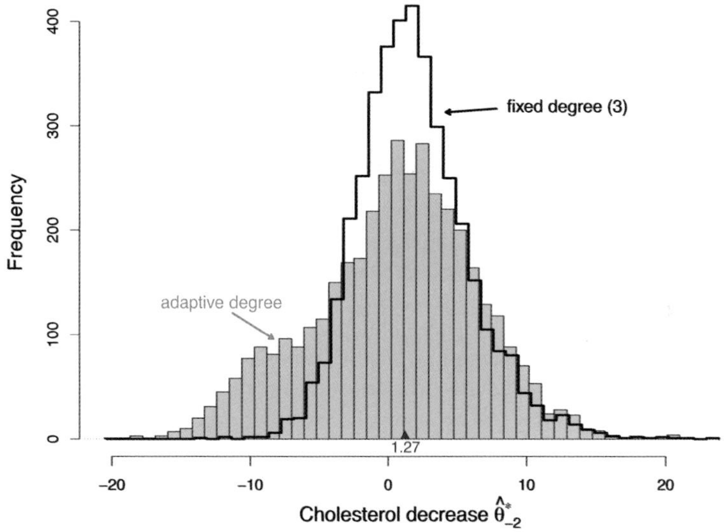

Figure 20.4 A histogram of 4000 nonparametric bootstrap replications for polynomial regression estimates of cholesterol decreases d at adjusted compliance $c = -2$. Solid histogram, adaptive estimator $\hat{\theta}_c^*$ (20.27), using full C_p/OLS algorithm for each bootstrap data set; line histogram, using OLS only with degree 3 for each bootstrap data set. Bootstrap standard errors are 5.98 and 3.97.

$B = 4000$ nonparametric bootstrap replications $t(\boldsymbol{x}^*)$ were generated.[3] Figure 20.4 shows the histogram of the 4000 $\hat{\theta}_c^*$ replications for $c = -2.0$. It is labeled "adaptive" to indicate that C_p model selection, as well as OLS fitting, was carred out anew for each $\boldsymbol{x}^*$. This is as opposed to the "fixed" histogram, where there was no C_p selection, cubic OLS regression always being used.

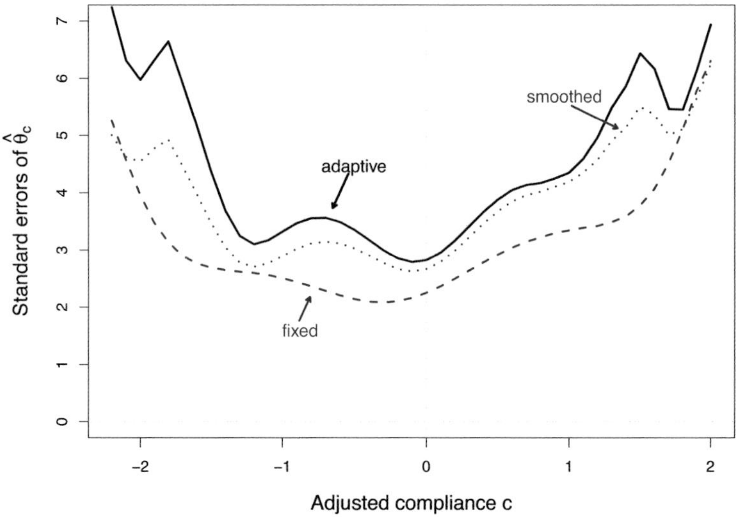

Figure 20.5 Bootstrap standard-error estimates of $\hat{\theta}_c$, for $-2.2 \le c \le 2$. Solid black curve, adaptive estimator (20.27) using full C_p/OLS model selection estimate; red dashed curve, using OLS only with polynomial degree fixed at 3; blue dotted curve, "bagged estimator" using bootstrap smoothing (20.28). Average standard-error ratios: adaptive/fixed $= 1.43$, adaptive/smoothed $= 1.14$.

The bootstrap estimate of standard error (10.16) obtained from the adaptive values $\hat{\theta}_c^*$ was 5.98, compared with 3.97 for the fixed values.[4] In this case, accounting for model selection ("adaptation") adds more than 50% to the standard error estimates. The same comparison was made at all values

[3] Ten times more than needed for assessing standard errors, but helpful for the comparisons that follow.

[4] The latter is not the usual OLS assessment, following (8.30), that would be appropriate for a parametric bootstrap comparison. Rather, it's the nonparametric one-sample bootstrap assessment, resampling pairs (x_i, y_i) as individual sample points.

of the adjusted compliance c. Figure 20.5 graphs the results: the adaptive standard errors averaged 43% greater than the fixed values. The standard 95% confidence intervals $\hat{\theta}_c \pm \widehat{\text{se}} \cdot 1.96$ would be roughly 43% too short if we ignored model selection in assessing $\widehat{\text{se}}$.

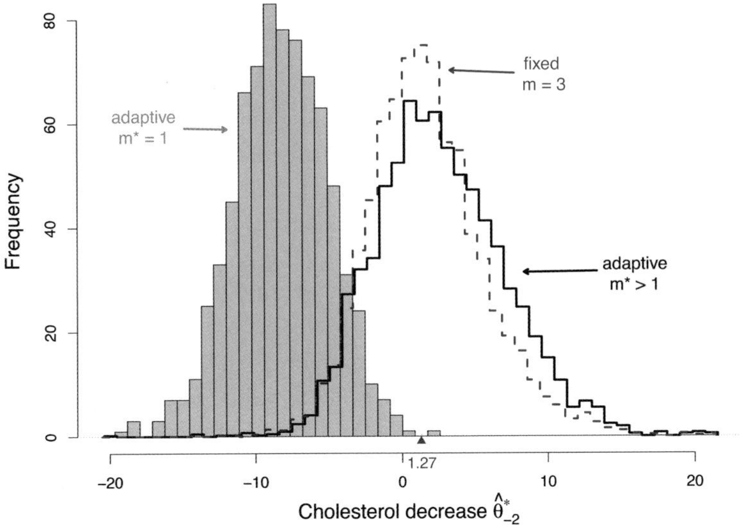

Figure 20.6 "Adaptive" histogram of Figure 20.4 now split into 19% of 4000 bootstrap replications where C_p selected linear regression ($m^* = 1$) as best, versus 81% having $m^* > 1$. $m^* = 1$ cases are shifted about 10 units downward. (The $m^* > 1$ cases resemble the "fixed" histogram in Figure 20.4.) Histograms are scaled to have equal areas.

Having an honest assessment of standard error doesn't mean that $t(c, \boldsymbol{x}^*)$ (20.27) is a good estimator. Model selection can induce an unpleasant "jumpiness" in an estimator, as the original data vector $\boldsymbol{x}$ crosses definitional boundaries. This happened in our example: for 19% of the 4000 bootstrap samples $\boldsymbol{x}^*$, the C_p algorithm selected linear regression, $m^* = 1$, as best, and in these cases $\hat{\theta}^*_{-2.0}$ tended toward smaller values. Figure 20.6 shows the $m^* = 1$ histogram shifted about 10 units down from the $m^* > 1$ histogram (which now resembles the "fixed" histogram in Figure 20.4).

Discontinuous estimators such as $t(c, \boldsymbol{x})$ can't be Bayesian, Bayes posterior expectations being continuous. They can also suffer frequentist difficulties,[†] including excess variability and overly long confidence intervals. †3

Bagging, or *bootstrap smoothing*, is a tactic for improving a discontinuous estimation rule by averaging (as in (12.80) and Chapter 17).

Suppose $t(\boldsymbol{x})$ is any estimator for which we have obtained bootstrap replications $\{t(\boldsymbol{x}^{*b}),\ b = 1, 2, \ldots, B\}$. The bagged version of $t(\boldsymbol{x})$ is the average

$$s(\boldsymbol{x}) = \frac{1}{B}\sum_{b=1}^{B} t(\boldsymbol{x}^{*b}). \tag{20.28}$$

The letter s here stands for "smooth." Small changes in $\boldsymbol{x}$, even ones that move across a model selection definitional boundary, produce only small changes in the bootstrap average $s(\boldsymbol{x})$.

Averaging over the 4000 bootstrap replications of $t(c, \boldsymbol{x}^*)$ (20.27) gave a bagged estimate $s_c(\boldsymbol{x})$ for each value of c. Bagging reduced the standard errors of the C_p/OLS estimates $t(c, \boldsymbol{x})$ by about 12%, as indicated by the green dotted curve in Figure 20.5.

Where did the green dotted curve come from? All 4000 bootstrap values $t(c, \boldsymbol{x}^{*b})$ were needed to produce the single value $s_c(\boldsymbol{x})$. It seems as if we would need to bootstrap the bootstrap in order to compute $\widehat{\text{se}}[s_c(\boldsymbol{x})]$. Fortunately, a more economical calculation is possible, one that requires only the original B bootstrap computations for $t(c, \boldsymbol{x})$.

Define

$$N_{bj} = \#\{\text{times } x_j \text{ occurs in } \boldsymbol{x}^{*b}\}, \tag{20.29}$$

for $b = 1, 2, \ldots, B$ and $j = 1, 2, \ldots, n$. For instance $N_{4000,7} = 3$ says that data point x_7 occurred three times in nonparametric bootstrap sample $\boldsymbol{x}^{4000}$. The B by n matrix $\{N_{bj}\}$ completely describes the B bootstrap samples. Also denote

$$t^{*b} = t(\boldsymbol{x}^{*b}) \tag{20.30}$$

and let cov_j indicate the covariance in the bootstrap sample between N_{bj} and t^{*b},

$$\text{cov}_j = \frac{1}{B}\sum_{b=1}^{B}(N_{bj} - N_{\cdot j})(t^{*b} - t^{*\cdot}), \tag{20.31}$$

where dots denote averaging over B: $N_{\cdot j} = \frac{1}{B}\sum_b N_{bj}$ and $t^{*\cdot} = \frac{1}{B}\sum_b t^{*b}$.

†4 **Theorem 20.2** † *The infinitesimal jackknife estimate of standard error*

(10.41) *for the bagged estimate* (20.28) *is*

$$\widehat{\text{se}}_{\text{IJ}}\left[s_c(\boldsymbol{x})\right] = \left(\sum_{j=1}^{n} \text{cov}_j^2 / n\right)^{1/2}. \tag{20.32}$$

Keeping track of N_{bj} *as we generate the bootstrap replications* t^{*b} *allows us to compute* cov_j *and* $\widehat{\text{se}}[s_c(\boldsymbol{x})]$ *without any additional computational effort.*

We expect averaging to reduce variability, and this is seen to hold true in Figure 20.5, the ratio of $\widehat{\text{se}}_{\text{IJ}}[s_c(\boldsymbol{x})]/\widehat{\text{se}}_{\text{boot}}[t(c,\boldsymbol{x})]$ averaging 0.88. In fact, we have the following general result.

Corollary *The ratio* $\widehat{\text{se}}_{\text{IJ}}[s_c(\boldsymbol{x})]/\widehat{\text{se}}_{\text{boot}}[t(c,\boldsymbol{x})]$ *is always* ≤ 1.

The savings due to bagging increase with the nonlinearity of $t(\boldsymbol{x}^*)$ as a function of the counts N_{bj} (or, in the language of Section 10.3, in the nonlinearity of $S(\boldsymbol{P})$ as a function of $\boldsymbol{P}$). Model-selection estimators such as the C_p/OLS rule tend toward greater nonlinearity and bigger savings.

Table 20.3 *Proportion of 4000 nonparametric bootstrap replications of* C_p*/OLS algorithm that selected degrees* $m = 1, 2, \ldots, 6$*; also infinitesimal jackknife standard deviations for proportions* (20.32)*, which mostly exceed the estimates themselves.*

	$m = 1$	2	3	4	5	6
proportion	.19	.12	.35	.07	.20	.06
$\widehat{\text{sd}}_{\text{IJ}}$	.24	.20	.24	.13	.26	.06

The first line of Table 20.3 shows the proportions in which the various degrees were selected in the 4000 cholesterol bootstrap replications, 19% for linear, 12% for quadratic, 35% for cubic, etc. With $B = 4000$, the proportions seem very accurate, the binomial standard error for 0.19 being just $(0.19 \cdot 0.81/4000)^{1/2} = 0.006$, for instance.

Theorem 20.2 suggests otherwise. Now let t^{*b} (20.30) indicate whether the bth bootstrap sample $\boldsymbol{x}^*$ made the C_p choice $m^* = 1$,

$$t^{*b} = \begin{cases} 1 & \text{if } m^{*b} = 1 \\ 0 & \text{if } m^{*b} > 1. \end{cases} \tag{20.33}$$

The bagged value of $\{t^{*b}, b = 1, 2, \ldots, B\}$ is the observed proportion

0.19. Applying the bagging theorem yielded $\widehat{\text{se}}_{\text{IJ}} = 0.24$, as seen in the second line of the table, with similarly huge standard errors for the other proportions.

The binomial standard errors are *internal*, saying how quickly the bootstrap resampling process is converging to its ultimate value as $B \to \infty$. The infinitesimal jackknife estimates are *external*: if we collected a new set of 164 data pairs (c_i, d_i) (20.2) the new proportion table might look completely different than the top line of Table 20.3.

Frequentist statistics has the advantage of being applicable to any algorithmic procedure, for instance to our C_p/OLS estimator. This has great appeal in an era of enormous data sets and fast computation. The drawback, compared with Bayesian statistics, is that we have no guarantee that our chosen algorithm is best in any way. Classical statistics developed a theory of *best* for a catalog of comparatively simple estimation and testing problems. In this sense, modern inferential theory has not yet caught up with modern problems such as data-based model selection, though techniques such as *model averaging* (e.g., bagging) suggest promising steps foward.

20.3 Selection Bias

Many a sports fan has been victimized by selection bias. Your team does wonderfully well and tops the league standings. But the next year, with the same players and the same opponents, you're back in the pack. This is the *winner's curse*, a more picturesque name for selection bias, the tendency of unusually good (or bad) comparative performances not to repeat themselves.

Modern scientific technology allows the simultaneous investigation of hundreds or thousands of candidate situations, with the goal of choosing the top performers for subsequent study. This is a setup for the heartbreak of selection bias. An apt example is offered by the prostate study data of Section 15.1, where we observe statistics z_i measuring patient–control differences for $N = 6033$ genes,

$$z_i \sim \mathcal{N}(\mu_i, 1), \qquad i = 1, 2, \ldots, N. \tag{20.34}$$

Here μ_i is the *effect size* for gene i, the true difference between the patient and control populations.

Genes with large positive or negative values of μ_i would be promising targets for further investigation. Gene number 610, with $z_{610} = 5.29$, attained the biggest z-value; (20.34) says that z_{610} is unbiased for μ_{610}. Can we believe the obvious estimate $\hat{\mu}_{610} = 5.29$?

"No" is the correct selection bias answer. Gene 610 has won a contest for bigness among 6033 contenders. In addition to being *good* (having a large value of μ) it has almost certainly been *lucky*, with the noise in (20.34) pushing z_{610} in the positive direction—or else it would not have won the contest. This is the essence of selection bias.

False-discovery rate theory, Chapter 15, provided a way to correct for selection bias in simultaneous hypothesis testing. This was extended to false-coverage rates in Section 20.1. Our next vignette concerns the realistic estimation of effect sizes μ_i in the face of selection bias.

We begin by assuming that an effect size μ has been obtained from a prior density $g(\mu)$ (which might include discrete atoms) and then $z \sim \mathcal{N}(\mu, \sigma^2)$ observed,

$$\mu \sim g(\cdot) \quad \text{and} \quad z|\mu \sim \mathcal{N}(\mu, \sigma^2) \tag{20.35}$$

(σ^2 is assumed known for this discussion). The marginal density of z is

$$f(z) = \int_{-\infty}^{\infty} g(\mu)\phi_\sigma(z-\mu)\, d\mu,$$
$$\text{where } \phi_\sigma(z) = (2\pi\sigma^2)^{-1/2} \exp\left(-\frac{1}{2}\frac{z^2}{\sigma^2}\right). \tag{20.36}$$

Tweedie's formula† is an intriguing expression for the Bayes expectation of μ given z. †5

Theorem 20.3 *In model* (20.35), *the posterior expectation of* μ *having observed* z *is*

$$E\{\mu|z\} = z + \sigma^2 l'(z) \qquad \text{with } l'(z) = \frac{d}{dz}\log f(z). \tag{20.37}$$

The especially convenient feature of Tweedie's formula is that $E\{\mu|z\}$ is expressed directly in terms of the marginal density $f(z)$. This is a setup for empirical Bayes estimation. We don't know $g(\mu)$, but in large-scale situations we can estimate the marginal density $f(z)$ from the observations $\boldsymbol{z} = (z_1, z_2, \ldots, z_N)$, perhaps by Poisson regression as in Table 15.1, yielding

$$\hat{E}\{\mu_i|z_i\} = z_i + \sigma^2 \hat{l}'(z_i) \qquad \text{with } \hat{l}'(z) = \frac{d}{dz}\log \hat{f}(z). \tag{20.38}$$

The solid curve in Figure 20.7 shows $\hat{E}\{\mu|z\}$ for the prostate study data,

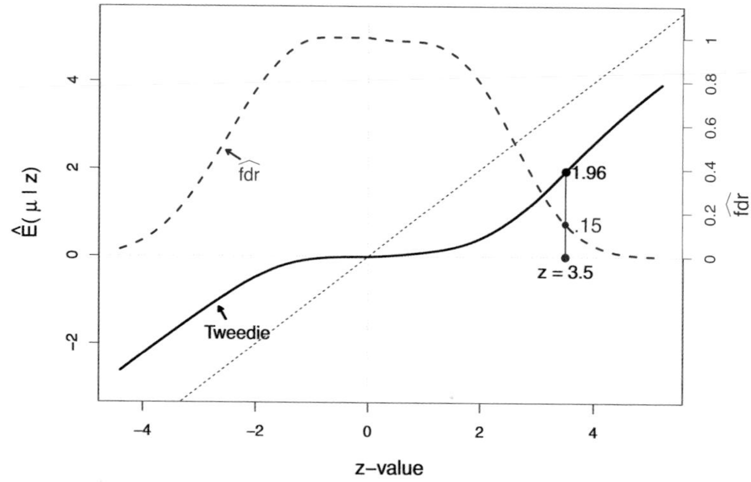

Figure 20.7 The solid curve is Tweedie's estimate $\hat{E}\{\mu|z\}$ (20.38) for the `prostate` study data. The dashed line shows the local false-discovery rate $\widehat{\text{fdr}}(z)$ from Figure 15.5 (red scale on right). At $z = 3.5$, $\hat{E}\{\mu|z\} = 1.96$ and $\widehat{\text{fdr}}(z) = 0.15$. For gene 610, with $z_{610} = 5.29$, Tweedie's estimate is 4.03.

with $\sigma^2 = 1$ and $\hat{f}(z)$ obtained using fourth-degree log polynomial regression as in Section 15.4. The curve has $E\{\mu|z\}$ hovering near zero for $|z_i| \leq 2$, agreeing with the local false-discovery rate curve $\widehat{\text{fdr}}(z)$ of Figure 15.5 that says these are mostly null genes.

$\hat{E}\{\mu|z\}$ increases for $z > 2$, equaling 1.96 for $z = 3.5$. At that point $\widehat{\text{fdr}}(z) = 0.15$. So even though $z_i = 3.5$ has a one-sided p-value of 0.0002, with 6033 genes to consider at once, it still is not a sure thing that gene i is non-null. About 85% of the genes with z_i near 3.5 will be non-null, and these will have effect sizes averaging about 2.31 ($= 1.96/0.85$). All of this nicely illustrates the combination of frequentist and Bayesian inference possible in large-scale studies, and also the combination of estimation and hypothesis-testing ideas in play.

If the prior density $g(\mu)$ in (20.35) is assumed to be normal, Tweedie's formula (20.38) gives (almost) the James–Stein estimator (7.13). The corresponding curve in Figure 20.7 in that case would be a straight line passing through the origin at slope 0.22. Like the James–Stein estimator, ridge regression, and the lasso of Chapter 16, Tweedie's formula is a shrinkage estimator. For $z_{610} = 5.29$, the most extreme observation, it gave

$\hat{\mu}_{629} = 4.03$, shrinking the maximum likelihood estimate more than one σ unit toward the origin.

Bayes estimators are immune to selection bias, as discussed in Sections 3.3 and 3.4. This offers some hope that Tweedie's empirical Bayes estimates might be a realistic cure for the winners' curse. A small simulation experiment was run as a test.

- A hundred data sets $\boldsymbol{z}$, each of length $N = 1000$, were generated according to a combination of exponential and normal sampling,

$$\mu_i \overset{\text{ind}}{\sim} e^{-\mu} \quad (\mu > 0) \quad \text{and} \quad z_i|\mu_i \overset{\text{ind}}{\sim} \mathcal{N}(\mu_i, 1), \tag{20.39}$$

  for $i = 1, 2, \ldots, 1000$.
- For each $\boldsymbol{z}$, $\hat{l}(z)$ was computed as in Section 15.4, now using a natural spline model with five degrees of freedom.
- This gave Tweedie's estimates

$$\hat{\mu}_i = z_i + \hat{l}'(z_i), \qquad i = 1, 2, \ldots, 1000, \tag{20.40}$$

  for that data set $\boldsymbol{z}$.
- For each data set $\boldsymbol{z}$, the 20 largest z_i values and the corresponding $\hat{\mu}_i$ and μ_i valus were recorded, yielding the

$$\begin{aligned} &\textit{uncorrected differences} \quad z_i - \mu_1 \\ \text{and} \quad &\textit{corrected differences} \quad \hat{\mu}_i - \mu_i, \end{aligned} \tag{20.41}$$

  the hope being that empirical Bayes shrinkage would correct the selection bias in the z_i values.
- Figure 20.8 shows the 2000 (100 data sets, 20 top cases each) uncorrected and corrected differences. Selection bias is quite obvious, with the uncorrected differences shifted one unit to the right of zero. In this case at least, the empirical Bayes corrections have worked well, the corrected differences being nicely centered at zero. Bias correction often adds variance, but in this case it hasn't.

Finally, it is worth saying that the "empirical" part of empirical Bayes is less the estimation of Bayesian rules from the aggregate data than the application of such rules to individual cases. For the prostate data we began with no definite prior opinions but arrived at strong (i.e., *not* "uninformative") Bayesian conclusions for, say, μ_{610} in the prostate study.

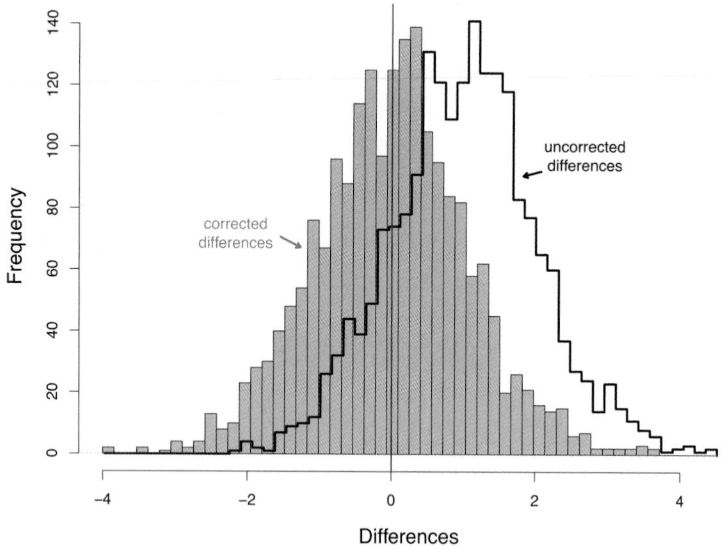

Figure 20.8 Corrected and uncorrected differences for 20 top cases in each of 100 simulations (20.39)–(20.41). Tweedie corrections effectively counteracted selection bias.

20.4 Combined Bayes–Frequentist Estimation

As mentioned previously, Bayes estimates are, at least theoretically, immune from selection bias. Let $\boldsymbol{z} = (z_1, z_2, \ldots, z_N)$ represent the prostate study data of the previous section, with parameter vector $\boldsymbol{\mu} = (\mu_1, \mu_2, \ldots, \mu_N)$. Bayes' rule (3.5)

$$g(\boldsymbol{\mu}|\boldsymbol{z}) = g(\boldsymbol{\mu}) f_{\boldsymbol{\mu}}(\boldsymbol{z}) / f(\boldsymbol{z}) \tag{20.42}$$

yields the posterior density of $\boldsymbol{\mu}$ given $\boldsymbol{z}$. A data-based model selection rule such as "estimate the $\boldsymbol{\mu}$ corresponding to the largest observation z_i" has no effect on the likelihood function $f_{\boldsymbol{\mu}}(\boldsymbol{z})$ (with $\boldsymbol{z}$ fixed) or on $g(\boldsymbol{\mu}|\boldsymbol{z})$. Having chosen a prior $g(\boldsymbol{\mu})$, our posterior estimate of μ_{610} is unaffected by the fact that $z_{610} = 5.29$ happens to be largest.

This same argument applies just as well to any data-based model selection procedure, for instance a preliminary screening of possible variables to include in a regression analysis—the C_p choice of a cubic regression in Figure 20.1 having no effect on its Bayes posterior accuracy.

There is a catch: the chosen prior $g(\boldsymbol{\mu})$ must apply to the entire parameter vector $\boldsymbol{\mu}$ and not just the part we are interested in (e.g., μ_{610}). This is feasible in one-parameter situations like the stopping rule example of Figure 3.3. It becomes difficult and possibly dangerous in higher dimensions. Empirical Bayes methods such as Tweedie's rule can be thought of as allowing the data vector $\boldsymbol{z}$ to assist in the choice of a high-dimensional prior, an effective collaboration between Bayesian and frequentist methodology.

Our chapter's final vignette concerns another Bayes–frequentist estimation technique. Dropping the boldface notation, suppose that $\mathcal{F} = \{f_\alpha(x)\}$ is a multi-dimensional family of densities (5.1) (now with α playing the role of μ), and that we are interested in estimating a particular parameter $\theta = t(\alpha)$. A prior $g(\alpha)$ has been chosen, yielding posterior expectation

$$\hat{\theta} = E\{t(\alpha)|x\}. \tag{20.43}$$

How accurate is $\hat{\theta}$? The usual answer would be calculated from the posterior distribution of θ given x. This is obviously the correct answer if $g(\alpha)$ is based on genuine prior experience. Most often though, and especially in high-dimensional problems, the prior reflects mathematical convenience and a desire to be uninformative, as in Chapter 13. There is a danger of circular reasoning in using a self-selected prior distribution to calculate the accuracy of its own estimator.

An alternate approach, discussed next, is to calculate the *frequentist* accuracy of $\hat{\theta}$; that is, even though (20.43) is a Bayes estimate, we consider $\hat{\theta}$ simply as a function of x, and compute its frequentist variability. The next theorem leads to a computationally efficient way of doing so. (The Bayes and frequentist standard errors for $\hat{\theta}$ operate in conceptually orthogonal directions as pictured in Figure 3.5. Here we are supposing that the prior $g(\cdot)$ is unavailable or uncertain, forcing more attention on frequentist calculations.)

For convenience, we will take the family $\mathcal{F}$ to be a p-parameter exponential family (5.50),

$$f_\alpha(x) = e^{\alpha' x - \psi(\alpha)} f_0(x), \tag{20.44}$$

now with α being the parameter vector called μ above. The $p \times p$ covariance matrix of x (5.59) is denoted

$$V_\alpha = \text{cov}_\alpha(x). \tag{20.45}$$

Let Cov_x indicate the posterior covariance given x between $\theta = t(\alpha)$, the parameter of interest, and α,

$$\text{Cov}_x = \text{cov}\{\alpha, t(\alpha)|x\}, \tag{20.46}$$

a $p \times 1$ vector. Cov_x leads directly to a frequentist estimate of accuracy for $\hat{\theta}$.

†6 **Theorem 20.4** † *The delta method estimate of standard error for* $\hat{\theta} = E\{t(\alpha)|x\}$ (10.41) *is*

$$\widehat{\mathrm{se}}_{\mathrm{delta}}\left\{\hat{\theta}\right\} = \left(\mathrm{Cov}_x' \, \boldsymbol{V}_{\hat{\alpha}} \, \mathrm{Cov}_x\right)^{1/2}, \tag{20.47}$$

where $\boldsymbol{V}_{\hat{\alpha}}$ *is* $\boldsymbol{V}_{\alpha}$ *evaluated at the MLE* $\hat{\alpha}$.

The theorem allows us to calculate the frequentist accuracy estimate $\widehat{\mathrm{se}}_{\mathrm{delta}}\{\hat{\theta}\}$ with hardly any additional computational effort beyond that required for $\hat{\theta}$ itself. Suppose we have used an MCMC or Gibbs sampling algorithm, Section 13.4, to generate a sample from the Bayes posterior distribution of α given x,

$$\alpha^{(1)}, \alpha^{(2)}, \ldots, \alpha^{(B)}. \tag{20.48}$$

These yield the usual estimate for $E\{t(\alpha)|x\}$,

$$\hat{\theta} = \frac{1}{B}\sum_{b=1}^{B} t\left(\alpha^{(b)}\right). \tag{20.49}$$

They also give a similar expression for $\mathrm{cov}\{\alpha, t(\alpha)|x\}$,

$$\mathrm{Cov}_x = \frac{1}{B}\sum_{b=1}^{B}\left(\alpha^{(b)} - \alpha^{(\cdot)}\right)\left(t^{(b)} - t^{(\cdot)}\right), \tag{20.50}$$

$t^{(b)} = t(\alpha^{(b)})$, $t^{(\cdot)} = \sum_b t^{(b)}/B$, and $\alpha^{(\cdot)} = \sum_b \alpha^{(b)}/B$, from which we can calculate[5] $\widehat{\mathrm{se}}_{\mathrm{delta}}(\hat{\theta})$ (20.47).

For an example of Theorem 20.4 in action we consider the `diabetes` data of Section 20.1, with x_i' the ith row of $\boldsymbol{X}$, the 442×10 matrix of prediction, so x_i is the vector of 10 predictors for patient i. The response vector $\boldsymbol{y}$ of progression scores has now been rescaled to have $\sigma^2 = 1$ in the normal regression model,[6]

$$\boldsymbol{y} \sim \mathcal{N}_n(\boldsymbol{X}\beta, \boldsymbol{I}). \tag{20.51}$$

The prior distribution $g(\beta)$ was taken to be

$$g(\beta) = c e^{-\lambda \|\beta\|_1}, \tag{20.52}$$

[5] $\boldsymbol{V}_{\hat{\alpha}}$ may be known theoretically, calculated by numerical differentiation in (5.57), or obtained from parametric bootstrap resampling—taking the empirical covariance matrix of bootstrap replications $\hat{\beta}_i^*$.

[6] By dividing the original data vector $\boldsymbol{y}$ by its estimated standard error from the linear model $E\{\boldsymbol{y}\} = \boldsymbol{X}\beta$.

with $\lambda = 0.37$ and c the constant that makes $g(\beta)$ integrate to 1. This is the "Bayesian lasso prior,"[†] so called because of its connection to the lasso, (7.42) and (16.1). (The lasso plays no part in what follows). †7

An MCMC algorithm generated B =10,000 samples (20.48) from the posterior distribution $g(\beta|\boldsymbol{y})$. Let

$$\theta_i = x_i'\beta, \tag{20.53}$$

the (unknown) expectation of the ith patient's response y_i. The Bayes posterior expectation of θ_i is

$$\hat{\theta}_i = \frac{1}{B}\sum_{b=1}^{B} x_i'\beta. \tag{20.54}$$

It has Bayes posterior standard error

$$\widehat{\text{se}}_{\text{Bayes}}\left(\hat{\theta}_i\right) = \left[\frac{1}{B}\sum_{b=1}^{B}\left(x_i'\beta^{(b)} - \hat{\theta}_i\right)^2\right]^{1/2}, \tag{20.55}$$

which we can compare with $\widehat{\text{se}}_{\text{delta}}(\hat{\theta}_i)$, the frequentist standard error (20.47).

Figure 20.9 shows the 10,000 MCMC replications $\hat{\theta}_i^{(b)} = x_i'\beta$ for patient $i = 322$. The point estimate $\hat{\theta}_i$ equaled 2.41, with Bayes and frequentist standard error estimates

$$\widehat{\text{se}}_{\text{Bayes}} = 0.203 \quad \text{and} \quad \widehat{\text{se}}_{\text{delta}} = 0.186. \tag{20.56}$$

The frequentist standard error is 9% smaller in this case; $\widehat{\text{se}}_{\text{delta}}$ was less than $\widehat{\text{se}}_{\text{Bayes}}$ for all 442 patients, the difference averaging a modest 5%.

Things can work out differently. Suppose we are interested in the posterior cdf of θ_{332} given $\boldsymbol{y}$. For any given value of c let

$$t\left(c, \beta^{(b)}\right) = \begin{cases} 1 & \text{if } x_{322}'\beta^{(b)} \le c \\ 0 & \text{if } x_{332}'\beta^{(b)} > c, \end{cases} \tag{20.57}$$

so

$$\text{cdf}(c) = \frac{1}{B}\sum_{b=1}^{B} t\left(c, \beta^{(b)}\right) \tag{20.58}$$

is our MCMC assessment of $\Pr\{\theta_{322} \le c|\boldsymbol{y}\}$. The solid curve in Figure 20.10 graphs cdf(c).

If we believe prior (20.52) then the curve *exactly* represents the posterior distribution of θ_{322} given $\boldsymbol{y}$ (except for the simulation error due to stopping at B =10,000 replications). Whether or not we believe the prior we can use

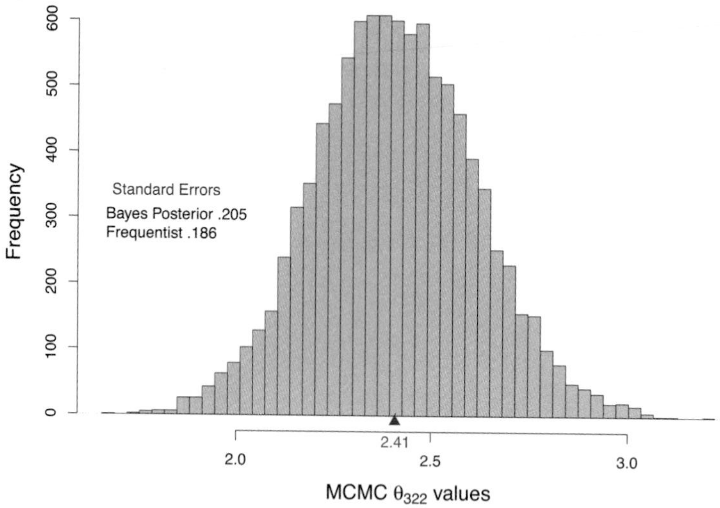

Figure 20.9 A histogram of 10,000 MCMC replications for posterior distribution of θ_{322}, expected progression for patient 322 in the `diabetes` study; model (20.51) and prior (20.52). The Bayes posterior expectation is 2.41. Frequentist standard error (20.47) for $\hat{\theta}_{322} = 2.41$ was 9% smaller than Bayes posterior standard error (20.55).

Theorem 20.4, with $t^{(b)} = t(c, \beta^{(b)})$ in (20.50), to evaluate the frequentist accuracy of the curve.

The dashed vertical red lines show cdf(c) plus or minus one $\widehat{\text{se}}_{\text{delta}}$ unit. The standard errors are disturbingly large, for instance 0.687 ± 0.325 at $c = 2.5$. The central 90% credible interval for θ_{322} (the c-values between cdf(c) 0.05 and 0.95),

$$(2.08, 2.73) \tag{20.59}$$

has frequentist standard errors about 0.185 for each endpoint—28% of the interval's length.

If we believe prior (20.52) then (2.08, 2.73) is an (almost) exact 90% credible interval for θ_{322}, and moreover is immune to any selection bias involved in our focus on θ_{322}. If not, the large frequentist standard errors are a reminder that calculation (20.59) might turn out much differently in a new version of the diabetes study, even ignoring selection bias.

To return to our main theme, Bayesian calculations encourage a disregard for model selection effects. This can be dangerous in objective Bayes

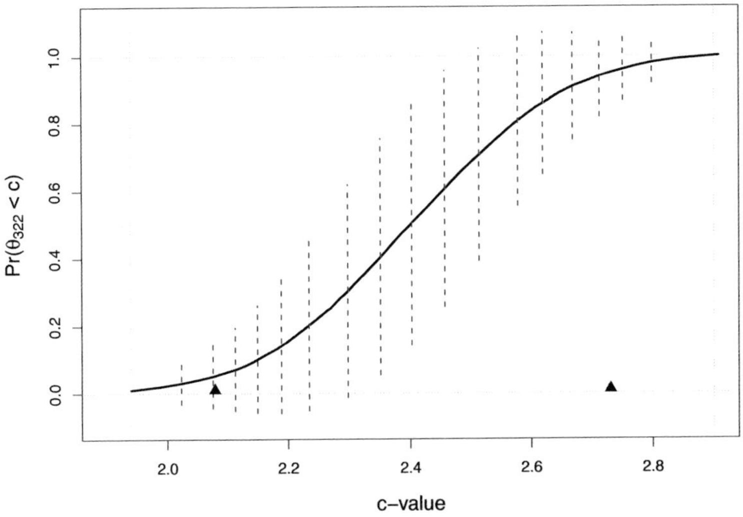

Figure 20.10 The solid curve is the posterior cdf of θ_{322}. Vertical red bars indicate $\pm$ one frequentist standard error, as obtained from Theorem 20.4. Black triangles are endpoints of the 0.90 central credible interval.

settings where one can't rely on genuine prior experience. Theorem 20.4 serves as a frequentist checkpoint, offering some reassurance as in Figure 20.9, or sounding a warning as in Figure 20.10.

20.5 Notes and Details

Optimality theories—statements of best possible results—are marks of maturity in applied mathematics. Classical statistics achieved two such theories: for unbiased or asymptotically unbiased estimation, and for hypothesis testing. Most of this book and all of this chapter venture beyond these safe havens. How far from *best* are the C_p/OLS bootstrap smoothed estimates of Section 20.2? At this time we can't answer such questions, though we can offer appealing methodologies in their pursuit, a few of which have been highlighted here.

The cholostyramine example comes from Efron and Feldman (1991) where it is discussed at length. Data for a control group is also analyzed there.

†1 [p. 398] *Scheffé intervals.* Scheffé's 1953 paper came at the beginning

of a period of healthy development in simultaneous inference techniques, mostly in classical normal theory frameworks. Miller (1981) gives a clear and thorough summary. The 1980s followed with a more computer-intensive approach, nicely developed in Westfall and Young's 1993 book, leading up to Benjamini and Hochberg's 1995 false-discovery rate paper (Chapter 15 here), and Benjamini and Yekutieli's (2005) false-coverage rate algorithm.

Scheffé's construction (20.15) is derived by transforming (20.6) to the case $\boldsymbol{V} = \boldsymbol{I}$ using the inverse square root of matrix $\boldsymbol{V}$,

$$\hat{\gamma} = \boldsymbol{V}^{-1/2}\hat{\beta} \quad \text{and} \quad \gamma = \boldsymbol{V}^{-1/2}\beta \tag{20.60}$$

($(\boldsymbol{V}^{-1/2})^2 = \boldsymbol{V}^{-1}$), which makes the ellipsoid of Figure 20.2 into a circle. Then $Q = \|\hat{\gamma} - \gamma\|^2/\hat{\sigma}^2$ in (20.10), and for a linear combination $\gamma_d = d'\gamma$ it is straightforward to see that $\Pr\{Q \le k_{p,q}^{(\alpha)2}\} = \alpha$ amounts to

$$\gamma_d \in \hat{\gamma}_d \pm \hat{\sigma}\,\|d\|\,k_{p,q}^{(\alpha)} \tag{20.61}$$

for all choices of d, the geometry of Figure 20.2 now being transparent. Changing coordinates back to $\hat{\beta} = \boldsymbol{V}^{1/2}\hat{\gamma}$, $\beta = \boldsymbol{V}^{1/2}\gamma$, and $c = \boldsymbol{V}^{1/2}d$ yields (20.15).

†2 [p. 399] *Restricting the catalog $\boldsymbol{C}$.* Suppose that all the sample sizes n_j in (20.16) take the same value n, and that we wish to set simultaneous confidence intervals for all pairwise differences $\beta_i - \beta_j$. Tukey's *studentized range* pivotal quantity (1952, unpublished)

$$T = \max_{i \ne j} \frac{\left|\left(\hat{\beta}_i - \hat{\beta}_j\right) - (\beta_i - \beta_j)\right|}{\hat{\sigma}} \tag{20.62}$$

has a distribution not depending on σ or $\boldsymbol{\beta}$. This implies that

$$\beta_i - \beta_j \in \hat{\beta}_i - \hat{\beta}_j \pm \frac{\hat{\sigma}}{\sqrt{n}} T^{(\alpha)} \tag{20.63}$$

is a set of simultaneous level-α confidence intervals for all pairwise differences $\beta_i - \beta_j$, where $T^{(\alpha)}$ is the αth quantile of T. (The factor $1/\sqrt{n}$ comes from $\hat{\beta}_j \sim \mathcal{N}(\beta_j, \sigma^2/n)$ in (20.16).)

Table 20.4 *Half-width of Tukey studentized range simultaneous 95% confidence intervals for pairwise differences $\beta_i - \beta_j$ (in units of $\hat{\sigma}/\sqrt{n}$) for $p = 2, 3, \ldots, 6$ and $n = 20$; compared with Scheffé intervals* (20.15).

p	2	3	4	5	6
Tukey	2.95	3.58	3.96	4.23	4.44
Scheffé	3.74	4.31	4.79	5.21	5.58

Reducing the catalog $\boldsymbol{C}$ from all linear combinations $c'\beta$ to only pairwise differences shortens the simultaneous intervals. Table 20.4 shows the comparison between the Tukey and Scheffé 95% intervals for $p = 2, 3, \ldots, 6$ and $n = 20$.

Calculating $T^{(\alpha)}$ was a substantial project in the early 1980s. Berk *et al.* (2013) now carry out the analogous computations for general catalogs of linear constraints. They discuss at length the inferential basis of such procedures.

†3 [p. 405] *Discontinuous estimators.* Looking at Figure 20.6 suggests that a confidence interval for $\theta_{-2.0}\ t(c, \boldsymbol{x})$ will move far left for data sets x where C_p selects linear regression ($m = 1$) as best. This kind of "jumpy" behavior lengthens the intervals needed to attain a desired coverage level. More seriously, intervals for $m = 1$ may give misleading inferences, another example of "accurate but incorrect" behavior. Bagging (20.28), in addition to reducing interval length, improves inferential correctness, as discussed in Efron (2014a).

†4 [p. 406] *Theorem 20.2 and its corollary.* Theorem 20.2 is proved in Section 3 of Efron (2014a), with a parametric bootstrap version appearing in Section 4. The corollary is a projection result illustrated in Figure 4 of that paper: let $\mathcal{L}(\boldsymbol{N})$ be the n-dimensional subspace of B-dimensional Euclidean space spanned by the columns of the $B \times n$ matrix (N_{bj}) (20.29) and $\boldsymbol{t}^*$ the B-vector with components $t^{*b} - t^{*\cdot}$; then

$$\widehat{\mathrm{se}}_{\mathrm{IJ}}(s) \big/ \widehat{\mathrm{se}}_{\mathrm{boot}}(t) = \left\|\hat{\boldsymbol{t}}^*\right\| \big/ \|\boldsymbol{t}^*\|, \tag{20.64}$$

where $\hat{\boldsymbol{t}}^*$ is the projection of $\boldsymbol{t}^*$ into $\mathcal{L}(\boldsymbol{N})$. In the language of Section 10.3, if $\hat{\theta}^* = S(\boldsymbol{P})$ is very nonlinear as a function of $\boldsymbol{P}$, then the ratio in (20.64) will be substantially less than 1.

†5 [p. 409] *Tweedie's formula.* For convenience, take $\sigma^2 = 1$ in (20.35). Bayes' rule (3.5) can then be arranged to give

$$g(\mu|z) = e^{\mu z - \psi(z)} g(\mu) e^{-\frac{1}{2}\mu^2} \big/ \sqrt{2\pi} \tag{20.65}$$

with

$$\psi(z) = \frac{1}{2}z + \log f(z). \tag{20.66}$$

This is a one-parameter exponential family (5.46) having natural parameter α equal to z. Differentiating ψ as in (5.55) gives

$$E\{\mu|z\} = \frac{d\psi}{dz} = z + \frac{d}{dz}\log f(z), \tag{20.67}$$

which is Tweedie's formula (20.37) when $\sigma^2 = 1$. The formula first appears in Robbins (1956), who credits it to a personal communication from M. K. Tweedie. Efron (2011) discusses general exponential family versions of Tweedie's formula, and its application to selection bias situations.

†6 [p. 414] *Theorem 20.4.* The delta method standard error approximation for a statistic $T(x)$ is

$$\widehat{\text{se}}_{\text{delta}} = \left[(\nabla T(x))' \, \hat{V} \, (\nabla T(x)) \right]^{1/2}, \tag{20.68}$$

where $\nabla T(x)$ is the gradient vector $(\partial T / \partial x_j)$ and $\hat{V}$ is an estimate of the covariance matrix of x. Other names include the "Taylor series method," as in (2.10), and "propagation of errors" in the physical sciences literature. The proof of Theorem 20.4 in Section 2 of Efron (2015) consists of showing that $\text{Cov}_x = T(x)$ when $T(x) = E\{t(\alpha)|x\}$. Standard deviations are only a first step in assessing the frequentist accuracy of $T(x)$. The paper goes on to show how Theorem 20.4 can be improved to give confidence intervals, correcting the impression in Figure 20.10 that cdf(c) can range outside [0, 1].

†7 [p. 415] *Bayesian lasso.* Applying Bayes' rule (3.5) with density (20.51) and prior (20.52) gives

$$\log g(\beta|\boldsymbol{y}) = -\left\{ \frac{\|\boldsymbol{y} - \boldsymbol{X}\beta\|^2}{2} + \lambda \|\beta\|_1 \right\}, \tag{20.69}$$

as discussed in Tibshirani (2006). Comparison with (7.42) shows that the maximizing value of β (the "MAP" estimate) agrees with the lasso estimate. Park and Casella (2008) named the "Bayesian lasso" and suggested an appropriate MCMC algorithm. Their choice $\lambda = 0.37$ was based on marginal maximum likelihood calculations, giving their analysis an empirical Bayes aspect ignored in their and our analyses.

21

Empirical Bayes Estimation Strategies

Classic statistical inference was focused on the analysis of individual cases: a single estimate, a single hypothesis test. The interpretation of direct evidence bearing on the case of interest—the number of successes and failures of a new drug in a clinical trial as a familiar example—dominated statistical practice.

The story of modern statistics very much involves indirect evidence, "learning from the experience of others" in the language of Sections 7.4 and 15.3, carried out in both frequentist and Bayesian settings. The computer-intensive prediction algorithms described in Chapters 16–19 use regression theory, the frequentist's favored technique, to mine indirect evidence on a massive scale. False-discovery rate theory, Chapter 15, collects indirect evidence for hypothesis testing by means of Bayes' theorem as implemented through empirical Bayes estimation.

Empirical Bayes methodology has been less studied than Bayesian or frequentist theory. As with the James–Stein estimator (7.13), it can seem to be little more than plugging obvious frequentist estimates into Bayes estimation rules. This conceals a subtle and difficult task: learning the equivalent of a Bayesian prior distribution from ongoing statistical observations. Our final chapter concerns the empirical Bayes learning process, both as an exercise in applied deconvolution and as a relatively new form of statistical inference. This puts us back where we began in Chapter 1, examining the two faces of statistical analysis, the algorithmic and the inferential.

21.1 Bayes Deconvolution

A familiar formulation of empirical Bayes inference begins by assuming that an unknown prior density $g(\theta)$, our object of interest, has produced a random sample of real-valued variates $\Theta_1, \Theta_2, \ldots, \Theta_N$,

$$\Theta_i \overset{\text{iid}}{\sim} g(\theta), \qquad i = 1, 2, \ldots, N. \tag{21.1}$$

(The "density" $g(\cdot)$ may include discrete atoms of probability.) The Θ_i are unobservable, but each yields an observable random variable X_i according to a known family of density functions

$$X_i \overset{\text{ind}}{\sim} p_i(X_i|\Theta_i). \tag{21.2}$$

From the observed sample $X_1, X_2, \ldots, X_N$ we wish to estimate the prior density $g(\theta)$.

A famous example has $p_i(X_i|\Theta_i)$ the Poisson family,

$$X_i \sim \text{Poi}(\Theta_i), \tag{21.3}$$

as in Robbins' formula, Section 6.1. Still more familiar is the normal model (3.28),

$$X_i \sim \mathcal{N}(\Theta_i, \sigma^2), \tag{21.4}$$

often with $\sigma^2 = 1$. A binomial model was used in the medical example of Section 6.3,

$$X_i \sim \text{Bi}(n_i, \Theta_i). \tag{21.5}$$

There the n_i differ from case to case, accounting for the need for the first subscript i in $p_i(X_i|\Theta_i)$ (21.2).

Let $f_i(X_i)$ denote the *marginal density* of X_i obtained from (21.1)–(21.2),

$$f_i(X_i) = \int_{\mathcal{T}} p_i(X_i|\theta_i) g(\theta_i)\, d\theta_i, \tag{21.6}$$

the integral being over the space $\mathcal{T}$ of possible Θ values. The statistician has only the marginal observations available,

$$X_i \overset{\text{ind}}{\sim} f_i(\cdot), \qquad i = 1, 2, \ldots, N, \tag{21.7}$$

from which he or she wishes to estimate the density $g(\cdot)$ in (21.6).

In the normal model (21.4), f_i is the convolution of the unknown $g(\theta)$ with a known normal density, denoted

$$f = g * \mathcal{N}(0, \sigma^2) \tag{21.8}$$

(now f_i not depending on i). Estimating g using a sample $X_1, X_2, \ldots, X_N$ from f is a problem in *deconvolution*. In general we might call the estimation of g in model (3.1)–(3.2) the "Bayes deconvolution problem."

An artificial example appears in Figure 21.1, where $g(\theta)$ is a mixture distribution: seven-eighths $\mathcal{N}(0, 0.5^2)$ and one-eighth uniform over the interval $[-3, 3]$. A normal sampling model $X_i \overset{\text{ind}}{\sim} \mathcal{N}(\Theta_i, 1)$ is assumed, yielding f by convolution as in (21.8). The convolution process makes f wider

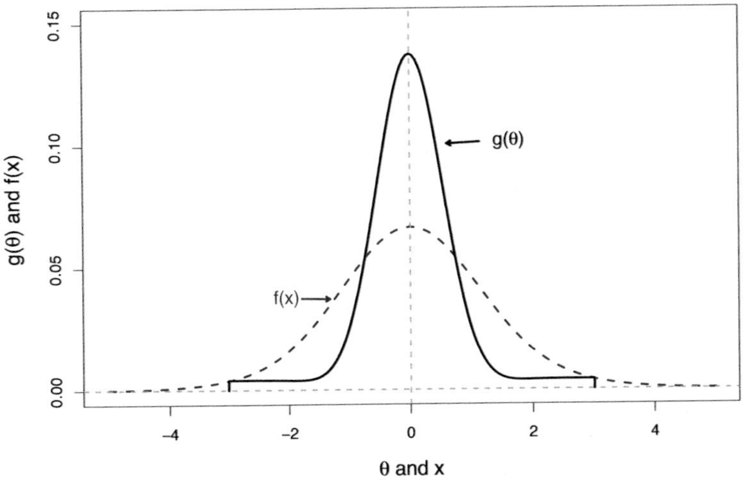

Figure 21.1 An artificial example of the Bayes deconvolution problem. The solid curve is $g(\theta)$, the prior density of Θ (21.1); the dashed curve is the density of an observation X from marginal distribution $f = g * \mathcal{N}(0, 1)$ (21.8). We wish to estimate $g(\theta)$ on the basis of a random sample $X_1, X_2, \dots, X_N$ from $f(x)$.

and smoother than g, as illustrated in the figure. Having observed a random sample from f, we wish to estimate the deconvolute g, which begins to look difficult in the figure's example.

Deconvolution has a well-deserved reputation for difficulty. It is the classic ill-posed problem: because of the convolution process (21.6), large changes in $g(\theta)$ are smoothed out, often yielding only small changes in $f(x)$. Deconvolution operates in the other direction, with small changes in the estimation of f disturbingly magnified on the g scale. Nevertheless, modern computation, modern theory, and most of all modern sample sizes, together can make empirical deconvolution a practical reality.

Why would we want to estimate $g(\theta)$? In the `prostate` data example (3.28) (where Θ is called μ) we might wish to know $\Pr\{\Theta = 0\}$, the probability of a *null* gene, ones whose effect size is zero; or perhaps $\Pr\{|\Theta| \geq 2\}$, the proportion of genes that are substantially non-null. Or we might want to estimate Bayesian posterior expectations like $E\{\Theta|X = x\}$ in Figure 20.7, or posterior densities as in Figure 6.5.

Two main strategies have developed for carrying out empirical Bayes estimation: modeling on the θ scale, called g-*modeling* here, and modeling on the x scale, called *f-modeling*. We begin in the next section with g-modeling.

21.2 g-Modeling and Estimation

There has been a substantial amount of work on the asymptotic accuracy of estimates $\hat{g}(\theta)$ in the empirical Bayes model (21.1)–(21.2), most often in the normal sampling framework (21.4). The results are discouraging, with the rate of convergence of $\hat{g}(\theta)$ to $g(\theta)$ as slow as $(\log N)^{-1}$. In our terminology, much of this work has been carried out in a nonparametric g-modeling framework, allowing the unknown prior density $g(\theta)$ to be virtually anything at all. More optimistic results are possible if the g-modeling is pursued parametrically, that is, by restricting $g(\theta)$ to lie within some parametric family of possibilities.

We assume, for the sake of simpler exposition, that the space $\mathcal{T}$ of possible Θ values is finite and discrete, say

$$\mathcal{T} = \left\{\theta_{(1)}, \theta_{(2)}, \ldots, \theta_{(m)}\right\}. \tag{21.9}$$

The prior density $g(\theta)$ is now represented by a vector $\boldsymbol{g} = (g_1, g_2, \ldots, g_m)'$, with components

$$g_j = \Pr\left\{\Theta = \theta_{(j)}\right\} \qquad \text{for } j = 1, 2, \ldots, m. \tag{21.10}$$

A p-parameter exponential family (5.50) for $\boldsymbol{g}$ can be written as

$$\boldsymbol{g} = \boldsymbol{g}(\alpha) = e^{\boldsymbol{Q}\alpha - \psi(\alpha)}, \tag{21.11}$$

where the p-vector α is the natural parameter and $\boldsymbol{Q}$ is a known $m \times p$ *structure matrix*. Notation (21.11) means that the jth component of $\boldsymbol{g}(\alpha)$ is

$$g_j(\alpha) = e^{Q'_j\alpha - \psi(\alpha)}, \tag{21.12}$$

with Q'_j the jth row of $\boldsymbol{Q}$; the function $\psi(\alpha)$ is the normalizer that makes $\boldsymbol{g}(\alpha)$ sum to 1,

$$\psi(\alpha) = \log\left(\sum_{j=1}^{m} e^{Q'_j\alpha}\right). \tag{21.13}$$

In the `nodes` example of Figure 6.4, the set of possible Θ values was $\mathcal{T} = \{0.01, 0.02, \ldots, 0.99\}$, and $\boldsymbol{Q}$ was a fifth-degree polynomial matrix,

$$\boldsymbol{Q} = \texttt{poly}(\mathcal{T}, 5) \tag{21.14}$$

in **R** notation, indicating a five-parameter exponential family for $\boldsymbol{g}$, (6.38)–(6.39).

In the development that follows we will assume that the kernel $p_i(\cdot|\cdot)$ in (21.2) does not depend on i, i.e., that X_i has the same family of conditional distributions $p(X_i|\Theta_i)$ for all i, as in the Poisson and normal situations (21.3) and (21.4), but not the binomial case (21.5). And moreover we assume that the sample space $\mathcal{X}$ for the X_i observations is finite and discrete, say

$$\mathcal{X} = \left\{x_{(1)}, x_{(2)}, \ldots, x_{(n)}\right\}. \tag{21.15}$$

None of this is necessary, but it simplifies the exposition.

Define

$$p_{kj} = \Pr\left\{X_i = x_{(k)}|\Theta_i = \theta_{(j)}\right\}, \tag{21.16}$$

for $k = 1, 2, \ldots, n$ and $j = 1, 2, \ldots, m$, and the corresponding $n \times m$ matrix

$$\boldsymbol{P} = (p_{kj}), \tag{21.17}$$

having kth row $P_k = (p_{k1}, p_{k2}, \ldots, p_{km})'$. The convolution-type formula (21.6) for the marginal density $f(x)$ now reduces to an inner product,

$$\begin{aligned} f_k(\alpha) &= \Pr_\alpha\left\{X_i = x_{(k)}\right\} = \textstyle\sum_{j=1}^m p_{kj}\, g_j(\alpha) \\ &= P_k' \boldsymbol{g}(\alpha). \end{aligned} \tag{21.18}$$

In fact we can write the entire marginal density $\boldsymbol{f}(\alpha) = (f_1(\alpha), f_2(\alpha), \ldots, f_n(\alpha))'$ in terms of matrix multiplication,

$$\boldsymbol{f}(\alpha) = \boldsymbol{P}\boldsymbol{g}(\alpha). \tag{21.19}$$

The vector of counts $\boldsymbol{y} = (y_1, y_2, \ldots, y_n)$, with

$$y_k = \#\left\{X_i = x_{(k)}\right\}, \tag{21.20}$$

is a sufficient statistic in the iid situation. It has a multinomial distribution (5.38),

$$\boldsymbol{y} \sim \mathrm{Mult}_n(N, \boldsymbol{f}(\alpha)), \tag{21.21}$$

indicating N independent draws for a density $\boldsymbol{f}(\alpha)$ on n categories.

All of this provides a concise description of the g-modeling probability model:

$$\alpha \to \boldsymbol{g}(\alpha) = e^{\boldsymbol{Q}\alpha - \psi(\alpha)} \to \boldsymbol{f}(\alpha) = \boldsymbol{P}\boldsymbol{g}(\alpha) \to \boldsymbol{y} \sim \mathrm{Mult}_n(N, \boldsymbol{f}(\alpha)). \tag{21.22}$$

The inferential task goes in the reverse direction,

$$y \to \hat{\alpha} \to f(\hat{\alpha}) \to g(\hat{\alpha}) = e^{Q\hat{\alpha}-\psi(\hat{\alpha})}. \tag{21.23}$$

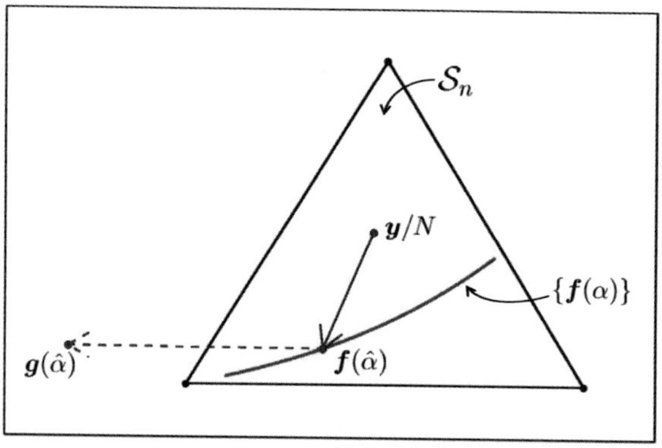

Figure 21.2 A schematic diagram of empirical Bayes estimation, as explained in the text. $\mathcal{S}_n$ is the n-dimensional simplex, containing the p-parameter family $\mathcal{F}$ of allowable probability distributions $f(\alpha)$. The vector of observed proportions y/N yields MLE $f(\hat{\alpha})$, which is then deconvolved to obtain estimate $g(\hat{\alpha})$.

A schematic diagram of the estimation process appears in Figure 21.2.

- The vector of observed proportions y/N is a point in $\mathcal{S}_n$, the simplex (5.39) of all possible probability vectors f on n categories; y/N is the usual nonparametric estimate of f.
- The parametric family of allowable f vectors (21.19)

$$\mathcal{F} = \{f(\alpha),\ \alpha \in A\}, \tag{21.24}$$

indicated by the red curve, is a curved p-dimensional surface in $\mathcal{S}_n$. Here A is the space of allowable vectors α in family (21.11).

- The nonparametric estimate y/N is "projected" down to the parametric estimate $f(\hat{\alpha})$; if we are using MLE estimation, $f(\hat{\alpha})$ will be the closest point in $\mathcal{F}$ to y/N measured according to a deviance metric, as in (8.35).
- Finally, $f(\hat{\alpha})$ is mapped back to the estimate $g(\hat{\alpha})$, by inverting mapping (21.19). (Inversion is not actually necessary with g-modeling since, having found $\hat{\alpha}$, $g(\hat{\alpha})$ is obtained directly from (21.11); the inversion step is more difficult for f-modeling, Section 21.6.)

The maximum likelihood estimation process for g-modeling is discussed in more detail in the next section, where formulas for its accuracy will be developed.

21.3 Likelihood, Regularization, and Accuracy[1]

Parametric g-modeling, as in (21.11), allows us to work in low-dimensional parametric families—just five parameters for the **nodes** example (21.14)—where classic maximum likelihood methods can be more confidently applied. Even here though, some regularization will be necessary for stable estimation, as discussed in what follows.

The g-model probability mechanism (21.22) yields a log likelihood for the multinomial vector $\boldsymbol{y}$ of counts as a function of α, say $l_{\boldsymbol{y}}(\alpha)$;

$$l_{\boldsymbol{y}}(\alpha) = \log\left(\prod_{k=1}^{n} f_k(\alpha)^{y_k}\right) = \sum_{k=1}^{n} y_k \log f_k(\alpha). \tag{21.25}$$

Its score function $\dot{l}_{\boldsymbol{y}}(\alpha)$, the vector of partial derivatives $\partial l_{\boldsymbol{y}}(\alpha)/\partial \alpha_h$ for $h = 1, 2, \ldots, p$, determines the MLE $\hat{\alpha}$ according to $\dot{l}_{\boldsymbol{y}}(\hat{\alpha}) = 0$. The $p \times p$ matrix of second derivatives $\ddot{l}_{\boldsymbol{y}}(\alpha) = (\partial^2 l_{\boldsymbol{y}}(\alpha)/\partial \alpha_h \partial \alpha_l)$ gives the *Fisher information matrix* (5.26)

$$\mathcal{I}(\alpha) = E\{-\ddot{l}_{\boldsymbol{y}}(\alpha)\}. \tag{21.26}$$

The exponential family model (21.11) yields simple expressions for $\dot{l}_{\boldsymbol{y}}(\alpha)$ and $\mathcal{I}(\alpha)$. Define

$$w_{kj} = g_j(\alpha)\left(\frac{p_{kj}}{f_k(\alpha)} - 1\right) \tag{21.27}$$

and the corresponding m-vector

$$W_k(\alpha) = (w_{k1}(\alpha), w_{k2}(\alpha), \ldots, w_{km}(\alpha))'. \tag{21.28}$$

Lemma 21.1 *The score function $\dot{l}_{\boldsymbol{y}}(\alpha)$ under model (21.22) is*

$$\dot{l}_{\boldsymbol{y}}(\alpha) = \boldsymbol{Q} W_+(\alpha), \qquad \textit{where } W_+(\alpha) = \sum_{k=1}^{n} W_k(\alpha) y_k \tag{21.29}$$

and $\boldsymbol{Q}$ is the $m \times p$ structure matrix in (21.11).

[1] The technical lemmas in this section are not essential to following the subsequent discussion.

Lemma 21.2 *The Fisher information matrix $\mathcal{I}(\alpha)$, evaluated at $\alpha = \hat{\alpha}$, is*

$$\mathcal{I}(\hat{\alpha}) = \boldsymbol{Q}' \left\{ \sum_{k=1}^{n} W_k(\hat{\alpha}) N f_k(\hat{\alpha}) W_k(\hat{\alpha})' \right\} \boldsymbol{Q}, \tag{21.30}$$

where $N = \sum_1^n y_k$ is the sample size in the empirical Bayes model (21.1)–(21.2).

†1 See the chapter endnotes† for a brief discussion of Lemmas 21.1 and 21.2. $\mathcal{I}(\hat{\alpha})^{-1}$ is the usual maximum likelihood estimate of the covariance matrix of $\hat{\alpha}$, but we will use a regularized version of the MLE that is less variable.

In the examples that follow, $\hat{\alpha}$ was found by numerical maximization.[2] Even though $\boldsymbol{g}(\alpha)$ is an exponential family, the marginal density $\boldsymbol{f}(\alpha)$ in (21.22) *is not.* As a result, some care is needed in avoiding local maxima of $l_{\boldsymbol{y}}(\alpha)$. These tend to occur at "corner" values of α, where one of its components goes to infinity. A small amount of regularization pulls $\hat{\alpha}$ away from the corners, decreasing its variance at the possible expense of increased bias.

Instead of maximizing $l_{\boldsymbol{y}}(\alpha)$ we maximize a *penalized likelihood*

$$m(\alpha) = l_{\boldsymbol{y}}(\alpha) - s(\alpha), \tag{21.31}$$

where $s(\alpha)$ is a positive penalty function. Our examples use

$$s(\alpha) = c_0 \|\alpha\| = c_0 \left(\sum_{h=1}^{p} \alpha_h^2 \right)^{1/2} \tag{21.32}$$

(with c_0 equal 1), which prevents the maximizer $\hat{\alpha}$ of $m(\alpha)$ from venturing too far into corners.

The following lemma is discussed in the chapter endnotes.

†2 **Lemma 21.3** †*The maximizer $\hat{\alpha}$ of $m(\alpha)$ has approximate bias vector and covariance matrix*

$$\begin{aligned} \text{Bias}(\hat{\alpha}) &= -\left(\mathcal{I}(\hat{\alpha}) + \ddot{s}(\hat{\alpha})\right)^{-1} \dot{s}(\hat{\alpha}) \\ \textit{and } \text{Var}(\hat{\alpha}) &= \left(\mathcal{I}(\hat{\alpha}) + \ddot{s}(\hat{\alpha})\right)^{-1} \mathcal{I}(\hat{\alpha}) \left(\mathcal{I}(\hat{\alpha}) + \ddot{s}(\hat{\alpha})\right)^{-1}, \end{aligned} \tag{21.33}$$

where $\mathcal{I}(\hat{\alpha})$ is given in (21.30).

With $s(\alpha) \equiv 0$ (no regularization) the bias is zero and $\text{Var}(\hat{\alpha}) = \mathcal{I}(\hat{\alpha})^{-1}$,

[2] Using the nonlinear maximizer `nlm` in R.

the usual MLE approximations: including $s(\alpha)$ reduces variance while introducing bias.

For $s(\alpha) = c_0\|\alpha\|$ we calculate

$$\dot{s}(\alpha) = c_0\alpha/\|\alpha\| \quad \text{and} \quad \ddot{s}(\alpha) = \frac{c_0}{\|\alpha\|}\left(\boldsymbol{I} - \frac{\alpha\alpha'}{\|\alpha\|^2}\right), \tag{21.34}$$

with $\boldsymbol{I}$ the $p \times p$ identity matrix. Adding the penalty $s(\alpha)$ in (21.31) pulls the MLE of α toward zero and the MLE of $\boldsymbol{g}(\alpha)$ toward a flat distribution over $\mathcal{T}$. Looking at $\text{Var}(\hat{\alpha})$ in (21.33), a measure of the regularization effect is

$$\text{tr}(\ddot{s}(\hat{\alpha}))/\,\text{tr}(\mathcal{I}(\hat{\alpha})), \tag{21.35}$$

which was never more than a few percent in our examples.

Most often we will be more interested in the accuracy of $\hat{\boldsymbol{g}} = \boldsymbol{g}(\hat{\alpha})$ than in that of $\hat{\alpha}$ itself. Letting

$$\boldsymbol{D}(\hat{\alpha}) = \text{diag}(\boldsymbol{g}(\hat{\alpha})) - \boldsymbol{g}(\hat{\alpha})\boldsymbol{g}(\hat{\alpha})', \tag{21.36}$$

the $m \times p$ derivative matrix $(\partial g_j/\partial\alpha_h)$ is

$$\partial\boldsymbol{g}/\partial\alpha = \boldsymbol{D}(\alpha)\boldsymbol{Q}, \tag{21.37}$$

with $\boldsymbol{Q}$ the structure matrix in (21.11). The usual first-order delta-method calculations then give the following theorem.

Theorem 21.4 *The penalized maximum likelihood estimate* $\hat{\boldsymbol{g}} = \boldsymbol{g}(\hat{\alpha})$ *has estimated bias vector and covariance matrix*

$$\begin{aligned} \text{Bias}(\hat{\boldsymbol{g}}) &= \boldsymbol{D}(\hat{\alpha})\boldsymbol{Q}\,\textit{Bias}(\hat{\alpha}) \\ \textit{and}\ \text{Var}(\hat{\boldsymbol{g}}) &= \boldsymbol{D}(\hat{\alpha})\boldsymbol{Q}\,\textit{Var}(\hat{\alpha})\boldsymbol{Q}'\boldsymbol{D}(\hat{\alpha}) \end{aligned} \tag{21.38}$$

with $\text{Bias}(\hat{\alpha})$ *and* $\text{Var}(\hat{\alpha})$ *as in* (21.33).[3]

The many approximations going into Theorem 21.4 can be short-circuited by means of the parametric bootstrap, Section 10.4. Starting from $\hat{\alpha}$ and $\boldsymbol{f}(\hat{\alpha}) = \boldsymbol{P}\boldsymbol{g}(\hat{\alpha})$, we resample the count vector

$$\boldsymbol{y}^* \sim \text{Mult}_n(N, \boldsymbol{f}(\hat{\alpha})), \tag{21.39}$$

and calculate[4] the penalized MLE $\hat{\alpha}^*$ based on $\boldsymbol{y}^*$, yielding $\hat{\boldsymbol{g}}^* = \boldsymbol{g}(\hat{\alpha}^*)$.

[3] Note that the bias treats model (21.11) as the true prior, and arises as a result of the penalization.

[4] Convergence of the `nlm` search process is speeded up by starting from $\hat{\alpha}$.

B replications $\hat{g}^{*1}, \hat{g}^{*2}, \ldots, \hat{g}^{*B}$ gives bias and covariance estimates

$$\widehat{\text{Bias}} = \hat{g}^{*\cdot} - \hat{g}$$
$$\text{and } \widehat{\text{Var}} = \sum_{b=1}^{B} (\hat{g}^{*b} - \hat{g}^{*\cdot})(\hat{g}^{*b} - \hat{g}^{*\cdot})' / (B-1), \tag{21.40}$$

and $\hat{g}^{*\cdot} = \sum_1^B \hat{g}^{*b}/B$.

Table 21.1 *Comparison of delta method* (21.38) *and bootstrap* (21.40) *standard errors and biases for the* `nodes` *study estimate of* **g** *in Figure 6.4. All columns except the first multiplied by 100.*

		Standard Error		Bias	
θ	$g(\theta)$	Delta	Boot	Delta	Boot
.01	12.048	.887	.967	−.518	−.592
.12	1.045	.131	.139	.056	.071
.23	.381	.058	.065	.025	.033
.34	.779	.096	.095	−.011	−.013
.45	1.119	.121	.117	−.040	−.049
.56	.534	.102	.100	.019	.027
.67	.264	.047	.051	.023	.027
.78	.224	.056	.053	.018	.020
.89	.321	.054	.048	.013	.009
.99	.576	.164	.169	−.008	.008

Table 21.1 compares the delta method of Theorem 20.4 with the parametric bootstrap ($B = 1000$ replications) for the surgical nodes example of Section 6.3. Both the standard errors—square roots of the diagonal elements of $\text{Var}(\hat{g})$—and biases are well approximated by the delta method formulas (21.38). The delta method also performed reasonably well on the two examples of the next section.

It did less well on the artificial example of Figure 21.1, where

$$g(\theta) = \frac{1}{8}\frac{I_{[-3,3]}(\theta)}{6} + \frac{7}{8}\frac{1}{\sqrt{2\pi\sigma^2}} e^{-\frac{1}{2}\frac{\theta^2}{\sigma^2}} \qquad (\sigma = 0.5) \tag{21.41}$$

(1/8 uniform on $[-3,3]$ and 7/8 $\mathcal{N}(0, 0.5^2)$). The vertical bars in Figure 21.3 indicate ± one standard error obtained from the parametric bootstrap, taking $\mathcal{T} = \{-3, -2.8, \ldots, 3\}$ for the sample space of Θ, and assuming a natural spline model in (21.11) with five degrees of freedom,

$$\boldsymbol{g}(\alpha) = e^{\boldsymbol{Q}\alpha - \psi(\alpha)}, \qquad \boldsymbol{Q} = \texttt{ns}(\mathcal{T}\texttt{,df=5}). \tag{21.42}$$

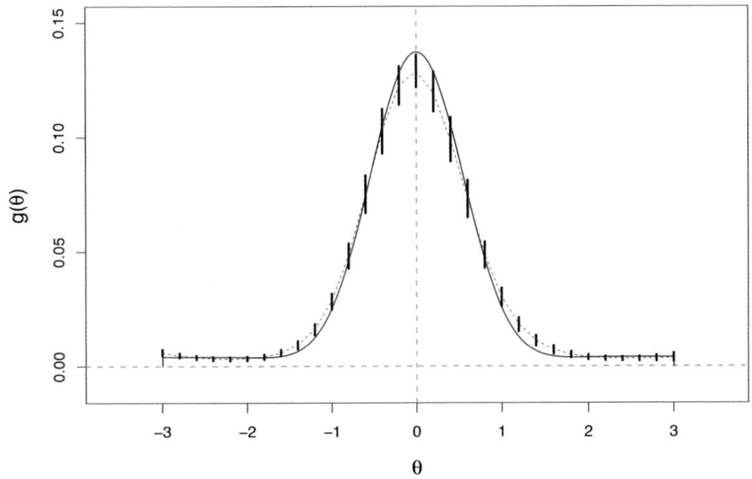

Figure 21.3 The red curve is $g(\theta)$ for the artificial example of Figure 21.1. Vertical bars are $\pm$ one standard error for g-model estimate $\boldsymbol{g}(\hat{\alpha})$; specifications (21.41)–(21.42), sample size $N = 1000$ observations $X_i \sim \mathcal{N}(\Theta_i, 1)$, using parametric bootstrap (21.40), $B = 500$. Tghe light dashed line follows bootstrap means $\hat{g}_j^*$. Some definitional bias is apparent.

The sampling model was $X_i \sim \mathcal{N}(\Theta_i, 1)$ for $i = 1, 2, \ldots, N = 1000$. In this case the delta method standard errors were about 25% too small.

The light dashed curve in Figure 21.3 traces $\bar{g}(\theta)$, the average of the $B = 500$ bootstrap replications $\boldsymbol{g}^{*b}$. There is noticeable bias, compared with $g(\theta)$. The reason is simple: the exponential family (21.42) for $\boldsymbol{g}(\alpha)$ does not include $g(\theta)$ (21.41). In fact, $\bar{g}(\theta)$ is (nearly) the closest member of the exponential family to $g(\theta)$. This kind of *definitional bias* is a disadvantage of parametric g-modeling.

Our g-modeling examples, and those of the next section, bring together a variety of themes from modern statistical practice: classical maximum likelihood theory, exponential family modeling, regularization, bootstrap methods, large data sets of parallel structure, indirect evidence, and a combination of Bayesian and frequentist thinking, all of this enabled by massive computer power. Taken together they paint an attractive picture of the range of inferential methodology in the twenty-first century.

21.4 Two Examples

We now reconsider two previous data sets from a g-modeling point of view. the first is the artificial microarray-type example (20.24) comprising N =10,000 independent observations

$$z_i \overset{\text{ind}}{\sim} \mathcal{N}(\mu_i, 1), \qquad i = 1, 2, \ldots, N = 10{,}000, \tag{21.43}$$

with

$$\mu_i \sim \begin{cases} 0 & \text{for } i = 1, 2, \ldots, 9000 \\ \mathcal{N}(-3, 1) & \text{for } i = 9001, \ldots, 10{,}000. \end{cases} \tag{21.44}$$

Figure 20.3 displays the points (z_i, μ_i) for $i = 9001, \ldots, 10,000$, illustrating the Bayes posterior 95% conditional intervals (20.26),

$$\mu_i \in (z_i - 3)/2 \pm 1.96/\sqrt{2}. \tag{21.45}$$

These required knowing the Bayes prior distribution $\mu_i \sim \mathcal{N}(-3, 1)$. We would like to recover intervals (21.45) using just the observed data z_i, $i = 1, 2, \ldots, 10,000$, without knowledge of the prior.

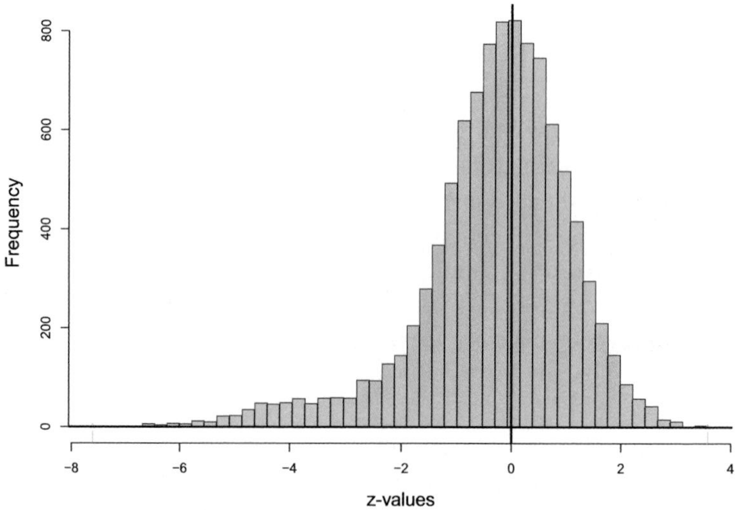

Figure 21.4 Histogram of observed sample of $N = 10{,}000$ values z_i from simulations (21.43)–(21.44).

A histogram of the 10,000 z-values is shown in Figure 21.4; g-modeling (21.9)–(21.11) was applied to them (now with μ playing the role of "Θ"

and z being "x"), taking $\mathcal{T} = (-6, -5.75, \ldots, 3)$. $\boldsymbol{Q}$ was composed of a delta function at $\mu = 0$ and a fifth-degree polynomial basis for the nonzero μ, again a family of spike-and-slab priors. The penalized MLE $\hat{\boldsymbol{g}}$ (21.31), (21.32), $c_0 = 1$, estimated the probability of $\mu = 0$ as

$$\hat{g}(0) = 0.891 \pm 0.006 \tag{21.46}$$

(using (21.38), which also provided bias estimate 0.001).

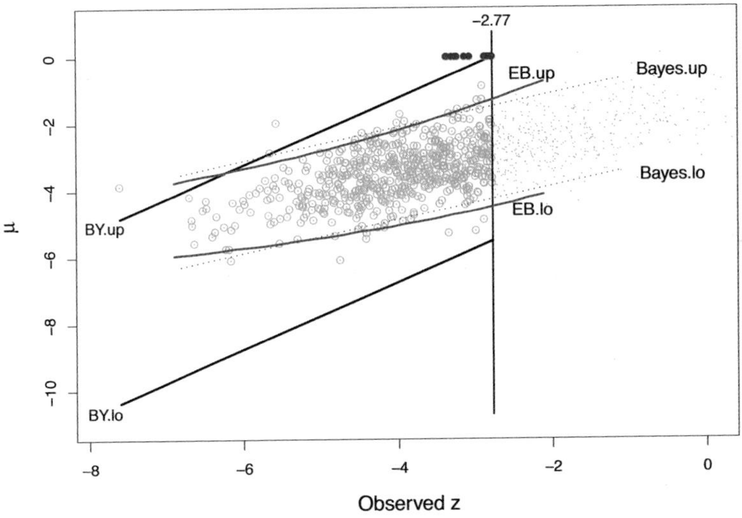

Figure 21.5 Purple curves show g-modeling estimates of conditional 95% credible intervals for μ given z in artificial microarray example (21.43)–(21.44). They are a close match to the actual Bayes intervals, dotted lines; cf. Figure 20.3.

The estimated posterior density of μ given z is

$$\hat{g}(\mu|z) = c_z \hat{g}(\mu)\phi(z - \mu), \tag{21.47}$$

$\phi(\cdot)$ the standard normal density and c_z the constant required for $\hat{g}(\mu|z)$ to integrate to 1. Let $q^{(\alpha)}(z)$ denote the αth quantile of $\hat{g}(\mu|z)$. The purple curves in Figure 21.5 trace the estimated 95% credible intervals

$$\left(q^{(.025)}(z), q^{(.975)}(z)\right). \tag{21.48}$$

They are a close match to the actual credible intervals (21.45).

The solid black curve in Figure 21.6 shows $\hat{g}(\mu)$ for $\mu \neq 0$ (the "slab" portion of the estimated prior). As an estimate of the actual slab density

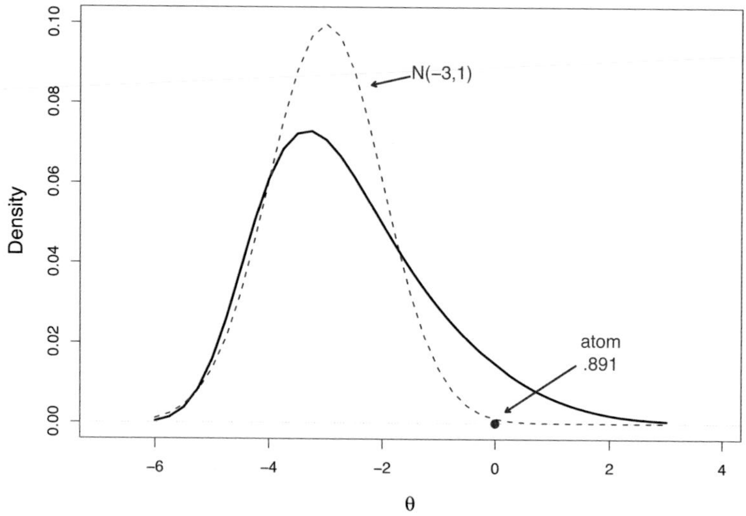

Figure 21.6 The heavy black curve is the g-modeling estimate of $g(\mu)$ for $\mu \neq 0$ in the artificial microarray example, supressing the atom at zero, $\hat{g}(0) = 0.891$. It is only a rough estimate of the actual nonzero density $\mathcal{N}(-3, 1)$.

$\mu \sim \mathcal{N}(-3, 1)$ it is only roughly accurate, but apparently still accurate enough to yield the reasonably good posterior intervals seen in Figure 21.5. The fundamental impediment to deconvolution—that large changes in $g(\theta)$ produce only small changes in $f(x)$—can sometimes operate in the statistician's favor, when only a rough knowledge of g suffices for applied purposes.

Our second example concerns the **prostate** study data, last seen in Figure 15.1: $n = 102$ men, 52 cancer patients and 50 normal controls, each have had their genetic activities measured on a microarray of $N = 6033$ genes; gene$_i$ yields a test statistic z_i comparing patients with controls,

$$z_i \sim \mathcal{N}(\mu_i, \sigma_0^2), \tag{21.49}$$

with μ_i the gene's effect size. (Here we will take the variance σ_0^2 as a parameter to be estimated, rather than assuming $\sigma_0^2 = 1$.) What is the prior density $g(\mu)$ for the effects?

The local false-discovery rate program **locfdr**, Section 15.5, was applied to the 6033 z_i values, as shown in Figure 21.7. **Locfdr** is an "f-modeling" method, where probability models are proposed directly for

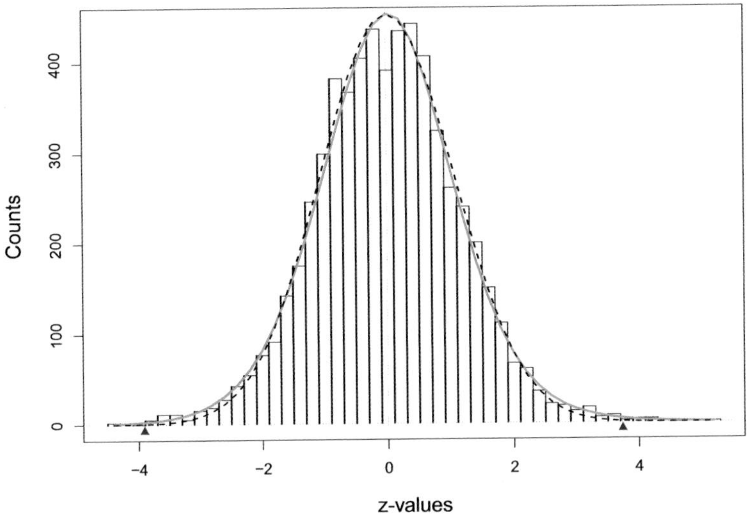

Figure 21.7 The green curve is a six-parameter Poisson regression estimate fit to counts of the observed z_i values for the `prostate` data. The dashed curve is the empirical null (15.48), $z_i \sim \mathcal{N}(0.00, 1.06^2)$. The f-modeling program `locfdr` estimated null probability $\Pr\{\mu = 0\} = 0.984$. Genes with z-values lying beyond the red triangles have estimated fdr values less than 0.20.

the marginal density $f(\cdot)$ rather than for the prior density $g(\cdot)$; see Section (21.6). Here we can compare `locfdr`'s results with those from g-modeling. The former gave[5]

$$\left(\hat{\delta}_0, \hat{\sigma}_0, \hat{\pi}_0\right) = (0.00, 1.06, 0.984) \tag{21.50}$$

in the notation of (15.50); that is, it estimated the null distribution as $\mu \sim \mathcal{N}(0, 1.06^2)$, with probability $\hat{\pi}_0 = 0.984$ of a gene being null ($\mu = 0$).

Only 22 genes were estimated to have local fdr values less than 0.20, the 9 with $z_i \leq -3.71$ and the 12 with $z_i \geq 3.81$. (These are more pessimistic results than in Figure 15.5, where we used the theoretical null $\mathcal{N}(0, 1)$ rather than the empirical null $\mathcal{N}(0, 1.06^2)$.)

The g-modeling approach (21.11) was applied to the `prostate` study data, assuming $z_i \sim \mathcal{N}(\mu_i, \sigma_0^2)$, $\sigma_0 = 1.06$ as suggested by (21.50). The

[5] Using a six-parameter Poisson regression fit to the z_i values, of the type employed in Section 10.4.

structure matrix $\boldsymbol{Q}$ in (21.11) had a delta function at $\mu = 0$ and a five-parameter natural spline basis for $\mu \neq 0$; $\mathcal{T} = (-3.6, -3.4, \ldots, 3.6)$ for the discretized Θ space (21.9). This gave a penalized MLE $\hat{\boldsymbol{g}}$ having null probability

$$\hat{g}(0) = 0.946 \pm 0.011. \tag{21.51}$$

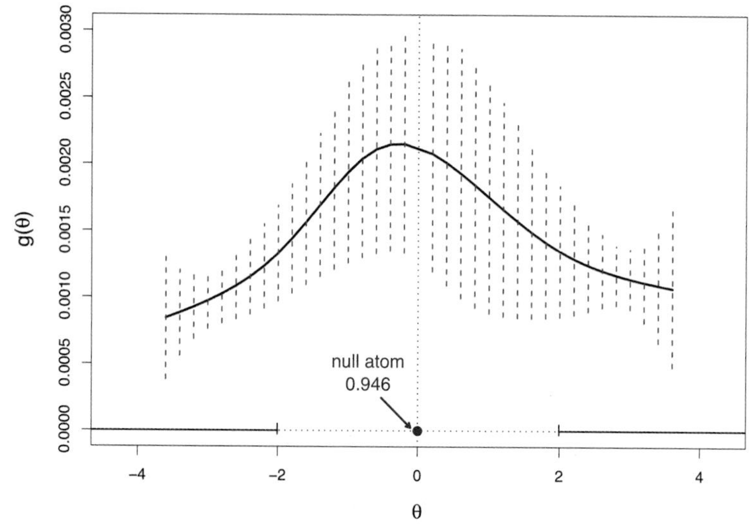

Figure 21.8 The g-modeling estimate for the non-null density $\hat{g}(\mu)$, $\mu \neq 0$, for the `prostate` study data, also indicating the null atom $\hat{g}(0) = 0.946$. About 2% of the genes are estimated to have effect sizes $|\mu_i| \geq 2$. The red bars show $\pm$ one standard error as computed from Theorem 21.4 (page 429).

The non-null distribution, $\hat{g}(\mu)$ for $\mu \neq 0$, appears in Figure 21.8, where it is seen to be modestly unimodal around $\mu = 0$. Dashed red bars indicate $\pm$ one standard error for the $\hat{g}(\theta_{(j)})$ estimates obtained from Theorem 21.4 (page 429). The accuracy is not very good. It is better for larger regions of the Θ space, for example

$$\widehat{\Pr}\{|\theta| \geq 2\} = 0.020 \pm 0.0014. \tag{21.52}$$

Here g-modeling estimated less prior null probability, 0.946 compared with 0.984 from f-modeling, but then attributed much of the non-null probability to small values of $|\mu_i|$.

Taking (21.52) literally suggests 121 ($= 0.020 \cdot 6033$) genes with true

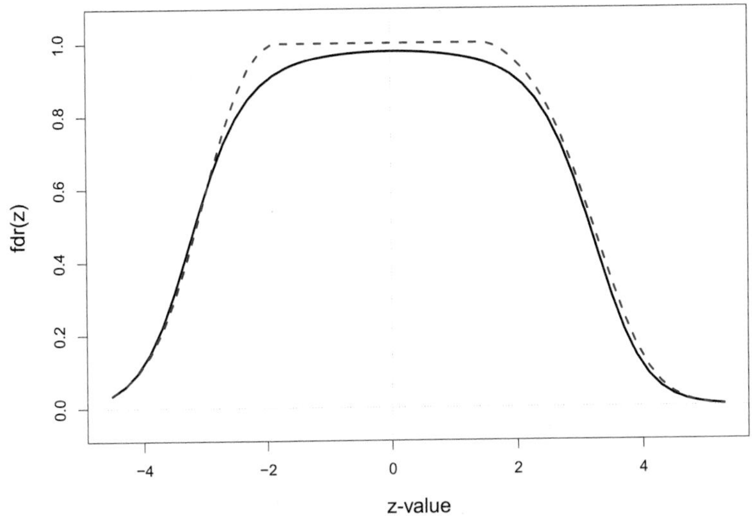

Figure 21.9 The black curve is the empirical Bayes estimated false-discovery rate $\widehat{\Pr}\{\mu = 0|z\}$ from g-modeling. For large values of $|z|$ it nearly matches the **locfdr** f-modeling estimate fdr(z), red curve.

effect sizes $|\mu_i| \geq 2$. That doesn't mean we can say with certainty *which* 121. Figure 21.9 compares the g-modeling empirical Bayes false-discovery rate

$$\widehat{\Pr}\{\mu = 0|z\} = c_z \hat{g}(0)\phi\left(\frac{z-\mu}{\hat{\sigma}_0}\right), \tag{21.53}$$

as in (21.47), with the f-modeling estimate $\widehat{\text{fdr}}(z)$ produced by **locfdr**. Where it counts, in the tails, they are nearly the same.

21.5 Generalized Linear Mixed Models

The g-modeling theory can be extended to the situation where each observation X_i is accompanied by an observed vector of covariates c_i, say of dimension d. We return to the generalized linear model setup of Section 8.2, where each X_i has a one-parameter exponential family density indexed by its own natural parameter λ_i,

$$f_{\lambda_i}(X_i) = \exp\{\lambda_i X_i - \gamma(\lambda_i)\} f_0(X_i) \tag{21.54}$$

in notation (8.20).

Our key assumption is that each λ_i is the sum of a deterministic component, depending on the covariates c_i, and a random term Θ_i,

$$\lambda_i = \Theta_i + c_i'\beta. \tag{21.55}$$

Here Θ_i is an unobserved realization from $g(\alpha) = \exp\{Q\alpha - \psi(\alpha)\}$ (21.11) and β is an unknown d-dimensional parameter. If $\beta = 0$ then (21.55) is a g-model as before,[6] while if all the $\Theta_i = 0$ then it is a standard GLM (8.20)–(8.22). Taken together, (21.55) represents a *generalized linear mixed model* (GLMM). The likelihood and accuracy calculations of Section 21.3 extend to GLMMs, as referenced in the endnotes, but here we will only discuss a GLMM analysis of the `nodes` study of Section 6.3.

In addition to n_i the number of `nodes` removed and X_i the number found `positive` (6.33), a vector of four covariates

$$c_i = (\texttt{age}_i,\ \texttt{sex}_i,\ \texttt{smoke}_i,\ \texttt{prog}_i) \tag{21.56}$$

was observed for each patient: a standardized version of `age` in years; `sex` being 0 for female or 1 for male; `smoke` being 0 for no or 1 for yes to long-term smoking; and `prog` being a post-operative prognosis score with large values more favorable.

GLMM model (21.55) was applied to the `nodes` data. Now λ_i was the logit $\log[\pi_i/(1-\pi_i)]$, where

$$X_i \sim \text{Bi}(n_i, \pi_i) \tag{21.57}$$

as in Table 8.4, i.e., π_i is the probability that any one node from patient i is positive. To make the correspondence with the analysis in Section 6.3 exact, we used a variant of (21.55)

$$\lambda_i = \text{logit}(\Theta_i) + c_i'\beta. \tag{21.58}$$

Now with $\beta = 0$, Θ_i is exactly the binomial probability π_i for the ith case. Maximum likelihood estimates were calculated for α in (21.11)—with $\mathcal{T} = (0.01, 0.02, \ldots, 0.99)$ and $Q = \texttt{poly}(\mathcal{T}, 5)$ (21.14)—and β in (21.58). The MLE prior $g(\hat{\alpha})$ was almost the same as that estimated without covariates in Figure 6.4.

Table 21.2 shows the MLE values $(\hat{\beta}_1, \hat{\beta}_2, \hat{\beta}_3, \hat{\beta}_4)$, their standard errors (from a parametric bootstrap simulation), and the z-values $\hat{\beta}_k/\widehat{\text{se}}_k$. `Sex` looks like it has a significant effect, with males tending toward larger values of π_i, that is, a greater number of positive nodes. The big effect though is `prog`, larger values of `prog` indicating smaller values of π_i.

[6] Here the setup is more specific; f is exponential family, and Θ_i is on the natural-parameter scale.

Table 21.2 *Maximum likelihood estimates $(\hat{\beta}_1, \hat{\beta}_2, \hat{\beta}_3, \hat{\beta}_4)$ for GLMM analysis of the* `nodes` *data, and standard errors from a parametric bootstrap simulation; large values of* $\texttt{prog}_i$ *predict low values of* π_i.

	`age`	`sex`	`smoke`	`prog`
MLE	−.078	.192	.089	−.698
Boot st err	.066	.070	.063	.077
***z*-value**	**−1.18**	**2.74**	**1.41**	**9.07**

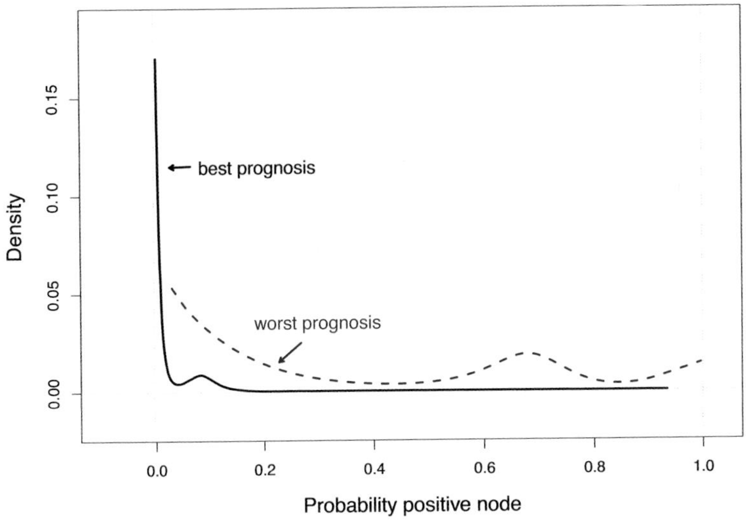

Figure 21.10 Distribution of π_i, individual probabilities of a positive node, for best and worst levels of factor `prog`; from GLMM analysis of `nodes` data.

Figure 21.10 displays the distribution of $\pi_i = 1/[1+\exp(-\lambda_i)]$ implied by the GLMM model for the best and worst values of prog (setting `age`, `sex`, and `smoke` to their average values and letting Θ have distribution $g(\hat{\alpha})$). The implied distribution is concentrated near $\pi = 0$ for the best-level `prog`, while it is roughly uniform over [0, 1] for the worst level.

The random effects we have called Θ_i are sometimes called *frailties*: a composite of unmeasured individual factors lumped together as an index of disease susceptibility. Taken together, Figures 6.4 and 21.10 show substantial frailty and covariate effects both at work in the `nodes` data. In the language of Section 6.1, we have amassed "indirect evidence" for each patient, using both Bayesian and frequentist methods.

21.6 Deconvolution and f-Modeling

Empirical Bayes applications have traditionally been dominated by f-modeling—not the g-modeling approach of the previous sections—where probability models for the marginal density $f(x)$, usually exponential families, are fit directly to the observed sample $X_1, X_2, \ldots, X_N$. We have seen several examples: Robbins' estimator in Table 6.1 (particularly the bottom line), `locfdr`'s Poisson regression estimates in Figures 15.6 and 21.7, and Tweedie's estimate in Figure 20.7.

Both the advantages and the disadvantages of f-modeling can be seen in the inferential diagram of Figure 21.2. For f-modeling the red curve now can represent an exponential family $\{\boldsymbol{f}(\alpha)\}$, whose concave log likelihood function greatly simplifies the calculation of $\boldsymbol{f}(\hat{\alpha})$ from $\boldsymbol{y}/N$. This comes at a price: the deconvolution step, from $\boldsymbol{f}(\hat{\alpha})$ to a prior distribution $\boldsymbol{g}(\hat{\alpha})$, is problematical, as discussed below.

This is only a problem if we want to know $\boldsymbol{g}$. The traditional applications of f-modeling apply to problems where the desired answer can be phrased directly in terms of $\boldsymbol{f}$. This was the case for Robbins' formula (6.5), the local false-discovery rate (15.38), and Tweedie's formula (20.37).

Nevertheless, f-modeling methodology for the estimation of the prior $g(\theta)$ does exist, an elegant example being the *Fourier method* described next. A function $f(x)$ and its Fourier transform $\phi(t)$ are related by

$$\phi(t) = \int_{-\infty}^{\infty} f(x)e^{itx}\,dx \quad \text{and} \quad f(x) = \frac{1}{2\pi}\int_{-\infty}^{\infty} \phi(t)e^{-itx}\,dt. \tag{21.59}$$

For the *normal case* where $X_i = \Theta_i + Z_i$ with $Z_i \sim \mathcal{N}(0,1)$, the Fourier transform of $f(x)$ is a multiple of that for $g(\theta)$,

$$\phi_f(t) = \phi_g(t)e^{-t^2/2}, \tag{21.60}$$

so, on the transform scale, estimating g from f amounts to removing the factor $\exp(t^2/2)$.

The Fourier method begins with the empirical density $\bar{f}(x)$ that puts probability $1/N$ on each observed value X_i, and then proceeds in three steps.

†3 1 $\bar{f}(x)$ is smoothed using the "sinc" kernel,†

$$\bar{f}(x) = \frac{1}{N\lambda} \sum_{i=1}^{N} \text{sinc}\left(\frac{X_i - x}{\lambda}\right), \qquad \text{sinc}(x) = \frac{\sin(x)}{x}. \tag{21.61}$$

2 The Fourier transform of $\bar{f}$, say $\bar{\phi}(t)$, is calculated.
3 Finally, $\hat{g}(\theta)$ is taken to be the inverse Fourier transform of $\bar{\phi}(t)e^{t^2/2}$, this last step eliminating the unwanted factor $e^{-t^2/2}$ in (21.60).

A pleasantly surprising aspect of the Fourier method is that $\hat{g}(\theta)$ can be expressed directly as a kernel estimate,

$$\hat{g}(\theta) = \frac{1}{N} \sum_{i=1}^{N} k_\lambda(X_i - \theta) = \int_{-\infty}^{\infty} k_\lambda(x - \theta)\bar{f}(x)\,dx, \tag{21.62}$$

where the kernel $k_\lambda(\cdot)$ is

$$k_\lambda(x) = \frac{1}{\pi} \int_0^{1/\lambda} e^{t^2/2} \cos(tx)\,dt. \tag{21.63}$$

Large values of λ smooth $\bar{f}(x)$ more in (21.61), reducing the variance of $\hat{g}(\theta)$ at the expense of increased bias.

Despite its compelling rationale, there are two drawbacks to the Fourier method. First of all, it applies only to situations $X_i = \Theta_i + Z_i$ where X_i is Θ_i plus iid noise. More seriously, the bias/variance trade-off in the choice of λ can be quite unfavorable.

This is illustrated in Figure 21.11 for the artificial example of Figure 21.1. The black curve is the standard deviation of the g-modeling estimate of $g(\theta)$ for θ in $[-3, 3]$, under specifications (21.41)–(21.42). The red curve graphs the standard deviation of the f-modeling estimate (21.62), with $\lambda = 1/3$, a value that produced roughly the same amount of bias as the g-modeling estimate (seen in Figure 21.3). The ratio of red to black standard deviations averages more than 20 over the range of θ.

This comparison is at least partly unfair: g-modeling is parametric while the Fourier method is almost nonparametric in its assumptions about $f(x)$ or $g(\theta)$. It can be greatly improved by beginning the three-step algorithm with a parametric estimate $\hat{f}(x)$ rather than $\bar{f}(x)$. The blue dotted curve in Figure 21.11 does this with $\hat{f}(x)$ a Poisson regression on the data $X_1, X_2, \ldots, X_N$—as in Figure 10.5 but here using a natural spline basis `ns(df=5)` —giving the estimate

$$\hat{g}(\theta) = \int_{-\infty}^{\infty} k_\lambda(x - \theta)\hat{f}(x)\,dx. \tag{21.64}$$

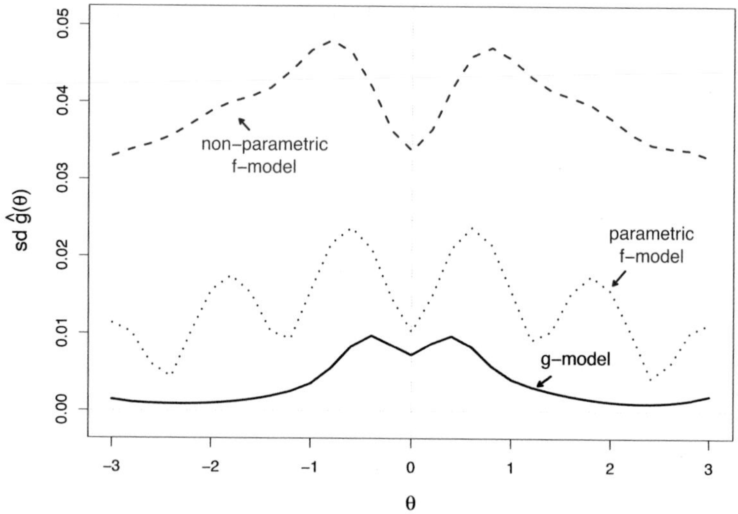

Figure 21.11 Standard deviations of estimated prior density $\hat{g}(\theta)$ for the artificial example of Figure 21.1, based on $N = 1000$ observations $X_i \sim \mathcal{N}(\Theta_i, 1)$; black curve using g-modeling under specifications (21.41)–(21.42); red curve nonparametric f-modeling (21.62), $\lambda = 1/3$; blue curve parametric f-modeling (21.64), with $\hat{f}(x)$ estimated from Poisson regression with a structure matrix having five degrees of freedom.

We see a substantial decrease in standard deviation, though still not attaining g-modeling rates.

As commented before, the great majority of empirical Bayes applications have been of the Robbins/fdr/Tweedie variety, where f-modeling is the natural choice. g-modeling comes into its own for situations like the `nodes` data analysis of Figures 6.4 and 6.5, where we really want an estimate of the prior $g(\theta)$. Twenty-first-century science is producing more such data sets, an impetus for the further development of g-modeling strategies.

Table 21.3 concerns the g-modeling estimation of $E_x = E\{\Theta | X = x\}$,

$$E_x = \int_{\mathcal{T}} \theta g(\theta) f_\theta(x)\, d\theta \Bigg/ \int_{\mathcal{T}} g(\theta) f_\theta(x)\, d\theta \tag{21.65}$$

for the artificial example, under the same specifications as in Figure 21.11. Samples of size $N = 1000$ of $X_i \sim \mathcal{N}(\Theta_i, 1)$ were drawn from model (21.41)–(21.42), yielding MLE $\hat{g}(\theta)$ and estimates $\hat{E}_x$ for x between -4

Table 21.3 *Standard deviation of* $\hat{E}\{\Theta|x\}$ *computed from parametric bootstrap simulations of* $\hat{g}(\theta)$. *The g-modeling is as in Figure 21.11, with* $N = 1000$ *observations* $X_i \sim \mathcal{N}(\Theta_i, 1)$ *from the artificial example for each simulation. The column "info" is the implied empirical Bayes information for estimating* $E\{\Theta|x\}$ *obtained from one "other" observation* X_i.

x	$E\{\Theta\|x\}$	sd($\hat{E}$)	info
−3.5	−2.00	.10	.11
−2.5	−1.06	.10	.11
−1.5	−.44	.05	.47
−.5	−.13	.03	.89
.5	.13	.04	.80
1.5	.44	.05	.44
2.5	1.06	.10	.10
3.5	2.00	.16	.04

and 4. One thousand such estimates $\hat{E}_x$ were generated, averaging almost exactly E_x, with standard deviations as shown. Accuracy is reasonably good, the coefficient of variation $\text{sd}(\hat{E}_x)/E_x$ being about 0.05 for large values of $|x|$. (Estimate (21.65) is a favorable case: results are worse for other conditional estimates† such as $E\{\Theta^2|X = x\}$.) †4

Theorem 21.4 (page 429) implies that, for large values of the sample size N, the variance of $\hat{E}_x$ decreases as $1/N$, say

$$\text{var}\left\{\hat{E}_x\right\} \doteq c_x/N. \qquad (21.66)$$

By analogy with the Fisher information bound (5.27), we can define the *empirical Bayes information* for estimating E_x in one observation to be

$$i_x = 1\Big/\left(N \cdot \text{var}\left\{\hat{E}_x\right\}\right), \qquad (21.67)$$

so that $\text{var}\{\hat{E}_x\} \doteq i_x^{-1}/N$.

Empirical Bayes inference leads us directly into the world of indirect evidence, *learning from the experience of others* as in Sections 6.4 and 7.4. So, if $X_i = 2.5$, each " other" observation X_j provides 0.10 units of information for learning $E\{\Theta|X_i = 2.5\}$ (compared with the usual Fisher information value $\mathcal{I} = 1$ for the direct estimation of Θ_i from X_i). This is a favorable case, as mentioned, and i_x is often much smaller. The main point, perhaps, is that assuming a Bayes prior is not a casual matter, and can amount to the assumption of an enormous amount of relevant *other* information.

21.7 Notes and Details

Empirical Bayes and James–Stein estimation, Chapters 6 and 7, exploded onto the statistics scene almost simultaneously in the 1950s. They represented a genuinely new branch of statistical inference, unlike the computer-based extensions of classical methodology reviewed in previous chapters. Their development as practical tools has been comparatively slow. The pace has quickened in the twenty-first century, with false-discovery rates, Chapter 15, as a major step forward. A practical empirical Bayes methodology for use beyond traditional f-modeling venues such as fdr is the goal of the g-modeling approach.

†1 [p. 428] *Lemmas 21.1 and 21.2.* The derivations of Lemmas 21.1 and 21.2 are straightforward but somewhat involved exercises in differential calculus, carried out in Remark B of Efron (2016). Here we will present just a sample of the calculations. From (21.18), the gradient vector $\dot{f}_k(\alpha) = (\partial f_k(\alpha)/\partial \alpha_l)$ with respect to α is

$$\dot{f}_k(\alpha) = \dot{\boldsymbol{g}}(\alpha)'\boldsymbol{P}_k, \tag{21.68}$$

where $\dot{\boldsymbol{g}}(\alpha)$ is the $m \times p$ derivative matrix

$$\dot{\boldsymbol{g}}(\alpha) = (\partial g_j(\alpha)/\partial \alpha_l) = \boldsymbol{DQ}, \tag{21.69}$$

with $\boldsymbol{D}$ as in (21.36), the last equality following, after some work, by differentiation of $\log \boldsymbol{g}(\alpha) = \boldsymbol{Q}\alpha - \phi(\alpha)$.

Let $l_k = \log f_k$ (now suppressing α from the notation). The gradient with respect to α of l_k is then

$$\dot{l}_k = \dot{f}_k/f_k = \boldsymbol{Q}'\boldsymbol{DP}_k/f_k. \tag{21.70}$$

The vector $\boldsymbol{DP}_k/f_k$ has components

$$(g_j p_{kj} - g_j f_k)/f_k = w_{kj} \tag{21.71}$$

(21.27), using $\boldsymbol{g}'\boldsymbol{P}_k = f_k$. This gives $\dot{l}_k = \boldsymbol{Q}'W_k(\alpha)$ (21.28). Adding up the independent score functions $\dot{l}_k$ over the full sample yields the overall score $\dot{l}_{\boldsymbol{y}}(\alpha) = \boldsymbol{Q}'\sum_1^n y_k W_k(\alpha)$, which is Lemma 21.1.

†2 [p. 428] *Lemma 2.* The penalized MLE $\hat{\alpha}$ satisfies

$$\boldsymbol{0} = \dot{m}(\hat{\alpha}) \doteq \dot{m}(\alpha_0) + \ddot{m}(\alpha_0)(\hat{\alpha} - \alpha_0), \tag{21.72}$$

where α_0 is the true value of α, or

$$\hat{\alpha} - \alpha_0 \doteq (-\dddot{m}(\alpha_0))^{-1}\dot{m}(\alpha_0)\left(-\ddot{l}_y(\alpha_0) + \ddot{s}(\alpha_0)\right)^{-1}\left(\dot{l}_y(\alpha_0) - \dot{s}(\alpha_0)\right). \tag{21.73}$$

Standard MLE theory shows that the random variable $\dot{l}_y(\alpha_0)$ has mean 0 and covariance Fisher information matrix $\mathcal{I}(\alpha_0)$, while $-\ddot{l}_y(\alpha_0)$ asymptotically approximates $\mathcal{I}(\alpha_0)$. Substituting in (21.73),

$$\hat{\alpha} - \alpha_0 \doteq (\mathcal{I}(\alpha_0) + \ddot{s}(\alpha_0))^{-1} Z, \tag{21.74}$$

where Z has mean $-\dot{s}(\alpha_0)$ and covariance $\mathcal{I}(\alpha_0)$. This gives $\text{Bias}(\hat{\alpha})$ and $\text{Var}(\hat{\alpha})$ as in Lemma 2. Note that the bias is with respect to a *true* parametric model (21.11), and is a consequence of the penalization.

†3 [p. 440] *The sinc kernel.* The Fourier transform $\phi_s(t)$ of the scaled sinc function $s(x) = \sin(x/\lambda)/(\pi x)$ is the indicator of the interval $[-1/\lambda, 1/\lambda]$, while that of $\bar{f}(x)$ is $(1/N)\sum_1^N \exp(itX_j)$. Formula (21.61) is the convolution $\bar{f} * s$, so $\tilde{f}$ has the product transform

$$\phi_{\tilde{f}}(t) = \left[\frac{1}{N}\sum_{j=1}^{N} e^{itX_j}\right] I_{[-1/\lambda,1/\lambda]}(t). \tag{21.75}$$

The effect of the sinc convolution is to censor the high-frequency (large t) components of $\bar{f}$ or $\phi_{\bar{f}}$. Larger λ yields more censoring. Formula (21.63) has upper limits $1/\lambda$ because of $\phi_s(t)$. All of this is due to Stefanski and Carroll (1990). Smoothers other than the sinc kernel have been suggested in the literature, but without substantial improvements on deconvolution performance.

†4 [p. 443] *Conditional expectation* (21.65). Efron (2014b) considers estimating $E\{\Theta^2 | X = x\}$ and other such conditional expectations, both for f-modeling and for g-modeling. $E\{\Theta | X = x\}$ is by far the easiest case, as might be expected from the simple form of Tweedie's estimate (20.37).

Epilogue

Something important changed in the world of statistics in the new millennium. Twentieth-century statistics, even after the heated expansion of its late period, could still be contained within the classic Bayesian–frequentist–Fisherian inferential triangle (Figure 14.1). This is not so in the twenty-first century. Some of the topics discussed in Part III—false-discovery rates, post-selection inference, empirical Bayes modeling, the lasso—fit within the triangle but others seem to have escaped, heading south from the frequentist corner, perhaps in the direction of computer science.

The escapees were the large-scale prediction algorithms of Chapters 17–19: neural nets, deep learning, boosting, random forests, and support-vector machines. Notably missing from their development were parametric probability models, the building blocks of classical inference. Prediction algorithms are the media stars of the big-data era. It is worth asking why they have taken center stage and what it means for the future of the statistics discipline.

The *why* is easy enough: prediction is commercially valuable. Modern equipment has enabled the collection of mountainous data troves, which the "data miners" can then burrow into, extracting valuable information. Moreover, prediction is the simplest use of regression theory (Section 8.4). It can be carried out successfully without probability models, perhaps with the assistance of nonparametric analysis tools such as cross-validation, permutations, and the bootstrap.

A great amount of ingenuity and experimentation has gone into the development of modern prediction algorithms, with statisticians playing an important but not dominant role.[1] There is no shortage of impressive success stories. In the absence of optimality criteria, either frequentist or Bayesian, the prediction community grades algorithmic excellence on per-

[1] All papers mentioned in this section have their complete references in the bibliography. Footnotes will identify papers not fully specified in the text.

formance within a catalog of often-visited examples such as the spam and digits data sets of Chapters 17 and 18.[2] Meanwhile, "traditional statistics" —probability models, optimality criteria, Bayes priors, asymptotics—has continued successfully along on a parallel track. Pessimistically or optimistically, one can consider this as a bipolar disorder of the field or as a healthy duality that is bound to improve both branches. There are historical and intellectual arguments favoring the optimists' side of the story.

The first thing to say is that the current situation is not entirely unprecedented. By the end of the nineteenth century there was available an impressive inventory of statistical methods—Bayes' theorem, least squares, correlation, regression, the multivariate normal distribution—but these existed more as individual algorithms than as a unified discipline. Statistics as a distinct intellectual enterprise was not yet well-formed.

A small but crucial step forward was taken in 1914 when the astrophysicist Arthur Eddington[3] claimed that mean absolute deviation was superior to the familiar root mean square estimate for the standard deviation from a normal sample. Fisher in 1919 showed that this was wrong, and moreover, in a clear mathematical sense, the root mean square was the *best possible estimate*. Eddington conceded the point while Fisher went on to develop the theory of sufficiency and optimal estimation.[4]

"Optimal" is the key word here. Before Fisher, statisticians didn't really understand estimation. The same can be said now about prediction. Despite their impressive performance on a raft of test problems, it might still be possible to do much better than neural nets, deep learning, random forests, and boosting—or perhaps they are coming close to some as-yet unknown theoretical minimum.

It is the job of statistical inference to connect "dangling algorithms" to the central core of well-understood methodology. The connection process is already underway. Section 17.4 showed how `Adaboost`, the original machine learning algorithm, could be restated as a close cousin of logistic regression. Purely empirical approaches like the Common Task Framework are ultimately unsatisfying without some form of principled justification. Our optimistic scenario has the big-data/data-science prediction world rejoining the mainstream of statistical inference, to the benefit of both branches.

[2] This empirical approach to optimality is sometimes codified as the *Common Task Framework* (Liberman, 2015 and Donoho, 2015).

[3] Eddington became world-famous for his 1919 empirical verification of Einstein's relativity theory.

[4] See Stigler (2006) for the full story.

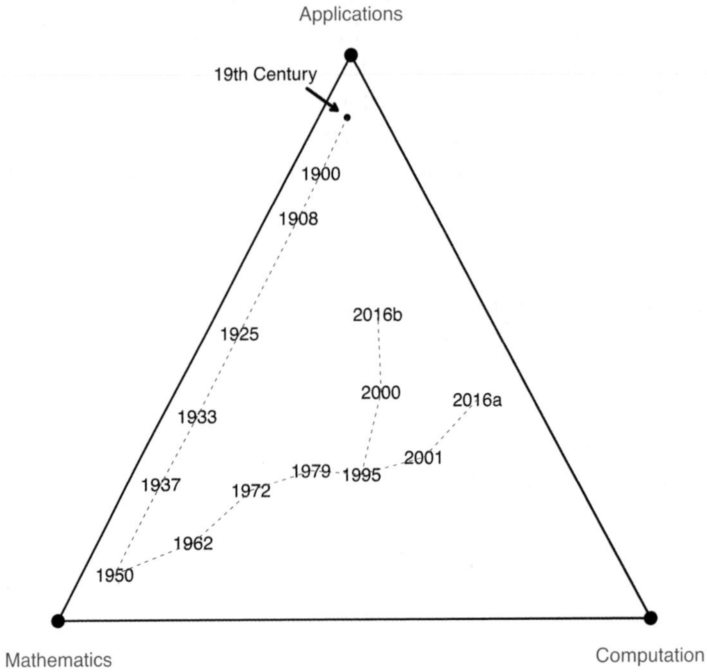

Development of the statistics discipline since the end of the nineteenth century, as discussed in the text.

Whether or not we can predict the future of statistics, we can at least examine the past to see how we've gotten where we are. The next figure does so in terms of a new triangle diagram, this time with the poles labeled *Applications*, *Mathematics*, and *Computation*. "Mathematics" here is shorthand for the mathematical/logical justification of statistical methods. "Computation" stands for the empirical/numerical approach.

Statistics is a branch of applied mathematics, and is ultimately judged by how well it serves the world of applications. Mathematical logic, *à la* Fisher, has been the traditional vehicle for the development and understanding of statistical methods. Computation, slow and difficult before the 1950s, was only a bottleneck, but now has emerged as a competitor to (or perhaps a seven-league boots enabler of) mathematical analysis. At any one time the discipline's energy and excitement is directed unequally toward the three poles. The figure attempts, in admittedly crude fashion, to track the changes in direction over the past 100+ years.

The tour begins at the end of the nineteenth century. Mathematicians of the caliber of Gauss and Laplace had contributed to the available methodology, but the subsequent development was almost entirely applications-driven. Quetelet[5] was especially influential, applying the Gauss–Laplace formulation to census data and his "Average Man." A modern reader will search almost in vain for any mathematical symbology in nineteenth-century statistics journals.

1900

Karl Pearson's chi-square paper was a bold step into the new century, applying a new mathematical tool, matrix theory, in the service of statistical methodology. He and Weldon went on to found *Biometrika* in 1901, the first recognizably modern statistics journal. Pearson's paper, and *Biometrika*, launched the statistics discipline on a fifty-year march toward the mathematics pole of the triangle.

1908

Student's t statistic was a crucial first result in small-sample "exact" inference, and a major influence on Fisher's thinking.

1925

Fisher's great estimation paper—a more coherent version of its 1922 predecessor. It introduced a host of fundamental ideas, including sufficiency, efficiency, Fisher information, maximum likelihood theory, and the notion of optimal estimation. Optimality is a mark of maturity in mathematics, making 1925 the year statistical inference went from a collection of ingenious techniques to a coherent discipline.

1933

This represents Neyman and Pearson's paper on optimal hypothesis testing. A logical completion of Fisher's program, it nevertheless aroused his strong antipathy. This was partly personal, but also reflected Fisher's concern that mathematization was squeezing intuitive correctness out of statistical thinking (Section 4.2).

1937

Neyman's seminal paper on confidence intervals. His sophisticated mathematical treatment of statistical inference was a harbinger of decision theory.

[5] Adolphe Quetelet was a tireless organizer, helping found the Royal Statistical Society in 1834, with the American Statistical Association following in 1839.

1950

The publication of Wald's *Statistical Decision Functions*. Decision theory completed the full mathematization of statistical inference. This date can also stand for Savage's and de Finetti's decision-theoretic formulation of Bayesian inference. We are as far as possible from the Applications corner of the triangle now, and it is fair to describe the 1950s as a nadir of the influence of the statistics discipline on scientific applications.

1962

The arrival of electronic computation in the mid 1950s began the process of stirring statistics out of its inward-gazing preoccupation with mathematical structure. Tukey's paper "The future of data analysis" argued for a more application- and computation-oriented discipline. Mosteller and Tukey later suggested changing the field's name to *data analysis*, a prescient hint of today's *data science*.

1972

Cox's proportional hazards paper. Immensely useful in its own right, it signaled a growing interest in biostatistical applications and particularly survival analysis, which was to assert its scientific importance in the analysis of AIDS epidemic data.

1979

The bootstrap, and later the widespread use of MCMC: electronic computation used for the extension of classic statistical inference.

1995

This stands for false-discovery rates and, a year later, the lasso.[6] Both are computer-intensive algorithms, firmly rooted in the ethos of statistical inference. They lead, however, in different directions, as indicated by the split in the diagram.

2000

Microarray technology inspires enormous interest in large-scale inference, both in theory and as applied to the analysis of microbiological data.

[6] Benjamini and Hochberg (1995) and Tibshirani (1996).

2001

Random forests; it joins boosting[7] and the resurgence of neural nets in the ranks of *machine learning* prediction algorithms.

2016a

Data science: a more popular successor to Tukey and Mosteller's "data analysis," at one extreme it seems to represent a statistics discipline without parametric probability models or formal inference. The Data Science Association defines a practitioner as one who "...uses scientific methods to liberate and create meaning from raw data." In practice the emphasis is on the algorithmic processing of large data sets for the extraction of useful information, with the prediction algorithms as exemplars.

2016b

This represents the traditional line of statistical thinking, of the kind that could be located within Figure 14.1, but now energized with a renewed focus on applications. Of particular applied interest are biology and genetics. Genome-wide association studies (GWAS) show a different face of big data. Prediction is important here,[8] but not sufficient for the scientific understanding of disease.

A cohesive inferential theory was forged in the first half of the twentieth century, but unity came at the price of an inwardly focused discipline, of reduced practical utility. In the century's second half, electronic computation unleashed a vast expansion of useful—and much used—statistical methodology. Expansion accelerated at the turn of the millennium, further increasing the reach of statistical thinking, but now at the price of intellectual cohesion.

It is tempting but risky to speculate on the future of statistics. What will the Mathematics–Applications–Computation diagram look like, say 25 years from now? The appetite for statistical analysis seems to be always increasing, both from science and from society in general. Data science has blossomed in response, but so has the traditional wing of the field. The data-analytic initiatives represented in the diagram by 2016a and 2016b are in actuality not isolated points but the centers of overlapping distributions.

[7] Breiman (1996) for random forests, Freund and Schapire (1997) for boosting.

[8] "Personalized medicine" in which an individual's genome predicts his or her optimal treatment has attracted grail-like attention.

A hopeful scenario for the future is one of an increasing overlap that puts data science on a solid footing while leading to a broader general formulation of statistical inference.

References

Abu-Mostafa, Y. 1995. Hints. *Neural Computation*, **7**, 639–671.

Achanta, R., and Hastie, T. 2015. *Telugu OCR Framework using Deep Learning*. Tech. rept. Statistics Department, Stanford University.

Akaike, H. 1973. Information theory and an extension of the maximum likelihood principle. Pages 267–281 of: *Second International Symposium on Information Theory (Tsahkadsor, 1971)*. Akadémiai Kiadó, Budapest.

Anderson, T. W. 2003. *An Introduction to Multivariate Statistical Analysis*. Third edn. Wiley Series in Probability and Statistics. Wiley-Interscience.

Bastien, F., Lamblin, P., Pascanu, R., Bergstra, J., Goodfellow, I. J., Bergeron, A., Bouchard, N., and Bengio, Y. 2012. *Theano: new features and speed improvements*. Deep Learning and Unsupervised Feature Learning NIPS 2012 Workshop.

Becker, R., Chambers, J., and Wilks, A. 1988. *The New S Language: A Programming Environment for Data Analysis and Graphics*. Pacific Grove, CA: Wadsworth and Brooks/Cole.

Bellhouse, D. R. 2004. The Reverend Thomas Bayes, FRS: A biography to celebrate the tercentenary of his birth. *Statist. Sci.*, **19**(1), 3–43. With comments and a rejoinder by the author.

Bengio, Y., Courville, A., and Vincent, P. 2013. Representation learning: a review and new perspectives. *IEEE Transactions on Pattern Analysis and Machine Intelligence*, **35**(8), 1798–1828.

Benjamini, Y., and Hochberg, Y. 1995. Controlling the false discovery rate: A practical and powerful approach to multiple testing. *J. Roy. Statist. Soc. Ser. B*, **57**(1), 289–300.

Benjamini, Y., and Yekutieli, D. 2005. False discovery rate-adjusted multiple confidence intervals for selected parameters. *J. Amer. Statist. Assoc.*, **100**(469), 71–93.

Berger, J. O. 2006. The case for objective Bayesian analysis. *Bayesian Anal.*, **1**(3), 385–402 (electronic).

Berger, J. O., and Pericchi, L. R. 1996. The intrinsic Bayes factor for model selection and prediction. *J. Amer. Statist. Assoc.*, **91**(433), 109–122.

Bergstra, J., Breuleux, O., Bastien, F., Lamblin, P., Pascanu, R., Desjardins, G., Turian, J., Warde-Farley, D., and Bengio, Y. 2010 (June). Theano: a CPU and GPU math expression compiler. In: *Proceedings of the Python for Scientific Computing Conference (SciPy)*.

Berk, R., Brown, L., Buja, A., Zhang, K., and Zhao, L. 2013. Valid post-selection inference. *Ann. Statist.*, **41**(2), 802–837.

Berkson, J. 1944. Application of the logistic function to bio-assay. *J. Amer. Statist. Assoc.*, **39**(227), 357–365.

Bernardo, J. M. 1979. Reference posterior distributions for Bayesian inference. *J. Roy. Statist. Soc. Ser. B*, **41**(2), 113–147. With discussion.

Birch, M. W. 1964. The detection of partial association. I. The 2×2 case. *J. Roy. Statist. Soc. Ser. B*, **26**(2), 313–324.

Bishop, C. 1995. *Neural Networks for Pattern Recognition*. Clarendon Press, Oxford.

Boos, D. D., and Serfling, R. J. 1980. A note on differentials and the CLT and LIL for statistical functions, with application to M-estimates. *Ann. Statist.*, **8**(3), 618–624.

Boser, B., Guyon, I., and Vapnik, V. 1992. A training algorithm for optimal margin classifiers. In: *Proceedings of COLT II*.

Breiman, L. 1996. Bagging predictors. *Mach. Learn.*, **24**(2), 123–140.

Breiman, L. 1998. Arcing classifiers (with discussion). *Annals of Statistics*, **26**, 801–849.

Breiman, L. 2001. Random forests. *Machine Learning*, **45**, 5–32.

Breiman, L., Friedman, J., Olshen, R. A., and Stone, C. J. 1984. *Classification and Regression Trees*. Wadsworth Statistics/Probability Series. Wadsworth Advanced Books and Software.

Carlin, B. P., and Louis, T. A. 1996. *Bayes and Empirical Bayes Methods for Data Analysis*. Monographs on Statistics and Applied Probability, vol. 69. Chapman & Hall.

Carlin, B. P., and Louis, T. A. 2000. *Bayes and Empirical Bayes Methods for Data Analysis*. 2 edn. Texts in Statistical Science. Chapman & Hall/CRC.

Chambers, J. M., and Hastie, T. J. (eds). 1993. *Statistical Models in S*. Chapman & Hall Computer Science Series. Chapman & Hall.

Cleveland, W. S. 1981. LOWESS: A program for smoothing scatterplots by robust locally weighted regression. *Amer. Statist.*, **35**(1), 54.

Cox, D. R. 1958. The regression analysis of binary sequences. *J. Roy. Statist. Soc. Ser. B*, **20**, 215–242.

Cox, D. R. 1970. *The Analysis of Binary Data*. Methuen's Monographs on Applied Probability and Statistics. Methuen & Co.

Cox, D. R. 1972. Regression models and life-tables. *J. Roy. Statist. Soc. Ser. B*, **34**(2), 187–220.

Cox, D. R. 1975. Partial likelihood. *Biometrika*, **62**(2), 269–276.

Cox, D. R., and Hinkley, D. V. 1974. *Theoretical Statistics*. Chapman & Hall.

Cox, D. R., and Reid, N. 1987. Parameter orthogonality and approximate conditional inference. *J. Roy. Statist. Soc. Ser. B*, **49**(1), 1–39. With a discussion.

Crowley, J. 1974. Asymptotic normality of a new nonparametric statistic for use in organ transplant studies. *J. Amer. Statist. Assoc.*, **69**(348), 1006–1011.

de Finetti, B. 1972. *Probability, Induction and Statistics. The Art of Guessing*. John Wiley & Sons, London-New York-Sydney.

Dembo, A., Cover, T. M., and Thomas, J. A. 1991. Information-theoretic inequalities. *IEEE Trans. Inform. Theory*, **37**(6), 1501–1518.

Dempster, A. P., Laird, N. M., and Rubin, D. B. 1977. Maximum likelihood from incomplete data via the EM algorithm. *J. Roy. Statist. Soc. Ser. B*, **39**(1), 1–38.

Diaconis, P., and Ylvisaker, D. 1979. Conjugate priors for exponential families. *Ann. Statist.*, **7**(2), 269–281.

DiCiccio, T., and Efron, B. 1992. More accurate confidence intervals in exponential families. *Biometrika*, **79**(2), 231–245.

Donoho, D. L. 2015. 50 years of data science. *R-bloggers*. `www.r-bloggers.com/50-years-of-data-science-by-david-donoho/`.

Edwards, A. W. F. 1992. *Likelihood*. Expanded edn. Johns Hopkins University Press. Revised reprint of the 1972 original.

Efron, B. 1967. The two sample problem with censored data. Pages 831–853 of: *Proc. 5th Berkeley Symp. Math. Statist. and Prob., Vol. 4*. University of California Press.

Efron, B. 1975. Defining the curvature of a statistical problem (with applications to second order efficiency). *Ann. Statist.*, **3**(6), 1189–1242. With discussion and a reply by the author.

Efron, B. 1977. The efficiency of Cox's likelihood function for censored data. *J. Amer. Statist. Assoc.*, **72**(359), 557–565.

Efron, B. 1979. Bootstrap methods: Another look at the jackknife. *Ann. Statist.*, **7**(1), 1–26.

Efron, B. 1982. *The Jackknife, the Bootstrap and Other Resampling Plans*. CBMS-NSF Regional Conference Series in Applied Mathematics, vol. 38. Society for Industrial and Applied Mathematics (SIAM).

Efron, B. 1983. Estimating the error rate of a prediction rule: Improvement on cross-validation. *J. Amer. Statist. Assoc.*, **78**(382), 316–331.

Efron, B. 1985. Bootstrap confidence intervals for a class of parametric problems. *Biometrika*, **72**(1), 45–58.

Efron, B. 1986. How biased is the apparent error rate of a prediction rule? *J. Amer. Statist. Assoc.*, **81**(394), 461–470.

Efron, B. 1987. Better bootstrap confidence intervals. *J. Amer. Statist. Assoc.*, **82**(397), 171–200. With comments and a rejoinder by the author.

Efron, B. 1988. Logistic regression, survival analysis, and the Kaplan–Meier curve. *J. Amer. Statist. Assoc.*, **83**(402), 414–425.

Efron, B. 1993. Bayes and likelihood calculations from confidence intervals. *Biometrika*, **80**(1), 3–26.

Efron, B. 1998. R. A. Fisher in the 21st Century (invited paper presented at the 1996 R. A. Fisher Lecture). *Statist. Sci.*, **13**(2), 95–122. With comments and a rejoinder by the author.

Efron, B. 2004. The estimation of prediction error: Covariance penalties and cross-validation. *J. Amer. Statist. Assoc.*, **99**(467), 619–642. With comments and a rejoinder by the author.

Efron, B. 2010. *Large-Scale Inference: Empirical Bayes Methods for Estimation, Testing, and Prediction*. Institute of Mathematical Statistics Monographs, vol. 1. Cambridge University Press.

Efron, B. 2011. Tweedie's formula and selection bias. *J. Amer. Statist. Assoc.*, **106**(496), 1602–1614.

Efron, B. 2014a. Estimation and accuracy after model selection. *J. Amer. Statist. Assoc.*, **109**(507), 991–1007.

Efron, B. 2014b. Two modeling strategies for empirical Bayes estimation. *Statist. Sci.*, **29**(2), 285–301.

Efron, B. 2015. Frequentist accuracy of Bayesian estimates. *J. Roy. Statist. Soc. Ser. B*, **77**(3), 617–646.

Efron, B. 2016. Empirical Bayes deconvolution methods. *Biometrika*, **101**(1), 1–20.

Efron, B., and Feldman, D. 1991. Compliance as an explanatory variable in clinical trials. *J. Amer. Statist. Assoc.*, **86**(413), 9–17.

Efron, B., and Gous, A. 2001. Scales of evidence for model selection: Fisher versus Jeffreys. Pages 208–256 of: *Model Selection*. IMS Lecture Notes Monograph Series, vol. 38. Beachwood, OH: Institute of Mathematics and Statististics. With discussion and a rejoinder by the authors.

Efron, B., and Hinkley, D. V. 1978. Assessing the accuracy of the maximum likelihood estimator: Observed versus expected Fisher information. *Biometrika*, **65**(3), 457–487. With comments and a reply by the authors.

Efron, B., and Morris, C. 1972. Limiting the risk of Bayes and empirical Bayes estimators. II. The empirical Bayes case. *J. Amer. Statist. Assoc.*, **67**, 130–139.

Efron, B., and Morris, C. 1977. Stein's paradox in statistics. *Scientific American*, **236**(5), 119–127.

Efron, B., and Petrosian, V. 1992. A simple test of independence for truncated data with applications to redshift surveys. *Astrophys. J.*, **399**(Nov), 345–352.

Efron, B., and Stein, C. 1981. The jackknife estimate of variance. *Ann. Statist.*, **9**(3), 586–596.

Efron, B., and Thisted, R. 1976. Estimating the number of unseen species: How many words did Shakespeare know? *Biometrika*, **63**(3), 435–447.

Efron, B., and Tibshirani, R. 1993. *An Introduction to the Bootstrap*. Monographs on Statistics and Applied Probability, vol. 57. Chapman & Hall.

Efron, B., and Tibshirani, R. 1997. Improvements on cross-validation: The .632+ bootstrap method. *J. Amer. Statist. Assoc.*, **92**(438), 548–560.

Efron, B., Hastie, T., Johnstone, I., and Tibshirani, R. 2004. Least angle regression. *Annals of Statistics*, **32**(2), 407–499. (with discussion, and a rejoinder by the authors).

Finney, D. J. 1947. The estimation from individual records of the relationship between dose and quantal response. *Biometrika*, **34**(3/4), 320–334.

Fisher, R. A. 1915. Frequency distribution of the values of the correlation coefficient in samples from an indefinitely large population. *Biometrika*, **10**(4), 507–521.

Fisher, R. A. 1925. Theory of statistical estimation. *Math. Proc. Cambridge Phil. Soc.*, **22**(7), 700–725.

Fisher, R. A. 1930. Inverse probability. *Math. Proc. Cambridge Phil. Soc.*, **26**(10), 528–535.

Fisher, R. A., Corbet, A., and Williams, C. 1943. The relation between the number of species and the number of individuals in a random sample of an animal population. *J. Anim. Ecol.*, **12**, 42–58.

Fithian, W., Sun, D., and Taylor, J. 2014. Optimal inference after model selection. *ArXiv e-prints*, Oct.

Freund, Y., and Schapire, R. 1996. Experiments with a new boosting algorithm. Pages 148–156 of: *Machine Learning: Proceedings of the Thirteenth International Conference*. Morgan Kauffman, San Francisco.

Freund, Y., and Schapire, R. 1997. A decision-theoretic generalization of online learning and an application to boosting. *Journal of Computer and System Sciences*, **55**, 119–139.

Friedman, J. 2001. Greedy function approximation: a gradient boosting machine. *Annals of Statistics*, **29**(5), 1189–1232.

Friedman, J., and Popescu, B. 2005. *Predictive Learning via Rule Ensembles*. Tech. rept. Stanford University.

Friedman, J., Hastie, T., and Tibshirani, R. 2000. Additive logistic regression: a statistical view of boosting (with discussion). *Annals of Statistics*, **28**, 337–307.

Friedman, J., Hastie, T., and Tibshirani, R. 2009. *glmnet: Lasso and elastic-net regularized generalized linear models*. R package version 1.1-4.

Friedman, J., Hastie, T., and Tibshirani, R. 2010. Regularization paths for generalized linear models via coordinate descent. *Journal of Statistical Software*, **33**(1), 1–22.

Geisser, S. 1974. A predictive approach to the random effect model. *Biometrika*, **61**, 101–107.

Gerber, M., and Chopin, N. 2015. Sequential quasi Monte Carlo. *J. Roy. Statist. Soc. B*, **77**(3), 509–580. with discussion, doi: 10.1111/rssb.12104.

Gholami, S., Janson, L., Worhunsky, D. J., Tran, T. B., Squires, Malcolm, I., Jin, L. X., Spolverato, G., Votanopoulos, K. I., Schmidt, C., Weber, S. M., Bloomston, M., Cho, C. S., Levine, E. A., Fields, R. C., Pawlik, T. M., Maithel, S. K., Efron, B., Norton, J. A., and Poultsides, G. A. 2015. Number of lymph nodes removed and survival after gastric cancer resection: An analysis from the US Gastric Cancer Collaborative. *J. Amer. Coll. Surg.*, **221**(2), 291–299.

Good, I., and Toulmin, G. 1956. The number of new species, and the increase in population coverage, when a sample is increased. *Biometrika*, **43**, 45–63.

Hall, P. 1988. Theoretical comparison of bootstrap confidence intervals. *Ann. Statist.*, **16**(3), 927–985. with discussion and a reply by the author.

Hampel, F. R. 1974. The influence curve and its role in robust estimation. *J. Amer. Statist. Assoc.*, **69**, 383–393.

Hampel, F. R., Ronchetti, E. M., Rousseeuw, P. J., and Stahel, W. A. 1986. *Robust Statistics: The approach based on influence functions*. Wiley Series in Probability and Mathematical Statistics. John Wiley & Sons.

Harford, T. 2014. Big data: A big mistake? *Significance*, **11**(5), 14–19.

Hastie, T., and Loader, C. 1993. Local regression: automatic kernel carpentry (with discussion). *Statistical Science*, **8**, 120–143.

Hastie, T., and Tibshirani, R. 1990. *Generalized Additive Models*. Chapman and Hall.

Hastie, T., and Tibshirani, R. 2004. Efficient quadratic regularization for expression arrays. *Biostatistics*, **5**(3), 329–340.

Hastie, T., Tibshirani, R., and Friedman, J. 2009. *The Elements of Statistical Learning. Data mining, Inference, and Prediction*. Second edn. Springer Series in Statistics. Springer.

Hastie, T., Tibshirani, R., and Wainwright, M. 2015. *Statistical Learning with Sparsity: the Lasso and Generalizations*. Chapman and Hall, CRC Press.

Hoeffding, W. 1952. The large-sample power of tests based on permutations of observations. *Ann. Math. Statist.*, **23**, 169–192.

Hoeffding, W. 1965. Asymptotically optimal tests for multinomial distributions. *Ann. Math. Statist.*, **36**(2), 369–408.

Hoerl, A. E., and Kennard, R. W. 1970. Ridge regression: Biased estimation for nonorthogonal problems. *Technometrics*, **12**(1), 55–67.

Huber, P. J. 1964. Robust estimation of a location parameter. *Ann. Math. Statist.*, **35**, 73–101.

Jaeckel, L. A. 1972. Estimating regression coefficients by minimizing the dispersion of the residuals. *Ann. Math. Statist.*, **43**, 1449–1458.

James, W., and Stein, C. 1961. Estimation with quadratic loss. Pages 361–379 of: *Proc. 4th Berkeley Symposium on Mathematical Statistics and Probability*, vol. I. University of California Press.

Jansen, L., Fithian, W., and Hastie, T. 2015. Effective degrees of freedom: a flawed metaphor. *Biometrika*, **102**(2), 479–485.

Javanmard, A., and Montanari, A. 2014. Confidence intervals and hypothesis testing for high-dimensional regression. *J. of Machine Learning Res.*, **15**, 2869–2909.

Jaynes, E. 1968. Prior probabilities. *IEEE Trans. Syst. Sci. Cybernet.*, **4**(3), 227–241.

Jeffreys, H. 1961. *Theory of Probability*. Third ed. Clarendon Press.

Johnson, N. L., and Kotz, S. 1969. *Distributions in Statistics: Discrete Distributions*. Houghton Mifflin Co.

Johnson, N. L., and Kotz, S. 1970a. *Distributions in Statistics. Continuous Univariate Distributions. 1*. Houghton Mifflin Co.

Johnson, N. L., and Kotz, S. 1970b. *Distributions in Statistics. Continuous Univariate Distributions. 2*. Houghton Mifflin Co.

Johnson, N. L., and Kotz, S. 1972. *Distributions in Statistics: Continuous Multivariate Distributions*. John Wiley & Sons.

Kaplan, E. L., and Meier, P. 1958. Nonparametric estimation from incomplete observations. *J. Amer. Statist. Assoc.*, **53**(282), 457–481.

Kass, R. E., and Raftery, A. E. 1995. Bayes factors. *J. Amer. Statist. Assoc.*, **90**(430), 773–795.

Kass, R. E., and Wasserman, L. 1996. The selection of prior distributions by formal rules. *J. Amer. Statist. Assoc.*, **91**(435), 1343–1370.

Kuffner, R., Zach, N., Norel, R., Hawe, J., Schoenfeld, D., Wang, L., Li, G., Fang, L., Mackey, L., Hardiman, O., Cudkowicz, M., Sherman, A., Ertaylan, G., Grosse-Wentrup, M., Hothorn, T., van Ligtenberg, J., Macke, J. H., Meyer, T., Scholkopf, B., Tran, L., Vaughan, R., Stolovitzky, G., and Leitner, M. L. 2015. Crowdsourced analysis of clinical trial data to predict amyotrophic lateral sclerosis progression. *Nat Biotech*, **33**(1), 51–57.

LeCun, Y., and Cortes, C. 2010. *MNIST Handwritten Digit Database*. http://yann.lecun.com/exdb/mnist/.

LeCun, Y., Bengio, Y., and Hinton, G. 2015. Deep learning. *Nature*, **521**(7553), 436–444.

Lee, J., Sun, D., Sun, Y., and Taylor, J. 2016. Exact post-selection inference, with application to the Lasso. *Annals of Statistics*, **44**(3), 907–927.

Lehmann, E. L. 1983. *Theory of Point Estimation*. Wiley Series in Probability and Mathematical Statistics. John Wiley & Sons.

Leslie, C., Eskin, E., Cohen, A., Weston, J., and Noble, W. S. 2003. Mismatch string kernels for discriminative pretein classification. *Bioinformatics*, **1**, 1–10.

Liaw, A., and Wiener, M. 2002. Classification and regression by randomForest. *R News*, **2**(3), 18–22.

Liberman, M. 2015 (April). *"Reproducible Research and the Common Task Method"*. Simons Foundation Frontiers of Data Science Lecture, April 1, 2015; video available.

Lockhart, R., Taylor, J., Tibshirani, R., and Tibshirani, R. 2014. A significance test for the lasso. *Annals of Statistics*, **42**(2), 413–468. With discussion and a rejoinder by the authors.

Lynden-Bell, D. 1971. A method for allowing for known observational selection in small samples applied to 3CR quasars. *Mon. Not. Roy. Astron. Soc.*, **155**(1), 95–18.

Mallows, C. L. 1973. Some comments on C_p. *Technometrics*, **15**(4), 661–675.

Mantel, N., and Haenszel, W. 1959. Statistical aspects of the analysis of data from retrospective studies of disease. *J. Natl. Cancer Inst.*, **22**(4), 719–748.

Mardia, K. V., Kent, J. T., and Bibby, J. M. 1979. *Multivariate Analysis*. Academic Press.

McCullagh, P., and Nelder, J. 1983. *Generalized Linear Models*. Monographs on Statistics and Applied Probability. Chapman & Hall.

McCullagh, P., and Nelder, J. 1989. *Generalized Linear Models*. Second edn. Monographs on Statistics and Applied Probability. Chapman & Hall.

Metropolis, N., Rosenbluth, A. W., Rosenbluth, M. N., Teller, A. H., and Teller, E. 1953. Equation of state calculations by fast computing machines. *J. Chem. Phys.*, **21**(6), 1087–1092.

Miller, Jr, R. G. 1964. A trustworthy jackknife. *Ann. Math. Statist*, **35**, 1594–1605.

Miller, Jr, R. G. 1981. *Simultaneous Statistical Inference*. Second edn. Springer Series in Statistics. New York: Springer-Verlag.

Nesterov, Y. 2013. Gradient methods for minimizing composite functions. *Mathematical Programming*, **140**(1), 125–161.

Neyman, J. 1937. Outline of a theory of statistical estimation based on the classical theory of probability. *Phil. Trans. Roy. Soc.*, **236**(767), 333–380.

Neyman, J. 1977. Frequentist probability and frequentist statistics. *Synthese*, **36**(1), 97–131.

Neyman, J., and Pearson, E. S. 1933. On the problem of the most efficient tests of statistical hypotheses. *Phil. Trans. Roy. Soc. A*, **231**(694-706), 289–337.

Ng, A. 2015. *Neural Networks*. `http://deeplearning.stanford.edu/wiki/index.php/Neural_Networks`. Lecture notes.

Ngiam, J., Chen, Z., Chia, D., Koh, P. W., Le, Q. V., and Ng, A. 2010. Tiled convolutional neural networks. Pages 1279–1287 of: Lafferty, J., Williams, C., Shawe-Taylor, J., Zemel, R., and Culotta, A. (eds), *Advances in Neural Information Processing Systems 23*. Curran Associates, Inc.

O'Hagan, A. 1995. Fractional Bayes factors for model comparison. *J. Roy. Statist. Soc. Ser. B*, **57**(1), 99–138. With discussion and a reply by the author.

Park, T., and Casella, G. 2008. The Bayesian lasso. *J. Amer. Statist. Assoc.*, **103**(482), 681–686.

Pearson, K. 1900. On the criterion that a given system of deviations from the probable in the case of a correlated system of variables is such that it can be reasonably supposed to have arisen from random sampling. *Phil. Mag.*, **50**(302), 157–175.

Pritchard, J., Stephens, M., and Donnelly, P. 2000. Inference of Population Structure using Multilocus Genotype Data. *Genetics*, **155**(June), 945–959.

Quenouille, M. H. 1956. Notes on bias in estimation. *Biometrika*, **43**, 353–360.

R Core Team. 2015. *R: A Language and Environment for Statistical Computing*. R Foundation for Statistical Computing, Vienna, Austria.

Ridgeway, G. 2005. *Generalized boosted models: A guide to the gbm package*. Available online.

Ridgeway, G., and MacDonald, J. M. 2009. Doubly robust internal benchmarking and false discovery rates for detecting racial bias in police stops. *J. Amer. Statist. Assoc.*, **104**(486), 661–668.

Ripley, B. D. 1996. *Pattern Recognition and Neural Networks*. Cambridge University Press.

Robbins, H. 1956. An empirical Bayes approach to statistics. Pages 157–163 of: *Proc. 3rd Berkeley Symposium on Mathematical Statistics and Probability*, vol. I. University of California Press.

Rosset, S., Zhu, J., and Hastie, T. 2004. Margin maximizing loss functions. In: Thrun, S., Saul, L., and Schölkopf, B. (eds), *Advances in Neural Information Processing Systems 16*. MIT Press.

Rubin, D. B. 1981. The Bayesian bootstrap. *Ann. Statist.*, **9**(1), 130–134.

Savage, L. J. 1954. *The Foundations of Statistics*. John Wiley & Sons; Chapman & Hill.

Schapire, R. 1990. The strength of weak learnability. *Machine Learning*, **5**(2), 197–227.

Schapire, R., and Freund, Y. 2012. *Boosting: Foundations and Algorithms*. MIT Press.

Scheffé, H. 1953. A method for judging all contrasts in the analysis of variance. *Biometrika*, **40**(1-2), 87–110.

Schölkopf, B., and Smola, A. 2001. *Learning with Kernels: Support Vector Machines, Regularization, Optimization, and Beyond (Adaptive Computation and Machine Learning)*. MIT Press.

Schwarz, G. 1978. Estimating the dimension of a model. *Ann. Statist.*, **6**(2), 461–464.

Senn, S. 2008. A note concerning a selection "paradox" of Dawid's. *Amer. Statist.*, **62**(3), 206–210.

Soric, B. 1989. Statistical "discoveries" and effect-size estimation. *J. Amer. Statist. Assoc.*, **84**(406), 608–610.

Spevack, M. 1968. *A Complete and Systematic Concordance to the Works of Shakespeare*. Vol. 1–6. Georg Olms Verlag.

Srivastava, N., Hinton, G., Krizhevsky, A., Sutskever, I., and Salakhutdinov, R. 2014. Dropout: a simple way to prevent neural networks from overfitting. *J. of Machine Learning Res.*, **15**, 1929–1958.

Stefanski, L., and Carroll, R. J. 1990. Deconvoluting kernel density estimators. *Statistics*, **21**(2), 169–184.

Stein, C. 1956. Inadmissibility of the usual estimator for the mean of a multivariate normal distribution. Pages 197–206 of: *Proc. 3rd Berkeley Symposium on Mathematical Statististics and Probability*, vol. I. University of California Press.

Stein, C. 1981. Estimation of the mean of a multivariate normal distribution. *Ann. Statist.*, **9**(6), 1135–1151.

Stein, C. 1985. On the coverage probability of confidence sets based on a prior distribution. Pages 485–514 of: *Sequential Methods in Statistics*. Banach Center Publication, vol. 16. PWN, Warsaw.

Stigler, S. M. 2006. How Ronald Fisher became a mathematical statistician. *Math. Sci. Hum. Math. Soc. Sci.*, **176**(176), 23–30.

Stone, M. 1974. Cross-validatory choice and assessment of statistical predictions. *J. Roy. Statist. Soc. B*, **36**, 111–147. With discussion and a reply by the author.

Storey, J. D., Taylor, J., and Siegmund, D. 2004. Strong control, conservative point estimation and simultaneous conservative consistency of false discovery rates: A unified approach. *J. Roy. Statist. Soc. B*, **66**(1), 187–205.

Tanner, M. A., and Wong, W. H. 1987. The calculation of posterior distributions by data augmentation. *J. Amer. Statist. Assoc.*, **82**(398), 528–550. With discussion and a reply by the authors.

Taylor, J., Loftus, J., and Tibshirani, R. 2015. Tests in adaptive regression via the Kac-Rice formula. *Annals of Statistics*, **44**(2), 743–770.

Thisted, R., and Efron, B. 1987. Did Shakespeare write a newly-discovered poem? *Biometrika*, **74**(3), 445–455.

Tibshirani, R. 1989. Noninformative priors for one parameter of many. *Biometrika*, **76**(3), 604–608.

Tibshirani, R. 1996. Regression shrinkage and selection via the lasso. *J. Roy. Statist. Soc. B*, **58**(1), 267–288.

Tibshirani, R. 2006. A simple method for assessing sample sizes in microarray experiments. *BMC Bioinformatics*, **7**(Mar), 106.

Tibshirani, R., Bien, J., Friedman, J., Hastie, T., Simon, N., Taylor, J., and Tibshirani, R. 2012. Strong rules for discarding predictors in lasso-type problems. *J. Roy. Statist. Soc. B*, **74**.

Tibshirani, R., Tibshirani, R., Taylor, J., Loftus, J., and Reid, S. 2016. *selectiveInference: Tools for Post-Selection Inference*. R package version 1.1.3.

Tukey, J. W. 1958. "Bias and confidence in not-quite large samples" in Abstracts of Papers. *Ann. Math. Statist.*, **29**(2), 614.

Tukey, J. W. 1960. A survey of sampling from contaminated distributions. Pages 448–485 of: *Contributions to Probability and Statistics: Essays in Honor of Harold Hotelling* (I. Olkin, et. al, ed.). Stanford University Press.

Tukey, J. W. 1962. The future of data analysis. *Ann. Math. Statist.*, **33**, 1–67.

Tukey, J. W. 1977. *Exploratory Data Analysis*. Behavioral Science Series. Addison-Wesley.

van de Geer, S., Bühlmann, P., Ritov, Y., and Dezeure, R. 2014. On asymptotically optimal confidence regions and tests for high-dimensional models. *Annals of Statistics*, **42**(3), 1166–1202.

Vapnik, V. 1996. *The Nature of Statistical Learning Theory*. Springer.

Wager, S., Wang, S., and Liang, P. S. 2013. Dropout training as adaptive regularization. Pages 351–359 of: Burges, C., Bottou, L., Welling, M., Ghahramani, Z., and Weinberger, K. (eds), *Advances in Neural Information Processing Systems 26*. Curran Associates, Inc.

Wager, S., Hastie, T., and Efron, B. 2014. Confidence intervals for random forests: the jacknife and the infintesimal jacknife. *J. of Machine Learning Res.*, **15**, 1625–1651.

Wahba, G. 1990. *Spline Models for Observational Data*. SIAM.

Wahba, G., Lin, Y., and Zhang, H. 2000. GACV for support vector machines. Pages 297–311 of: Smola, A., Bartlett, P., Schölkopf, B., and Schuurmans, D. (eds), *Advances in Large Margin Classifiers*. MIT Press.

Wald, A. 1950. *Statistical Decision Functions*. John Wiley & Sons; Chapman & Hall.

Wedderburn, R. W. M. 1974. Quasi-likelihood functions, generalized linear models, and the Gauss–Newton method. *Biometrika*, **61**(3), 439–447.

Welch, B. L., and Peers, H. W. 1963. On formulae for confidence points based on integrals of weighted likelihoods. *J. Roy. Statist. Soc. B*, **25**, 318–329.

Westfall, P., and Young, S. 1993. *Resampling-based Multiple Testing: Examples and Methods for p-Value Adjustment*. Wiley Series in Probability and Statistics. Wiley-Interscience.

Xie, M., and Singh, K. 2013. Confidence distribution, the frequentist distribution estimator of a parameter: A review. *Int. Statist. Rev.*, **81**(1), 3–39. with discussion.

Ye, J. 1998. On measuring and correcting the effects of data mining and model selection. *J. Amer. Statist. Assoc.*, **93**(441), 120–131.

Zhang, C.-H., and Zhang, S. 2014. Confidence intervals for low-dimensional parameters with high-dimensional data. *J. Roy. Statist. Soc. B*, **76**(1), 217–242.

Zou, H., Hastie, T., and Tibshirani, R. 2007. On the "degrees of freedom" of the lasso. *Ann. Statist.*, **35**(5), 2173–2192.

Author Index

Abu-Mostafa, Y. 372
Achanta, R. 372
Akaike, H. 231
Anderson, T. W. 69

Bastien, F. 374
Becker, R. 128
Bellhouse, D. R. 36
Bengio, Y. 372, 374
Benjamini, Y. 294, 418, 450
Berger, J. O. 36, 261
Bergeron, A. 374
Bergstra, J. 374
Berk, R. 323, 419
Berkson, J. 128
Bernardo, J. M. 261
Bibby, J. M. 37, 69
Bien, J. 322
Birch, M. W. 128
Bishop, C. 371
Bloomston, M. 89
Boos, D. D. 180
Boser, B. 390
Bouchard, N. 374
Breiman, L. 129, 348, 451
Breuleux, O. 374
Brown, L. 323, 419
Bühlmann, P. 323
Buja, A. 323, 419

Carlin, B. P. 89, 261
Carroll, R. J. 445
Casella, G. 420
Chambers, J. 128
Chen, Z. 372
Chia, D. 372
Cho, C. S. 89
Chopin, N. 261
Cleveland, W. S. 11
Cohen, A. 393
Cortes, C. 372
Courville, A. 372
Cover, T. M. 52
Cox, D. R. 52, 128, 152, 153, 262
Crowley, J. 153
Cudkowicz, M. 349

de Finetti, B. 261
Dembo, A. 52
Dempster, A. P. 152
Desjardins, G. 374
Dezeure, R. 323
Diaconis, P. 262
DiCiccio, T. 204
Donnelly, P. 261
Donoho, D. L. 447

Edwards, A. W. F. 37
Efron, B. 11, 20, 37, 51, 52, 69, 89, 90, 105, 106, 130, 152, 154, 177–179, 204, 206, 207, 231, 232, 262, 263, 267, 294–297, 322, 323, 348, 417, 419, 420, 444, 445
Ertaylan, G. 349
Eskin, E. 393

Fang, L. 349
Feldman, D. 417
Fields, R. C. 89
Finney, D. J. 262
Fisher, R. A. 184, 204, 449
Fithian, W. 323
Freund, Y. 348, 451
Friedman, J. 128, 129, 231, 321, 322, 348–350, 371

Geisser, S. 231
Gerber, M. 261
Gholami, S. 89
Good, I. 88
Goodfellow, I. J. 374

Gous, A. 262, 263
Grosse-Wentrup, M. 349
Guyon, I. 390

Haenszel, W. 152
Hall, P. 204
Hampel, F. R. 179
Hardiman, O. 349
Harford, T. 232
Hastie, T. 128, 231, 321–323, 348–350, 371, 372, 392, 393
Hawe, J. 349
Hinkley, D. V. 52, 69
Hinton, G. 372
Hochberg, Y. 294, 418, 450
Hoeffding, W. 129, 296
Hoerl, A. E. 105
Hothorn, T. 349
Huber, P. J. 179

Jaeckel, L. A. 178
James, W. 104
Jansen, L. 323
Janson, L. 89
Javanmard, A. 323
Jaynes, E. 261
Jeffreys, H. 261
Jin, L. X. 89
Johnson, N. L. 36
Johnstone, I. 231, 322, 323

Kaplan, E. L. 152
Kass, R. E. 261–263
Kennard, R. W. 105
Kent, J. T. 37, 69
Koh, P. W. 372
Kotz, S. 36
Krizhevsky, A. 372
Kuffner, R. 349

Laird, N. M. 152
Lamblin, P. 374
Le, Q. V. 372
LeCun, Y. 372
Lee, J. 323
Lehmann, E. L. 52
Leitner, M. L. 349
Leslie, C. 393
Levine, E. A. 89
Li, G. 349
Liang, P. S. 372, 373
Liaw, A. 348
Liberman, M. 447
Lin, Y. 391
Loader, C. 393
Lockhart, R. 323
Loftus, J. 323
Louis, T. A. 89, 261
Lynden-Bell, D. 150

MacDonald, J. M. 294
Macke, J. H. 349
Mackey, L. 349
Maithel, S. K. 89
Mallows, C. L. 231
Mantel, N. 152
Mardia, K. V. 37, 69
McCullagh, P. 128, 322
Meier, P. 152
Metropolis, N. 261
Meyer, T. 349
Miller, R. G., Jr 177, 294, 418
Montanari, A. 323
Morris, C. 105

Nelder, J. 128, 322
Nesterov, Y. 372
Neyman, J. 20, 204, 449
Ng, A. 372
Ngiam, J. 372
Noble, W. S. 393
Norel, R. 349
Norton, J. A. 89

O'Hagan, A. 261
Olshen, R. A. 129, 348

Park, T. 420
Pascanu, R. 374
Pawlik, T. M. 89
Pearson, E. S. 449
Pearson, K. 449
Peers, H. W. 37, 207, 261
Pericchi, L. R. 261
Petrosian, V. 130
Popescu, B. 348
Poultsides, G. A. 89
Pritchard, J. 261

Quenouille, M. H. 177

R Core Team 128
Raftery, A. E. 262
Reid, N. 262
Reid, S. 323
Ridgeway, G. 294, 348
Ripley, B. D. 371
Ritov, Y. 323
Robbins, H. 88, 104, 420
Ronchetti, E. M. 179

Rosenbluth, A. W. 261
Rosenbluth, M. N. 261
Rosset, S. 392
Rousseeuw, P. J. 179
Rubin, D. B. 152, 179

Salakhutdinov, R. 372
Savage, L. J. 261
Schapire, R. 348, 451
Scheffé, H. 417
Schmidt, C. 89
Schoenfeld, D. 349
Schölkopf, B. 390
Schwarz, G. 263
Senn, S. 37
Serfling, R. J. 180
Sherman, A. 349
Siegmund, D. 294
Simon, N. 322
Singh, K. 51, 207
Smola, A. 390
Soric, B. 294
Spevack, M. 89
Spolverato, G. 89
Squires, I., Malcolm 89
Srivastava, N. 372
Stahel, W. A. 179
Stefanski, L. 445
Stein, C. 104, 106, 178, 232, 261
Stephens, M. 261
Stigler, S. M. 447
Stolovitzky, G. 349
Stone, C. J. 129, 348
Stone, M. 231
Storey, J. D. 294
Sun, D. 323
Sun, Y. 323
Sutskever, I. 372

Tanner, M. A. 263
Taylor, J. 294, 322, 323
Teller, A. H. 261
Teller, E. 261
Thisted, R. 89
Thomas, J. A. 52
Tibshirani, R. 128, 179, 207, 231, 232, 261, 321–323, 348–350, 371, 392, 420, 450
Toulmin, G. 88
Tran, L. 349
Tran, T. B. 89
Tukey, J. W. 11, 177, 179, 450
Turian, J. 374

van de Geer, S. 323
van Ligtenberg, J. 349
Vapnik, V. 390
Vaughan, R. 349
Vincent, P. 372
Votanopoulos, K. I. 89

Wager, S. 348, 372, 373
Wahba, G. 391, 392
Wainwright, M. 321–323
Wald, A. 450
Wang, L. 349
Wang, S. 372, 373
Warde-Farley, D. 374
Wasserman, L. 261, 263
Weber, S. M. 89
Wedderburn, R. W. M. 128
Welch, B. L. 37, 207, 261
Westfall, P. 294, 418
Weston, J. 393
Wiener, M. 348
Wilks, A. 128
Wong, W. H. 263
Worhunsky, D. J. 89

Xie, M. 51, 207

Ye, J. 231
Yekutieli, D. 418
Ylvisaker, D. 262
Young, S. 294, 418

Zach, N. 349
Zhang, C.-H. 323
Zhang, H. 391
Zhang, K. 323, 419
Zhang, S. 323
Zhao, L. 323, 419
Zhu, J. 392
Zou, H. 231, 322

Subject Index

abc method, 194, 204
Accelerated gradient descent, 359
Acceleration, 192, 205
Accuracy, 14
  after model selection, 402–408
Accurate but not correct, 402
Activation function, 355, 361
  leaky rectified linear, 362
  rectified linear, 362
  ReLU, 362
  tanh, 362
Active set, 302, 308
`adaboost` algorithm, 341–345, 447
Adaboost.M1, 342
Adaptation, 404
Adaptive estimator, 404
Adaptive rate control, 359
Additive model, 324
  adaptive, 346
Adjusted compliance, 404
Admixture modeling, 256–260
AIC, *see* Akaike information criterion
Akaike information criterion, 208, 218, 226, 231, 246, 267
Allele frequency, 257
American Statistical Association, 449
Ancillary, 44, 46, 139
Apparent error, 211, 213, 219
arcsin transformation, 95
Arthur Eddington, 447
Asymptotics, xvi, 119, 120
Autoencoder, 362–364

Backfitting, 346
Backpropagation, 356–358
Bagged estimate, 404, 406
Bagging, 226, 327, 406, 408, 419
Balance equations, 256
Barycentric plot, 259
Basis expansion, 375
Bayes
  deconvolution, 421–424
  factor, 244, 285
  false-discovery rate, 279
  posterior distribution, 254
  posterior probability, 280
  shrinkage, 212
  t-statistic, 255
  theorem, 22
Bayes–frequentist estimation, 412–417
Bayesian
  inference, 22–37
  information criterion, 246
  lasso, 420
  lasso prior, 415
  model selection, 244
  trees, 349
Bayesian information criterion, 267
Bayesianism, 3
BCa
  accuracy and correctness, 206
  confidence density, 202, 207, 237, 242, 243
  interval, 202
  method, 192
Benjamini and Hochberg, 276
Benjamini–Yekutieli, 400
Bernoulli, 338
Best-approximating linear subspace, 363
Best-subset selection, 299
Beta
  distribution, 54, 239
BH_q, 276
Bias, 14, 352
Bias-corrected, 330
  and accelerated, *see* BCa method
  confidence intervals, 190–191
  percentile method, 190

Bias-correction value, 191
Biased estimation, 321
BIC, *see* Bayesian information criterion
Big-data era, xv, 446
Binomial, 109, 117
  distribution, 54, 117, 239
  log-likelihood, 380
  standard deviation, 111
Bioassay, 109
Biometrika, 449
Bivariate normal, 182
Bonferroni bound, 273
Boole's inequality, 274
Boosting, 320, 324, 333–350
Bootstrap, 7, 155–180, 266, 327
  Baron Munchausen, 177
  Bayesian, 168, 179
  cdf, 187
  confidence intervals, 181–207
  ideal estimate, 160, 179
  jackknife after, 179
  moving blocks, 168
  multisample, 167
  nonparametric, 159–162, 217
  out of bootstrap, 232
  packages, 178
  parametric, 169–173, 223, 312, 429
  probabilities, 164
  replication, 159
  sample, 159
  sample size, 179, 205
  smoothing, 226, 404, 406
  t, 196
  t intervals, 195–198
Bound form, 305
Bounding hyperplane, 398
Burn-in, 260
BY_q algorithm, 400

Causal inference, xvi
Censored
  data, 134–139
  not truncated, 150
Centering, 107
Central limit theorem, 119
Chain rule for differentiation, 356
Classic statistical inference, 3–73
Classification, 124, 209
Classification accuracy, 375
Classification error, 209
Classification tree, 348
Cochran–Mantel–Haenszel test, 131
Coherent behavior, 261
Common task framework, 447
Compliance, 394
Computational bottleneck, 128
Computer age, xv
Computer-intensive, 127
  inference, 189, 267
  statistics, 159
Conditional, 58
Conditional distribution
  full, 253
Conditional inference, 45–48, 139, 142
  lasso, 318
Conditionality, 44
Confidence
  density, 200, 201, 235
  distribution, 198–203
  interval, 17
  region, 397
Conjugate, 253, 259
  prior, 238
  priors, 237
Convex optimization, 304, 308, 321, 323, 377
Convolution, 422, 445
  filters, 368
  layer, 367
Corrected differences, 411
Correlation effects, 295
Covariance
  formula, 312
  penalty, 218–226
Coverage, 181
Coverage level, 274
Coverage matching prior, 236–237
Cox model, *see* proportional hazards model
C_p, 217, 218, 221, 231, 267, 300, 394, 395, 403
Cramér–Rao lower bound, 44
Credible interval, 198, 417
Cross-validation, 208–232, 267, 335
  10-fold, 326
  estimate, 214
  K-fold, 300
  leave one out, 214, 231
Cumulant generating function, 67
Curse of dimensionality, 387

Dark energy, 210, 231
Data analysis, 450
Data science, xvii, 450, 451

Data sets
ALS, 334
AML, *see* leukemia
cholesterol, 395, 402, 403
CIFAR-100, 365
diabetes, 209, 396, 414, 416
dose-response, 109
handwritten digits (MNIST), 353
head/neck cancer, 135
human ancestry, 257
kidney function, 157, 222
leukemia, 176, 196, 377
NCOG, 134
nodes, 424, 427, 430, 438, 439, 442
pediatric cancer, 143
police, 287
prostate, 249, 272, 289, 408, 410, 423, 434–436
protein classification, 385
spam, 113, 127, 209, 215, 300–302, 325
student score, 173, 181, 186, 202, 203
supernova, 210, 212, 217, 221, 224
vasoconstriction, 240, 241, 246, 252
Data snooping, 398
De Finetti, B., 35, 36, 251, 450
De Finetti–Savage school, 251
Debias, 318
Decision rule, 275
Decision theory, xvi
Deconvolution, 422
Deep learning, 351–374
Definitional bias, 431
Degrees of freedom, 221, 231, 312–313
Delta method, 15, 414, 420
Deviance, 112, 118, 119, 301
Deviance residual, 123
Diffusion tensor imaging, 291
Direct evidence, 105, 109, 421
Directional derivatives, 158
Distribution
beta, 54, 239
binomial, 54, 117, 239
gamma, 54, 117, 239
Gaussian, 54
normal, 54, 117, 239
Poisson, 54, 117, 239
Divide-and-conquer algorithm, 325
Document retrieval, 298
Dose–response, 109
Dropout learning, 368, 372
DTI, *see* diffusion tensor imaging

Early computer-age, xvi, 75–268
Early stopping, 362
Effect size, 272, 288, 399, 408
Efficiency, 44, 120
Eigenratio, 162, 173, 194
Elastic net, 316, 356
Ellipsoid, 398
EM algorithm, 146–150
missing data, 266
Empirical Bayes, 75–90, 93, 264
estimation strategies, 421–445
information, 443
large-scale testing, 278–282
Empirical null, 286
estimation, 289–290
maximum-likelihood estimation, 296
Empirical probability distribution, 160
Ensemble, 324, 334
Ephemeral predictors, 227
Epoch, 359
Equilibrium distribution, 256
Equivariant, 106
Exact inferences, 119
Expectation parameter, 118
Experimental design, xvi
Exponential family, 53–72, 225
p-parameter, 117, 413, 424
curved, 69
one-parameter, 116

F distribution, 397
F tests, 394
f-modeling, 424, 434, 440–444
Fake-data principle, 148, 154, 266
False coverage
control, 399
False discovery, 275
control, 399
control theorem, 294
proportion, 275
rate, 271–297
False-discovery
rate, 9
Family of probability densities, 64
Family-wise error rate, 274
FDR, *see* false-discovery rate
Feed-forward, 351
Fiducial, 267

constructions, 199
density, 200
inference, 51
Fisher, 79
Fisher information, 29, 41, 59
bound, 41
matrix, 236, 427
Fisherian correctness, 205
Fisherian inference, 38–52, 235
Fixed-knot regression splines, 345
Flat prior, 235
Forward pass, 357
Forward-stagewise, 346
fitting, 320
Forward-stepwise, 298–303
computations, 322
logistic regression, 322
regression, 300
Fourier
method, 440
transform, 440
Frailties, 439
Frequentism, 3, 12–22, 30, 35, 51, 146, 267
Frequentist, 413
inference, 12–21
strongly, 218
Fully connected layer, 368
Functional gradient descent, 340
FWER, *see* family-wise error rate

g-modeling, 423
Gamma, 117
distribution, 54, 117, 239
General estimating equations, xvi
General information criterion, 248
Generalized
linear mixed model, 437–440
linear model, 108–123, 266
ridge problem, 384
Genome, 257
Genome-wide association studies, 451
Gibbs sampling, 251–260, 267, 414
GLM, *see* generalized linear model
GLMM, *see* generalized linear mixed model
Google flu trends, 230, 232
Gradient boosting, 338–341
Gradient descent, 354, 356
Gram matrix, 381
Gram-Schmidt orthogonalization, 322
Graphical lasso, 321
Graphical models, xvi
Greenwood's formula, 137, 151
Group lasso, 321

Hadamard product, 358
Handwritten digits, 353
Haplotype estimation, 261
Hazard rate, 131–134
parametric estimate, 138
Hidden layer, 351, 352, 354
High-order interaction, 325
Hinge loss, 380
Hints
learning with, 369
Hoeffding's lemma, 118
Holm's procedure, 274, 294
Homotopy path, 306
Hypergeometric distribution, 141, 152

Imputation, 149
Inadmissable, 93
Indirect evidence, 102, 109, 266, 290, 421, 440, 443
Inductive inference, 120
Inference, 3
Inference after model selection, 394–420
Inferential triangle, 446
Infinitesimal forward stagewise, 320
Infinitesimal jackknife, 167
estimate, 406
standard deviations, 407
Influence function, 174–177
empirical, 175
Influenza outbreaks, 230
Input distortion, 369, 373
Input layer, 355
Insample error, 219
Inverse chi-squared, 262
Inverse gamma, 239, 262
IRLS, *see* iteratively reweighted least squares
Iteratively reweighted least squares, 301, 322

Jackknife, 155–180, 266, 330
estimate of standard error, 156
standard error, 178
James–Stein
estimation, 91–107, 282, 305, 410
ridge regression, 265
Jeffreys
prior, 237
Jeffreys'

prior, 28–30, 36, 198, 203, 236
prior, multiparameter, 242
scale, 285
Jumpiness of estimator, 405
Kaplan–Meier, 131, 134, 136, 137
estimate, 134–139, 266
Karush–Kuhn–Tucker optimality conditions, 308
Kernel
function, 382
logistic regression, 386
method, 375–393
SVM, 386
trick, 375, 381–383, 392
Kernel smoothing, 375, 387–390
Knots, 309
Kullback–Leibler distance, 112
ℓ_1 regularization, 321
Lagrange
dual, 381
form, 305, 308
multiplier, 391
Large-scale
hypothesis testing, 271–297
testing, 272–275
Large-scale prediction algorithms, 446
Lasso, 101, 210, 217, 222, 231, 298–323
modification, 312
path, 312
penalty, 356
Learning from the experience of others, 104, 280, 290, 421, 443
Learning rate, 358
Least squares, 98, 112, 299
Least-angle regression, 309–313, 321
Least-favorable family, 262
Left-truncated, 150
Lehmann alternative, 294
Life table, 131–134
Likelihood function, 38
concavity, 118
Limited-translation rule, 293
Lindsey's method, 68
Linearly separable, 375
Link function, 237, 340
Local false-discovery rate, 280, 282–286
Local regression, 387–390, 393
Local translation invariance, 368
Log polynomial regression, 410
Log-rank statistic, 152
Log-rank test, 131, 139–142, 152, 266
Logic of inductive inference, 185, 205
Logistic regression, 109–115, 139, 214, 299, 375
multiclass, 355
Logit, 109
Loss plus penalty, 385
Machine learning, 208, 267, 375
Mallows' C_p, *see* C_p
Mantel–Haenzel test, 131
MAP, 101
MAP estimate, 420
Margin, 376
Marginal density, 409, 422
Markov chain Monte Carlo, *see* MCMC
Markov chain theory, 256
Martingale theory, 294
Matching prior, 198, 200
Matlab, 271
Matrix completion, 321
Max pool layer, 366
Maximized a-posteriori probability, *see* MAP
Maximum likelihood, 299
Maximum likelihood estimation, 38–52
MCMC, 234, 251–260, 267, 414
McNemar test, 341
Mean absolute deviation, 447
Median unbiased, 190
Memory-based methods, 390
Meter reader, 30
Meter-reader, 37
Microarrays, 227, 271
Minitab, 271
Misclassification error, 302
Missing data, 146–150, 325
EM algorithm, 266
Missing-species problem, 78–84
Mixed features, 325
Mixture density, 279
Model averaging, 408
Model selection, 243–250, 398
criteria, 250
Monotone lasso, 320
Monotonic increasing function, 184
Multinomial
distribution, 61–64, 425
from Poisson, 63
Multiple testing, 272
Multivariate
analysis, 119
normal, 55–59

n-gram, 385
N-P complete, 299
Nadaraya–Watson estimator, 388
Natural parameter, 116
Natural spline model, 430
NCOG, *see* Northern California Oncology Group
Nested models, 299
Neural Information Processing Systems, 372
Neural network, 351–374
  adaptive tuning, 360
  number of hidden layers, 361
Neurons, 351
Neyman's construction, 181, 183, 193, 204
Neyman–Pearson, 18, 19, 293
Non-null, 272
Noncentral chi-square variable, 207
Nonlinear transformations, 375
Nonlinearity, 361
Nonparameteric
  regression, 375
Nonparametric, 53, 127
  MLE, 150, 160
  percentile interval, 187
Normal
  correlation coefficient, 182
  distribution, 54, 117, 239
  multivariate, 55–59
  regression model, 414
  theory, 119
Northern California Oncology Group, 134
Nuclear norm, 321
Nuisance parameters, 142, 199

Objective Bayes, 36, 267
  inference, 233–263
  intervals, 198–203
  prior distribution, 234–237
OCR, *see* optical character recognition
Offset, 349
OLS
  algorithm, 403
  estimation, 395
  predictor, 221
One-sample nonparametric bootstrap, 161
One-sample problems, 156
OOB, *see* out-of-bag error
Optical character recognition, 353
Optimal separating hyperplane, 375–377
Optimal-margin classifier, 376
Optimality, 18
Oracle, 275
Orthogonal parameters, 262
Out-of-bag error, 232, 327, 329–330
Out-the-box learning algorithm, 324
Output layer, 352
Outsample error, 219
Over parametrized, 298
Overfitting, 304
Overshrinks, 97

p-value, 9, 282
Package/program
  `gbm`, 335, 348
  `glmnet`, 214, 315, 322, 348
  `h2o`, 372
  `lars`, 312, 320
  `liblineaR`, 381
  `locfdr`, 289–291, 296, 437
  `lowess`, 6, 222, 388
  `nlm`, 428
  `randomForest`, 327, 348
  `selectiveInference`, 323
Pairwise inner products, 381
Parameter space, 22, 29, 54, 62, 66
Parametric bootstrap, 242
Parametric family, 169
Parametric models, 53–72
Partial likelihood, 142, 145, 151, 153, 266, 341
Partial logistic regression, 152
Partial residual, 346
Path-wise coordinate descent, 314
Penalized
  least squares, 101
  likelihood, 101, 428
  logistic regression, 356
  maximum likelihood, 226, 307
Percentile method, 185–190
  central interval, 187
Permutation null, 289, 296
Permutation test, 49–51
Phylogenetic tree, 261
Piecewise
  linear, 313
  nonlinear, 314
Pivotal
  argument, 183
  quantity, 196, 198
  statistic, 16

.632 rule, 232
Poisson, 117, 193
distribution, 54, 117, 239
regression, 120–123, 249, 284, 295, 435
Poisson regression, 171
Polynomial kernel, 382, 392
Positive-definite function, 382
Post-selection inference, 317, 394–420
Posterior density, 235, 238
Posterior distribution, 416
Postwar era, 264
Prediction
errors, 216
rule, 208–213
Predictors, 124, 208
Principal components, 362
Prior distribution, 234–243
beta, 239
conjugate, 237–243
coverage matching, 236–237
gamma, 239
normal, 239
objective Bayes, 234
proper, 239
Probit analysis, 112, 120, 128
Propagation of errors, 420
Proper prior, 239
Proportional hazards model, 131, 142–146, 266
Proximal-Newton, 315

q-value, 280
QQ plot, 287
QR decomposition, 311, 322
Quadratic program, 377
Quasilikelihood, 266
Quetelet, Adolphe, 449

R, 178, 271
Random forest, 209, 229, 324–332, 347–350
adaptive nearest-neighbor estimator, 328
leave-one-out cross-validated error, 329
Monte Carlo variance, 330
sampling variance, 330
standard error, 330–331
Randomization, 49–51
Rao–Blackwell, 227, 231
Rate annealing, 360
Rectified linear, 359
Regression, 109
Regression rule, 219
Regression to the mean, 33
Regression tree, 124–128, 266, 348
Regularization, 101, 173, 298, 379, 428
path, 306
Relevance, 290–293
Relevance function, 293
Relevance theory, 297
Reproducing kernel Hilbert space, 375, 384, 392
Resampling, 162
plans, 162–169
simplex, 164, 169
vector, 163
Residual deviance, 283
Response, 124, 208
Ridge regression, 97–102, 209, 304, 327, 332, 372, 381
James–Stein, 265
Ridge regularization, 368
logistic regression, 392
Right-censored, 150
Risk set, 144
RKHS, *see* reproducing-kernel Hilbert space
Robbins' formula, 75, 77, 422, 440
Robust estimation, 174–177
Royal Statistical Society, 449

S language, 271
Sample correlation coefficient, 182
Sample size coherency, 248
Sampling distribution, 312
SAS, 271
Savage, L. J., 35, 36, 51, 199, 233, 251, 450
Scale of evidence
Fisher, 245
Jeffreys, 245
Scheffé
interval, 396, 397, 417
theorem, 398
Score function, 42
Score tests, 301
Second-order accuracy, 192–195
Selection bias, 33, 408–411
Self-consistent, 149
Separating hyperplane, 375
geometry, 390
Seven-league boots, 448
Shrinkage, 115, 316, 338

estimator, 59, 91, 94, 96, 410
Sigmoid function, 352
Significance level, 274
Simulation, 155–207
Simultaneous confidence intervals, 395–399
Simultaneous inference, 294, 418
Sinc kernel, 440, 445
Single-nucleotide polymorphism, *see* SNP
Smoothing operator, 346
SNP, 257
Soft margin classifier, 378–379
Soft-threshold, 315
Softmax, 355
Spam filter, 115
Sparse
  models, 298–323
  principal components, 321
Sparse matrix, 316
Sparsity, 321
Split-variable randomization, 327, 332
SPSS, 271
Squared error, 209
Standard candles, 210, 231
Standard error, 155
  external, 408
  internal, 408
Standard interval, 181
Stein's
  paradox, 105
  unbiased risk estimate, 218, 231
Stepwise selection, 299
Stochastic gradient descent, 358
Stopping rule, 32, 413
Stopping rules, 243
String kernel, 385, 386
Strong rules, 316, 322
Structure, 261
Structure matrix, 97, 424
Student *t*
  confidence interval, 396
  distribution, 196, 272
  statistic, 449
    two-sample, 8, 272
Studentized range, 418
Subgradient
  condition, 308
  equation, 312, 315
Subjective prior distribution, 233
Subjective probability, 233
Subjectivism, 35, 233, 243, 261
Sufficiency, 44
Sufficient
  statistic, 66, 112, 116
  vector, 66
Supervised learning, 352
Support
  set, 377, 378
  vector, 377
  vector classifiers, 381
  vector machine, 319, 375–393
SURE, *see* Stein's unbiased risk estimate
Survival analysis, 131–154, 266
Survival curve, 137, 279
SVM
  Lagrange dual, 391
  Lagrange primal, 391
  loss function, 391

Taylor series, 157, 420
Theoretical null, 286
Tied weights, 368
Time series, xvi
Training set, 208
Transformation invariance, 183–185, 236
Transient episodes, 228
Trees
  averaging, 348
  best-first, 333
  depth, 335
  terminal node, 126
Tricube kernel, 388, 389
Trimmed mean, 175
Triple-point, xv
True error rate, 210
True-discovery rates, 286
Tukey, J. W., 418, 450
Tukey, J. W., 418
Tweedie's formula, 409, 419, 440
Twenty-first-century methods, xvi, 271–446
Two-groups model, 278

Uncorrected differences, 411
Uninformative prior, 28, 169, 233, 261
Universal approximator, 351
Unlabeled images, 365
Unobserved covariates, 288

Validation set, 213
Vapnik, V., 390
Variable-importance plot, 331–332, 336
Variance, 14

Variance reduction, 324
Velocity vector, 360
Voting, 333
Warm starts, 314, 363
Weak learner, 333, 342
Weight
  decay, 356
  regularization, 361, 362
  sharing, 352, 367
Weighted exponential loss, 345
Weighted least squares, 315
Weighted majority vote, 341
Weights, 352
Wide data, 298, 321
Wilks' likelihood ratio statistic, 246
Winner's curse, 33, 408
Winsorized mean, 175
Working response, 315, 322
$z^{(\alpha)}$, 188
Zero set, 296